Important Formulas

1 The **distance** $d(P_1, P_2)$ between $P_1(x_1, y_1)$ and $P_2(x_2, y_2)$:

$$d(P_1, P_2) = \sqrt{(x_2 - x_1)^2 + (y_2 - y_1)^2}$$

2 The **equation of a circle** with center $C(h, k)$ and radius r:

$$(x - h)^2 + (y - k)^2 = r^2$$

3 The **slope** m of the line through $P_1(x_1, y_1)$ and $P_2(x_2, y_2)$:

$$m = \frac{y_2 - y_1}{x_2 - x_1}$$

4 The **point-slope form** for the equation of a line through $P_1(x_1, y_1)$ with slope m:

$$y - y_1 = m(x - x_1)$$

5 The **slope-intercept form** for the equation of a line with slope m and y-intercept b:

$$y = mx + b$$

6 **Linear function** f: $f(x) = ax + b$

7 **Quadratic function** f: $f(x) = ax^2 + bx + c$

8 **Polynomial function** f: $f(x) = a_n x^n + a_{n-1} x^{n-1} + \cdots + a_1 x + a_0$

9 **Rational function** f: $f(x) = \dfrac{h(x)}{g(x)}$, where h and g are polynomial functions

10 **Exponential function** f with base $a > 0$: $f(x) = a^x$

11 **Logarithmic function** f with base $a > 0$: $f(x) = \log_a x$, where $y = \log_a x$ if and only if $a^y = x$

12 **Geometric formulas** for areas, circumference of a circle, volumes, and curved surface area are as follows, where r = radius, h = altitude, b (or a) = length of base, A = area, C = circumference, V = volume, and S = curved surface area:

Triangle	$A = \frac{1}{2}bh$
Circle	$A = \pi r^2, \quad C = 2\pi r$
Parallelogram	$A = bh$
Trapezoid	$A = \frac{1}{2}(a + b)h$
Right circular cylinder	$V = \pi r^2 h, \quad S = 2\pi rh$
Right circular cone	$V = \frac{1}{3}\pi r^2 h, \quad S = \pi r \sqrt{r^2 + h^2}$
Sphere	$V = \frac{4}{3}\pi r^3, \quad S = 4\pi r^2$

EARL W. SWOKOWSKI

MARQUETTE UNIVERSITY

FUNDAMENTALS OF ALGEBRA AND TRIGONOMETRY

FIFTH EDITION

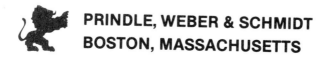

PRINDLE, WEBER & SCHMIDT
BOSTON, MASSACHUSETTS

PWS PUBLISHERS

Prindle, Weber & Schmidt · 🐝 · Willard Grant Press · **wg** · Duxbury Press · ♠
Statler Office Building · 20 Providence Street · Boston, Massachusetts 02116

PWS Publishers is a division of Wadsworth, Inc.

Library of Congress Cataloging in Publication Data

Swokowski, Earl William
    Fundamentals of algebra and trigonometry.

    Includes index.
    1. Algebra.   2.   Trigonometry, Plane.   I.   Title.
QA152.2.S967   1981   512'.13   80-26359
ISBN 0-87150-307-7

ISBN 0-87150-307-7

Printed in the United States of America.
10 9 8 7 6 5 4 3 — 86 85 84 83 82

Cover design by Novum, Inc. Text design by Nancy Blodget. Text composition in Monophoto Times Roman and Helvetica by Syntax International PTE. LTD. Cover printed by New England Book Components, Inc. Text printed and bound by Rand McNally & Company.

PREFACE

The fifth edition of *Fundamentals of Algebra and Trigonometry* reflects the continuing change in the needs and abilities of students who enroll in precalculus mathematics courses. The goal of this new edition is to maintain the mathematical soundness of earlier editions, but to make some of the discussions less formal by rewriting, placing more emphasis on graphing and applications, and adding new examples, figures, and exercises.

Chapter 1 consists of topics that are fundamental for the study of algebra. These include properties of real numbers, exponents, radicals, and simplification of algebraic expressions. Equations and inequalities are considered in Chapter 2. The technique of solving inequalities involving polynomials has been simplified by introducing the notion of *test value*.

Basic properties of functions and graphs are discussed in Chapter 3. A noteworthy change from earlier editions is the additional emphasis on symmetry, horizontal or vertical shifts, and stretching of graphs.

Chapter 4 contains a rather detailed examination of polynomial and rational functions. Once again the use of test values has made it possible to simplify and unify this material. Section 4.4, on partial fractions, is new to this edition. The chapter ends with a discussion of conic sections.

Exponential and logarithmic functions are discussed in Chapter 5. Computational aspects of logarithms appear in the last two sections, and may be omitted without interrupting the continuity of the book.

In Chapter 6, an accessible introduction to trigonometric functions using a unit circle is managed by emphasizing examples and keeping the discussion relatively brief. Angles are introduced early in the chapter, and the unit circle approach is then supplemented by an angular description of the trigonometric functions. Later, examples are again emphasized to help students learn to sketch graphs rapidly. A new section on harmonic motion was added at the end of the chapter to provide some non-triangular applications of trigonometry.

Most of Chapter 7 consists of work with trigonometric identities, equations, and solutions of oblique triangles. The last section of this chapter contains an introduction to geometric and algebraic properties of vectors in two dimensions.

Matrices are introduced in Chapter 8 as a tool for solving systems of linear equations. The chapter concludes with a study of determinants and a brief introduction to linear programming.

Complex numbers are discussed in Chapter 9. Chapter 10 contains material on zeros of polynomials. Several additional theorems about the location and bounds of zeros were added to this edition. Chapter 11 has sections on mathematical induction, the Binomial Theorem, summation notation, sequences, and probability.

There is a review section at the end of each chapter consisting of a list of important topics and pertinent exercises. The review exercises are similar in scope to those which appear throughout the chapter and may be used to prepare for examinations. Answers to odd-numbered exercises are given at the end of the text. An answer booklet for the even-numbered exercises may be obtained from the publisher.

This revision has benefited from comments and suggestions of users of previous editions. I wish to thank the following individuals, who reviewed all or parts of the manuscript and offered many helpful suggestions: Ben P. Bockstege, Jr. (Broward Community College), Gary L. Ebert (University of Delaware), Leonard E. Fuller (Kansas State University), F. Cecilia Hakeem (Southern Illinois University), Douglas W. Hall (Michigan State University), Arthur M. Hobbs (Texas A&M University), Adam J. Hulin (University of New Orleans), William B. Jones (University of Colorado), Jimmie Lawson (Louisiana State University), Burnett Meyer (University of Colorado), Eldon L. Miller (University of Mississippi), Robert W. Owens (Lewis and Clark College), Anthony L. Peressini (University of Illinois), Jack W. Pope (University of San Diego), Larry H. Potter (Memphis State University), Sandra M. Powers (The College of Charleston), Billie Ann Rice (Dekalb Community College), John Riner (Ohio State University), Mary W. Scott (University of Florida), William F. Stearns (University of Maine), Janet T. Vasak (University of Wisconsin-Milwaukee), Richard G. Vinson (University of South Alabama), T. Perrin Wright, Jr. (Florida State University), and Paul M. Young (Kansas State University).

I am also grateful to the staff of Prindle, Weber & Schmidt, for their painstaking work in the production of this book. In particular, Mary Lu Walsh, John Martindale, and Nancy Blodget gave valuable assistance with the revision.

Special thanks are due to my wife Shirley and the members of our family: Mary, Mark, John, Steve, Paul, Tom, Bob, Nancy, Judy, and Jay. All have had an influence on the book—either directly, through working exercises, proofreading, or typing, or indirectly, through continued interest and moral support.

To all of the people named here, and to the many unnamed students and teachers who have helped shape my views about precalculus mathematics, I express my sincere appreciation.

Earl W. Swokowski

TABLE OF CONTENTS

v

Attention Students: A Programmed Guide by Roy A. Dobyns which gives you assistance in understanding the major concepts discussed in this text and practice for chapter tests is available. Check for it in your bookstore, or talk with your instructor or bookstore manager about ordering copies.

COMMENTS ABOUT HANDHELD CALCULATORS

There is disagreement among mathematics educators as to the extent that electronic handheld calculators should be employed in courses in algebra and trigonometry. Those who favor their use point out that, in addition to reducing the time spent on numerical computations, calculators are useful for illustrating and reinforcing mathematical concepts. It is unlikely that any teacher would recommend that a calculator be used as a tool for solving the quadratic equation $x^2 + 2x - 3 = 0$; however, the process of approximating the solutions of $241x^2 + 527x - 73 = 0$ *without* a calculator is pure drudgery, adding little to a student's mathematical knowledge. On the theoretical side, some claim that calculating $10^{\log x}$ for various values of x will help beginning students accept, and remember, the identity $10^{\log x} = x$. The extent to which this is true is, of course, difficult to measure.

Teachers who oppose the use of calculators in algebra and trigonometry courses argue that they distract from the mainstream of study, and that classroom time is better spent on theoretical and manipulative aspects which are needed in advanced courses such as calculus. Clearly, it is essential to understand basic results about equations before using a calculator to approximate solutions. As another illustration, working with logarithmic and trigonometric tables has some worthwhile side-effects concerning the variation of functions and intermediate values.

In this text an attempt is made to satisfy both those instructors who wish to use calculators as a teaching aid, and those who prefer not to use them. Specifically, optional problems labeled *Calculator Exercises* have been included in appropriate sections of the text, following the regular exercises. They may either be omitted or assigned, depending on course objectives. In order to make proper use of these exercises, it is necessary to have access to a scientific calculator which has the capability to deal with exponential, logarithmic, trigonometric, and inverse trigonometric expressions.

If calculators are used, students should be aware of the fact that inaccuracies may occur due to round off, truncation, or algorithms which the calculator employs to compute certain numerical values. For example, to use a calculator to demonstrate that $(\frac{1}{3}) \cdot 3 = 1$, we may enter 3, press the reciprocal key $\boxed{1/x}$, and then multiply by 3.

However, if the decimal approximation 0.33333333 is entered, then multiplying by 3 results in 0.99999999. Similarly, to demonstrate that $\sqrt{2}\sqrt{3} = \sqrt{6}$ we may, on the one hand, calculate $\sqrt{2}\sqrt{3}$ and $\sqrt{6}$ separately, and verify that the same decimal approximation is obtained in each case. On the other hand, calculating $\sqrt{6} - \sqrt{2}\sqrt{3}$ with a typical calculator we may obtain 0.000000000013 (this number may vary, depending on the type of calculator). If we agree to round off answers to eight decimal places, then this result may be considered as a calculator demonstration that $\sqrt{6} - \sqrt{2}\sqrt{3} = 0$, and hence that $\sqrt{6} = \sqrt{2}\sqrt{3}$.

Other discrepencies may occur. For example, given a calculator which displays a maximum of eight digits, if we enter 10,000,000 and then add 0.1, we obtain 10,000,000, a result not in keeping with laws of algebra! Similarly, laws such as $a(b + c) = ab + ac$ may not hold if certain values are entered for a, b, and c.

It should also be noted that if calculators are used to solve the *regular* exercises involving logarithmic or trigonometric functions, some answers may differ from those given in the text, since the latter were obtained using the four-place tables in the Appendix, whereas calculators produce a higher degree of accuracy.

Finally, instructions on how to operate a calculator are not provided in this text. The large variety of calculators on the market make it impractical to do so. Owner manuals should be consulted for information and illustrations.

1

FUNDAMENTAL CONCEPTS OF ALGEBRA

The material in this chapter is basic to the study of algebra. We begin by discussing properties of real numbers. Next we turn our attention to exponents and radicals, and how they may be used to simplify complicated algebraic expressions.

1.1 ALGEBRA: A POWERFUL LANGUAGE AND TOOL

A good foundation in algebra is essential for advanced courses in mathematics, the natural sciences, and engineering. It is also required for problems which arise in business, industry, statistics, and many other fields of endeavor. Indeed, every situation which makes use of numerical processes is a candidate for algebraic methods.

Algebra evolved from the operations and rules of arithmetic. The study of arithmetic begins with addition, multiplication, subtraction, and division of numbers, such as

$$4 + 7, \quad (37)(681) \quad 79 - 22, \quad \text{and} \quad 40 \div 8.$$

In *algebra* we introduce symbols or letters a, b, c, d, x, y, etc., to denote *arbitrary* numbers and, instead of special cases, we often consider *general* statements, such as

$$a + b, \quad cd, \quad x - y, \quad \text{and} \quad x \div a.$$

This *language of algebra* serves a two-fold purpose. First, it may be used as a short-hand, to abbreviate and simplify long or complicated statements. Second, it is a convenient means of generalizing many specific statements. To illustrate, at an early

1

age, children learn that

$$2 + 3 = 3 + 2, \qquad 4 + 7 = 7 + 4, \qquad 5 + 9 = 9 + 5, \qquad 1 + 8 = 8 + 1,$$

and so on. In words, this property may be phrased "if two numbers are added, then the order of addition is immaterial; that is, the same result is obtained whether the second number is added to the first, or the first number is added to the second." This lengthy description can be shortened, and at the same time made easier to understand by means of the algebraic statement

$$a + b = b + a,$$

where a and b denote arbitrary numbers.

Illustrations of the generality of algebra may be found in formulas used in science and industry. For example, if an airplane flies at a constant rate of 300 mph (miles per hour) for two hours, then the distance it travels is given by

$$(300)(2), \quad \text{or} \quad 600 \text{ miles.}$$

If the rate is 250 mph and the elapsed time is 3 hours, then the distance traveled is

$$(250)(3), \quad \text{or} \quad 750 \text{ miles.}$$

If we introduce symbols and let r denote the constant rate, t the elapsed time, and d the distance traveled, then the two illustrations we have given are special cases of the general algebraic formula

$$d = rt.$$

When specific numerical values for r and t are given, the distance d may be found readily by an appropriate substitution in the formula. Moreover, the formula may also be used to solve related problems. For example, suppose the distance between two cities is 645 miles, and we wish to find the constant rate which would enable an airplane to cover that distance in 2 hours and 30 minutes. Thus, we are given

$$d = 645 \text{ miles}, \quad \text{and} \quad t = 2.5 \text{ hours,}$$

and the problem is to find r. Since $d = rt$, it follows that

$$r = \frac{d}{t}$$

and hence, for our special case,

$$r = \frac{645}{2.5} = 258 \text{ mph.}$$

That is, if an airplane flies at a constant rate of 258 mph, then it will travel 645 miles in 2 hours and 30 minutes. In like manner, given r, the time t required to travel a distance d may be found by means of the formula

$$t = \frac{d}{r}.$$

Note how the introduction of a general algebraic formula not only allows us to solve special problems conveniently, but also to enlarge the scope of our knowledge by suggesting new problems that can be considered.

We have given several elementary illustrations of the value of algebraic methods. There are an unlimited number of situations where a symbolic approach may lead to insights and solutions that would be impossible to obtain using only numerical processes. As you proceed through this text and go on to either more advanced courses in mathematics or fields which employ mathematics, you will become further aware of the importance and the power of algebraic techniques.

1.2 REAL NUMBERS

Real numbers are used in all phases of mathematics and you are undoubtedly well acquainted with symbols which are used to represent them, such as

$$1, \quad 73, \quad -5, \quad \frac{49}{12}, \quad \sqrt{2}, \quad 0, \quad \sqrt[3]{-85}, \quad 0.33333\ldots, \quad \text{and} \quad 596.25.$$

The real numbers are said to be **closed** relative to operations of addition (denoted by $+$) and multiplication (denoted by $\cdot$). This means that, to every pair a, b of real numbers, there corresponds a real number $a + b$ called the **sum** of a and b and a unique real number $a \cdot b$ (also written ab) called the **product** of a and b. These operations have the following properties, where all lower-case letters denote arbitrary real numbers, and where 0 and 1 are special real numbers referred to as **zero** and **one**, respectively.

Commutative Properties

$$a + b = b + a, \qquad ab = ba$$

Associative Properties

$$a + (b + c) = (a + b) + c, \qquad a(bc) = (ab)c$$

Identities

$$a + 0 = a = 0 + a, \qquad a \cdot 1 = a = 1 \cdot a$$

Inverses For every real number a, there is a real number denoted by $-a$ such that

$$a + (-a) = 0 = (-a) + a.$$

For every real number $a \neq 0$, there is a real number denoted by $1/a$ such that

$$a\left(\frac{1}{a}\right) = 1 = \left(\frac{1}{a}\right)a.$$

Distributive Properties

$$a(b + c) = ab + ac, \qquad (a + b)c = ac + bc$$

The equals sign, $=$, indicates that the expressions immediately to the right and left of the sign represent the same real number. The real numbers 0 and 1 are sometimes referred to as the **additive identity** and **multiplicative identity**, respectively. We call $-a$ the **additive inverse** of a (or the **negative** of a). If $a \neq 0$, then $1/a$ is called the **multiplicative inverse** of a (or the **reciprocal** of a). The symbol a^{-1} is often used in place of $1/a$. Thus, by definition,

Definition of a^{-1}

$$a^{-1} = \frac{1}{a}$$

Example 1 Verify the following special cases of the Associative and Distributive Properties.

(a) $2 + (3 + 4) = (2 + 3) + 4$

(b) $2 \cdot (3 \cdot 4) = (2 \cdot 3) \cdot 4$

(c) $2 \cdot (3 + 4) = 2 \cdot 3 + 2 \cdot 4$

(d) $(2 + 3) \cdot 4 = 2 \cdot 4 + 3 \cdot 4$

Solutions   To verify each of parts (a)–(d) we perform the operations indicated on opposite sides of the equals sign and observe that the resulting numbers are identical. Thus,

(a)

$$2 + (3 + 4) = 2 + 7 = 9$$
$$(2 + 3) + 4 = 5 + 4 = 9$$

(b)

$$2 \cdot (3 \cdot 4) = 2 \cdot 12 = 24$$
$$(2 \cdot 3) \cdot 4 = 6 \cdot 4 = 24$$

(c)

$$2 \cdot (3 + 4) = 2 \cdot 7 = 14$$
$$2 \cdot 3 + 2 \cdot 4 = 6 + 8 = 14$$

(d)

$$(2 + 3) \cdot 4 = 5 \cdot 4 = 20$$
$$2 \cdot 4 + 3 \cdot 4 = 8 + 12 = 20 \qquad \blacksquare$$

Since $a + (b + c)$ and $(a + b) + c$ are always equal, we may, without ambiguity, use the symbol $a + b + c$ to denote this real number. Similarly, the notation abc is used for either $a(bc)$ or $(ab)c$. An analogous situation exists if four real numbers $a, b, c,$ and d are added. For example, we could consider

$$(a + b) + (c + d), \qquad a + [(b + c) + d], \qquad [(a + b) + c] + d,$$

and so on. It can be shown that, regardless of how the four numbers are grouped, the same result is obtained, and consequently it is customary to write $a + b + c + d$ for any of these expressions. Furthermore, it follows from the Commutative Property for addition that the numbers can be interchanged in any way. For example,

$$a + b + c + d = a + d + c + b = a + c + d + b.$$

A similar situation exists for multiplication, where the expression $abcd$ is used to denote the product of four real numbers.

The Distributive Properties are useful for finding products of many different types of expressions. The next example provides an illustration. Others will be found in the exercises.

Example 2   If a, b, c, and d denote real numbers, show that

$$(a + b)(c + d) = ac + bc + ad + bd.$$

Solution   Using the two Distributive Properties,

$$\begin{aligned}
(a + b)(c + d) &= (a + b)c + (a + b)d \\
&= (ac + bc) + (ad + bd) \\
&= ac + bc + ad + bd.
\end{aligned}$$ ∎

If $a = b$ and $c = d$, then, since a and b are merely different names for the same real number, and likewise for c and d, it follows that $a + c = b + d$ and $ac = bd$. This is often called the **substitution principle**, since we may think of replacing a by b and c by d in the expressions $a + c$ and ac. As a special case, using the fact that $c = c$ gives us the following rules.

> If $a = b$, then $a + c = b + c$.
> If $a = b$, then $ac = bc$.

We sometimes refer to these rules by the statements "any number c may be added to both sides of an equality" and "both sides of an equality may be multiplied by the same number c." We shall make heavy use of them in Chapter 2, in conjunction with solving equations.

The following results can be proved. (See Exercises 34 and 35.)

> $a \cdot 0 = 0$ for every real number a.
> If $ab = 0$, then either $a = 0$ or $b = 0$.

The last two statements imply that $ab = 0$ *if and only if* either $a = 0$ or $b = 0$. The phrase "if and only if," which is used throughout mathematics, always has a two-fold character. Here it means that if $ab = 0$, then $a = 0$ or $b = 0$ and, *conversely*, if $a = 0$ or $b = 0$, then $ab = 0$. Consequently, if both $a \neq 0$ and $b \neq 0$, then $ab \neq 0$; that is, *the product of two nonzero real numbers is always nonzero*.

The following can also be proved. (See Exercise 39.)

Properties of Negatives

> $-(-a) = a$
> $(-a)b = -(ab) = a(-b)$
> $(-a)(-b) = ab$
> $(-1)a = -a$

The operation of **subtraction** (denoted by $-$) is defined by

**Definition of
Subtraction**

$$a - b = a + (-b)$$

The next example indicates that the Distributive Properties hold for subtraction.

Example 3   If a, b, and c are real numbers, show that

$$a(b - c) = ab - ac.$$

Solution   We shall list reasons after each step, as follows.

$$
\begin{aligned}
a(b - c) &= a[b + (-c)] \quad &&\text{(Definition of Subtraction)} \\
&= ab + a(-c) \quad &&\text{(Distributive Property)} \\
&= ab + [-(ac)] \quad &&\text{(Property of Negatives)} \\
&= ab - ac \quad &&\text{(Definition of Subtraction)}
\end{aligned}
$$

∎

If $b \neq 0$, then **division** (denoted by $\div$) is defined by

**Definition of
Division**

$$a \div b = a\left(\frac{1}{b}\right) = ab^{-1}$$

The symbol a/b is often used in place of $a \div b$, and we refer to it as the **quotient of a by b** or the **fraction a over b**. The numbers a and b are called the **numerator** and **denominator**, respectively. It is important to note that, since 0 has no multiplicative inverse, a/b is not defined if $b = 0$; that is, *division by zero is not permissible*. Also note that

$$1 \div b = \frac{1}{b} = b^{-1}.$$

The following properties of quotients may be established, where all denominators are nonzero real numbers.

**Properties
of Quotients**

$$\frac{a}{b} = \frac{c}{d} \quad \text{if and only if} \quad ad = bc$$

$$\frac{a}{b} = \frac{ad}{bd}, \qquad \frac{a}{-b} = \frac{-a}{b} = -\frac{a}{b}$$

$$\frac{a}{b} + \frac{c}{b} = \frac{a + c}{b}, \qquad \frac{a}{b} + \frac{c}{d} = \frac{ad + bc}{bd}$$

$$\frac{a}{b} \cdot \frac{c}{d} = \frac{ac}{bd}, \qquad \frac{a}{b} \div \frac{c}{d} = \frac{a}{b} \cdot \frac{d}{c} = \frac{ad}{bc}$$

Example 4 Find (a) $\dfrac{2}{3} + \dfrac{9}{5}$ (b) $\dfrac{2}{3} \cdot \dfrac{9}{5}$ (c) $\dfrac{2}{3} \div \dfrac{9}{5}$.

Solutions Using the Properties of Quotients we have

(a) $$\frac{2}{3} + \frac{9}{5} = \frac{(2 \cdot 5) + (3 \cdot 9)}{3 \cdot 5} = \frac{10 + 27}{15} = \frac{37}{15}$$

(b) $$\frac{2}{3} \cdot \frac{9}{5} = \frac{2 \cdot 9}{3 \cdot 5} = \frac{18}{15} = \frac{6 \cdot 3}{5 \cdot 3} = \frac{6}{5}$$

(c) $$\frac{2}{3} \div \frac{9}{5} = \frac{2}{3} \cdot \frac{5}{9} = \frac{2 \cdot 5}{3 \cdot 9} = \frac{10}{27}$$ ∎

The **positive integers** 1, 2, 3, 4, ... may be obtained by adding the real number 1 successively to itself. The negatives, $-1, -2, -3, -4, \ldots$, of the positive integers are referred to as **negative integers**. The **integers** consist of the totality of positive and negative integers together with the real number 0.

Observe that, by the Distributive Properties, if a is a real number, then

$$a + a = (1 + 1)a = 2a$$

and

$$a + a + a = (1 + 1 + 1)a = 3a.$$

Similarly, the sum of four a's is $4a$, the sum of five a's is $5a$, and so on.

If a, b, and c are integers and $c = ab$, then a and b are called **factors**, or **divisors**, of c. For example, the integer 6 may be written as

$$6 = 2 \cdot 3 = (-2)(-3) = 1 \cdot 6 = (-1)(-6).$$

Hence $1, -1, 2, -2, 3, -3, 6$, and -6 are factors of 6.

A positive integer p different from 1 is **prime** if its only positive factors are 1 and p. The first few primes are 2, 3, 5, 7, 11, 13, 17, and 19. One of the reasons for the importance of prime numbers is that every positive integer a different from 1 can be expressed in one and only one way (except for order of factors) as a product of primes. The proof of this result, which is called The **Fundamental Theorem of Arithmetic**, will not be given in this book. As examples, we have

$$12 = 2 \cdot 2 \cdot 3, \quad 126 = 2 \cdot 3 \cdot 3 \cdot 7, \quad \text{and} \quad 540 = 2 \cdot 2 \cdot 3 \cdot 3 \cdot 3 \cdot 5.$$

A real number is called a **rational number** if it can be written in the form a/b, where a and b are integers and $b \neq 0$. Real numbers that are not rational are called **irrational**. The ratio of the circumference of a circle to its diameter is an irrational real number and is denoted by π. It is often approximated by the decimal 3.1416 or by the rational number 22/7. We use the notation $\pi \approx 3.1416$ to indicate that π is *approximately equal* to 3.1416. To cite another example, a real number a, such that $a^2 = 2$, where a^2 denotes $a \cdot a$, is not rational. There are two such irrational numbers denoted by the symbols $\sqrt{2}$ and $-\sqrt{2}$.

Real numbers may be represented by decimal expressions. Decimal representations for rational numbers either terminate or are nonterminating and repeating. For example, it can be shown by long division that a decimal representation for 7434/2310 is 3.2181818 . . . , where the digits 1 and 8 repeat indefinitely. The rational number 5/4 has the terminating decimal representation 1.25. Decimal representations for irrational numbers may also be obtained; however, they are always nonterminating and nonrepeating. The process of finding decimal representations for irrational numbers is usually difficult. Often some method of successive approximation is employed. For example, the device learned in arithmetic for extracting square roots can be used to find a decimal representation for $\sqrt{2}$. Using this technique, we successively obtain the approximations 1, 1.4, 1.41, 1.414, 1.4142, and so on.

Sometimes, it is convenient to use the notation and terminology of sets. A **set** may be thought of as a collection of objects of some type. The objects are called **elements** of the set. Capital letters A, B, C, R, S, ... will often be used to denote sets. Lower-case letters a, b, x, y, . . . will represent elements of sets. The notation $x \in S$ is used to specify that x is an element of the set S. Throughout our work, $\mathbb{R}$ will denote the set of real numbers, and $\mathbb{Z}$ the set of integers. If every element of a set S is also an element of a set T, then S is called a **subset** of T, written $S \subseteq T$. For example, $\mathbb{Z} \subseteq \mathbb{R}$. Two sets S and T are said to be equal, written $S = T$, if S and T contain precisely the same elements. The notation $S \neq T$ means that S and T are not equal.

EXERCISES 1.2

In each of Exercises 1–10, justify the equality by stating only one of the following properties:
(a) commutative, (b) associative, (c) identity, (d) inverse, (e) distributive.

1 $(4 \cdot 5) \cdot 4 = 4 \cdot (4 \cdot 5)$

2 $3 \cdot (4 + 5) = (4 + 5) \cdot 3$

3 $(4 \cdot 5) \cdot 4 = 4 \cdot (5 \cdot 4)$

4 $(4 + 5) \cdot 3 = 4 \cdot 3 + 5 \cdot 3$

5 $3 \cdot (5 + 0) = 3 \cdot 5$

6 $3 + (-3) = 0$

7 $1 \cdot (2 + 3) = 2 + 3$

8 $(1 + 2) + 1 = 1 + (1 + 2)$

9 $(1/4)4 = 1$

10 $0 \cdot 1 = 0$

Use properties of real numbers to find the products in Exercises 11–20, where all letters represent real numbers.

11 $a(b + 3) + 2(b + 3)$

12 $c(d + 1) + 5(d + 1)$

13 $(a + 2)(b + 3)$

14 $(c + 5)(d + 1)$

15 $2x(y + 2) - 3(y + 2)$

16 $2p(6q + 5) - 3(6q + 5)$

17 $(2x - 3)(y + 2)$

18 $(2p - 3)(6q + 5)$

19 $(4r + 5)(3s + 6)$

20 $(3u + 7)(4v - 1)$

In each of Exercises 21–30, write the expression as a rational number with least positive numerator.

21 $\dfrac{3}{4} + \dfrac{5}{3}$

22 $\dfrac{1}{5} + \dfrac{3}{2}$

23 $\dfrac{5}{6} \cdot \dfrac{2}{3}$

24 $\dfrac{4}{9}\left(\dfrac{1}{2} + \dfrac{3}{4}\right)$

25 $\dfrac{1}{2} + \dfrac{1}{4} + \dfrac{1}{6}$

26 $\dfrac{3}{-2} \cdot \dfrac{-5}{6}$

27 $\dfrac{13}{4} \div \left(\dfrac{3}{2} + 1\right)$

28 $\dfrac{2}{5} + \dfrac{7}{4} + \dfrac{3}{2}$

29 $\dfrac{5}{7}\left(\dfrac{3}{2} - \dfrac{5}{6}\right)$

30 $\dfrac{10}{11} \cdot \dfrac{11}{10}$

31 Show, by means of examples, that the operation of subtraction on $\mathbb{R}$ is neither commutative nor associative.

32 Show, by means of examples, that the operation of division, as applied to nonzero real numbers, is neither commutative nor associative.

Prove the rules in Exercises 33–38, where all letters denote real numbers and no denominators are zero.

33 $a = -a$ if and only if $a = 0$.

34 $a \cdot 0 = 0$ (*Hint*: Write $a \cdot 0 = a \cdot (0 + 0) = a \cdot 0 + a \cdot 0$ and then add $-(a \cdot 0)$ to both sides.)

35 If $ab = 0$, then either $a = 0$ or $b = 0$

36 $\dfrac{-a}{-b} = \dfrac{a}{b}$

37 $\left(\dfrac{a}{b}\right)^{-1} = \dfrac{b}{a}$

38 $\dfrac{a}{b} + \dfrac{c}{b} = \dfrac{a + c}{b}$ (*Hint*: Write $\dfrac{a}{b} + \dfrac{c}{b} = ab^{-1} + cb^{-1}$ and use the Distributive Properties.)

39 Prove each of the properties of negatives.

1.3 COORDINATE LINES

It is possible to associate the set of real numbers with the set of all points on a line l in such a way that for each real number a there corresponds one and only one point, and conversely, to each point P on l there corresponds precisely one real number. Such an association between two sets is referred to as a **one-to-one correspondence**. We first choose an arbitrary point O, called the **origin**, and associate with it the real number 0. Points associated with the integers are then determined by laying off successive line segments of equal length on either side of O as illustrated in Figure 1.1. The points corresponding to rational numbers such as 23/5 and $-1/2$ are obtained by subdividing the equal line segments. Points associated with certain irrational numbers, such as $\sqrt{2}$, can be found by geometric construction. For other irrational numbers such as π, no construction is possible. However, the point corresponding to π can be approximated to any degree of accuracy by locating successively the points corresponding to 3, 3.1, 3.14, 3.141, 3.1415, 3.14159, etc.

Figure 1.1

The number a that is associated with a point A on l is called the **coordinate** of A. An assignment of coordinates to points on l is called a **coordinate system** for l, and l is called a **coordinate line**, or a **real line**. A direction can be assigned to l by taking the **positive direction** along l to the right and the **negative direction** to the left. The positive direction is noted by placing an arrowhead on l as shown in Figure 1.1.

The numbers which correspond to points to the right of O in Figure 1.1 are called **positive real numbers**, whereas those which correspond to points to the left of O are **negative real numbers**. The real number 0 is neither positive nor negative. The set of positive real numbers is **closed** relative to addition and multiplication; that is, if a and b are positive, then so is the sum $a + b$ and the product ab.

Note that if a is positive, then $-a$ is negative. Similarly, if $-a$ is positive, then $-(-a) = a$ is negative. A common error is to think that $-a$ is always a negative number; however, this is not necessarily the case. For example, if $a = -3$, then $-a = -(-3) = 3$, which is positive. Indeed, whenever a is negative the number $-a$ is positive.

If a and b are real numbers, and $a - b$ is positive, we say that a **is greater than** b and write $a > b$. An equivalent statement is b **is less than** a, written $b < a$. The symbols $>$ or $<$ are called **inequality signs** and expressions such as $a > b$ or $b < a$ are called **inequalities**. From the manner in which we constructed the coordinate line l in Figure 1.1, we see that if A and B are points with coordinates a and b, respectively, then $a > b$ (or $b < a$) *if and only if A lies to the right of B*. The following definition is stated for reference, where a and b denote real numbers.

Definition of
> and <

> $a > b$ means $a - b$ is positive.
>
> $b < a$ means $a - b$ is positive.

Note that $a > b$ and $b < a$ have exactly the same meaning. As illustrations,

$5 > 3$ since $5 - 3 = 2$ is positive;

$-6 < -2$ since $-2 - (-6) = -2 + 6 = 4$ is positive;

$-\sqrt{2} < 1$ since $1 - (-\sqrt{2}) = 1 + \sqrt{2}$ is positive;

$2 > 0$ since $2 - 0 = 2$ is positive; and

$-5 < 0$ since $0 - (-5) = 5$ is positive.

The last two illustrations are special cases of the following general properties.

> $a > 0$ if and only if a is positive.
>
> $a < 0$ if and only if a is negative.

These properties are consequences of the definitions of $>$ and $<$.

It should be clear from our discussion that if a and b are real numbers, then *one and only one of the following is true*:

$$a = b, \quad a > b, \quad \text{or} \quad a < b.$$

We refer to the **sign** of a real number as being positive or negative according as the number is positive or negative. We also say that two real numbers *have the same sign* if both are positive or both are negative. They have *opposite signs* if one is positive and the other is negative. The next result about the signs of products and quotients of two real numbers a and b can be proved using the Properties of Negatives and Quotients listed in Section 1.2.

Laws of Signs

> (i) If a and b have the same sign, then ab and a/b are positive.
>
> (ii) If a and b have opposite signs, then ab and a/b are negative.

As illustrations, note that

$$(-2)(3) = -(2 \cdot 3) = -6$$
$$(2)(-3) = -(2 \cdot 3) = -6$$
$$(-2)(-3) = 2 \cdot 3 = 6.$$

$$\frac{-2}{3} = -\frac{2}{3}$$

$$\frac{-2}{-3} = -\frac{2}{-3} = -\left(-\frac{2}{3}\right) = \frac{2}{3}.$$

The converses of the Laws of Signs are also true. For example, if a quotient is negative, then the numerator and denominator have opposite signs.

There are several other useful symbols which involve inequality signs. In particular, $a \geq b$, which is read **a is greater than or equal to b**, means that either $a > b$ or $a = b$ (but not both). As an illustration, we have $a^2 \geq 0$ for every real number a. The symbol $a \leq b$ is read **a is less than or equal to b** and means that either $a < b$ or $a = b$. The expression $a < b < c$ means that both $a < b$ and $b < c$, in which case we say that **b is between a and c**. We may also write $c > b > a$. For instance,

$$1 < 5 < \frac{11}{2}, \qquad -4 < \frac{2}{3} < \sqrt{2}, \qquad 3 > -6 > -10.$$

Other variations of the inequality notation are used. For example, $a < b \leq c$ means both $a < b$ and $b \leq c$. Similarly, $a \leq b < c$ means both $a \leq b$ and $b < c$. Finally, $a \leq b \leq c$ means both $a \leq b$ and $b \leq c$.

If a is a real number, then it is the coordinate of some point A on a coordinate line l, and the symbol $|a|$ is used to denote the number of units (or distance) between A and the origin, without regard to direction. This nonnegative number $|a|$ is called the **absolute value** of a. Referring to Figure 1.2, we see that, for the point with coordinate -4, we have $|-4| = 4$. Similarly, $|4| = 4$. In general, if a is negative we change its sign to find $|a|$, whereas if a is nonnegative, then $|a| = a$. The next definition summarizes this discussion.

Figure 1.2

Definition

> If a is a real number, then the **absolute value** $|a|$ of a is defined by
>
> $$|a| = \begin{cases} a & \text{if } a \geq 0. \\ -a & \text{if } a < 0. \end{cases}$$

Example 1   Find $|3|$, $|-3|$, $|0|$, $|\sqrt{2} - 2|$, and $|2 - \sqrt{2}|$.

Solution   Since 3, $2 - \sqrt{2}$, and 0 are nonnegative, we have

$$|3| = 3, \quad |2 - \sqrt{2}| = 2 - \sqrt{2}, \quad \text{and} \quad |0| = 0.$$

Since -3 and $\sqrt{2} - 2$ are negative, we use the formula $|a| = -a$ to obtain

$$|-3| = -(-3) = 3 \quad \text{and} \quad |\sqrt{2} - 2| = -(\sqrt{2} - 2) = 2 - \sqrt{2}. \quad \blacksquare$$

Note that in Example 1, $|-3| = |3|$ and $|2 - \sqrt{2}| = |\sqrt{2} - 2|$. It can be shown that

$$\boxed{|a| = |-a| \text{ for every real number } a.}$$

We shall use the concept of absolute value to define the distance between any two points on a coordinate line. Let us begin by noting that the distance between the points with coordinates 2 and 7 shown in Figure 1.3 equals 5 units on l. This distance is the difference, $7 - 2$, obtained by subtracting the smaller coordinate from the larger. If we employ absolute values, then, since $|7 - 2| = |2 - 7|$, it is unnecessary to be concerned about the order of subtraction. We shall use this as our motivation for the next definition.

Figure 1.3

Definition   Let a and b be the coordinates of two points A and B, respectively, on a coordinate line l. The **distance between A and B**, denoted by $d(A, B)$, is defined by

$$d(A, B) = |b - a|.$$

The number $d(A, B)$ is also called the **length of the line segment AB**.
Observe that, since $d(B, A) = |a - b|$ and $|b - a| = |a - b|$, we may write

$$d(A, B) = d(B, A).$$

Also note that the distance between the origin O and the point A is

$$d(O, A) = |a - 0| = |a|,$$

which agrees with the geometric interpretation of absolute value illustrated in Figure 1.2. The formula $d(A, B) = |b - a|$ is true regardless of the signs of a and b, as illustrated in the next example.

Example 2   Let A, B, C, and D have coordinates -5, -3, 1, and 6 respectively on a coordinate line l (see Figure 1.4). Find $d(A, B)$, $d(C, B)$, $d(O, A)$, and $d(C, D)$.

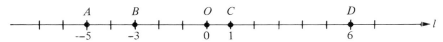

Figure 1.4

Solution   By definition,

$$d(A, B) = \left| -3 - (-5) \right| = \left| -3 + 5 \right| = \left| 2 \right| = 2$$
$$d(C, B) = \left| -3 - 1 \right| = \left| -4 \right| = 4$$
$$d(O, A) = \left| -5 - 0 \right| = \left| -5 \right| = 5$$
$$d(C, D) = \left| 6 - 1 \right| = \left| 5 \right| = 5.$$

These answers can be checked geometrically by referring to Figure 1.4.   ■

The concept of absolute value has uses other than that of finding distances between points. Generally, it is employed whenever we are interested in the "magnitude" or "numerical value" of a real number without regard to its sign.

EXERCISES 1.3

In Exercises 1–4, replace the □ with either $<$, $>$, or $=$.

1 (a) $-7 \,\square\, -4$ (b) $3 \,\square\, -1$
    (c) $1 + 3 \,\square\, 6 - 2$

2 (a) $-3 \,\square\, -5$ (b) $-6 \,\square\, 2$
    (c) $1/4 \,\square\, 0.25$

3 (a) $1/3 \,\square\, 0.33$ (b) $125/57 \,\square\, 2.193$
    (c) $22/7 \,\square\, \pi$

4 (a) $1/7 \,\square\, 0.143$
    (b) $(3/4) + (2/3) \,\square\, 19/12$
    (c) $\sqrt{2} \,\square\, 1.4$

In each of Exercises 5–16, express the given statement in terms of inequalities.

5 -8 is less than -5.

6 2 is greater than 1.9.

7 0 is greater than -1.

8 $\sqrt{2}$ is less than π.

9 x is negative.

10 y is positive.

11 a is between 5 and 3.

12 b is between $1/10$ and $1/3$.

13 b is greater than or equal to 2.

14 x is less than or equal to -5.

15 c is not greater than 1.

16 d is nonnegative.

Rewrite the numbers in Exercises 17–20 without using symbols for absolute value.

17 (a) $\left| 4 - 9 \right|$ (b) $\left| -4 \right| - \left| -9 \right|$
    (c) $\left| 4 \right| + \left| -9 \right|$

18 (a) $\left| 3 - 6 \right|$ (b) $\left| 0.2 - (1/5) \right|$
    (c) $\left| -3 \right| - \left| -4 \right|$

19 (a) $3 - \left| -3 \right|$ (b) $\left| \pi - 4 \right|$
    (c) $(-3)/\left| -3 \right|$

20 (a) $\left| 8 - 5 \right|$ (b) $-5 + \left| -7 \right|$
    (c) $(-2)\left| -2 \right|$

In each of Exercises 21–24, the given numbers are coordinates of three points A, B, and C (in that order) on a coordinate line l. For each, find: (a) $d(A, B)$ (b) $d(B, C)$ (c) $d(C, B)$ (d) $d(A, C)$.

21 $-6, \ -2, \ 4$ **22** $3, \ 7, \ -5$

23 $8, \ -4, \ -1$ **24** $-9, \ 1, \ 10$

25 Are the positive real numbers closed relative to:

 (a) subtraction?   (b) division?   Explain.

26 Are the negative real numbers closed relative to:
 (a) addition?   (b) multiplication?
 (c) subtraction?   Explain.

Rewrite the following without using symbols for absolute value.

27 $|5 - x|$ if $x > 5$ **28** $|x^2 + 1|$

29 $|-4 - x^2|$ **30** $|a - b|$ if $a < b$

1.4 INTEGRAL EXPONENTS

Throughout the remainder of this chapter, all symbols for elements will denote real numbers. We have had occasion to use the notation a^2 for the real number $a \cdot a$. Similarly, we write

$$a^3 = a \cdot a \cdot a, \qquad a^4 = a \cdot a \cdot a \cdot a, \qquad a^5 = a \cdot a \cdot a \cdot a \cdot a,$$

and so on. In general we have the following.

Definition

> If n is any positive integer, then
>
> $$a^n = \underbrace{a \cdot a \cdot \cdots \cdot a}_{n \text{ factors}}$$
>
> The integer n is called an **exponent**.

We sometimes refer to a^n as "a raised to the nth power," or "a to the nth power," or simply "a to the n." Note that $a^1 = a$. Some numerical examples are

$$\left(\frac{1}{2}\right)^5 = \frac{1}{2} \cdot \frac{1}{2} \cdot \frac{1}{2} \cdot \frac{1}{2} \cdot \frac{1}{2} = \frac{1}{32}$$

$$(-3)^3 = (-3)(-3)(-3) = -27$$

$$(\sqrt{2})^4 = \sqrt{2}\sqrt{2}\sqrt{2}\sqrt{2} = (\sqrt{2})^2(\sqrt{2})^2 = (2)(2) = 4.$$

It is important to remember that if n is a positive integer, then an expression such as $3a^n$ means $3(a^n)$ and not $(3a)^n$. The real number 3 is called the **coefficient** of a^n in the expression $3a^n$. Similarly, $-3a^n$ means $(-3)a^n$, not $(-3a)^n$. For example, we have

$$5 \cdot 2^3 = 5 \cdot 8 = 40 \quad \text{and} \quad -5 \cdot 2^3 = -5 \cdot 8 = -40.$$

Example 1   Express each of the following in the form a^n, where n is a positive integer.

(a) a^3a^4 (b) $(a^4)^3$ (c) $\dfrac{a^5}{a^2}$,   where $a \neq 0$

Solutions   By the definition of exponent,

(a) $$a^3a^4 = (aaa)(aaaa) = a^7$$

(b) $$(a^4)^3 = (a^4)(a^4)(a^4) = (aaaa)(aaaa)(aaaa) = a^{12}$$

(c) $$\frac{a^5}{a^2} = \frac{aaaaa}{aa} = \frac{aa}{aa} \cdot \frac{aaa}{1} = aaa = a^3$$ ■

Example 2   Express the following in forms which involve only powers of a and b:
(a) $(ab)^4$ (b) $(a/b)^3$,   where $b \neq 0$.

Solutions

(a) Using the definition of exponent and the Commutative and Associative Properties,

$$(ab)^4 = (ab)(ab)(ab)(ab) = (aaaa)(bbbb) = a^4b^4.$$

(b) Using the definition of exponent and Properties of Quotients,

$$\left(\frac{a}{b}\right)^3 = \frac{a}{b} \cdot \frac{a}{b} \cdot \frac{a}{b} = \frac{aaa}{bbb} = \frac{a^3}{b^3}.$$ ■

The solutions of Examples 1 and 2 are tedious and boring, to say the least. If the exponents were larger, the methods of solution would be even more aggravating. Fortunately, the following laws make such tasks less painful.

Laws of Exponents

If a and b are real numbers and m and n are positive integers, then

(i) $a^m a^n = a^{m+n}$;

(ii) $(a^m)^n = a^{mn}$;

(iii) $(ab)^n = a^n b^n$;

(iv) $\left(\dfrac{a}{b}\right)^n = \dfrac{a^n}{b^n}$,   if $b \neq 0$;

(v)   If $a \neq 0$, then

$$\frac{a^m}{a^n} = a^{m-n} \quad \text{for } m > n$$

$$\frac{a^m}{a^n} = \frac{1}{a^{n-m}} \quad \text{for } n > m$$

$$\frac{a^m}{a^n} = 1 \quad \text{for } m = n.$$

Let us illustrate Laws (i), (ii), and (v) by reworking Example 1. Thus,

$$a^3a^4 = a^{3+4} = a^7$$
$$(a^4)^3 = a^{4 \cdot 3} = a^{12}$$
$$\frac{a^5}{a^2} = a^{5-2} = a^3.$$

Example 2 could be solved in similar fashion using Laws (iii) and (iv).

Complete proofs of (i)–(v) require the method of mathematical induction discussed in Section 11.1. However, if we are allowed to count an arbitrary number of factors, then it is easy to supply arguments which establish the laws. To prove Law (i) we have

$$a^m a^n = \underbrace{a \cdot a \cdots \cdot a}_{m \text{ factors}} \cdot \underbrace{a \cdot a \cdots \cdot a}_{n \text{ factors}}.$$

Since the total number of factors a on the right is $m + n$, this expression is equal to a^{m+n}.

Similarly, we could write

$$(a^m)^n = \underbrace{a^m \cdot a^m \cdots \cdot a^m}_{n \text{ factors } a^m}$$

and count the number of times a appears as a factor on the right-hand side. Since $a^m = a \cdot a \cdots \cdot a$ where a occurs as a factor m times, and since the number of such groups of m factors is n, the total number of factors is $m \cdot n$. This gives us Law (ii).

Laws (iii) and (iv) can be obtained in similar fashion and are left to the reader. Law (v) is clear if $m = n$. For the case $m > n$, the integer $m - n$ is positive and we may write

$$\frac{a^m}{a^n} = \frac{a^n a^{m-n}}{a^n} = \frac{a^n}{a^n} \cdot a^{m-n} = 1 \cdot a^{m-n} = a^{m-n}.$$

A similar argument can be used if $n > m$.

We can, of course, use symbols for real numbers other than a and b when applying the Laws of Exponents, as in the following illustrations.

$$x^5 x^6 = x^{5+6} = x^{11} \qquad (y^5)^7 = y^{5 \cdot 7} = y^{35}$$

$$(rs)^7 = r^7 s^7 \qquad \left(\frac{p}{q}\right)^{10} = \frac{p^{10}}{q^{10}}$$

$$\frac{c^8}{c^3} = c^{8-3} = c^5 \qquad \frac{u^3}{u^8} = \frac{1}{u^{8-3}} = \frac{1}{u^5}$$

The Laws of Exponents can be extended to rules such as $a^m a^n a^p = a^{m+n+p}$ and $(abc)^n = a^n b^n c^n$.

Example 3   Simplify each of the following:

(a) $(3x^3y^4)(4xy^5)$ (b) $(2a^2b^3c)^4$ (c) $\left(\dfrac{2r^3}{s}\right)^2\left(\dfrac{s}{r^3}\right)^3$

Solutions   We may proceed as follows (supply reasons for each step).

(a) $(3x^3y^4)(4xy^5) = (3)(4)x^3xy^4y^5 = 12x^4y^9$

(b) $(2a^2b^3c)^4 = 2^4(a^2)^4(b^3)^4c^4 = 16a^8b^{12}c^4$

(c) $\left(\dfrac{2r^3}{s}\right)^2\left(\dfrac{s}{r^3}\right)^3 = \left(\dfrac{2^2r^6}{s^2}\right)\left(\dfrac{s^3}{r^9}\right) = 2^2\left(\dfrac{r^6}{r^9}\right)\left(\dfrac{s^3}{s^2}\right) = 4\left(\dfrac{1}{r^3}\right)(s) = \left(\dfrac{4s}{r^3}\right)$ ∎

It is possible to extend our work to exponents which are negative integers or 0. If we want (i) of the Laws of Exponents to be true for $n = 0$, then $a^m \cdot a^0 = a^{m+0} = a^m$ and, if $a \neq 0$, multiplication by $1/(a^m)$ leads to $a^0 = 1$. Thus, in order to be consistent with our previous development, we introduce the following definition.

**Definition of
Zero
Exponent**

> If a is any nonzero real number, then $a^0 = 1$.

Some special cases are

$$3^0 = 1, \quad (-14)^0 = 1, \quad \text{and} \quad (5 - \sqrt{2})^0 = 1.$$

The symbol 0^0 will be left undefined. Thus, whenever we write $x^0 = 1$, it is always assumed that x represents a *nonzero* real number.

Let us next turn our attention to negative exponents. In Section 1.2 we used the notation $a^{-1} = 1/a$ if $a \neq 0$. If (i) of the Laws of Exponents is to be true in this situation, then we must have

$$a^{-2} = a^{-1} \cdot a^{-1} = \frac{1}{a} \cdot \frac{1}{a} = \frac{1}{a^2}$$

$$a^{-3} = a^{-2} \cdot a^{-1} = \frac{1}{a^2} \cdot \frac{1}{a} = \frac{1}{a^3}.$$

It is natural, therefore, to define negative integral exponents as follows.

**Definition of
Negative
Exponents**

> If a is a nonzero real number and n is a positive integer, then $a^{-n} = \dfrac{1}{a^n}$.

It is possible to show that the Laws of Exponents are valid for all integers m and n, whether positive, negative, or zero. If negative exponents are allowed, then Law (v) may be abbreviated

$$\frac{a^m}{a^n} = a^{m-n}, \quad \text{for } all \text{ integers } m \text{ and } n.$$

For example,

$$\frac{a^2}{a^5} = a^{2-5} = a^{-3} = \frac{1}{a^3} = \frac{1}{a^{5-2}}.$$

In the future it will be assumed that the Laws of Exponents are true for all integral exponents.

To simplify statements, we shall assume that, in all examples and exercises involving exponents, symbols which appear in denominators represent *nonzero* real numbers.

Example 4   Eliminate negative exponents and simplify:

(a) $(a^{-2}b^3)^{-3}$ (b) $\dfrac{8x^3y^{-5}}{4x^{-1}y^2}.$

Solutions   We may proceed as follows (supply reasons for each step).

(a)
$$(a^{-2}b^3)^{-3} = (a^{-2})^{-3}(b^3)^{-3}$$
$$= a^6 b^{-9}$$
$$= a^6 \cdot \frac{1}{b^9}$$
$$= \frac{a^6}{b^9}$$

(b)
$$\frac{8x^3y^{-5}}{4x^{-1}y^2} = \frac{8}{4}\frac{x^3}{x^{-1}}\frac{y^{-5}}{y^2}$$
$$= 2x^{3-(-1)}y^{-5-2}$$
$$= 2x^4 y^{-7}$$
$$= \frac{2x^4}{y^7}$$

Another method for simplifying (b) is to multiply numerator and denominator of the given fraction by $y^5 x$, thereby eliminating negative exponents. Thus,

$$\frac{8x^3y^{-5}}{4x^{-1}y^2} = \frac{8x^3y^{-5}}{4x^{-1}y^2} \cdot \frac{y^5 x}{y^5 x}$$
$$= \frac{8}{4}\frac{x^4 y^0}{x^0 y^7}$$
$$= 2\frac{x^4(1)}{(1)y^7}$$
$$= \frac{2x^4}{y^7}.$$ ∎

EXERCISES 1.4

In Exercises 11–50, the word "simplify" means to replace the given expression by one in which letters representing real numbers appear only once, and no negative exponents occur.

Express the numbers in Exercises 1–10 in the form a/b, where a and b are integers.

1 $\left(-\dfrac{3}{5}\right)^3$ **2** $(-2)^4$

3 $3^2 + (2 \cdot 3)$ **4** $(-3)^2 - 2^3$

5 $\left(\dfrac{4}{5}\right)^{-2}\left(\dfrac{2}{5}\right)^3$ **6** $\dfrac{5^{-2}}{2^{-5}}$

7 $3^0 + 0^3$ **8** $2^{-4} + (-4)^2$

9 $\dfrac{3^2 \cdot 5^{-3}}{5 \cdot 3^{-1}}$ **10** $(15,678)^0$

Simplify the following expressions.

11 $(4a^3)(3a^5)$ **12** $(-3b^4)(4b^2)(\tfrac{1}{6}b^7)$

13 $\dfrac{(2c^7)(3c^2)}{12c^{12}}$ **14** $\dfrac{18c^{11}}{(6c^4)^2}$

15 $(4x^2)^2(3x^4)$ **16** $(5y^2)(-2y^3)^3$

17 $\dfrac{(4b^2)^3}{(2b^3)^2}$ **18** $\dfrac{(6z^2)(2z^2)^4}{(3z^2)^5}$

19 $(8t^3u^5)(2^{-1}tu^{-3})$ **20** $(m^4)(-4m^3)(3m^{-2})$

21 $(-2a^2bc^{-4})(5a^{-1}b^3c^5)$

22 $(8x^2y^3)(2x^4y^{-2})(\tfrac{1}{4}x^{-3}y)$

23 $(-4x^3)^2(8x^4)^{-1}$ **24** $(3z^3y^4w^{-9})^0$

25 $\dfrac{(-6a^3)^2}{9a^4}$ **26** $\dfrac{(4c^{-3})(-3c^7)}{6c^2}$

27 $(4a^2b^{-1})^3$ **28** $\left(\dfrac{3x^2y^3}{4x^3y^2}\right)\left(\dfrac{2xy^4}{6x^5}\right)$

29 $(-2u^{-1}y^2)^3(5u^2y)^{-1}$ **30** $(3x^{-2}yz^3)^4$

31 $(-3ab^2)^3\left(\dfrac{a^5}{9b}\right)^2$ **32** $\dfrac{(10r^{-2}s)^3}{(50r^3s^{-1})^2}$

33 $\left(\dfrac{5t^2v}{4tv^3}\right)^2\left(\dfrac{2tv^3}{t^2v}\right)^4$ **34** $\left(\dfrac{3p}{q^3}\right)^{-2}\left(\dfrac{-q}{2p^2}\right)^3$

35 $\left(\dfrac{16a^2b^{-4}c}{a^5bc^7}\right)^0$ **36** $(a+b)^2(a+b)^{-2}$

37 $(3x^2y^{-3})^{-2}$ **38** $(-4a^3b^{-2})^{-2}$

39 $\dfrac{8u^2v^{-1}}{2u^{-1}v^3} \div \dfrac{u^3}{v^2}$

40 $(7x^2y^{-5} + 3x^{-1}y^2)^0$

41 $(3a+2b)^3(3a+2b)^{-2}$

42 $(cd^{-1})^{-1}$

43 $[(x^2y^{-1})^2]^{-3}$ **44** $\dfrac{(x^2y^{-3}z)^{-2}}{(x^{-2}yz^3)^{-1}}$

45 $[(a^{-1})^{-1}]^{-1}$ **46** $(p^{-1}q^{-1})^{-1}$

47 $\dfrac{(-2rs^2vp^{-1})^3}{(5r^{-3}sv^2p)^{-2}}$ **48** $\left(\dfrac{y^{-1}}{2x^{-1}}\right)\left(\dfrac{2x}{y}\right)^{-1}$

49 $\dfrac{(xy)^{-1}}{x^{-1}y^{-1}}$ **50** $\left(\dfrac{a^{-2}}{b^{-2}}\right)\left(\dfrac{b}{a}\right)^2$

1.5 RADICALS

We have used the symbol $\sqrt{2}$ to denote the positive real number whose square is 2. In general, if $a > 0$ and $b > 0$, we write

Definition of $\sqrt{b}$

$$\sqrt{b} = a \quad \text{if and only if} \quad a^2 = b$$

and call a the **principal square root of b**, or simply the **square root of b**.

Example 1   Find $\sqrt{9}$ and $\sqrt{25/49}$.

Solutions   We have

$$\sqrt{9} = 3 \quad \text{since} \quad 3^2 = 9,$$
$$\sqrt{25/49} = 5/7 \quad \text{since} \quad (5/7)^2 = 25/49.$$ ■

Note that if $b > 0$ and $a^2 = b$, then also $(-a)^2 = b$, and hence there are always two real numbers, one positive and one negative, that have b as their square. It is very important to note that $\sqrt{b}$ denotes only the *positive* real number whose square is b.

We also write

Definition of $\sqrt[3]{b}$

$$\boxed{\sqrt[3]{b} = a \quad \text{if and only if} \quad a^3 = b.}$$

In this case, if $b > 0$, then $a > 0$, whereas if $b < 0$, then $a < 0$. The number a is called the **principal cube root of b**, or simply the **cube root of b**.

Example 2   Find $\sqrt[3]{8}$, $\sqrt[3]{-8}$, and $\sqrt[3]{27/64}$.

Solutions   We have

$$\sqrt[3]{8} = 2 \quad \text{since} \quad 2^3 = 8,$$
$$\sqrt[3]{-8} = -2 \quad \text{since} \quad (-2)^3 = 8,$$
$$\sqrt[3]{27/64} = 3/4 \quad \text{since} \quad (3/4)^3 = 27/64.$$ ■

The **principal nth root** $\sqrt[n]{b}$ of b is defined as follows.

Definition of $\sqrt[n]{b}$

If both a and b are nonnegative real numbers and n is a positive integer, or if both a and b are negative and n is an odd positive integer, then

$$\sqrt[n]{b} = a \quad \text{if and only if} \quad a^n = b.$$

For brevity, we usually drop the adjective "principal." Of course, if $n = 2$, we write $\sqrt[2]{b} = \sqrt{b}$.

Example 3   Find $\sqrt[5]{-243}$, $\sqrt[4]{625/81}$, and $\sqrt[6]{64}$.

Solutions   Using the definition of $\sqrt[n]{b}$,

$$\sqrt[5]{-243} = -3 \quad \text{since} \quad (-3)^5 = -243,$$
$$\sqrt[4]{625/81} = 5/3 \quad \text{since} \quad (5/3)^4 = 625/81,$$
$$\sqrt[6]{64} = 2 \quad \text{since} \quad 2^6 = 64.$$ ■

Note that we have not defined $\sqrt[n]{b}$ if $b < 0$ and n is an *even* positive integer. The reason is that if n is even, then $a^n \geq 0$ for every real number a. We shall use the terminology "$\sqrt[n]{b}$ exists" if there is a real number a such that $a^n = b$. In the future, whenever we use the symbols $\sqrt[n]{b}$, $\sqrt[m]{c}$, $\sqrt[p]{x}$, etc., it is assumed that these roots exist. To complete our terminology, the symbol $\sqrt[n]{b}$ is called a **radical**, the number b is called the **radicand**, and n is the **index** of the radical. The $\sqrt{}$ is called a **radical sign**.

If $\sqrt{b} = a$, then $a^2 = b$; that is, $(\sqrt{b})^2 = b$. Similarly, if $\sqrt[3]{b} = a$, then $a^3 = b$; that is, $(\sqrt[3]{b})^3 = b$. In general, if n is any positive integer, then

$$(\sqrt[n]{b})^n = b.$$

It also follows that if $b > 0$, or if $b < 0$ and n is an odd positive integer, then

$$\sqrt[n]{b^n} = b.$$

For example,

$$\sqrt{5^2} = 5, \quad \sqrt[3]{(-2)^3} = -2, \quad \text{and} \quad \sqrt[4]{3^4} = 3.$$

In the following three laws m and n denote positive integers and it is assumed that the indicated roots exist.

Laws of Radicals

$$\text{(i)} \quad \sqrt[n]{bc} = \sqrt[n]{b}\,\sqrt[n]{c}; \qquad \text{(ii)} \quad \sqrt[n]{\frac{b}{c}} = \frac{\sqrt[n]{b}}{\sqrt[n]{c}}; \qquad \text{(iii)} \quad \sqrt[mn]{b} = \sqrt[m]{\sqrt[n]{b}}$$

Proof

To prove Law (i) we note that, by definition,

$$\sqrt[n]{bc} = a \quad \text{if and only if} \quad a^n = bc.$$

Consequently,

$$\sqrt[n]{bc} = \sqrt[n]{b}\,\sqrt[n]{c} \quad \text{if and only if} \quad (\sqrt[n]{b}\,\sqrt[n]{c})^n = bc.$$

The last relationship can be proved as follows (supply reasons):

$$(\sqrt[n]{b}\,\sqrt[n]{c})^n = (\sqrt[n]{b})^n(\sqrt[n]{c})^n = bc.$$

Law (ii) is proved in similar fashion.

To prove Law (iii) we note that, by definition,

$$\sqrt[mn]{b} = a \quad \text{if and only if} \quad a^{mn} = b,$$

and hence,

$$\sqrt[mn]{b} = \sqrt[m]{\sqrt[n]{b}} \quad \text{if and only if} \quad (\sqrt[m]{\sqrt[n]{b}})^{mn} = b.$$

To establish the last relationship we first note that if $\sqrt[m]{x}$ exists for a number x, then $(\sqrt[m]{x})^m = x$. In particular, letting $x = \sqrt[n]{b}$.

$$(\sqrt[m]{\sqrt[n]{b}})^m = \sqrt[n]{b}.$$

Raising both sides of the last equation to the nth power gives us

$$(\sqrt[m]{\sqrt[n]{b}})^{mn} = (\sqrt[n]{b})^n = b,$$

which is what we wished to prove. ∎

Example 4 Show that (a) $\sqrt{50} = 5\sqrt{2}$;   (b) $\sqrt[3]{-108} = -3\sqrt[3]{4}$;
(c) $\sqrt[4]{5/16} = \sqrt[4]{5}/2$.

Solutions   (Supply reasons.)

(a)
$$\sqrt{50} = \sqrt{25 \cdot 2} = \sqrt{25}\sqrt{2} = 5\sqrt{2}$$
(b)
$$\sqrt[3]{-108} = \sqrt[3]{(-27)(4)} = \sqrt[3]{-27}\sqrt[3]{4} = -3\sqrt[3]{4}$$

(c)
$$\sqrt[4]{\frac{5}{16}} = \frac{\sqrt[4]{5}}{\sqrt[4]{16}} = \frac{\sqrt[4]{5}}{2}$$ ∎

Example 5 Express each of the following in terms of a single radical.
(a) $\sqrt[3]{\sqrt[4]{b}}$ (b) $\sqrt[3]{\sqrt{64}}$

Solutions

(a) Using Law (iii) with $m = 3$ and $n = 4$,

$$\sqrt[3]{\sqrt[4]{b}} = \sqrt[12]{b}.$$

(b) Applying Law (iii) with $m = 3$ and $n = 2$,

$$\sqrt[3]{\sqrt{64}} = \sqrt[6]{64} = 2.$$

As a check, we have

$$\sqrt[3]{\sqrt{64}} = \sqrt[3]{8} = 2.$$ ∎

An important type of simplification involving radicals is that of **rationalizing a denominator**. The process involves beginning with a quotient which contains a radical in the denominator and then multiplying numerator and denominator by some expression so that the resulting denominator contains no radicals. The next example illustrates this technique.

Example 6 Rationalize the denominator in each of the following.

(a) $\dfrac{1}{\sqrt{5}}$ (b) $\sqrt{\dfrac{2}{3}}$ (c) $\sqrt[3]{\dfrac{x}{y}}$

Solutions   We may proceed as follows (supply reasons):

(a)
$$\frac{1}{\sqrt{5}} = \frac{1}{\sqrt{5}} \cdot \frac{\sqrt{5}}{\sqrt{5}} = \frac{\sqrt{5}}{(\sqrt{5})^2} = \frac{\sqrt{5}}{5}$$

(b)
$$\sqrt{\frac{2}{3}} = \sqrt{\frac{2}{3} \cdot \frac{3}{3}} = \sqrt{\frac{6}{3^2}} = \frac{\sqrt{6}}{\sqrt{3^2}} = \frac{\sqrt{6}}{3}$$

(c)
$$\sqrt[3]{\frac{x}{y}} = \sqrt[3]{\frac{x}{y} \cdot \frac{y^2}{y^2}} = \frac{\sqrt[3]{xy^2}}{\sqrt[3]{y^3}} = \frac{\sqrt[3]{xy^2}}{y}$$

∎

Theorem

If x is any real number, then $\sqrt{x^2} = |x|$.

Proof

The theorem is clearly true if $x = 0$. If $x > 0$, then $\sqrt{x^2} = x = |x|$. Finally, if $x < 0$, then $-x > 0$, and since $\sqrt{b^2} = b$ for all $b > 0$, we have $\sqrt{(-x)^2} = -x$. However, $|x| = -x$ if $x < 0$. This gives us $\sqrt{x^2} = \sqrt{(-x)^2} = -x = |x|$, and the theorem is proved.

∎

As an illustration of the preceding theorem,

$$\sqrt{(-4)^2} = |-4| = 4.$$

Of course, this equality could also be obtained by writing $\sqrt{(-4)^2} = \sqrt{16} = 4$. Incidentally, we have shown that *it is not always true that* $\sqrt{x^2} = x$. Indeed, if x is negative, then $\sqrt{x^2} = -x$.

If c is a real number and c^n occurs as a factor in a radical of index n, then c can be removed from the radicand provided the sign of c is taken into account. For example,

$$\sqrt[n]{c^n d} = \sqrt[n]{c^n}\sqrt[n]{d} = c\sqrt[n]{d},$$

where we have assumed that the sign of c is such that $\sqrt[n]{c^n} = c$. If, in the preceding equalities, c is negative and $n = 2$, then, by the preceding theorem,

$$\sqrt{c^2 d} = \sqrt{c^2}\sqrt{d} = |c|\sqrt{d}.$$

A similar situation exists if n is *any* even integer. In order to avoid considering positive and negative cases separately in the examples and exercises to follow, *we shall assume that a, b, c, d, etc., always represent positive real numbers.*

The technique we have discussed is useful for radicals of the form $\sqrt[n]{b}$ where b is an integer. In this case we first obtain the prime factorization of b. If a positive prime power p^n appears as a factor, then p may be taken out from under the radical sign.

If we use the term "simplify" when referring to a radical, we mean to proceed as above until the radicand contains no factors with exponent greater than or equal to the index of the radical. Moreover, no fractions should appear under the final radical sign and denominators should be rationalized. The index n should also be as low as possible.

Example 7 Simplify (a) $\sqrt[3]{320}$; (b) $\sqrt{1400}$; (c) $\sqrt{x^7}$; (d) $\sqrt[4]{x^9 y^6}$.

Solutions

(a) $\sqrt[3]{320} = \sqrt[3]{4^3 \cdot 5} = \sqrt[3]{4^3}\sqrt[3]{5} = 4\sqrt[3]{5}$

(b) $\sqrt{1400} = \sqrt{2^3 5^2 7} = \sqrt{(2 \cdot 5)^2 \cdot 2 \cdot 7} = \sqrt{(2 \cdot 5)^2}\sqrt{14} = 10\sqrt{14}$

(c) $\sqrt{x^7} = \sqrt{x^6 x} = \sqrt{(x^3)^2 x} = \sqrt{(x^3)^2}\sqrt{x} = x^3\sqrt{x}$

(d) $\sqrt[4]{x^9 y^6} = \sqrt[4]{(x^8 y^4)(xy^2)} = \sqrt[4]{(x^2 y)^4 (xy^2)} = \sqrt[4]{(x^2 y)^4}\sqrt[4]{xy^2} = x^2 y\sqrt[4]{xy^2}$

∎

Example 8 Simplify (a) $\sqrt[3]{16x^3y^8z^4}$;   (b) $\sqrt{3a^2b^3}\sqrt{6a^5b}$;   (c) $\sqrt{\dfrac{27x^3}{8y^5}}$.

Solutions   We may proceed as follows (supply reasons):

(a)
$$\sqrt[3]{16x^3y^8z^4} = \sqrt[3]{(2^3x^3y^6z^3)(2y^2z)}$$
$$= \sqrt[3]{(2xy^2z)^3(2y^2z)}$$
$$= \sqrt[3]{(2xy^2z)^3}\sqrt[3]{2y^2z}$$
$$= 2xy^2z\sqrt[3]{2y^2z}$$

(b)
$$\sqrt{3a^2b^3}\sqrt{6a^5b} = \sqrt{18a^7b^4}$$
$$= \sqrt{(9a^6b^4)(2a)}$$
$$= \sqrt{(3a^3b^2)^2(2a)}$$
$$= \sqrt{(3a^3b^2)^2}\sqrt{2a}$$
$$= 3a^3b^2\sqrt{2a}$$

(c)
$$\sqrt{\frac{27x^3}{8y^5}} = \sqrt{\frac{27x^3}{8y^5}\cdot\frac{2y}{2y}}$$
$$= \sqrt{\frac{54x^3y}{16y^6}}$$
$$= \sqrt{\frac{(9x^2)(6xy)}{(4y^3)^2}}$$
$$= \sqrt{\left(\frac{3x}{4y^3}\right)^2(6xy)}$$
$$= \sqrt{\left(\frac{3x}{4y^3}\right)^2}\sqrt{6xy}$$
$$= \frac{3x}{4y^3}\sqrt{6xy}$$ ∎

Radicals with the same index can sometimes be combined by using the Distributive Properties, as illustrated in the next example.

Example 9 Simplify $\sqrt{12} - \sqrt{27} + \sqrt{4x^2y} + \sqrt{y^3}$.

Solution
$$\sqrt{12} - \sqrt{27} + \sqrt{4x^2y} + \sqrt{y^3} = \sqrt{2^2(3)} - \sqrt{3^3} + \sqrt{(2x)^2(y)} + \sqrt{(y^2)(y)}$$
$$= 2\sqrt{3} - 3\sqrt{3} + 2x\sqrt{y} + y\sqrt{y}$$
$$= -\sqrt{3} + (2x + y)\sqrt{y}.$$ ∎

EXERCISES 1.5

Simplify each of the expressions in Exercises 1–48.
(All letters denote positive real numbers.)

1 $\sqrt{81}$

2 $\sqrt[3]{-125}$

3 $\sqrt[5]{-64}$

4 $\sqrt[4]{256}$

5 $\dfrac{1}{\sqrt[3]{2}}$

6 $\sqrt{\dfrac{1}{7}}$

7 $\sqrt{192}$

8 $\sqrt[3]{135}$

9 $5\sqrt{20} - \sqrt{45} + 2\sqrt{80}$

10 $\sqrt[4]{162} + \sqrt[4]{32} - \sqrt[4]{2}$

11 $\sqrt{9x^{-4}y^6}$

12 $\sqrt{16a^8b^{-2}}$

13 $\sqrt[3]{8a^6b^{-3}}$

14 $\sqrt[4]{81r^5s^8}$

15 $\sqrt[3]{\dfrac{54a^7}{b^2}}$

16 $\sqrt[5]{\dfrac{-96x^7}{y^3}}$

17 $\sqrt{\dfrac{1}{3u^3v}}$

18 $\sqrt[3]{\dfrac{1}{4x^5y^2}}$

19 $\sqrt[4]{(3x^5y^{-2})^4}$

20 $\sqrt[6]{(2u^{-3}v^4)^6}$

21 $\sqrt[5]{\dfrac{8x^3}{y^4}}\sqrt[5]{\dfrac{4x^4}{y^2}}$

22 $\sqrt{5xy^7}\sqrt{10x^3y^3}$

23 $\sqrt[3]{3t^4v^2}\sqrt[3]{-9t^{-1}v^4}$

24 $\sqrt[3]{(2r-s)^3}$

25 $\sqrt[6]{32p^4y^5z^3}\sqrt[6]{2p^2yz^3}$

26 $(\sqrt{27a^3b^2c})^2$

27 $\sqrt[3]{\dfrac{1}{3x^2}}$

28 $\sqrt[4]{125d\sqrt{25d^6}}$

29 $\sqrt{\dfrac{12x^5}{y^3}}$

30 $\dfrac{\sqrt{6a^5b^3}}{\sqrt{12a^6b}}$

31 $\sqrt{(a-b)^2}$

32 $(\sqrt{8x^5y})^3$

33 $\sqrt[5]{(x^2-5y^3z)^0}$

34 $\sqrt[3]{\dfrac{(3w^{-2}z)^4}{(2wz^{-2})^3}}$

35 $\sqrt[4]{\left(\dfrac{4r^{-2}}{s^8}\right)^{-3}}$

36 $\sqrt{3ab^3c}\sqrt{2a^2bc^4}\sqrt{6a^3b^4c^3}$

37 $\sqrt{\sqrt{81a^4b^2}}$

38 $\sqrt{\sqrt[3]{64a^7}}$

39 $\left(\sqrt[7]{\dfrac{x^8y^9}{z^2}}\right)^2$

40 $\left(\sqrt[4]{\dfrac{16a^7}{b^3}}\right)^4$

41 $\sqrt{10a^3b}\sqrt{2ab^2}$

42 $\sqrt[3]{9xy}\sqrt[3]{6x^2y^4}$

43 $\sqrt[3]{\sqrt{a^{12}b}}$

44 $\sqrt[3]{\sqrt[4]{(x^7y^5)^2}}$

45 $\dfrac{1}{\sqrt{a}}\dfrac{1}{\sqrt{b}}$

46 $\dfrac{1}{\sqrt{x^2+1}}$

47 $\sqrt{9x^3} + \sqrt{25x^5} - \sqrt{x}$

48 $\sqrt[3]{27a^4b^5} + \sqrt[3]{-8ab^8}$

49 Prove that, for all real numbers a and b, $|ab| = |a|\,|b|$. (*Hint:* Write $|ab| = \sqrt{(ab)^2} = \sqrt{a^2b^2} = \sqrt{a^2}\sqrt{b^2}$ and use the fact that $\sqrt{x^2} = |x|$.)

50 Prove that $|a/b| = |a|/|b|$ if $b \neq 0$.

CALCULATOR EXERCISES 1.5

In Exercises 1–4 demonstrate that the given equality is true by expressing each side as a decimal (do not use Laws of Radicals).

1 $\sqrt{5}\sqrt{3} = \sqrt{15}$

2 $\sqrt[3]{\sqrt{7}} = \sqrt[6]{7}$

3 $\sqrt[3]{5}/\sqrt[3]{4} = \sqrt[3]{5/4}$

4 $1/\sqrt{2} = \sqrt{2}/2$

In each of Exercises 5–10 use decimal approximations to determine whether $\square$ may be replaced by $<$, $>$, or $=$.

5 $\sqrt{2} + \sqrt{3} \ \square \ \sqrt{2+3}$

6 $\sqrt{7} - \sqrt{5} \ \square \ \sqrt{7-5}$

7 $\sqrt{2}\sqrt[3]{5} \ \square \ \sqrt[6]{10}$

8 $\sqrt[4]{523} \ \square \ \sqrt{22.8}$

9 $(1/\sqrt{2}) + (1/\sqrt{3}) \ \square \ 1/(\sqrt{2} + \sqrt{3})$

10 $\sqrt{\pi} \ \square \ \sqrt{154/7}$

1.6   RATIONAL EXPONENTS

Radicals can be used to introduce rational exponents. Let us begin by defining $b^{1/n}$, where n is a positive integer and $b > 0$. This must be done in a manner consistent with our previous work. In particular, if (ii) of the Laws of Exponents is to be true, then

$$(b^{1/n})^n = b^{(1/n)n} = b.$$

According to the definition of $\sqrt[n]{b}$, this implies that $b^{1/n} = \sqrt[n]{b}$. We are led, therefore, to the following definition of $b^{1/n}$.

Definition of $b^{1/n}$

> If b is a real number, n a positive integer, and $\sqrt[n]{b}$ exists, then $b^{1/n} = \sqrt[n]{b}$.

Whenever we use a symbol of the form $b^{1/n}$ in the future, it will be assumed that b and n are chosen so that $\sqrt[n]{b}$ exists.

Let us now consider $b^{m/n}$, where m and n are integers with $n > 0$. It is always possible to express a rational number m/n in this way since, for example, we can write $-(m/n)$ as $(-m)/n$. Let us prove that $(b^{1/n})^m = (b^m)^{1/n}$. If $m > 0$, then

$$(b^{1/n})^m = (b^{1/n})(b^{1/n}) \cdots \cdots (b^{1/n})$$

where there are m factors $b^{1/n}$ on the right. Hence,

$$
\begin{aligned}
(b^{1/n})^m &= \sqrt[n]{b}\sqrt[n]{b} \cdots \cdots \sqrt[n]{b} \\
&= \sqrt[n]{b \cdot b \cdots \cdots b} \\
&= \sqrt[n]{b^m} \\
&= (b^m)^{1/n}.
\end{aligned}
$$

The cases $m \le 0$ are left to the reader. Since $(\sqrt[n]{b})^m = \sqrt[n]{b^m}$, the following definition of rational exponents is meaningful.

Definition of Rational Exponents

> If m/n is a rational number, where n is a positive integer, and if b is a real number such that $\sqrt[n]{b}$ exists, then
>
> $$b^{m/n} = (\sqrt[n]{b})^m = \sqrt[n]{b^m}.$$

Note that we may also write

$$b^{m/n} = (b^{1/n})^m = (b^m)^{1/n}.$$

It can be shown that the Laws of Exponents are true for all rational exponents.

Example 1   Simplify:

(a) $(-27)^{2/3}(4)^{-5/2}$

(b) $(4a^{1/3})(2a^{1/2})$

(c) $(r^2 s^6)^{1/3}$

(d) $\left(\dfrac{2x^{2/3}}{y^{1/2}}\right)^2 \left(\dfrac{3x^{-5/6}}{y^{1/3}}\right)$

Solutions   (Supply reasons.)

(a) $(-27)^{2/3}(4)^{-5/2} = (\sqrt[3]{-27})^2(\sqrt{4})^{-5} = (-3)^2(2)^{-5} = \frac{9}{32}$

(b) $(4a^{1/3})(2a^{1/2}) = 8a^{1/3+1/2} = 8a^{5/6}$

(c) $(r^2s^6)^{1/3} = (r^2)^{1/3}(s^6)^{1/3} = r^{2/3}s^2$

(d) $\left(\dfrac{2x^{2/3}}{y^{1/2}}\right)^2\left(\dfrac{3x^{-5/6}}{y^{1/3}}\right) = \left(\dfrac{4x^{4/3}}{y}\right)\left(\dfrac{3x^{-5/6}}{y^{1/3}}\right) = \dfrac{12x^{1/2}}{y^{4/3}}$ ∎

Rational exponents are often useful for simplifying expressions which involve radicals. The technique is to use either the definition of $b^{1/n}$ or $b^{m/n}$ to transform radicals into expressions with rational exponents, then simplify, and finally change back to radical form. We shall illustrate the procedure in the following example. The reader should supply the reason for each equality.

Example 2   Simplify: (a) $\sqrt[4]{b^2}$;   (b) $\dfrac{\sqrt[3]{4}}{\sqrt{2}}$,   (c) $\dfrac{\sqrt[4]{a^3}}{\sqrt[3]{a^2}}$.   (d) $\sqrt{xy}\sqrt[3]{x^2y}$.

Solutions

(a) $\sqrt[4]{b^2} = (b^2)^{1/4} = b^{1/2} = \sqrt{b}$

(b) $\dfrac{\sqrt[3]{4}}{\sqrt{2}} = \dfrac{4^{1/3}}{2^{1/2}} = \dfrac{4^{2/6}}{2^{3/6}} = \left(\dfrac{4^2}{2^3}\right)^{1/6} = \left(\dfrac{16}{8}\right)^{1/6} = 2^{1/6} = \sqrt[6]{2}$

(c) $\dfrac{\sqrt[4]{a^3}}{\sqrt[3]{a^2}} = \dfrac{a^{3/4}}{a^{2/3}} = a^{3/4-2/3} = a^{1/12} = \sqrt[12]{a}$

(d) $\sqrt{xy}\sqrt[3]{x^2y} = (xy)^{1/2}(x^2y)^{1/3}$

$= x^{1/2}y^{1/2}x^{2/3}y^{1/3}$

$= x^{7/6}y^{5/6}$

$= xx^{1/6}y^{5/6}$

$= x(xy^5)^{1/6}$

$= x\sqrt[6]{xy^5}$ ∎

In Chapter 4 we shall extend our work to irrational exponents such as $3^{\sqrt{2}}$, 5^π, etc.

In scientific areas it is not unusual to work with numbers that are very large or very small. To simplify matters it is customary to write such numbers in the form $a \cdot 10^n$, where a is expressed as a decimal between 1 and 10 and n is an integer. This is referred to as the **scientific form** for real numbers. In applications it is customary to use $\times$ for the multiplication symbol and write $a \times 10^n$ instead of $a \cdot 10^n$. As an illustration, the distance a ray of light travels in one year is approximately 5,900,000,000,000 miles. This may be written in scientific form as 5.9×10^{12}. The positive exponent 12 indicates that the decimal point should be moved 12 places to the *right*. The notation works equally well for small numbers. To illustrate, it is estimated that the weight of an oxygen molecule is 0.000000000000000000000053 g or, in scientific form, 5.3×10^{-23} g. The negative exponent indicates that the decimal point should be moved 23 places to the *left*. Some other illustrations of scientific form are

$$513 = 5.13 \times 10^2, \qquad 20{,}700 = 2.07 \times 10^4, \qquad 92{,}000{,}000 = 9.2 \times 10^7$$
$$0.000648 = 6.48 \times 10^{-4}, \qquad \text{and} \qquad 0.00000000043 = 4.3 \times 10^{-10}.$$

There are a number of advantages to scientific notation. It is compact, and it enables the reader to see quickly (without counting zeros) the relative magnitudes of large or small quantities. It may also be used to simplify calculations which involve quantities such as those illustrated in the next two examples.

Example 3   Calculate $(0.000000004)(320{,}000{,}000{,}000)$.

Solution   Introducing scientific form, the given product may be found as follows:

$$(4 \times 10^{-9})(3.2 \times 10^{11}) = (4)(3.2) \times 10^{-9+11}$$
$$= 12.8 \times 10^2$$
$$= 1.28 \times 10^3 \qquad \blacksquare$$

Example 4   Calculate $\dfrac{(1{,}100{,}000)^2 \sqrt{0.00000004}}{(8{,}000{,}000{,}000)^{2/3}}$.

Solution   Writing each number in scientific form and using the Laws of Exponents, we have

$$\frac{(1.1 \times 10^6)^2 (4 \times 10^{-8})^{1/2}}{(8 \times 10^9)^{2/3}} = \frac{[(1.1)^2 \times 10^{12}][4^{1/2} \times 10^{-4}]}{8^{2/3} \times 10^6}$$
$$= \frac{[1.21 \times 10^{12}][2 \times 10^{-4}]}{4 \times 10^6}$$
$$= \frac{(1.21)(2)}{4} \times 10^{12-4-6}$$
$$= (0.605) \times 10^2$$
$$= 6.05 \times 10. \qquad \blacksquare$$

Hand-held calculators often employ scientific notation in their display panels. In this case the number 10 used in the scientific notation $a \times 10^n$ is suppressed and only the exponent is shown. For example, if we wanted to find $(4{,}500{,}000)^2$ on a typical calculator, we would enter the integer 4,500,000 and press the x^2 (or squaring) key.

The display panel would show

which is translated as 2.025×10^{13}. Thus,

$$(4,500,000)^2 = 20,250,000,000,000.$$

Many calculators also allow the entry of a number in scientific form. The owner of a calculator should consult the user's manual for details.

EXERCISES 1.6

Rewrite the expressions in Exercises 1–6 in terms of fractional exponents.

1 $\sqrt[4]{x^3}$

2 $\sqrt[3]{x^5}$

3 $\sqrt[3]{(a+b)^2}$

4 $\sqrt{a} + \sqrt{b}$

5 $\sqrt{x^2 + y^2}$

6 $\sqrt[3]{r^3 - s^3}$

Rewrite the expressions in Exercises 7 and 8 in terms of radicals.

7 (a) $4x^{3/2}$ (b) $(4x)^{3/2}$
   (c) $4 + x^{3/2}$ (d) $(4 + x)^{3/2}$

8 (a) $8y^{1/3}$ (b) $(8y)^{1/3}$
   (c) $8 - y^{1/3}$ (d) $(8 - y)^{1/3}$

Simplify the expressions in Exercises 9–38.

9 $(-64)^{2/3}$

10 $16^{3/2}$

11 $(0.0009)^{3/2}$

12 $(0.027)^{-1/3}$

13 $(256)^{-3/4}$

14 $(-243)^{-2/5}$

15 $(3x^{2/3})(5x^{4/3})$

16 $(2u^{5/2})(6u^{1/2})$

17 $(2a^{1/3})(9a^{1/2})$

18 $y^{1/6}(8y^{1/3})$

19 $(16t^4)^{-3/2}$

20 $(-8w^6)^{2/3}$

21 $(3c^{2/3})(4c)^{3/2}$

22 $(6x^{1/3}y^{3/2})^2$

23 $\left(\dfrac{16a^8}{p^{-4}}\right)^{3/4}$

24 $\left(\dfrac{w^{-1/3}}{w^{3/2}}\right)^6$

25 $(p^{-3}s^9r^{-6})^{-2/3}$

26 $\dfrac{(x^{-4}y)^{-1/2}}{(x^2y^3)^{-1/3}}$

27 $\left(\dfrac{r^{1/2}}{s^2}\right)^4\left(\dfrac{s^{-1/3}}{r^{2/3}}\right)^3$

28 $\left(\dfrac{1}{4x}\right)^{1/2}$

29 $u^{1/3}u^{1/4}u^{1/6}$

30 $a^{1/2}a^{1/3}a^{1/6}$

31 $\left(\dfrac{x^{-2/3}z^{4/3}}{x^{4/5}z^{-1/4}}\right)^0$

32 $\dfrac{(a^{-1/5}b^{5/2})^{10}}{(a^{2/3}b^{-3/5})^{15}}$

33 $(4a^2b^{2/3}c^{-1})^{1/2}(3a^{-2}b^{1/3}c^3)^2$

34 $\left(\dfrac{p^{-2/3}q^{5/6}}{p^{3/4}q^{7/2}}\right)^{12}$

35 $(5x^{3/2}(xy^{-2})^{1/2})^4$

36 $((x^4y^{-3})^{1/5})^{-10}$

37 $a^{1/2}(a^{1/2} + a^{-1/2})$

38 $s^{1/3}(s^{2/3} - s^{5/3})$

Write the expressions in Exercises 39–52 as a radical with least possible index.

39 $\sqrt[6]{x^3}$

40 $\sqrt[4]{u^2}$

41 $\sqrt[4]{c^2d^2}$

42 $\sqrt[9]{c^3}$

43 $\sqrt[8]{a^4b^2}$

44 $\sqrt[6]{r^3s^9}$

45 $\sqrt[3]{4}\sqrt{2}$

46 $\sqrt{5}\sqrt[3]{2}$

47 $\sqrt[3]{x^2y}\sqrt{xy}$

48 $\sqrt[3]{xy^2}\sqrt[4]{x^3y}$

49 $\dfrac{\sqrt[3]{u^2v}}{\sqrt[6]{u^5v^4}}$

50 $\dfrac{\sqrt{ab^3}}{\sqrt[3]{a^2b}}$

51 $\sqrt[3]{b^2\sqrt{b}}$

52 $\sqrt{z\sqrt[3]{z}}$

53 The mass of a hydrogen atom is approximately 0.00000000000000000000000017 g. Express this number in scientific form.

54 The mass of an electron is approximately 9.1×10^{-31} kg. Express this number in decimal form.

Express the numbers in Exercises 55 and 56 in scientific form.

55 (a) 427,000   (b) 0.000000098
     (c) 810,000,000

56 (a) 85,200   (b) 0.0000055
     (c) 24,900,000

Express the numbers in Exercises 57 and 58 in decimal form.

57 (a) 8.3×10^5 (b) 2.9×10^{-12}
     (c) 5.63×10^8

58 (a) 2.3×10^7 (b) 7.01×10^{-9}
     (c) 1.23×10^{10}

Use scientific form to find the numbers in Exercises 59–64. If a calculator is available, use it to check your answers.

59 $(2,100,000,000)(0.0000000033)$

60 $(840,000,000)(0.00000021)$

61 $\dfrac{\sqrt{81,000,000}}{(0.0000002)^3}$

62 $\dfrac{(300,000)^4}{(8,000,000)^{2/3}}$

63 $\dfrac{\sqrt[5]{(0.00243)^2}(20,000)^3}{\sqrt[3]{8,000,000}}$

64 $\dfrac{(160,000)^{3/2}(12,100,000)}{(0.0000011)^2}$

65 In astronomy, distances to stars are measured in light years, where 1 light year is the distance a ray of light travels in one year. If the speed of light is 186,000 miles per second, approximate 1 light year.

66 (a) It is estimated that the Milky Way galaxy contains 100 billion stars. Express this number in scientific form.
     (b) The diameter d of the Milky Way galaxy is estimated as 100,000 light years. Express d in miles (refer to Exercise 65).

CALCULATOR EXERCISES 1.6

In Exercises 1–4 demonstrate that the given equality is true by expressing each side as a decimal.

1 $7^{0.5} = \sqrt{7}$ **2** $\sqrt[5]{(-7)^3} = (\sqrt[5]{-7})^3$

3 $10^{0.25} = \sqrt{\sqrt{10}}$ **4** $(\sqrt[4]{2.1})^5 = \sqrt[4]{(2.1)^5}$

Express the numbers in Exercises 4–8 as decimals, rounded to the nearest thousandth.

5 $(-6.413)^{3/5}$ **6** $4^{0.999}$

7 $(4.17)^{-1.38}$ **8** $(3671)^{4316/6123}$

In Exercises 9–12 use decimal approximations to determine whether $\square$ may be replaced by $<$, $>$, or $=$.

9 $10^{2.3} \;\square\; 10^{2.4}$ **10** $4^{3.62} \;\square\; 4^{3.61}$

11 $(0.7)^{2.64} \;\square\; (0.7)^{2.65}$

12 $(437)^{2/3} \;\square\; 437^{0.6666}$

1.7 POLYNOMIALS AND ALGEBRAIC EXPRESSIONS

We frequently make use of symbols to denote arbitrary elements of a set. For example, we may use x to denote a real number, although no *particular* real number is specified. A letter which is used to represent any element of a given set is sometimes called a **variable**. Throughout this text, unless otherwise specified, variables will represent real numbers. In some cases it is necessary to restrict the numbers which are represented by a variable. We shall use the following terminology.

Definition

> The **domain of a variable** is the set of real numbers represented by the variable.

The domain of a variable x is often referred to as the set of "permissible" or "allowable" values for x. To illustrate, given the expression $\sqrt{x}$, in order to obtain a real number we must have $x \geq 0$, and hence in this case the domain of x is assumed to be the set of nonnegative real numbers. Similarly, when considering the expression $1/(x - 2)$, we must exclude $x = 2$ (Why?) and consequently we take the domain of x as the set of all real numbers different from 2.

If we begin with any collection of variables and real numbers, then an **algebraic expression** is the result obtained by applying additions, subtractions, multiplications, divisions, or the taking of roots. The following are examples of algebraic expressions:

$$x^3 - 2x + \frac{3^{1/9}}{\sqrt{2x}}, \quad \frac{2xy + 3x}{y - 1}, \quad \frac{4yz^{-2} + \left(\dfrac{-7}{x + w}\right)^5}{\sqrt[3]{y^2} + \sqrt{5z}},$$

where x, y, z, and w are variables. If specific numbers are substituted for the variables in an algebraic expression, the resulting real number is called the **value** of the expression for these numbers. For example, the value of the second expression above when $x = -2$ and $y = 3$ is

$$\frac{2(-2)(3) + 3(-2)}{3 - 1} = \frac{-12 - 6}{2} = -9.$$

When we work with algebraic expressions, it will be assumed that the domains are chosen so that variables do not represent numbers which make the expressions meaningless. Thus, it is assumed that denominators do not vanish, roots always exist, and so on. To simplify our work, we shall not always state the domains of variables. If meaningless expressions occur when certain numbers are substituted for the variables, then these numbers are *not* in the domains of the variables.

Certain algebraic expressions are given special names. If x is a variable, then a **monomial** in x is an expression of the form ax^n, where a is a real number and n is a nonnegative integer. The number a is called the **coefficient** of x^n.

A *polynomial in x* is any finite sum of monomials in x. Another way of stating this is as follows.

Definition

> A **polynomial in x** is an expression of the form
>
> $$a_n x^n + a_{n-1} x^{n-1} + \cdots + a_1 x + a_0$$
>
> where n is a nonnegative integer and each a_i is a real number.

In the preceding definition, each of the expressions $a_k x^k$ in the sum is called a **term** of the polynomial. If a coefficient a_i is zero, we usually delete the term $a_i x^i$. The coefficient a_n of the highest power of x is the **leading coefficient** of the polynomial and, if $a_n \neq 0$, we say that the polynomial has **degree n**. If *all* the coefficients of a polynomial are zero, it is called the **zero polynomial** and is denoted by 0. It is customary not to assign a degree to the zero polynomial.

The following are some examples of polynomials.

$$3x^4 + 5x^3 + (-7)x + 4 \qquad \text{(degree 4)}$$

$$x^8 + 9x^2 + (-2)x \qquad \text{(degree 8)}$$

$$5x^2 + 1 \qquad \text{(degree 2)}$$

$$7x + 2 \qquad \text{(degree 1)}$$

$$5 \qquad \text{(degree 0)}$$

If some of the coefficients in a polynomial are negative, then for convenience we often use minus signs between appropriate terms. To illustrate, instead of $3x^2 + (-5)x + (-7)$, we write $3x^2 - 5x - 7$ for this polynomial of degree 2. Polynomials in other variables may also be considered. For example, $\frac{2}{5}z^2 - 3z^7 + 8 - \sqrt{5}z^4$ is a polynomial in z of degree 7. We ordinarily arrange the terms in order of decreasing powers of the variable and write $-3z^7 - \sqrt{5}z^4 + \frac{2}{5}z^2 + 8$.

A polynomial in x may be thought of as an algebraic expression obtained by employing only additions, subtractions, and multiplications involving x. In particular, the expressions

$$\frac{1}{x} + 3x, \quad \frac{x-5}{x^2+2}, \quad \text{and} \quad 3x^2 + \sqrt{x} - 2$$

are not polynomials since they also involve divisions by variables, or roots involving variables.

Since polynomials, and the monomials which make up polynomials, are symbols representing real numbers, all of the rules given in Section 1.2 can be applied. If additions, multiplications, and subtractions are carried out with polynomials, we may simplify the result by using various properties of real numbers.

Example 1   Find the sum of the polynomials $x^3 + 2x^2 - 5x + 7$ and $4x^3 - 5x^2 + 3$.

Solution   Rearranging terms and using the Distributive Properties gives us

$$(x^3 + 2x^2 - 5x + 7) + (4x^3 - 5x^2 + 3)$$
$$= x^3 + 4x^3 + 2x^2 - 5x^2 - 5x + 3 + 7$$
$$= (1 + 4)x^3 + (2 - 5)x^2 + (-5)x + (3 + 7)$$
$$= 5x^3 - 3x^2 - 5x + 10. \qquad \blacksquare$$

Example 1 illustrates the fact that the sum of any two polynomials in x can be obtained by adding coefficients of like powers of x. In order to keep track of the coefficients it is sometimes convenient to use the following scheme for finding the sum.

$$x^3 + 2x^2 - 5x + 7$$
$$\underline{4x^3 - 5x^2 + 3}$$
$$5x^3 - 3x^2 - 5x + 10$$

The difference of two polynomials is found by subtracting coefficients of like powers, as indicated by the next example.

Example 2 Subtract $4x^3 - 5x^2 + 3$ from $x^3 + 2x^2 - 5x + 7$.

Solution We could use the scheme shown previously to subtract, proceeding as follows:

$$(x^3 + 2x^2 - 5x + 7) - (4x^3 - 5x^2 + 3)$$
$$= x^3 + 2x^2 - 5x + 7 - 4x^3 + 5x^2 - 3$$
$$= x^3 - 4x^3 + 2x^2 + 5x^2 - 5x + 7 - 3$$
$$= (1 - 4)x^3 + (2 + 5)x^2 - 5x + (7 - 3)$$
$$= -3x^3 + 7x^2 - 5x + 4.$$ ■

The intermediate steps in the previous solution were used for completeness. After you become proficient with such manipulations, these may be omitted. In order to multiply two polynomials, we use the Distributive Laws together with the Laws of Exponents and combine like terms, as in the next example.

Example 3 Find the product of $x^2 + 5x - 4$ and $2x^3 + 3x - 1$.

Solution

$$(x^2 + 5x - 4)(2x^3 + 3x - 1)$$
$$= x^2(2x^3 + 3x - 1) + 5x(2x^3 + 3x - 1) - 4(2x^3 + 3x - 1)$$
$$= 2x^5 + 3x^3 - x^2 + 10x^4 + 15x^2 - 5x - 8x^3 - 12x + 4$$
$$= 2x^5 + 10x^4 + (3 - 8)x^3 + (-1 + 15)x^2 + (-5 - 12)x + 4$$
$$= 2x^5 + 10x^4 - 5x^3 + 14x^2 - 17x + 4$$

For convenience, this procedure may be arranged as follows.

$$
\begin{array}{l}
2x^3 + 3x - 1 \\
\underline{x^2 + 5x - 4} \\
2x^5 \qquad\quad + 3x^3 - \quad x^2 \qquad\qquad\quad = \ x^2(2x^3 + 3x - 1) \\
\qquad 10x^4 \qquad\quad + 15x^2 - \ 5x \qquad = \ 5x(2x^3 + 3x - 1) \\
\underline{\qquad\qquad - 8x^3 \qquad\quad - 12x + 4} = -4(2x^3 + 3x - 1) \\
2x^5 + 10x^4 - 5x^3 + 14x^2 - 17x + 4 = \quad \text{sum of above}
\end{array}
$$
■

We may also consider polynomials in more than one variable. For example, a polynomial in two variables x and y is a sum of terms, each of the form $ax^m y^k$ for some real number a and nonnegative integers m and k. An example is

$$3x^4y + 2x^3y^5 + 7x^2 - 4xy + 8y - 5.$$

In like manner, we may consider polynomials in three variables x, y, z or, for that matter, in *any* number of variables. Addition, subtraction, and multiplication are performed using properties of real numbers. A simple illustration is given in the next example.

Example 4   Find the product of $x^2 + xy + y^2$ and $x - y$.

Solution
$$(x^2 + xy + y^2)(x - y) = (x^2 + xy + y^2)x - (x^2 + xy + y^2)y$$
$$= x^3 + x^2y + xy^2 - x^2y - xy^2 - y^3$$
$$= x^3 - y^3$$

We could also arrange our work as follows:

$$x^2 + xy + y^2$$
$$x - y$$
$$\overline{x^3 + x^2y + xy^2} \qquad = x(x^2 + xy + y^2)$$
$$\underline{- x^2y - xy^2 - y^3} = (-y)(x^2 + xy + y^2)$$
$$x^3 \qquad\qquad - y^3 = \text{sum of the above}$$ ■

Division by a monomial is relatively easy, as seen in the next example.

Example 5   Divide $6x^2y^3 + 4x^3y^2 - 10xy$ by $2xy$.

Solution   Using the Properties of Quotients and the Laws of Exponents we obtain

$$\frac{6x^2y^3 + 4x^3y^2 - 10xy}{2xy} = \frac{6x^2y^3}{2xy} + \frac{4x^3y^2}{2xy} - \frac{10xy}{2xy}$$
$$= 3xy^2 + 2x^2y - 5.$$ ■

Certain products occur so frequently in algebra that they deserve special attention. We list some of these next, where the letters represent real numbers. The reader should check the validity of each formula by actually carrying out the multiplications.

Product Formulas

(i) $(x + y)(x - y) = x^2 - y^2$
(ii) $(ax + b)(cx + d) = acx^2 + (ad + bc)x + bd$
(iii) $(x + y)^2 = x^2 + 2xy + y^2$
(iv) $(x - y)^2 = x^2 - 2xy + y^2$
(v) $(x + y)^3 = x^3 + 3x^2y + 3xy^2 + y^3$
(vi) $(x - y)^3 = x^3 - 3x^2y + 3xy^2 - y^3$

Since the symbols x and y used in these formulas represent real numbers, they may be replaced by algebraic expressions, as illustrated in the next example.

Example 6   Find the products:

(a) $(2r^2 - \sqrt{s})(2r^2 + \sqrt{s})$ (b) $\left(\sqrt{c} + \dfrac{1}{\sqrt{c}}\right)^2$ (c) $(2a - 5b)^3$

Solutions

(a) Using Product Formula (i) with $x = 2r^2$ and $y = \sqrt{s}$,

$$(2r^2 - \sqrt{s})(2r^2 + \sqrt{s}) = (2r^2)^2 - (\sqrt{s})^2$$
$$= 4r^4 - s.$$

(b) Using Product Formula (iii) with $x = \sqrt{c}$ and $y = 1/\sqrt{c}$,

$$\left(\sqrt{c} + \frac{1}{\sqrt{c}}\right)^2 = (\sqrt{c})^2 + 2\sqrt{c} \cdot \frac{1}{\sqrt{c}} + \left(\frac{1}{\sqrt{c}}\right)^2$$

$$= c + 2 + \frac{1}{c}.$$

(c) Applying Product Formula (vi) with $x = 2a$ and $y = 5b$,

$$(2a - 5b)^3 = (2a)^3 - 3(2a)^2(5b) + 3(2a)(5b)^2 - (5b)^3$$
$$= 8a^3 - 60a^2b + 150ab^2 - 125b^3.$$ ∎

1 – 37 ODD)

EXERCISES 1.7

In Exercises 1–22 perform the indicated operations and express the result as a polynomial.

1 $(4x^3 + 2x^2 - x + 5) + (x^3 - 3x^2 - 5x + 1)$

2 $(x^4 - 3x^2 + 7x + 4) + (x^3 + 3x^2 - 4x - 3)$

3 $(5x^4 - 6x^2 + 9x) - (2x^3 + 3x^2 - 8x + 4)$

4 $(2x^2 - 11x + 13) - (3x^4 + 9x^2 - 10)$

5 $(5y^3 - 6y^2 + y - 7) - (5y^3 + 6y^2 + y + 2)$

6 $(4z^2 - 4z + 1) - (2z + 1)^2$

7 $(2a^4 - 3a^2 + 5) + a(a^3 + 3a - 4)$

8 $(3u + 1)(2u - 3) + 6u(u + 5)$

9 $(3x - 4)(2x^2 + x - 5)$

10 $(4x^3 - x^2 - 7)(3x + 2)$

11 $(r^2 + 2r + 3)(3r^2 - 2r + 4)$

12 $(s + t)(s^2 - st + t^2)$

13 $(6x^3 - 3x^2 - x + 7)(2x^2 + 4x + 5)$

14 $(2x^2 - xy + y^2)(3x - y)$

15 $(r - t)(r^2 + rt + t^2)$

16 $(x + y)(x^3 - x^2y + xy^2 - y^3)$

17 $(3x + 1)(2x^2 - x + 2)(x^2 + 4)$

18 $(3c + 1)(2c^2 + 5)(c^3 + 4)$

19 $\dfrac{8x^2y^3 - 10x^3y}{2x^2y}$

20 $\dfrac{6a^3b^3 - 9a^2b^2 + 3ab^4}{3ab^2}$

21 $\dfrac{3u^3v^4 - 2u^5v^2 + (u^2v^2)^2}{u^3v^2}$

22 $\dfrac{6x^2yz^3 - xy^2z}{xyz}$

Use the Product Formulas to express the following as polynomials.

23 $(x - 3)(2x + 1)$

24 $(3x + 2)(3x - 5)$

25 $(2s - 7t)(4s - 5t)$

26 $(8n - 6p)(7n - 10p)$

27 $(5x^2 + 2y)(3x^2 - 7y)$

28 $(x + 9y^2)(3x - 4y^2)$

29 $(6t - 5v)(6t + 5v)$

30 $(8u + 3)^2$

31 $(3r + 10s)^2$

32 $(4v - 3w)(4v + 3w)$

33 $(4x^2 - 5y^2)^2$

34 $(10p^2 + 7q^2)^2$

35 $(x - x^{-1})^2$

36 $(b^3 - b^{-3})^2$

37 $(\sqrt{a} + \sqrt{b})(\sqrt{a} - \sqrt{b})$

38 $(3\sqrt{s} + \sqrt{t})^2$

39 $(x - 2y)^3$

40 $(4x - y)^3$

41 $(3r + 4s)^3$

42 $(2a + 5b)^3$

43 $(x^2 + y^2)^3$

44 $(u^2 - 3v)^3$

45 $(a^{1/3} - b^{1/3})^3$

46 $(a + b)^2(a - b)^2$

47 $(x + y + z)(x + y - z)$

48 $(2a - b + 3c)(2a - b - 3c)$

49 $(3x + 2y + z)^2$

50 $(x^2 + y^2 + z^2)^2$

1.8 FACTORING

If a polynomial is written as a product of other polynomials, then each polynomial in the product is called a **factor** of the original polynomial. The process of expressing a polynomial as a product is called **factoring**. For example, since $x^2 - 9 = (x + 3)(x - 3)$, we see that $x + 3$ and $x - 3$ are factors of $x^2 - 9$.

Factoring plays a major role in numerous mathematical applications, since it may be used to reduce the study of a complicated expression to the study of several simpler expressions. For example, important properties of the polynomial $x^2 - 9$ can be determined by examining the factors $x + 3$ and $x - 3$. As another illustration, it can be shown that

$$6x^3 + 37x^2 + 47x - 20 = (2x + 5)(3x - 1)(x + 4).$$

Various properties of the indicated third-degree polynomial can then be found by studying the factors $2x + 5$, $3x - 1$, and $x + 4$. In future chapters you will encounter many specific problems which can be simplified by factoring. This will be especially true when you are finding solutions of equations and inequalities, or studying functions.

We shall be interested primarily in **nontrivial factors** of polynomials; that is, factors which contain polynomials of degree greater than zero. An exception to this rule is that if the coefficients are restricted to *integers*, then it is customary to remove a common integral factor from each term of the polynomial. This can be done by means of the distributive law, as in the factorization

$$4x^2y + 8z^3 = 4(x^2y + 2z^3).$$

Before carrying out factorizations of polynomials it is necessary to specify the system from which the coefficients of the factors are to be chosen. In this chapter we shall use the rule that *if a polynomial with integer coefficients is given, then the factors should be polynomials with integer coefficients.* For example,

$$x^2 + x - 6 = (x + 3)(x - 2).$$
$$4x^2 - 9y^2 = (2x - 3y)(2x + 3y).$$

An integer $a > 1$ is prime if it cannot be written as a product of two positive integers greater than 1. A polynomial is said to be **prime** or **irreducible** if it cannot be written as a product of two polynomials of positive degree. When factoring a polynomial the objective is to express it as a product of prime polynomials or powers of prime polynomials. According to the rule given above, $x^2 - 2$ is prime since it cannot be factored as a product of two polynomials of positive degree which have *integral* coefficients. If we allow the factors to have *real* coefficients, then $x^2 - 2$ is not prime, since

$$x^2 - 2 = (x + \sqrt{2})(x - \sqrt{2}).$$

However, as we have said, we shall not allow irrational coefficients in the factorizations of this chapter. Later, in our work with solutions of equations, this rule will be rescinded.

In general, it is very difficult to factor polynomials of degree greater than 2. In simple cases some of the Product Formulas given in Section 1.7 are useful. One of the most important is Formula (i). Let us restate it for reference as follows.

Difference of Two Squares

$$\boxed{a^2 - b^2 = (a + b)(a - b)}$$

Example 1   Factor each of the following.
   (a) $25r^2 - 49s^2$ (b) $81x^4 - y^4$ (c) $16x^4 - (y - 2z)^2$

Solutions

   (a) Applying the Difference of Two Squares Formula with $a = 5r$ and $b = 7s$ gives us

$$25r^2 - 49s^2 = (5r)^2 - (7s)^2 = (5r + 7s)(5r - 7s).$$

   (b) We make two applications of the formula as follows:

$$81x^4 - y^4 = (9x^2)^2 - (y^2)^2$$
$$= (9x^2 + y^2)(9x^2 - y^2)$$
$$= (9x^2 + y^2)(3x + y)(3x - y).$$

   (c) If we write $16x^4 = (4x^2)^2$, then we may use the formula with $a = 4x^2$ and $b = y - 2z$ as follows:

$$16x^4 - (y - 2z)^2 = (4x^2)^2 - (y - 2z)^2$$
$$= [(4x^2) + (y - 2z)][(4x^2) - (y - 2z)]$$
$$= (4x^2 + y - 2z)(4x^2 - y + 2z). \qquad \blacksquare$$

A factorization of a second-degree polynomial $px^2 + qx + r$, where p, q, and r are integers, must be of the form $(ax + b)(cx + d)$, where a, b, c, and d are integers. It follows that $ac = p$, $bd = r$, and $ad + bc = q$. Evidently there is only a limited number of choices for a, b, c, and d which satisfy these conditions. If none of the choices work, then $px^2 + qx + r$ is prime. This method is also applicable to polynomials of the form $px^2 + qxy + ry^2$.

Example 2   Factor the following:

    (a) $6x^2 - 7x - 3$ (b) $4x^2 - 12xy + 9y^2$ (c) $4x^4y - 11x^3y^2 + 6x^2y^3$

Solutions

    (a)   If we write

$$6x^2 - 7x - 3 = (ax + b)(cx + d),$$

then the product of a and c is 6; the product of b and d is -3; and $ad + bc = -7$. Trying various possibilities, we arrive at the factorization

$$6x^2 - 7x - 3 = (2x - 3)(3x + 1).$$

    (b)   If a factorization as a product of two first-degree polynomials exists, then it must be of the form

$$4x^2 - 12xy + 9y^2 = (ax + by)(cx + dy).$$

By trial we obtain

$$4x^2 - 12xy + 9y^2 = (2x - 3y)(2x - 3y) = (2x - 3y)^2.$$

    (c)   Since each term has x^2y as a factor, we begin by writing

$$4x^4y - 11x^3y^2 + 6x^2y^3 = x^2y(4x^2 - 11xy + 6y^2).$$

Then by trial we obtain the following factorization:

$$4x^4y - 11x^3y^2 + 6x^2y^3 = x^2y(4x - 3y)(x - 2y). \qquad \blacksquare$$

In some cases, if terms in a sum are grouped in suitable fashion, then a factorization can be found by means of the Distributive Properties, as illustrated in the next example.

Example 3   Factor the following:

    (a) $a(x^2 - y) + 2b(x^2 - y)$ (b) $4ac + 2bc - 2ad - bd$
    (c) $3x^3 + 2x^2 - 12x - 8$

Solutions

    (a)   The given expression is of the form $ac + 2bc$ with $c = x^2 - y^2$. Since $ac + 2bc = (a + 2b)c$, we have

$$a(x^2 - y) + 2b(x^2 - y) = (a + 2b)(x^2 - y).$$

(b) We group the first two terms and the last two terms, and then use the Distributive Properties as follows:

$$
\begin{aligned}
4ac + 2bc - 2ad - bd &= (4ac + 2bc) - (2ad + bd) \\
&= 2c(2a + b) - d(2a + b) \\
&= (2c - d)(2a + b).
\end{aligned}
$$

(c) We again begin by grouping the first two terms and the last two terms of the given expression as follows:

$$3x^3 + 2x^2 - 12x - 8 = x^2(3x + 2) - 4(3x + 2).$$

The right-hand side has the form $x^2c - 4c$, where $c = 3x + 2$. Since $x^2c - 4c = (x^2 - 4)c$, we have

$$3x^3 + 2x^2 - 12x - 8 = (x^2 - 4)(3x + 2).$$

Finally, using the Difference of Two Squares Formula, we obtain the factorization

$$3x^3 + 2x^2 - 12x - 8 = (x + 2)(x - 2)(3x + 2). \qquad \blacksquare$$

The technique illustrated in Example 3 is called **factorization by grouping**. Each of the next two formulas may be verified by multiplying the two factors on the right-hand side of the equation. Pay careful attention to where the plus or minus signs occur.

Sum of Two Cubes

$$x^3 + y^3 = (x + y)(x^2 - xy + y^2)$$

Difference of Two Cubes

$$x^3 - y^3 = (x - y)(x^2 + xy + y^2)$$

Example 4 Factor the following:

(a) $a^3 + 64b^3$ (b) $8x^6 - 27y^9$

Solutions

(a) Using the Sum of Two Cubes Formula, with $x = a$ and $y = 4b$,

$$
\begin{aligned}
a^3 + 64b^3 &= a^3 + (4b)^3 \\
&= (a + 4b)[a^2 - a(4b) + (4b)^2] \\
&= (a + 4b)(a^2 - 4ab + 16b^2).
\end{aligned}
$$

(b) Using the Difference of Two Cubes Formula,

$$
\begin{aligned}
8x^6 - 27y^9 &= (2x^2)^3 - (3y^3)^3 \\
&= (2x^2 - 3y^3)[(2x^2)^2 + (2x^2)(3y^3) + (3y^3)^2] \\
&= (2x^2 - 3y^3)(4x^4 + 6x^2y^3 + 9y^6). \qquad \blacksquare
\end{aligned}
$$

EXERCISES 1.8

1-47 ODD

Factor each of the following

1 $rs + 4st$

2 $4u^2 - 2uv$

3 $3a^2b^2 - 6a^2b$

4 $10xy + 15xy^2$

5 $9x^2y^2 + 15xy^4$

6 $-8p^4qr^2 - 4p^3q^3r^2$

7 $4x^2 + 5x - 6$

8 $21x^2 + 29x - 10$

9 $4x^2 - 4x - 15$

10 $6x^2 + x - 5$

11 $16y^2 - 14y - 15$

12 $12m^2 - 17m - 14$

13 $9z^2 - 12z + 4$

14 $4x^2 + 12x + 9$

15 $6c^2 + 25c + 24$

16 $20y^2 - 41y + 20$

17 $4r^2 - 25t^2$

18 $36a^2 - 49b^2$

19 $50x^2 + 45xy - 18y^2$

20 $45x^2 + 38xy + 8y^2$

21 $16w^4 - 9s^2$

22 $25p^2 - 16v^4$

23 $36z^2 + 60z + 25$

24 $64y^2 + 112y + 49$

25 $27x^3 - y^3$

26 $x^3 + 8y^3$

27 $27a^3 + 64b^3$

28 $8r^3 - 27s^3$

29 $8x^6 - 125$

30 $216 - y^6$

31 $6ax - 3ay + 2bx - by$

32 $5ru + 10vr + 2ut + 4vt$

33 $40zw + 8x^2w - 35z - 7x^2$

34 $18ck + 4dk + 9cj + 2dj$

35 $a^3 - a^2b + ab^2 - b^3$

36 $6w^8 + 17w^4 + 12$

37 $a^6 - b^6$

38 $x^8 - 16$

39 $x^4 + 25$

40 $a^2 + a + 1$

41 $6x^2 + 42x + 60$

42 $16x^2 + 40x - 24$

43 $4x^2 - 24x + 36$

44 $60x^2 - 85x + 30$

45 $75x^2 + 120x + 48$

46 $64x^2 - 16$

47 $36 - 9x^2$

48 $18x^2 - 50$

49 $2x^2y + xy - 2xz - z$

50 $3ac - 6bd + 3ad - 6bc$

51 $12x^2z + 8y^2z - 15x^2w - 10y^2w$

52 $4x^3 + 6x^2y - 4y^2x - 6y^3$

53 $y^6 + 7y^3 - 8$

54 $8c^6 + 19c^3 - 27$

55 $(x + y)^3 - 27$

56 $(a + b)^4 - 1$

57 $x^2 + 2x + 5$

58 $4x^3 + 4x^2 + x$

59 $x^{16} - 1$

60 $x^{16} + 1$

1.9 FRACTIONAL EXPRESSIONS

Quotients of algebraic expressions are called **fractional expressions**. As a special case, a quotient of two polynomials is called a **rational expression**. Some examples are

$$\frac{x^2 - 5x + 1}{x^3 + 7}, \quad \frac{z^2x^4 - 3yz}{5z}, \quad \text{and} \quad \frac{1}{4xy}.$$

Many problems in mathematics involve combining rational expressions and then simplifying the result. Since rational expressions are quotients containing symbols which represent real numbers, the Properties of Quotients listed in Section 1.2 may be used. Of course, the letters a, b, c, and d will now be replaced by polynomials. Of particular importance in simplification problems is the formula

$$\boxed{\frac{ad}{bd} = \frac{a}{b}}$$

which is obtained by multiplying the numerator and denominator of the given fraction by $1/d$. This rule is sometimes phrased "a common factor in the numerator and denominator may be *canceled* from the quotient." To use this technique in problems, we factor both the numerator and denominator of the given rational expression into prime factors and then cancel common factors which occur in the numerator and denominator. We refer to the resulting expression as being *simplified*, or *reduced to lowest terms.*

Example 1   Simplify $\dfrac{3x^2 - 5x - 2}{x^2 - 4}$.

Solution   Factoring the numerator and denominator and canceling common factors gives us

$$\frac{3x^2 - 5x - 2}{x^2 - 4} = \frac{(3x + 1)(x - 2)}{(x + 2)(x - 2)} = \frac{3x + 1}{x + 2}. \qquad \blacksquare$$

In the preceding example we canceled the common factor $x - 2$; that is, we divided numerator and denominator by $x - 2$. This simplification is valid only if $x - 2 \neq 0$, that is, $x \neq 2$. However, 2 is not in the domain of x, since it leads to a zero denominator when substituted in the original expression. Hence, our manipulations are valid. We shall always assume such restrictions when simplifying rational expressions.

Example 2   Simplify $\dfrac{2 - x - 3x^2}{6x^2 - x - 2}$.

Solution

$$\frac{2 - x - 3x^2}{6x^2 - x - 2} = \frac{(1 + x)(2 - 3x)}{(2x + 1)(3x - 2)} = \frac{-(1 + x)}{2x + 1}$$

The fact that $(2 - 3x) = -(3x - 2)$ accounts for the minus sign in the final answer, since $(2 - 3x)/(3x - 2) = -(3x - 2)/(3x - 2) = -1$. Another method of attack is to change the form of the numerator as follows:

$$\frac{2 - x - 3x^2}{6x^2 - x - 2} = \frac{-(3x^2 + x - 2)}{6x^2 - x - 2}$$

$$= -\frac{(3x - 2)(x + 1)}{(3x - 2)(2x + 1)}$$

$$= -\frac{x + 1}{2x + 1}. \qquad \blacksquare$$

Multiplication and division are performed using rules for quotients and then simplifying, as illustrated in the next example.

Example 3   Perform the indicated operations and simplify.

(a) $\dfrac{x^2 - 6x + 9}{x^2 - 1} \cdot \dfrac{2x - 2}{x - 3}$ (b) $\dfrac{x + 2}{2x - 3} \div \dfrac{x^2 - 4}{2x^2 - 3x}$

Solutions

(a)
$$\frac{x^2 - 6x + 9}{x^2 - 1} \cdot \frac{2x - 2}{x - 3} = \frac{(x - 3)^2}{(x + 1)(x - 1)} \cdot \frac{2(x - 1)}{x - 3}$$

$$= \frac{2(x - 3)^2(x - 1)}{(x + 1)(x - 1)(x - 3)}$$

$$= \frac{2(x - 3)}{x + 1}$$

(b)
$$\frac{x + 2}{2x - 3} \div \frac{x^2 - 4}{2x^2 - 3x} = \frac{x + 2}{2x - 3} \cdot \frac{2x^2 - 3x}{x^2 - 4}$$

$$= \frac{(x + 2)x(2x - 3)}{(2x - 3)(x + 2)(x - 2)}$$

$$= \frac{x}{x - 2}$$ ∎

When adding or subtracting two rational expressions, it is customary to find a common denominator and use the rules

$$\frac{a}{d} + \frac{c}{d} = \frac{a + c}{d}; \qquad \frac{a}{d} - \frac{c}{d} = \frac{a - c}{d}.$$

If different denominators occur, a common denominator may be introduced by multiplying numerator and denominator of each of the given fractions by suitable polynomials. It is usually desirable to use the **least common denominator (lcd)** of the two fractions. The lcd can be found by obtaining the prime factorization for each denominator and then forming the product of the different prime factors, using the *highest* exponent which appears with each prime factor. Let us begin with a numerical example of this technique.

Example 4   Express as a rational number in lowest terms:

$$\frac{7}{24} + \frac{5}{18}.$$

Solution   The prime factorizations of the denominators are $24 = (2^3)(3)$ and $18 = (2)(3^2)$. To find the lcd we form the product of the different prime factors, using the highest exponent associated with each factor. This gives us $(2^3)(3^2)$, or 72. We now change each fraction to an equal fraction with denominator 72 and add as follows:

$$\frac{7}{24} + \frac{5}{18} = \frac{7}{24} \cdot \frac{3}{3} + \frac{5}{18} \cdot \frac{4}{4}$$

$$= \frac{21}{72} + \frac{20}{72} = \frac{41}{72}.$$

If we use the addition formula for quotients stated in Section 1.2,

$$\frac{7}{24} + \frac{5}{18} = \frac{(7)(18) + (24)(5)}{(24)(18)}$$

$$= \frac{126 + 120}{432} = \frac{246}{432},$$

which reduced to 41/72. ∎

The method for finding the lcd for rational expressions is analogous to that illustrated in the solution of Example 4. The only difference is that we use factorizations of polynomials instead of integers.

Example 5 Change the following to a rational expression in lowest terms.

$$\frac{6}{x(3x - 2)} + \frac{5}{3x - 2} - \frac{2}{x^2}$$

Solution The denominators are already in factored form. Evidently the lcd is $x^2(3x - 2)$. In order to obtain three fractions having that denominator, we multiply numerator and denominator of the first fraction by x, of the second by x^2, and of the third by $3x - 2$. This gives us

$$\frac{6}{x(3x - 2)} + \frac{5}{3x - 2} - \frac{2}{x^2} = \frac{6x}{x^2(3x - 2)} + \frac{5x^2}{x^2(3x - 2)} - \frac{2(3x - 2)}{x^2(3x - 2)}$$

$$= \frac{6x + 5x^2 - (6x - 4)}{x^2(3x - 2)}$$

$$= \frac{5x^2 + 4}{x^2(3x - 2)}$$ ∎

Example 6 Simplify $\dfrac{2x + 5}{x^2 + 6x + 9} + \dfrac{x}{x^2 - 9} + \dfrac{1}{x - 3}$.

Solution We begin by factoring denominators as follows:

$$\frac{2x + 5}{x^2 + 6 + 9} + \frac{x}{x^2 - 9} + \frac{1}{x - 3} = \frac{2x + 5}{(x + 3)^2} + \frac{x}{(x + 3)(x - 3)} + \frac{1}{x - 3}.$$

Since the lcd is $(x + 3)^2(x - 3)$ we multiply numerator and denominator of the first fraction by $x - 3$, of the second by $x + 3$, and of the third by $(x + 3)^2$, and add as follows:

$$\frac{(2x + 5)(x - 3)}{(x + 3)^2(x - 3)} + \frac{x(x + 3)}{(x + 3)^2(x - 3)} + \frac{(x + 3)^2}{(x + 3)^2(x - 3)}$$

$$= \frac{(2x^2 - x - 15) + (x^2 + 3x) + (x^2 + 6x + 9)}{(x + 3)^2(x - 3)}$$

$$= \frac{4x^2 + 8x - 6}{(x + 3)^2(x - 3)} = \frac{2(2x^2 + 4x - 3)}{(x + 3)^2(x - 3)} \qquad \blacksquare$$

It is sometimes necessary to simplify quotients in which the numerator and denominator are not polynomials, as illustrated in the next two examples.

Example 7   Simplify $\dfrac{1 - \dfrac{2}{x + 1}}{x - \dfrac{1}{x}}$.

Solution

$$\frac{1 - \dfrac{2}{x + 1}}{x - \dfrac{1}{x}} = \frac{\dfrac{(x + 1) - 2}{x + 1}}{\dfrac{x^2 - 1}{x}} = \frac{\dfrac{x - 1}{x + 1}}{\dfrac{x^2 - 1}{x}}$$

$$= \frac{x - 1}{x + 1} \cdot \frac{x}{x^2 - 1} = \frac{(x - 1)x}{(x + 1)(x + 1)(x - 1)}$$

$$= \frac{x}{(x + 1)^2} \qquad \blacksquare$$

Example 8   Simplify $\dfrac{x^{-2} + y^{-2}}{(xy)^{-1}}$.

Solution

$$\frac{x^{-2} + y^{-2}}{(xy)^{-1}} = \frac{\dfrac{1}{x^2} + \dfrac{1}{y^2}}{\dfrac{1}{xy}} = \frac{\dfrac{y^2 + x^2}{x^2 y^2}}{\dfrac{1}{xy}}$$

$$= \frac{y^2 + x^2}{x^2 y^2} \cdot \frac{xy}{1} = \frac{y^2 + x^2}{xy}$$

Another method of attacking this problem is to multiply numerator and denominator of the original expression by $x^2 y^2$ and simplify. $\qquad \blacksquare$

The denominators of certain fractional expressions contain sums or differences involving radicals. In some cases the denominators can be rationalized, as illustrated in the next example.

Example 9   Simplify $\dfrac{1}{\sqrt{x} - \sqrt{y}}$.

Solution   Multiplying numerator and denominator by $\sqrt{x} + \sqrt{y}$ we obtain

$$\frac{1}{\sqrt{x} - \sqrt{y}} = \frac{1}{\sqrt{x} - \sqrt{y}} \cdot \frac{\sqrt{x} + \sqrt{y}}{\sqrt{x} + \sqrt{y}}$$

$$= \frac{\sqrt{x} + \sqrt{y}}{(\sqrt{x})^2 - \sqrt{y}\sqrt{x} + \sqrt{x}\sqrt{y} - (\sqrt{y})^2}$$

$$= \frac{\sqrt{x} + \sqrt{y}}{x - y}. \qquad ■$$

EXERCISES 1.9

P45 3-48 Every 3RD

Simplify each of the following.

1 $\dfrac{6x^2 + 7x - 10}{6x^2 + 13x - 15}$

2 $\dfrac{10x^2 + 29x - 21}{5x^2 - 23x + 12}$

3 $\dfrac{12y^2 + 3y}{20y^2 + 9y + 1}$

4 $\dfrac{4z^2 + 12z + 9}{2z^2 + 3z}$

5 $\dfrac{6 - 7a - 5a^2}{10a^2 - a - 3}$

6 $\dfrac{6y - 5y^2}{25y^2 - 36}$

7 $\dfrac{4x^3 - 9x}{10x^4 + 11x^3 - 6x^2}$

8 $\dfrac{16x^4 + 8x^3 + x^2}{4x^3 + 25x^2 + 6x}$

9 $\dfrac{6r}{3r - 1} - \dfrac{4r}{2r + 5}$

10 $\dfrac{3s}{s^2 + 1} - \dfrac{6}{2s - 1}$

11 $\dfrac{2x + 1}{2x - 1} - \dfrac{x - 1}{x + 1}$

12 $\dfrac{3u + 2}{u - 4} + \dfrac{4u + 1}{5u + 2}$

13 $\dfrac{9t - 6}{8t^3 - 27} \cdot \dfrac{4t^2 - 9}{12t^2 + 10t - 12}$

14 $\dfrac{a^2 + 4a + 3}{3a^2 + a - 2} \cdot \dfrac{3a^2 - 2a}{2a^2 + 13a + 21}$

15 $\dfrac{5a^2 + 12a + 4}{a^4 - 16} \div \dfrac{25a^2 + 20a + 4}{a^2 - 2a}$

16 $\dfrac{x^3 - 8}{x^2 - 4} \div \dfrac{x}{x^3 + 8}$

17 $\dfrac{2}{3x + 1} - \dfrac{9}{(3x + 1)^2}$

18 $\dfrac{4}{(5x - 2)^2} + \dfrac{x}{5x - 2}$

19 $\dfrac{1}{c} - \dfrac{c + 2}{c^2} + \dfrac{3}{c^3}$

20 $\dfrac{6}{3t} + \dfrac{t + 5}{t^3} + \dfrac{1 - 2t^2}{t^4}$

21 $\dfrac{5}{x - 1} + \dfrac{8}{(x - 1)^2} - \dfrac{3}{(x - 1)^3}$

22 $\dfrac{8}{x} + \dfrac{3}{2x - 4} + \dfrac{7x}{x^2 - 4}$

23 $\dfrac{2}{x} + \dfrac{7}{x^2} + \dfrac{5}{2x - 3} + \dfrac{1}{(2x - 3)^2}$

24 $\dfrac{4}{x} + \dfrac{3x^2 + 5}{x^3} - \dfrac{6}{2x + 1}$

25 $\dfrac{p^4 + 3p^3 - 8p - 24}{p^3 - 2p^2 - 9p + 18}$

26 $\dfrac{2ac + bc - 6ad - 3bd}{6ac + 2ad + 3bc + bd}$

27 $\left(\dfrac{1}{2x + 2h + 3} - \dfrac{1}{2x + 3}\right) \div h$

28 $\left(\dfrac{7}{5x + 5h - 2} - \dfrac{7}{5x - 2}\right) \div h$

29 $\dfrac{(x + h)^{-3} - x^{-3}}{h}$

30 $\dfrac{(x + h)^{-2} - x^{-2}}{h}$

31 $\dfrac{5}{7x - 3} - \dfrac{2}{2x + 1} + \dfrac{4x}{14x^2 + x - 3}$

32 $2 + \dfrac{3}{x} + \dfrac{7x}{3x + 10}$

33 $\dfrac{\dfrac{a}{b} - \dfrac{b}{a}}{\dfrac{1}{a} + \dfrac{1}{b}}$

34 $\dfrac{\dfrac{1}{x + 1} - 5}{\dfrac{1}{x} - x}$

35 $\dfrac{\dfrac{x}{y^2} - \dfrac{y}{x^2}}{\dfrac{1}{y^2} - \dfrac{1}{x^2}}$

36 $\dfrac{(r/s) + (s/r)}{(r^2/s^2) - (s^2/r^2)}$

37 $\dfrac{\dfrac{5}{x + 1} + \dfrac{2x}{x + 3}}{\dfrac{x}{x + 1} + \dfrac{7}{x + 3}}$

38 $\dfrac{\dfrac{3}{w} - \dfrac{6}{2w + 1}}{\dfrac{5}{w} + \dfrac{8}{2w + 1}}$

39 $\dfrac{a^2 - b^3 a^{-1}}{b^2 a^{-1} + a + b}$

40 $\dfrac{(c + d)^{-1} - (c - d)^{-1}}{(c + d)^{-1} + (c - d)^{-1}}$

41 $\dfrac{1}{\sqrt{a} + \sqrt{b}}$

42 $\dfrac{\sqrt{a} + \sqrt{b}}{\sqrt{a} - \sqrt{b}}$

43 $\dfrac{1}{\sqrt{x}} + \dfrac{1}{\sqrt{y}}$

44 $\dfrac{r}{\sqrt{s}} - \dfrac{s}{\sqrt{r}}$

45 $\dfrac{c}{1 - \sqrt{c}}$

46 $\dfrac{3}{\sqrt{t} + 4}$

47 $\dfrac{1}{a + b + \sqrt{c}}$

48 $\dfrac{1}{x + \sqrt{y + z}}$

49 $\dfrac{1}{h}\left(\dfrac{1}{x + h} - \dfrac{1}{x}\right)$

50 $\left(\dfrac{1}{\sqrt{x} + \sqrt{y} + \sqrt{z}}\right)^0$

51 $\dfrac{(3x + 1)^6}{(2x - 5)^{1/2}} + 2(2x - 5)^{1/2}(3x + 1)^5$

52 $\dfrac{1}{3}\dfrac{(x^2 + 9)^4}{(x + 6)^{2/3}} + 8x(x^2 + 9)^3(x + 6)^{1/3}$

53 $\dfrac{(x^2 + 4)^{1/3} - (x/3)(x^2 + 4)^{-2/3}(2x)}{(x^2 + 4)^{2/3}}$

54 $\dfrac{2x\sqrt{1 - x^2} - \dfrac{x^3}{\sqrt{1 - x^2}}}{1 - x^2}$

1.10   REVIEW

Concepts

Define or explain each of the following.

1 The Commutative Properties of real numbers

2 The Associative Properties

3 The Distributive Properties

4 Rational and irrational numbers

5 Integers

6 Prime number

7 Coordinate line

8 A number a is greater than a number b.

9 A number a is less than a number b.

10 The absolute value of a real number

11 The distance between points on a coordinate line

12 Integral exponents

13 The Laws of Exponents

14   Principal nth root

15   The radical notation

16   Rationalizing a denominator

17   Rational exponents

18   Scientific form for real numbers

19   Variable

20   Domain of a variable

21   Algebraic expression

22   Monomial

23   Polynomial

24   Degree of a polynomial

25   Prime polynomial

26   Irreducible polynomial

27   Factoring

28   Rational expression

Exercises

1   Express each of the following as a rational number with least positive numerator.

(a) $\left(\dfrac{2}{3}\right)\left(-\dfrac{5}{8}\right)$ (b) $\dfrac{3}{4}+\dfrac{6}{5}$

(c) $\dfrac{5}{8}-\dfrac{6}{7}$ (d) $\dfrac{3}{4}\div\dfrac{6}{5}$

2   Replace the $\square$ with either $<$, $>$, or $=$.

(a) $-0.1\ \square\ -0.01$ (b) $\sqrt{9}\ \square\ -3$

(c) $1/6\ \square\ 0.166$

3   Express in terms of inequalities:

(a) x is negative.

(b) a is between $1/2$ and $1/3$.

(c) The absolute value of x is not greater than 4.

4   Rewrite without using the absolute value symbol:

(a) $|-7|$ (b) $|-5|/(-5)$

(c) $|3^{-1}-2^{-1}|$

5   If points A, B, and C on a coordinate line have coordinates $-8, 4$, and -3, respectively, find the following:

(a) $d(A,C)$ (b) $d(C,A)$

(c) $d(B,C)$

6   Prove or disprove each of the following:

(a) $(x+y)^2=x^2+y^2$

(b) $\dfrac{1}{\sqrt{x+y}}=\dfrac{1}{\sqrt{x}}+\dfrac{1}{\sqrt{y}}$

(c) $\dfrac{1}{\sqrt{c}-\sqrt{d}}=\dfrac{\sqrt{c}+\sqrt{d}}{c-d}$

Simplify the expressions in Exercises 7–32.

7 $(3a^2b)^2(2ab^3)$

8 $\dfrac{6r^3y^2}{2r^5y}$

9 $\dfrac{(3x^2y^{-3})^{-2}}{x^{-5}y}$

10 $\left(\dfrac{a^{2/3}b^{3/2}}{a^2b}\right)^6$

11 $(-2p^2q)^3(p/4q^2)^2$

12 $c^{-4/3}c^{3/2}c^{1/6}$

13 $\left(\dfrac{xy^{-1}}{\sqrt{z}}\right)^4\div\left(\dfrac{x^{1/3}y^2}{z}\right)^3$

14 $\left(\dfrac{-64x^3}{z^6y^9}\right)^{2/3}$

15 $((a^{2/3}b^{-2})^3)^{-1}$

16 $(4x^3y^2z)^{-0}$

17 $\dfrac{r^{-1}+s^{-1}}{(rs)^{-1}}$

18 $(u+v)^3(u+v)^{-2}$

19 $\sqrt[3]{(x^4y^{-1})^6}$

20 $\sqrt[3]{8x^5y^3z^4}$

21 $\dfrac{1}{\sqrt[3]{4}}$

22 $\sqrt{\dfrac{a^2b^3}{c}}$

23 $\sqrt[3]{4x^2y}\,\sqrt[3]{2x^5y^2}$

24 $\sqrt[4]{(-4a^3b^2c)^2}$

25 $\dfrac{1}{\sqrt{t}}\left(\dfrac{1}{\sqrt{t}}-1\right)$

26 $\sqrt{\sqrt[3]{(c^3d^6)^4}}$

27 $\dfrac{\sqrt{12x^4y}}{\sqrt{3x^2y^5}}$

28 $\sqrt[3]{(a+2b)^3}$

29 $\sqrt[3]{x^2y}\,\sqrt[4]{x^3y^2}$

30 $\sqrt{x}-\sqrt{9x^3}$

31 $\dfrac{1}{\sqrt{a}-\sqrt{b}}$

32 $(\sqrt[3]{m^3+n^3})^3$

33   Express in scientific form:
   (a)   93,700,000,000   (b)   0.00000402

34   Express as a decimal:
   (a) 6.8×10^7 (b) 7.3×10^{-4}

In each of Exercises 35–42, perform the indicated operations.

35 $(3x^3 - 4x^2 + x - 7) + (x^4 - 2x^3 + 3x^2 + 5)$

36 $(4z^4 - 3z^2 + 1) - z(z^3 + 4z^2 - 4)$

37 $(x + 4)(x + 3) - (2x - 1)(x - 5)$

38 $(4x - 5)(2x^2 + 3x - 7)$

39 $(3y^3 - 2y^2 + y + 4)(y^2 - 3)$

40 $(3x + 2)(x - 5)(5x + 4)$

41 $(a - b)(a^3 + a^2b + ab^2 + b^3)$

42 $\dfrac{9p^4q^3 - 6p^2q^4 + 5p^3q^2}{3p^2q^2}$

Find the products in Exercises 43–50.

43 $(3a - 5b)(2a + 7b)$

44 $(4r^2 - 3s)^2$

45 $(13a^2 + 4b)(13a^2 - 4b)$

46 $(a^3 - a^{-3})^2$

47 $(2a + b)^3$

48 $(c^2 - d^2)^3$

49 $(3x + 2y)^2(3x - 2y)^2$

50 $(a + b + c + d)^2$

Factor the expressions in Exercises 51–62.

51 $60xw + 70w$ **52** $2r^4s^3 - 8r^2s^5$

53 $28x^2 + 4x - 9$

54 $16a^4 + 24a^2b^2 + 9b^4$

55 $2wy + 3yx - 8wz - 12zx$

56 $2c^3 - 12c^2 + 3c - 18$

57 $8x^3 + 64y^3$ **58** $u^3v^4 - u^6v$

59 $p^8 - q^8$ **60** $x^4 - 8x^3 + 16x^2$

61 $w^6 + 1$ **62** $3x + 6$

Simplify the expressions in Exercises 63–70.

63 $\dfrac{6x^2 - 7x - 5}{4x^2 + 4x + 1}$ **64** $\dfrac{r^3 - t^3}{r^2 - t^2}$

65 $\dfrac{6x^2 - 5x - 6}{x^2 - 4} \div \dfrac{2x^2 - 3x}{x + 2}$

66 $\dfrac{2}{4x - 5} - \dfrac{5}{10x + 1}$

67 $\dfrac{7}{x + 2} + \dfrac{3x}{(x + 2)^2} - \dfrac{5}{x}$

68 $\dfrac{x + x^{-2}}{1 + x^{-2}}$

69 $\dfrac{1}{x} - \dfrac{2}{x^2 + x} - \dfrac{3}{x + 3}$

70 $(a^{-1} + b^{-1})^{-1}$

2

EQUATIONS
AND
INEQUALITIES

For hundreds of years one of the main concerns in algebra has been the solutions of equations. More recently the study of inequalities has reached the same level of importance. Both of these topics are used extensively in applications of mathematics. In this chapter we shall discuss the rudiments of solving equations and inequalities. All variables will represent real numbers. In a later chapter we shall study equations that involve complex numbers.

2.1 LINEAR EQUATIONS

If x is a variable, then expressions such as

$$x + 3 = 0, \quad x^2 - 5 = 4x, \quad \text{or} \quad (x^2 - 9)\sqrt[3]{x + 1} = 0$$

are called **equations** in x. If certain numbers are substituted for x in these equations, true statements are obtained, whereas other numbers produce false statements. For example, the equation $x + 3 = 0$ leads to a false statement for every value of x except -3. If 2 is substituted for x in the equation $x^2 - 5 = 4x$, we obtain $4 - 5 = 8$, or $-1 = 8$, a false statement. However, if we let $x = 5$, then we obtain $(5)^2 - 5 = 4 \cdot 5$, or $20 = 20$, which is true. Given any equation in x, if a true statement is obtained when x is replaced by some real number a from the domain of x, then a is called a **solution** or a **root** of the equation. We also say that a **satisfies** the equation. To **solve** an equation means to find all the solutions.

Sometimes every number in the domain of the variable is a solution of an equation. In this case the equation is called an **identity**. For example,

$$\frac{1}{x^2 - 4} = \frac{1}{(x + 2)(x - 2)}$$

is an identity since it is true for every number in the domain of x. If there are numbers in the domain of x which are not solutions, then the equation is called a **conditional equation**.

The solutions of an equation depend on the system of numbers under consideration. For example, if we demand that solutions be rational numbers, then the equation $x^2 = 2$ has no solutions, since there is no rational number whose square is 2. However, if we allow *real* numbers, then the solutions are $-\sqrt{2}$ and $\sqrt{2}$. Similarly, the equation $x^2 = -1$ has no real solutions; however, we shall see later that this equation has solutions if *complex* numbers are allowed.

Two equations are said to be **equivalent** if they have exactly the same solutions. For example, the equations

$$x - 1 = 2, \quad x = 3, \quad 5x = 15, \quad \text{and} \quad 2x + 1 = 7$$

are all equivalent.

One method of solving an equation is to replace it by a chain of equivalent equations, each in some sense simpler than the preceding one and terminating in an equation for which the solutions are obvious. This is often accomplished by using various properties of real numbers. For example, we may add the same expression to both sides of an equation without changing the solutions. Similarly, we may subtract the same expression from both sides of an equation. We can also multiply or divide both sides of an equation by an expression which represents a nonzero real number. The following example illustrates these remarks.

Example 1   Solve the equation $2x - 5 = 3$.

Solution   The following is a chain of equations, each of which is equivalent to the preceding one. (Why?)

$$2x - 5 = 3$$
$$(2x - 5) + 5 = 3 + 5$$
$$2x = 8$$
$$\tfrac{1}{2}(2x) = \tfrac{1}{2}(8)$$
$$x = 4$$

Since the last equation has only one solution, 4, and since the last equation is equivalent to the first, it follows that 4 is the only solution of $2x - 5 = 3$.   ■

In order to guard against errors in manipulations and simplifications, answers should be checked by substitution in the original equation. If we apply a check in Example 1 we obtain $2(4) - 5 = 3$, which reduces to $3 = 3$, a true statement.

We sometimes say that an equation such as that given in Example 1 *has the solution* $x = 4$, in the sense that substitution of 4 for x produces a true statement. As another illustration, the equation $x^2 = 9$ is said to have solutions $x = 3$ and $x = -3$.

If we inadvertently multiply both sides of an equation by an expression which equals zero for some value of x, then an equivalent equation is not always obtained. The following example illustrates a manner in which this may happen.

Example 2   Solve the equation $\dfrac{3x}{x - 2} = 1 + \dfrac{6}{x - 2}$.

Solution   Multiplying both sides by $x - 2$ and simplifying leads to

$$\left(\frac{3x}{x - 2}\right)(x - 2) = 1(x - 2) + \left(\frac{6}{x - 2}\right)(x - 2)$$

$$3x = (x - 2) + 6$$

$$2x = 4$$

$$x = 2.$$

Let us check to see whether 2 is a solution of the given equation. Substituting 2 for x we obtain

$$\frac{3(2)}{2 - 2} = 1 + \frac{6}{2 - 2}, \quad \text{or} \quad \frac{6}{0} = 1 + \frac{6}{0}.$$

Since division by 0 is not permissible, 2 is not a solution. Actually, the given equation has no solutions, for we have shown that if the equation is true for some value of x, then that value must be 2. However, as we have seen, 2 is not a solution. ∎

The preceding example indicates that *it is essential to check answers obtained after multiplying both sides of an equation by an expression containing variables.*

In this section we shall restrict our efforts to equations which are equivalent to equations of the form

$$ax + b = 0,$$

where a and b are real numbers and $a \neq 0$. An equation of this type is called a **linear equation** in x. To solve the equation we first add $-b$ to both sides and then multiply by $1/a$ as follows:

$$(ax + b) + (-b) = 0 + (-b)$$

$$ax = -b$$

$$\frac{1}{a}(ax) = \frac{1}{a}(-b)$$

$$x = -\frac{b}{a}.$$

This shows that *if ax + b = 0 has a solution, then it must be* $-b/a$. To verify that $-b/a$ is actually a solution, we substitute it for x in $ax + b = 0$, obtaining

$$a\left(-\frac{b}{a}\right) + b = 0 \quad \text{or} \quad (-b) + b = 0,$$

which is a true statement. We have established the following fact.

Theorem

> If $a \neq 0$, then the equation $ax + b = 0$ has precisely one solution, $x = -b/a$.

In Section 2.3 we shall consider equations having more than one solution.

Example 3   Solve $(8x - 2)(3x + 4) = (4x + 3)(6x - 1)$.

Solution   The following equations are all equivalent (supply reasons).

$$(8x - 2)(3x + 4) = (4x + 3)(6x - 1)$$
$$24x^2 + 26x - 8 = 24x^2 + 14x - 3$$
$$26x - 14x = -3 + 8$$
$$12x = 5$$
$$x = \frac{5}{12}$$

Hence, the solution of the given equation is 5/12.   ∎

Example 4   Solve $\dfrac{3}{2x - 4} - \dfrac{5}{x + 3} = \dfrac{2}{x - 2}$.

Solution   If we rewrite the equation as

$$\frac{3}{2(x - 2)} - \frac{5}{x + 3} = \frac{2}{x - 2}$$

we see that the lcd of the three fractions is $2(x - 2)(x + 3)$. Multiplying both sides by this lcd leads to

$$3[(x + 3)] - 5[2(x - 2)] = 2[2(x + 3)].$$

Simplifying, we obtain

$$3x + 9 - 10x + 20 = 4x + 12$$
$$3x - 10x - 4x = 12 - 9 - 20$$
$$-11x = -17$$
$$x = \frac{17}{11}.$$

Since we multiplied by an expression involving x, we must check this result in the given equation. Substituting for x we have

$$\frac{3}{2(17/11) - 4} - \frac{5}{(17/11) + 3} = \frac{2}{(17/11) - 2}.$$

We leave it to the reader to show that this reduces to $-22/5 = -22/5$, a true statement. Hence, the given equation has the solution $17/11$. ∎

Students sometimes have difficulty in determining whether an equation is conditional or an identity. An identity will often be indicated when, after applying properties of real numbers, an equation of the form $p = p$ is obtained, where p is some expression. To illustrate, if we multiply both sides of the equation

$$\frac{x}{x^2 - 4} = \frac{x}{(x + 2)(x - 2)}$$

by $x^2 - 4$, we obtain $x = x$. This alerts us to the fact that we may have an identity on our hands; however, it does not prove anything. A standard method for verifying that an equation is an identity is to show, using properties of real numbers, that the expression which appears on one side of the given equation can be transformed into the expression which appears on the other side of the given equation. That is easy to do in the above illustration, since we know that $x^2 - 4 = (x + 2)(x - 2)$. Of course, to show that an equation is not an identity, we need only find one real number in the domain of the variable that fails to satisfy the original equation.

EXERCISES 2.1

P53 3 - 48 EVR 3RD

Solve the equations in Exercises 1–38.

1 $4x + 7 = 0$

2 $3x + 16 = 0$

3 $\sqrt{2}x - 5 = 0$

4 $\sqrt{3}x - 2 = 0$

5 $5x + 3 = 7x - 2$

6 $8x - 5 = 6x + 4$

7 $3(7y - 2) = 2(4y + 1)$

8 $2(9z + 2) - 5(z - 8) = 0$

9 $\frac{2}{3}t + 4 = 2 - \frac{5}{2}t$

10 $\frac{3}{4}u - 1 = 2 + \frac{1}{5}u$

11 $0.2(4 - 3x) + 0.3x = 2.6$

12 $0.7x - 1.2 = 0.3(2x - 1)$

13 $\dfrac{12 - 7w}{6} = \dfrac{2w + 1}{9}$

14 $\dfrac{3r + 2}{8} = 1 - \dfrac{r}{12}$

15 $\dfrac{18 - 5p}{3p + 2} = \dfrac{7}{3}$

16 $\dfrac{6}{5v - 2} = \dfrac{9}{7v + 3}$

17 $4 - \dfrac{3}{x} = 6 - \dfrac{5}{x}$

18 $\dfrac{5}{q} + \dfrac{2}{q} - \dfrac{10}{q} = 12$

19 $(6x - 5)^2 = (4x + 3)(9x - 2)$

20 $(x - 7)^2 - 4 = (x + 1)^2$

21 $(4x - 3)(2x + 3) - 8x(x - 4) = 0$

22 $(2x + 3)(3x - 2) = 6x^2 + 1$

23 $\dfrac{6s + 7}{4s - 1} = \dfrac{3s + 8}{2s - 4}$

24 $\dfrac{8t + 5}{10t - 7} = \dfrac{4t - 3}{5t + 7}$

25 $\dfrac{1}{3} + \dfrac{2}{6x + 3} = \dfrac{3}{2x + 1}$

26 $\dfrac{-3}{4x - 2} + \dfrac{2}{2x - 1} = \dfrac{7}{2}$

27 $\dfrac{5}{4a - 2} - \dfrac{1}{6a - 3} = \dfrac{4}{5}$

28 $\dfrac{8}{3b + 6} - \dfrac{1}{2b + 4} = \dfrac{3}{4}$

29 $\dfrac{1}{2x - 1} = \dfrac{3}{4x - 2}$

30 $\dfrac{6}{2x + 11} + 5 = 5$

31 $2 - \dfrac{5}{3x - 7} = 2$

32 $\dfrac{4}{5x + 2} - \dfrac{7}{15x + 6} = 0$

33 $\dfrac{7}{y^2 - 4} - \dfrac{4}{y + 2} = \dfrac{5}{y - 2}$

34 $\dfrac{4}{2u - 3} + \dfrac{10}{4u^2 - 9} = \dfrac{1}{2u + 3}$

35 $(x + 3)^3 - (3x - 1)^2 = x^3 + 4$

36 $(x - 1)^3 = (x + 1)^3 - 6x^2$

37 $\dfrac{9x}{3x - 1} = 2 + \dfrac{3}{3x - 1}$

38 $\dfrac{2x}{2x + 3} + \dfrac{6}{4x + 6} = 5$

Prove that the equations in Exercises 39–44 are identities.

39 $(3x + 1)^2 - 9x^2 = 6x + 1$

40 $(4x - 5)(3x + 2) + 7x = 12x^2 - 10$

41 $\dfrac{x^2 - 1}{x - 1} = x + 1$

42 $\dfrac{3x - 2}{x} = 3 - \dfrac{2}{x}$

43 $\dfrac{5x^2 + 2}{x} = \dfrac{2}{x} + 5x$

44 $\dfrac{9x^2 - 16}{3x + 4} = 3x - 4$

45 For what value of c is -3 a solution of the equation $3x + 1 - 5c = 2c + x - 10$?

46 For what value of b is 8 a solution of the equation $4x + 3b = 7$?

47 Determine values for a and b such that $5/3$ is a solution of the equation $ax + b = 0$. Are these the only possible values for a and b? Explain.

48 In each of the following, determine whether the two given equations are equivalent.
(a) $x^2 = 4$, $x = 2$
(b) $x = \sqrt{4}$, $x = 2$
(c) $2x = 4$, $x = 2$

49 (a) Find an equation in the following chain which is not equivalent to the preceding equation.

$$x^2 - x - 2 = x^2 - 4$$
$$(x + 1)(x - 2) = (x + 2)(x - 2)$$
$$x + 1 = x + 2$$
$$1 = 2$$

(b) Find the solutions of the first equation in part (a).

50 Find an equation in the following chain which is not equivalent to the preceding equation.

$$x + 3 = 0$$
$$5x - 4x = -3$$
$$5x + 6 = 4x + 3$$
$$x^2 + 5x + 6 = x^2 + 4x + 3$$
$$(x + 3)(x + 1) = (x + 2)(x + 3)$$
$$x + 1 = x + 2$$
$$0 = 1$$

2.2 APPLICATIONS

Formulas or equations involving variables are used in all fields which deal with numbers. For certain applications it is necessary to solve for a particular variable in terms of the remaining variables which appear in the formula. This is done by treating the equation as if the desired variable were the only one present and transforming the original equation into an equivalent equation in which that variable is isolated on one side, as illustrated in the next three examples.

Example 1   If a sum of money P (the principal) is invested at a simple interest rate of r percent per year, then the interest I at the end of t years is given by $I = Prt$. Solve for r in terms of the remaining variables.

Solution   We begin by writing

$$Prt = I.$$

In order to isolate r we multiply both sides by $1/Pt$, obtaining

$$\frac{1}{Pt} \cdot Prt = \frac{1}{Pt} \cdot I.$$

It follows that

$$r = \frac{I}{Pt}. \quad \blacksquare$$

Example 2   The relationship between the temperature F on the Fahrenheit scale and the temperature C on the Celsius scale is given by

$$C = \tfrac{5}{9}(F - 32).$$

Solve for F in terms of C.

Solution   We may proceed as follows (supply reasons):

$$C = \tfrac{5}{9}(F - 32)$$
$$\tfrac{9}{5}C = F - 32$$
$$\tfrac{9}{5}C + 32 = F$$
$$F = \tfrac{9}{5}C + 32. \quad \blacksquare$$

Example 3   The formula $R = \dfrac{R_1 R_2}{R_1 + R_2}$ is used in electrical theory to find the total resistance R when two resistances R_1 and R_2 are connected in parallel, as illustrated in Figure 2.1. Solve for R_1 in terms of R and R_2.

Figure 2.1

Solution   The following equations are equivalent to the given equation.

$$(R_1 + R_2)R = (R_1 + R_2)\left(\frac{R_1 R_2}{R_1 + R_2}\right)$$

$$R_1 R + R_2 R = R_1 R_2$$

$$R_1 R - R_1 R_2 = -R_2 R$$

$$R_1(R - R_2) = -R_2 R$$

$$R_1 = \frac{-R_2 R}{R - R_2}$$

$$R_1 = \frac{R_2 R}{R_2 - R} \qquad\blacksquare$$

Problems often occur in everyday life which can be solved by means of equations or other mathematical tools. Some problems are spoken, from one person to another. Others are stated using written words, as in textbooks. For this reason they are often called "word problems" by students and teachers of mathematics. They may also be referred to as "practical problems." We shall use the terminology "applied problem" for any problem which involves an application of mathematics to some other field.

Due to the unlimited variety of applied problems, it is difficult to state specific rules for finding solutions. However, it is possible to develop a general strategy for attacking such problems. Below are listed some guidelines which may be helpful, provided the problem can be formulated in terms of an equation in one variable. In a later chapter we shall introduce techniques which involve several variables.

Guidelines for Solving Applied Problems

1. If the problem is stated in written words, read it carefully several times and think about the given facts, together with the unknown quantity that is to be found.

2. Introduce a letter to denote the unknown quantity. This is one of the most crucial steps in the solution! Phrases containing words such as "what," "find," "how much," "how far," or "when" should alert you to the unknown quantity.

3. If possible, draw a picture and label it appropriately.

4. Make a list of known facts, together with any relationships involving the unknown quantity. A relationship may often be described by means of an equation in which written statements, instead of letters or numbers, appear on either one or both sides of the equals sign.

5. After analyzing the list in step 4, and perhaps rereading the problem several times, formulate an equation which describes precisely what is stated in words.

6. Solve the equation formulated in step 5.

7. Check the solutions obtained in step 6 by referring to the original statement of the problem. Carefully note whether the solution agrees with the stated conditions.

8. Don't become discouraged if you are unable to solve a given problem. It takes a great deal of effort and practice to become proficient in solving applied problems. Keep trying!

Example 4 A student has test scores of 64 and 78. What score on a third test will give the student an average of 80?

Solution   We shall follow the Guidelines which precede this example. Reading the problem carefully, as suggested in step 1, we note that the unknown quantity is the score on the third test. Accordingly, as in step 2, we introduce a letter as follows:

$$x = \text{score on the third test.}$$

Drawing a picture, as mentioned in step 3, is inappropriate for this problem, so we go on to step 4 and look for relationships involving x. Since the average of the three scores is found by adding them and dividing by 3, we may write

$$\frac{64 + 78 + x}{3} = \text{average of the three scores 64, 78, and } x.$$

From the statement of the problem we obtain

$$80 = \text{average desired.}$$

Consequently, x must satisfy the equation

$$\frac{64 + 78 + x}{3} = 80.$$

The last equation is the one referred to in step 5 of the Guidelines. We next solve the equation (see step 6) as follows:

$$64 + 78 + x = 240$$
$$142 + x = 240$$
$$x = 240 - 142$$
$$x = 98.$$

Step 7 tells us to check by referring to the original statement. If three test scores are 64, 78, and 98, then the average is

$$\frac{64 + 78 + 98}{3} = \frac{240}{3} = 80.$$

Hence, a score of 98 on the third test will give the student an average of 80. Happily, we can ignore step 8. ∎

In the remaining examples we shall not point out the explicit Guidelines which are used in the solution. The reader should be able to determine that without being told.

Example 5   A store holding a clearance sale advertises that all prices have been discounted 20%. If a certain article is on sale for $28, what was its price before the sale?

Solution   We begin by noting that the unknown quantity is the presale price. It is convenient to arrange our work as follows, where the quantities are measured in dollars.

$$x = \text{presale price}$$
$$0.20x = \text{discount}$$
$$28 = \text{sale price}$$

The sale price is determined as follows:

$$(\text{presale price}) - (\text{discount}) = (\text{sale price}).$$

This leads to the equation

$$x - 0.20x = 28$$

which we solve as follows:

$$x - \tfrac{1}{5}x = 28$$
$$\tfrac{4}{5}x = 28$$
$$x = (\tfrac{5}{4})28 = 35.$$

Hence, the price before the sale was $35.

To check this answer we note that if a $35 article is discounted 20%, then the discount (in dollars) is $(0.20)(35) = 7$, and the selling price is $35 - 7$, or $28.

∎

Example 6 A man has $15,000 to invest. He plans to deposit part of it in a savings account paying 5% simple interest and the remainder in an investment fund yielding 8% simple interest. How much should he invest in each to obtain a 7% return on his money after one year?

Solution The simple interest formula $I = Prt$ was given in Example 1. In the present example, $t = 1$ and hence, the interest is given by $I = Pr$. If we let x denote the amount deposited in the savings account, then the remainder, $15,000 - x$, will be put into the investment fund. This leads to the following equalities:

$$x = \text{amount invested at } 5\%$$
$$15,000 - x = \text{amount invested at } 8\%$$
$$0.05x = \text{interest on } x \text{ dollars at } 5\%$$
$$0.08(15,000 - x) = \text{interest on } 15,000 - x \text{ dollars at } 8\%$$
$$0.07(15,000) = \text{total interest desired.}$$

Since the total interest must equal the combined interest from the two investments, we see that

$$0.05x + 0.08(15000 - x) = 0.07(15000)$$
$$0.05x + 1200 - 0.08x = 1050$$
$$-0.03x = -150$$
$$x = \frac{-150}{-0.03} = 5000.$$

Consequently, $5,000 should be deposited in the savings account and $10,000 in the investment fund.

Checking, we see that if $5,000 is placed in the savings account, then the interest obtained is $(0.05)(\$5,000) = \$250$. If $10,000 is placed in the investment fund, then the interest is $(0.08)(\$10,000) = \$800$. Hence, the total interest is $1,050, which is 7% of $15,000.

∎

An important type of problem which may be solved using linear equations involves mixing two substances to obtain a prescribed mixture. For such problems it is often helpful to draw a picture, as illustrated in the next two examples.

Example 7 A chemist has 10 ml of a solution which contains a 30% concentration of acid. How many ml of pure acid must be added in order to increase the concentration to 50%?

Solution   Since we wish to find the amount of pure acid to add, we let

$$x = \text{ml of acid to be added.}$$

Figure 2.2 is self-explanatory.

| Original 30% mixture | | Pure acid | | New 50% mixture |
| 3 ml acid | | x ml acid | | $0.5(10 + x)$ ml acid |

Figure 2.2

This leads to the equation

$$3 + x = 0.5(10 + x),$$

or equivalently,

$$3 + x = 5 + 0.5x$$
$$0.5x = 2$$
$$\tfrac{1}{2}x = 2$$
$$x = 4.$$

Hence, 4 ml of the acid should be added to the original solution.

To check, we note that if 4 ml of acid is added to the given solution, then the new solution contains 14 ml, 7 of which are acid. This is the desired 50% concentration. ∎

Example 8   A radiator contains 8 quarts of a mixture of water and antifreeze. If 40% of the mixture is antifreeze, how much of the mixture should be drained and replaced by pure antifreeze in order that the resultant mixture will contain 60% antifreeze?

Solution   Let

$$x = \text{the number of qts to be drained.}$$

Since there were 8 quarts in the original 40% mixture, we may picture the problem as shown in Figure 2.3. This gives us the equation

$$0.4(8 - x) + x = 0.6(8),$$

Original 40% mixture, less amount drained

Pure antifreeze

New 60% mixture

8−x qts   +   x qts   =   8 qts

0.4(8 − x) qts antifreeze

x qts antifreeze

0.6(8) qts antifreeze

Figure 2.3

or equivalently,

$$3.2 - 0.4x + x = 4.8$$
$$-0.4x + x = 4.8 - 3.2$$
$$0.6x = 1.6$$
$$x = \frac{1.6}{0.6} = \frac{8}{3}.$$

Thus, 8/3 quarts should be drained from the original mixture.

To check, let us first note that the amount of antifreeze in the original 8-quart mixture was 0.4(8), or 3.2 quarts. In draining 8/3 quarts of the original 40% mixture, we lose 0.4(8/3) quarts of antifreeze, and hence, there remain (3.2) − 0.4(8/3) quarts of antifreeze. If we then add 8/3 quarts of pure antifreeze, the amount of antifreeze in the final mixture is (3.2) − 0.4(8/3) + 8/3 quarts. This reduces to 4.8, which is 60% of 8. ■

Many applied problems have to do with objects that move at a constant, or uniform, rate. If an object travels at a uniform (or average) rate r, then the distance d traversed in time t is given by $d = rt$. Of course, we assume that the units are properly chosen; that is, if r is in feet per second, then t is in seconds, and so on.

Example 9 Two cities, A and B, are connected by means of a highway 150 miles long. An automobile leaves A at 1:00 P.M. and travels at a uniform rate of 40 miles per hour toward B. Thirty minutes later, another automobile leaves A and travels toward B at a uniform rate of 55 miles per hour. At what time will the second car overtake the first car?

Solution   Let t denote the time, in hours, *after* 1:30 P.M. At 1:30 P.M. the first automobile has already traveled 20 miles. Hence, at time t *after* 1:30 P.M. the distance it has traveled is $20 + 40t$ miles. Since the second automobile starts the trip at 1:30 P.M., the distance it has traveled at time t is $55t$ miles. We wish to find the time t at which the distances traveled by the two automobiles are equal. This will be true when

$$55t = 20 + 40t.$$

Solving for t, we obtain

$$55t - 40t = 20$$
$$15t = 20$$
$$t = \tfrac{4}{3}.$$

Consequently, $t = 1\tfrac{1}{3}$ hours, or equivalently, 1 hour and 20 minutes. Since this is the amount of time after 1:30 P.M., it follows that the second car overtakes the first at 2:50 P.M.

To check our answer we note that, at 2:50 P.M. the first car has traveled for $1\tfrac{5}{6}$ hours and its distance from A is $40(11/6) = 220/3$ miles. At 2:50 P.M. the second car has traveled for $1\tfrac{1}{3}$ hours and is $55(4/3) = 220/3$ miles from A. Hence, they are together at 2:50 P.M.   ■

Example 10   Tom can do a certain job in 3 hours, and Bob can do the same job in 4 hours. If they work together, how long will it take them to do the job?

Solution   Let

$$t = \text{number of hours to do the job together.}$$

It is convenient to introduce the *part* of the job done in one hour as follows:

$$\frac{1}{3} = \text{part of job done by Tom in one hour}$$

$$\frac{1}{4} = \text{part of job done by Bob in one hour}$$

$$\frac{1}{t} = \text{part of job done working together in one hour.}$$

We have assumed that both Tom and Bob work at a steady pace throughout the job, and that there is no gain or loss in efficiency when they work together. Although this is probably not the case, such assumptions are often made when a real-world situation is represented mathematically. A representation of this type is called a **mathematical model** for the situation. Since

$$\begin{pmatrix} \text{part done by} \\ \text{Tom in one hour} \end{pmatrix} + \begin{pmatrix} \text{part done by} \\ \text{Bob in one hour} \end{pmatrix} = \begin{pmatrix} \text{part done working} \\ \text{together in one hour} \end{pmatrix}$$

we obtain

$$\frac{1}{3} + \frac{1}{4} = \frac{1}{t}, \quad \text{or} \quad \frac{7}{12} = \frac{1}{t}.$$

Solving for t gives us

$$7t = 12, \quad \text{or} \quad t = \frac{12}{7}.$$

Thus, by working together, Tom and Bob can do the job in 1 and $\tfrac{5}{7}$ hours, or approximately 1 hour and 43 minutes.   ■

EXERCISES 2.2

The formulas in Exercises 1–30 occur in mathematics and its applications. Solve each for the indicated variable in terms of the remaining variables.

1 $I = Prt$ for P

2 $d = rt$ for t

3 $A = \frac{1}{2}bh$ for h

4 $C = 2\pi r$ for r

5 $P = 2l + 2w$ for w

6 $A = P + Prt$ for r

7 $ax + by + cz = d$ for z

8 $ax + by + c = 0$ for y

9 $R = \dfrac{E}{I}$ for I

10 $K = \dfrac{mv^2}{2g}$ for g

11 $V = \frac{1}{3}\pi r^2 h$ for h

12 $F = g\dfrac{m_1 m_2}{d^2}$ for m_1

13 $S = \dfrac{a}{1 - r}$ for r

14 $m = \dfrac{y_2 - y_1}{x_2 - x_1}$ for x_1

15 $\dfrac{1}{R} = \dfrac{1}{R_1} + \dfrac{1}{R_2} + \dfrac{1}{R_3}$ for R_2

16 $\dfrac{x}{a} + \dfrac{y}{b} = 1$ for y

17 $S = P + Prt$ for P

18 $F = \frac{9}{5}C + 32$ for C

19 $V = \frac{1}{3}\pi h^2(3r - h)$ for r

20 $s = \frac{1}{2}gt^2 + v_0 t$ for v_0

21 $S = \dfrac{a - rl}{l - r}$ for r

22 $S = a + (n - 1)d$ for n

23 $Ft = mv_1 - mv_2$ for m

24 $\dfrac{1}{f} = \dfrac{1}{f_1} + \dfrac{1}{f_2}$ for f_1

25 $A = \frac{1}{2}(b_1 + b_2)h$ for b_1

26 $A = 2\pi r(r + h)$ for h

27 $a = \dfrac{v_2 - v_1}{t}$ for v_1

28 $l = l_0(1 + ct)$ for c

29 $R = \dfrac{nE - rI}{nI}$ for n

30 $E = \dfrac{T_1 - T_2}{T_1}$ for T_1

31 A newspaper boy collects \$13.45 in dimes and quarters. If there are 70 coins in all, how many quarters does he have?

32 A girl has 125 coins consisting of nickels and pennies. If the total amount is \$4.25, how many coins of each type does she have?

33 Find four consecutive integers whose sum is 550.

34 Find two consecutive integers such that the difference of their squares is 133.

35 The relationship between the temperature F on the Fahrenheit scale and the temperature C on the Celsius scale is given by

$$C = \frac{5}{9}(F - 32).$$

Find the temperature at which the reading is the same on both scales.

36 Refer to Exercise 35. When will the Celsius reading be twice the Fahrenheit reading?

37 A student in an algebra course has test scores of 75, 82, 71, and 84. What score on the next test will raise the student's average to 80?

38 Going into a final exam, a student has test scores of 72, 80, 65, 78, and 60. If the final exam counts as 1/3 of the final grade, what score must the student receive in order to end up with an average of 76?

39 A businesswoman wishes to invest $30,000 in two different funds which yield annual profits of 13% and $15\frac{1}{2}$%, respectively. How much should she invest in each in order to realize a profit of $4,350 after one year?

40 A banker plans to lend part of $24,000 at a simple interest rate of 16% and the remainder at 19%. How should he allot the loans in order to obtain a return of $18\frac{1}{2}$% after one year?

41 A college student has $3,000 in two different savings accounts, paying interest at the rates of $5\frac{1}{2}$% and 6%, respectively. If the total yearly interest is $174.10, how much is deposited in each account?

42 A man has $4,000 more invested at $7\frac{1}{2}$% simple interest than he has at 6%. If his total yearly interest is $467.40, how much is invested at each rate?

43 Two boys who are 224 meters apart start walking toward each other at the same time. If they walk at rates of 1.5 and 2 meters per second, respectively, when will they meet? How far will each have walked?

44 A jogger starts from a certain point and runs at a constant rate of 6 mph. Five minutes later a second jogger begins at the same point, running at a rate of 8 mph and following the same course. How long will it take the second jogger to catch up with the first?

45 If the radius of a circle is increased by 2 cm, its area increases by 16π cm². What is the original radius of the circle?

46 A rectangle is twice as long as it is wide. If the length and width are decreased by 2 cm and 3 cm, respectively, the area is decreased by 30 cm². Find the original dimensions.

47 After playing 100 games, a major league baseball team has a record of 0.650. If it wins only 50% of its games for the remainder of the season, when will its record be 0.600?

48 A projectile is fired horizontally at a target, and the sound of its impact is heard 1.5 seconds later. If the speed of the projectile is 3300 ft/sec and the speed of sound is 1100 ft/sec, how far away is the target?

49 Twenty liters of a solution contains 20% of a certain chemical. How much water should be added so that the resulting solution contains 15% of the chemical?

50 How many grams of an alloy containing 25% silver should be melted with 50 grams of an alloy containing 60% silver in order to obtain an alloy containing 50% silver?

51 How much water should be added to 1 liter of pure acid in order to obtain a solution which is 25% acid?

52 An automobile 20 feet long overtakes a truck that is 40 feet long which is traveling at 50 mph. At what constant speed must the automobile travel in order to pass the truck in 5 seconds?

53 A bus traveled from one city to another at an average rate of 50 mph. On the return trip the average rate was 45 mph and the elapsed time was 15 minutes longer. What was the total distance traveled?

54 At a concert certain tickets sold for $5.50 and others sold for $3.75. If a total of 520 tickets were sold for $2,448.75, how many of each kind were sold?

55 It takes a boy 90 minutes to mow his father's yard, but his sister can do it in 60 minutes. How long would it take them to mow the lawn if they worked together, using two lawnmowers?

56 Using water from one hose, a swimming pool can be filled in 8 hours. A second, larger, hose used alone can fill the pool in 5 hours. How long would it take to fill the pool if both hoses were used simultaneously?

57 How much water must be evaporated from 500 grams of a 10% salt solution in order to obtain a solution containing 15% salt?

58 A chemist has two acid solutions, the first containing 20% acid and the second 35% acid. How many ml of each should be mixed to obtain 50 ml of a solution containing 30% acid?

59 Generalize Exercise 57 as follows: Let a, b, and d be positive real numbers with $a < b < 100$. Given a solution of d grams of salt water that is a% salt, how much water must be evaporated in order that the resulting mixture will be $b\%$ salt? Express the answer in terms of a, b, and d.

60 Let a, b, c, and d be positive real numbers where $a < c < b < 100$. Generalize Exercise 58 to the case where the given solutions contain $a\%$ and $b\%$ acid, respectively, and where d ml of a solution containing $c\%$ acid is desired. Express the answer in terms of a, b, c, and d.

61 A salesperson purchased an automobile which was advertised as averaging 25 miles per gallon in the city and 40 miles per gallon on the highway. On a certain sales trip it required 51 gallons of gasoline to cover 1800 miles. Assuming that the advertised mileage estimates were correct, how many miles were driven in the city?

62 Two boys own Walky-Talky radios which have a maximum range of 2 miles. One of the boys leaves a certain point at 1:00 P.M., walking due north at a rate of 4 mph. The second boy leaves the same point at 1:15 P.M., traveling due south at 6 mph. When will they be unable to communicate with one another?

63 A farmer plans to use 180 feet of fencing to enclose a rectangular region, using part of a straight river bank instead of fencing as one side of the rectangle. Find the area of the region if the length of the side parallel to the river bank is
(a) twice the length of an adjacent side.
(b) one-half the length of an adjacent side.
(c) the same as the length of an adjacent side.

64 A consulting engineer charges $60 per hour, and his assistant receives $20 per hour. For a certain job a customer received a bill for $580. If the assistant worked 5 hours less than the engineer, how much time did each spend on the job?

65 It takes a boy 45 minutes to deliver the newspapers on his route; however, if his sister helps, it takes only 20 minutes. How long would it take his sister to deliver the newspapers by herself?

66 A water tank can be emptied by using one pump for 5 hours. A second, smaller, pump can empty the tank in 8 hours. If the larger pump is started at 1:00 P.M., at what time should the smaller pump be started so that the tank will be emptied at 5:00 P.M.?

67 A boy can row a boat at a rate of 5 mph in still water. If he rows upstream for 15 minutes, and then rows downstream, returning to his starting point in another 12 minutes, find:
(a) the rate of the current.
(b) the total distance traveled.

68 A woman begins jogging at 3:00 P.M., running due north at a 6-minute mile pace. Later she reverses direction and runs due south at a 7-minute mile pace. If she returns to her starting point at 3:45 P.M., find the total number of miles she ran.

69 The Intelligence Quotient (IQ) is determined by multiplying the quotient of a person's mental age and chronological age by 100.
(a) Find the IQ of a 12-year-old child whose mental age is 15.
(b) Find the mental age of a person 15 years old whose IQ is 140.

70 Ohm's Law in electrical theory states that $I = E/R$, where I is the current (in amperes), E is the electromotive force (in volts), and R is the resistance (in ohms). In a certain circuit $E = 110$ and $R = 50$. If E and R are to be changed by the same numerical amount, what change in them will cause I to double?

2.3 QUADRATIC EQUATIONS

Definition

A **quadratic equation** is an equation of the form

$$ax^2 + bx + c = 0$$

where a, b, and c are real numbers with $a \neq 0$.

An important technique for solving quadratic equations makes use of the fact that if p and q are real numbers such that $pq = 0$, then either $p = 0$ or $q = 0$. It follows that if $ax^2 + bx + c$ can be expressed as a product of two first-degree polynomials, then solutions can be found by setting each factor equal to 0, as illustrated in the next example.

Example 1   Solve the equation $3x^2 + x - 10 = 0$.

Solution   The equation may be written in the form

$$(3x - 5)(x + 2) = 0,$$

and hence, either

$$3x - 5 = 0 \quad \text{or} \quad x + 2 = 0.$$

The solutions of these linear equations are

$$x = \tfrac{5}{3} \quad \text{and} \quad x = -2.$$

This shows that if a real number satisfies the quadratic equation, then it must equal either 5/3 or -2. The fact that both numbers actually are roots may be seen by direct substitution in the given equation.   ■

The method used for solving the equation in Example 1 is called the **method of factoring**.

Example 2   Solve $x^2 - 8x + 16 = 0$.

Solution   Factoring, we see that the equation is equivalent to

$$(x - 4)^2 = 0 \quad \text{or} \quad (x - 4)(x - 4) = 0.$$

Setting each factor equal to zero, we obtain $x - 4 = 0$ and $x - 4 = 0$. Hence, the given equation has one solution, $x = 4$.   ■

Since $x - 4$ appears as a factor twice in the previous solution, the real number 4 is called a **double root**, or **root of multiplicity 2**, of the equation $x^2 - 8x + 16 = 0$.

Given a quadratic equation of the form $x^2 = d$, where $d \geq 0$, we may write $x^2 - d = 0$, or equivalently,

$$(x + \sqrt{d})(x - \sqrt{d}) = 0.$$

Setting each factor equal to zero gives us the solutions $-\sqrt{d}$ and $\sqrt{d}$. We frequently use $\pm\sqrt{d}$ as an abbreviation for solutions of this type. Thus, for $d \geq 0$, we have proved the following.

Theorem

$$\boxed{\text{If} \quad x^2 = d, \quad \text{then} \quad x = \pm\sqrt{d}.}$$

The process of solving $x^2 = d$ as indicated in the box may be referred to by the phrase "take the square root of both sides of the equation." Note that, in so doing, we introduce a positive and a negative square root, and not only the principal square root defined in Section 1.5.

Example 3 Solve the equation $x^2 = 5$.

Solution Taking the square root of both sides gives us $x = \pm\sqrt{5}$. Thus, the solutions are $\sqrt{5}$ and $-\sqrt{5}$. ∎

We shall now derive a general formula for solving any equation of the form

$$ax^2 + bx + c = 0$$

where $a \neq 0$. We begin by dividing both sides by a, obtaining

$$x^2 + \frac{b}{a}x + \frac{c}{a} = 0,$$

or equivalently,

$$x^2 + \frac{b}{a}x = -\frac{c}{a}.$$

Next we shall add a certain number to both sides so that the resulting expression on the left may be written in the form $(x + d)^2$ for some real number d. This can be accomplished by adding the square of half the coefficient of x as follows:

$$x^2 + \frac{b}{a}x + \left(\frac{b}{2a}\right)^2 = \left(\frac{b}{2a}\right)^2 - \frac{c}{a}.$$

We may now write

$$\left(x + \frac{b}{2a}\right)^2 = \frac{b^2 - 4ac}{4a^2}$$

$$x + \frac{b}{2a} = \pm\sqrt{\frac{b^2 - 4ac}{4a^2}}$$

$$x = -\frac{b}{2a} \pm \sqrt{\frac{b^2 - 4ac}{4a^2}}.$$

We may change the radical on the right-hand side of the last equation as follows:

$$\pm\sqrt{\frac{b^2 - 4ac}{4a^2}} = \pm\frac{\sqrt{b^2 - 4ac}}{\sqrt{(2a)^2}} = \pm\frac{\sqrt{b^2 - 4ac}}{|2a|}.$$

If $a > 0$, then $|2a| = 2a$ and the expression equals

$$\pm\frac{\sqrt{b^2 - 4ac}}{2a}.$$

If $a < 0$, then $|2a| = -2a$ and we obtain

$$\pm \frac{\sqrt{b^2 - 4ac}}{2a}.$$

Consequently, whether a is positive or negative,

$$x = -\frac{b}{2a} \pm \frac{\sqrt{b^2 - 4ac}}{2a}.$$

We have shown that if the quadratic equation $ax^2 + bx + c = 0$ has roots, then they are given by the two numbers in the last formula. Moreover, it can be shown by direct substitution that the two numbers do satisfy the equation. This gives us the following very important fact.

Quadratic Formula

> If $a \neq 0$, then the roots of the equation $ax^2 + bx + c = 0$ are given by
>
> $$x = \frac{-b \pm \sqrt{b^2 - 4ac}}{2a}.$$

The number $b^2 - 4ac$, which appears under the radical sign in the quadratic formula, is called the **discriminant** of the quadratic equation. It can be used to determine the nature of the roots of the equation as follows, where a, b, and c are real numbers.

> (i)   If $b^2 - 4ac = 0$, the equation has one root of multiplicity 2.
> (ii) If $b^2 - 4ac > 0$, the equation has two real and unequal solutions.
> (iii) If $b^2 - 4ac < 0$, the equation has no real solutions.

Example 4   Solve $4x^2 + x - 3 = 0$.

Solution   Letting $a = 4$, $b = 1$, and $c = -3$ in the Quadratic Formula, we obtain

$$x = \frac{-1 \pm \sqrt{1 - 4(4)(-3)}}{2(4)}$$

$$= \frac{-1 \pm \sqrt{49}}{8}$$

$$= \frac{-1 \pm 7}{8}.$$

Hence, the solutions are

$$x = \frac{-1 + 7}{8} = \frac{3}{4} \quad \text{and} \quad x = \frac{-1 - 7}{8} = -1. \qquad \blacksquare$$

Example 4 could also have been solved by factoring. If we write $(4x - 3)(x + 1) = 0$ and set each factor equal to zero, we obtain the solutions $x = 3/4$ and $x = -1$.

Example 5   Solve $2x^2 - 6x + 3 = 0$.

Solution   Letting $a = 2$, $b = -6$, and $c = 3$ in the Quadratic Formula, we obtain

$$x = \frac{6 \pm \sqrt{(-6)^2 - 4(2)(3)}}{2(2)}$$

$$= \frac{6 \pm \sqrt{12}}{4}$$

$$= \frac{6 \pm 2\sqrt{3}}{4}$$

$$= \frac{3 \pm \sqrt{3}}{2}.$$

Hence, the solutions are $x = (3 \pm \sqrt{3})/2$.   ∎

The following example illustrates the case of a double root.

Example 6   Solve the equation $9x^2 - 30x + 25 = 0$.

Solution   Letting $a = 9$, $b = -30$, $c = 25$ in the Quadratic Formula gives us

$$x = \frac{30 \pm \sqrt{(-30)^2 - 4(9)(25)}}{2(9)}$$

$$= \frac{30 \pm \sqrt{900 - 900}}{18}$$

$$= \frac{30 \pm 0}{18} = \frac{5}{3}.$$

Consequently, there is only one solution, $5/3$.   ∎

There are many applied problems which lead to quadratic equations. One is illustrated in the following example.

Example 7   A box with a square base and no top is to be made from a square piece of tin by cutting out 3-inch squares from each corner and folding up the sides. If the box is to hold 48 cubic inches, what size piece of tin should be used?

Solution   If we let x denote the length of the side of the piece of tin, then the length of the base of the box is $x - 6$ (see Figure 2.4). Since the area of the base is $(x - 6)^2$ and the height is 3, the volume of the box is $3(x - 6)^2$. Moreover, since the box is to hold 48 cubic inches,

$$3(x - 6)^2 = 48.$$

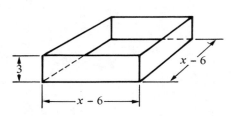

Figure 2.4

Solving for x, we obtain

$$(x - 6)^2 = 16$$
$$x - 6 = \pm 4$$
$$x = 6 \pm 4.$$

Consequently, either $x = 10$ or $x = 2$.

Let us now check each of these numbers. Referring to Figure 2.4, we see that 2 is unacceptable since no box is possible in this case. (Why?) However, if we begin with a 10-inch square of tin, cut out 3-inch corners, and fold, we obtain a box having dimensions 4 inches, 4 inches, and 3 inches. The box has the desired volume of 48 cubic inches. Thus, 10 inches is the answer to the problem. ∎

As illustrated in Example 7, even though an equation is formulated correctly, it is possible, owing to the physical nature of a given problem, to arrive at meaningless solutions. These solutions should be discarded. For example, we would not accept the answer -7 years for the age of an individual, nor $\sqrt{50}$ for the number of automobiles in a parking lot.

EXERCISES 2.3

Solve the equations in Exercises 1–10 by factoring.

Use the quadratic formula to solve the equations in Exercises 11–24.

1 $6x^2 + 11x - 10 = 0$

2 $15x^2 + x - 6 = 0$

3 $20x^2 - 33x + 7 = 0$

4 $4x^2 + 16x + 15 = 0$

5 $4y^2 + 29y + 30 = 0$

6 $8z^2 + 19z - 27 = 0$

7 $16t^2 - 24t + 9 = 0$

8 $9u^2 - 30u + 25 = 0$

9 $54r^2 - 9r - 30 = 0$

10 $60s^2 - 85s - 35 = 0$

11 $2x^2 - x - 3 = 0$

12 $3x^2 - 2x - 8 = 0$

13 $u^2 + 2u - 6 = 0$

14 $v^2 + 3v - 5 = 0$

15 $2x^2 - 4x - 5 = 0$

16 $3x^2 - 6x + 2 = 0$

17 $4y^2 - 20y + 25 = 0$

18 $9t^2 + 6t + 1 = 0$

19 $5s^2 + 9s + 3 = 0$

20 $3z^2 - 8z - 2 = 0$

21 $\dfrac{x+1}{3x+2} = \dfrac{x-2}{2x-3}$

22 $\dfrac{5}{w^2} - \dfrac{10}{w} + 2 = 0$

23 $9x^2 + 4x = 0$

24 $16x^2 - 9 = 0$

The formulas in Exercises 25–34 occur in mathematics and its applications. Solve for the indicated variable in terms of the remaining variables (all letters denote positive numbers).

25 $V = \frac{1}{3}\pi r^2 h$ for r

26 $K = \frac{1}{2}mv^2$ for v

27 $F = g\dfrac{m_1 m_2}{d^2}$ for d

28 $V = \frac{4}{3}\pi a^2 b$ for a

29 $s = \frac{1}{2}gt^2 + v_0 t$ for t

30 $A = 2\pi r(r + h)$ for r

31 $\dfrac{x^2}{a^2} - \dfrac{y^2}{b^2} = 1$ for y

32 $s = \sqrt{(r_1 - r_2)^2 + h^2}$ for r_2

33 $c = \sqrt{4h(2R - h)}$ for h

34 $\dfrac{x^2}{a^2} + \dfrac{y^2}{b^2} + \dfrac{z^2}{c^2} = 1$ for z

35 Find two consecutive odd integers whose product is 255.

36 Find two consecutive even integers, the sum of whose squares is 1060.

37 The diameter of a circle is 10 cm. What change in the radius will decrease the area by 16π cm^2?

38 The hypotenuse of a right triangle is 5 cm long. Find the lengths of the two legs if their sum is 6 cm.

39 A rectangular plot of ground having dimensions 26 ft by 30 ft is surrounded by a walk of uniform width. If the area of the walk is 240 ft^2, what is its width?

40 It can be shown by mathematical induction (see Chapter 11), that the sum of the first n positive integers $1, 2, 3, \ldots, n$ equals $\frac{1}{2}n(n + 1)$. For what value of n will the sum equal 276?

41 An airplane flying north at 200 mph passed over a point on the ground at 2:00 P.M. Another airplane at the same altitude passed over the point at 2:30 P.M., flying east at 400 mph.
  (a) If t denotes the time (in hours) after 2:30 P.M., express the distance d between the planes in terms of t.
  (b) At what time were the airplanes 500 miles apart?

42 A box with an open top is to be constructed by cutting out 3-inch squares from a rectangular sheet of tin whose length is twice its width. What size sheet will produce a box having a volume of 60 in^3?

43 A piece of wire 100 inches long is cut into two pieces, and then each piece is bent into the shape of a square. If the sum of the enclosed areas is 397 in^2, find the lengths of each piece of wire.

44 Generalize Exercise 43 to the case where the wire is l inches long and the area is A in^2. Express the lengths of the two pieces in terms of l and A.

45 A manufacturer of tin cans wishes to construct a cylindrical can of height 20 cm and capacity 3000 cc. Find the inner radius of the can.

46 A man plans to use 6 cubic yards of concrete to pour a rectangular slab for a patio. If the length of the patio is to be twice the width, and the thickness of the slab 3 inches, find the dimensions of the patio.

47 Two boys with Walky-Talky radios leave the same point at 9:00 A.M., one walking due south at 4 mph and the other due west at 3 mph. How long can they communicate with one another if each radio has a maximum range of 2 miles?

48 A farmer plans to enclose a rectangular region, using part of his barn for one side and fencing in the other three sides. If he wants the side parallel to the barn to be twice the length of an adjacent side, and the area of the region to be 128 square feet, how many feet of fencing should he purchase?

49 The speed of the current in a stream is 5 mph. It takes a girl 30 minutes longer to paddle a canoe 1.2 miles upstream than the same distance downstream. What is her rate in still water?

50   Water from two hoses can fill a swimming pool in 4 hours. If used separately it takes 6 hours longer for the smaller hose to fill the pool than the larger hose. How long would it take to fill the pool if only the smaller hose were used?

51   A 24×36 inch sheet of paper is to be used for a poster, with the smaller side at the bottom. The margins at the sides and top are to have the same width, and the bottom margin is to be twice as wide as the other margins. Find the width of the margins if the printed area is to be 661.5 square inches.

52   A manufacturer sells a certain article to dealers at a rate of \$20 each if less than 50 are ordered. If 50 or more are ordered (up to 600), the price per article is reduced at a rate of 2 cents times the number ordered. How many articles can a dealer purchase for \$4200?

53   A projectile is fired straight upward with an initial speed of 800 ft/sec. The number of feet s above the ground after t seconds is given by $s = -16t^2 + 800t$.
(a)   When will the projectile be 3200 feet above the ground?
(b)   When will it hit the ground?
(c)   What is its maximum height?

54   If a projectile is fired upward from a height of s_0 feet above the ground with an initial speed of v_0 feet per second, then its height above the ground at time t is given by $s = -16t^2 + v_0 t + s_0$. Solve for t in terms of s, v_0, and s_0. When will the projectile hit the ground? Express your answer in terms of v_0 and s_0.

55   If one solution of the equation $kx^2 + 7x + 3k = 0$ is -5, find the other solution.

56   If a, b, and c are rational numbers, can the equation $ax^2 + bx + c = 0$ have one rational and one irrational root? Explain.

57   If r_1 and r_2 are the two roots of $ax^2 + bx + c = 0$, show that $r_1 + r_2 = -b/a$ and $r_1 r_2 = c/a$.

58   Prove that the roots of $ax^2 + bx + c = 0$ are numerically equal but opposite in sign to the roots of $ax^2 - bx + c = 0$.

59   Given the equation $4x^2 - 4xy + 1 - y^2 = 0$, use the Quadratic Formula to solve for
(a) x in terms of y,   (b) y in terms of x.

60   Given the equation $2x^2 - xy = 3y^2 + 1$, use the Quadratic Formula to solve for
(a) x in terms of y,   (b) y in terms of x.

CALCULATOR EXERCISES 2.3

Approximate the solutions of the following equations to the nearest thousandth.

1 $47x^2 + 183x + 29 = 0$

2 $912x^2 - 3174x - 713 = 0$

3 $2.13x^2 + 6.04x - 5.71 = 0$

4 $6143x^2 + 5417x - 1439 = 0$

2.4   MISCELLANEOUS EQUATIONS

The method of factoring used in the previous section can also be applied to certain nonquadratic equations. Specifically, if an equation can be expressed in factored form, *with zero on one side*, then solutions may often be obtained by setting each factor equal to zero, as illustrated in the next two examples.

Example 1   Solve $x^3 + 2x^2 - x - 2 = 0$.

Solution   The left side may be factored by grouping, as follows:

$$x^3 + 2x^2 - x - 2 = 0$$
$$x^2(x + 2) - (x + 2) = 0$$
$$(x^2 - 1)(x + 2) = 0$$
$$(x + 1)(x - 1)(x + 2) = 0.$$

Setting each factor equal to 0 gives us

$$x + 1 = 0, \quad x - 1 = 0, \quad x + 2 = 0,$$

or equivalently,

$$x = -1, \quad x = 1, \quad x = -2.$$

That these three numbers are solutions may be checked by substitution in the original equation.   ■

Example 2   Solve $x^{3/2} = x^{1/2}$.

Solution   We may proceed as follows:

$$x^{3/2} - x^{1/2} = 0$$
$$x^{1/2}(x - 1) = 0.$$

Hence, $x^{1/2} = 0$ or $x - 1 = 0$, from which we obtain the solutions $x = 0$ and $x = 1$.   ■

A common error, when an equation such as $x^{3/2} = x^{1/2}$ is given (see Example 2), is to divide both sides by $x^{1/2}$, obtaining $x = 1$. Observe that this leads to the loss of the solution $x = 0$.

Certain equations that are not linear or quadratic can be transformed into one of those forms by a suitable manipulation. If equations involve radicals or fractional exponents, the method of raising both sides to a positive integral power is often used. When this is done, the solutions of the new equation always contain the solutions of the original equation. For example, the solutions of

$$2x - 3 = \sqrt{x + 6}$$

are also solutions of

$$(2x - 3)^2 = (\sqrt{x + 6})^2.$$

In some cases the new equation has *more* solutions than the original equation. To illustrate, if we start with the equation $x = 3$ and square both sides, we obtain $x^2 = 9$. Note that the given equation has only one solution, 3, whereas the new equation has two solutions, 3 and -3. Any solution of the new equation which is not a solution of the original equation is called an **extraneous solution**. Since extraneous solutions may arise, it is *absolutely essential* to check all solutions obtained after raising both sides of an equation to some power.

Example 3   Solve $\sqrt[3]{x^2 - 1} = 2$.

Solution   If we cube both sides, then the solutions of the given equation are included among the solutions of the following:

$$(\sqrt[3]{x^2 - 1})^3 = 2^3$$
$$x^2 - 1 = 8$$
$$x^2 = 9.$$

Hence, the only possible solutions of $\sqrt[3]{x^2 - 1} = 2$ are 3 or -3.

We next check each of these numbers by substitution in $\sqrt[3]{x^2 - 1} = 2$. Substituting 3 for x in the equation we obtain $\sqrt[3]{3^2 - 1} = 2$, or $\sqrt[3]{8} = 2$, which is a true statement. Thus, 3 is a solution. Similarly, -3 is a solution. Hence, the solutions of the given equation are 3 and -3.   ∎

Example 4   Solve $3 + \sqrt{3x + 1} = x$.

Solution   We begin by isolating the radical on one side as follows:

$$\sqrt{3x + 1} = x - 3.$$

Next we square both sides and simplify, obtaining

$$(\sqrt{3x + 1})^2 = (x - 3)^2$$
$$3x + 1 = x^2 - 6x + 9$$
$$x^2 - 9x + 8 = 0$$
$$(x - 1)(x - 8) = 0.$$

Since the last equation has solutions 1 and 8, it follows that 1 and 8 are the only possible solutions of the original equation.

We now check each of these by substitution in $3 + \sqrt{3x + 1} = x$. Letting $x = 1$ gives us

$$3 + \sqrt{4} = 1 \quad \text{or} \quad 5 = 1$$

which is false. Consequently, 1 is not a solution. Letting $x = 8$ in the given equation we obtain

$$3 + \sqrt{25} = 8 \quad \text{or} \quad 3 + 5 = 8$$

which is true. Hence, the equation $3 + \sqrt{3x + 1} = x$ has only one solution, $x = 8$.   ∎

For certain equations involving radicals it is necessary to use the process of raising sides to powers several times, as illustrated in the next example.

Example 5   Solve $\sqrt{2x - 3} - \sqrt{x + 7} + 2 = 0$.

Solution Let us begin by writing

$$\sqrt{2x - 3} = \sqrt{x + 7} - 2.$$

Squaring both sides we obtain

$$2x - 3 = (x + 7) - 4\sqrt{x + 7} + 4$$

which simplifies to

$$x - 14 = -4\sqrt{x + 7}.$$

Squaring both sides of the last equation and simplifying gives us

$$x^2 - 28x + 196 = 16(x + 7)$$
$$x^2 - 28x + 196 = 16x + 112$$
$$x^2 - 44x + 84 = 0$$
$$(x - 42)(x - 2) = 0.$$

Hence, the only possible solutions of the given equation are 42 and 2.

We next check each of these by substitution in the original equation. Substituting $x = 42$ gives us

$$\sqrt{84 - 3} - \sqrt{42 + 7} + 2 = 0$$

or

$$9 - 7 + 2 = 0$$

which is false. Hence, 42 is not a solution. If we substitute $x = 2$ we obtain

$$\sqrt{4 - 3} - \sqrt{2 + 7} + 2 = 0$$

or

$$1 - 3 + 2 = 0$$

which is true. Hence, the given equation has one solution, $x = 2$. ■

An equation is said to be of **quadratic type** if it can be written in the form

$$au^2 + bu + c = 0$$

where $a \neq 0$ and u is an expression in some variable. If we find the solutions in terms of u, then the solutions of the original equation can be obtained by referring to the specific form of u. The technique is illustrated in the following example.

Example 6 Solve $x^{2/3} + x^{1/3} - 6 = 0$.

Solution If we let $u = x^{1/3}$, then the equation can be written

$$u^2 + u - 6 = 0 \quad \text{or} \quad (u + 3)(u - 2) = 0$$

which has solutions

$$u = -3 \quad \text{and} \quad u = 2.$$

Since $u = x^{1/3}$, we have

$$x^{1/3} = -3 \quad \text{or} \quad x^{1/3} = 2.$$

Cubing gives us

$$x = -27 \quad \text{or} \quad x = 8.$$

We next check each of these by substitution. Letting $x = -27$ in the given equation we obtain

$$[(-27)^{1/3}]^2 + (-27)^{1/3} - 6 = 9 - 3 - 6 = 0.$$

Thus, -27 is a solution. Similarly, it can be shown that 8 is a solution. Hence, the solutions of the given equations are -27 and 8.   ∎

Example 7   Solve $x^4 - 3x^2 + 1 = 0$.

Solution   Letting $u = x^2$ gives us

$$u^2 - 3u + 1 = 0.$$

Using the quadratic formula, we obtain

$$u = \frac{3 \pm \sqrt{9 - 4}}{2} = \frac{3 \pm \sqrt{5}}{2}.$$

Since $u = x^2$, we have

$$x^2 = \frac{3 \pm \sqrt{5}}{2} \quad \text{or} \quad x = \pm \sqrt{\frac{3 \pm \sqrt{5}}{2}}.$$

Thus, there are four possible solutions:

$$\sqrt{\frac{3 + \sqrt{5}}{2}}, \quad -\sqrt{\frac{3 + \sqrt{5}}{2}}, \quad \sqrt{\frac{3 - \sqrt{5}}{2}}, \quad -\sqrt{\frac{3 - \sqrt{5}}{2}}.$$

By checking it can be shown that each is a solution of the original equation.   ∎

EXERCISES 2.4

Solve the equations in Exercises 1–36.

1   $4x^3 + 12x^2 - 9x - 27 = 0$

2   $2x^3 + 5x^2 - 8x - 20 = 0$

3   $6x^5 + 10x^4 = 3x^3 + 5x^2$

4   $25z^4 + 5z = 125z^3 + z^2$

5   $y^{3/2} = 4y$

6   $2x^3 = 5x^2$

7 $\dfrac{1}{x} = \dfrac{1}{x^2}$

8 $\sqrt{x}\sqrt[3]{x} - 3\sqrt[3]{x} - 2\sqrt{x} + 6 = 0$

9 $\sqrt{7 - 5x} = 8$

10 $\sqrt{2x - 9} = 3^{-1}$

11 $2 + \sqrt[3]{1 - 5t} = 0$

12 $\sqrt[3]{6 - s^2} + 5 = 0$

13 $\sqrt[5]{2x^2 + 1} - 2 = 0$

14 $\sqrt[4]{2x^2 - 1} = x$

15 $3\sqrt{2x - 3} + 2\sqrt{7 - x} = 11$

16 $\sqrt{2x + 15} - 2 = \sqrt{6x + 1}$

17 $\sqrt{7 - 2x} - \sqrt{5 + x} = \sqrt{4 + 3x}$

18 $4\sqrt{1 + 3x} + \sqrt{6x + 3} = \sqrt{-6x - 1}$

19 $\sqrt{11 + 8x} + 1 = \sqrt{9 + 4x}$

20 $2\sqrt{x} - \sqrt{x - 3} = \sqrt{5 + x}$

21 $\sqrt{2\sqrt{x} + 1} = \sqrt{3x - 5}$

22 $\sqrt{5\sqrt{x}} = \sqrt{2x - 3}$

23 $\sqrt{1 + 4\sqrt{x}} = \sqrt{x} + 1$

24 $\sqrt{x + 1} = \sqrt{x - 1}$

25 $4x^4 - 37x^2 + 9 = 0$

26 $2x^4 - 9x^2 + 4 = 0$

27 $3z^4 - 5z^2 + 1 = 0$

28 $2y^4 + y^2 - 5 = 0$

29 $3x^{2/3} + 8x^{1/3} - 3 = 0$

30 $2t^{1/3} - 5t^{1/6} + 2 = 0$

31 $3w - 19\sqrt{w} + 20 = 0$

32 $2x^{-2/3} + 7x^{-1/3} - 4 = 0$

33 $\dfrac{2}{(x - 1)^2} + \dfrac{3}{x - 1} - 2 = 0$

34 $\dfrac{4}{(x^2 - 1)^2} - \dfrac{5}{x^2 - 1} + 1 = 0$

35 $\left(\dfrac{t}{t + 1}\right)^2 + \dfrac{2t}{t + 1} - 15 = 0$

36 $6u^{-1/2} - 17u^{-1/4} + 5 = 0$

The formulas in Exercises 37–42 occur in mathematics and its applications. Solve for the indicated variable in terms of the remaining variables (all letters denote positive real numbers).

37 $S = \pi r \sqrt{r^2 + h^2}$ for h

38 $d = \frac{1}{2}\sqrt{4R^2 - C^2}$ for C

39 $y = \dfrac{b}{a}\sqrt{a^2 - x^2}$ for x

40 $T = 2\pi \sqrt{\dfrac{l}{g}}$ for l

41 $x^{2/3} + y^{2/3} = a^{2/3}$ for y

42 $y = (\sqrt[3]{a} - \sqrt[3]{x})^3$ for x

43 As sand leaks out of a certain container it forms a pile which has the shape of a right circular cone whose height is always one-half the diameter of the base. What is the diameter of the base at the instant that 144 cubic centimeters has leaked out?

44 If the volume of a spherical balloon increases from $10\frac{2}{3}$ cubic feet to 36 cubic feet, how much does the diameter increase?

CALCULATOR EXERCISES 2.4

Approximate the solutions of the following equations to the nearest hundredth.

1 $x^{6/7} - 28x^{1/7} = 0$

2 $45x^4 - 63x^2 + 14 = 0$

3 $62x^{2/3} - 75x^{1/3} - 23 = 0$

4 $12x^{-2/5} - 19x^{-1/5} - 11 = 0$

2.5 INEQUALITIES

If a and b are real numbers, then either of the inequalities $a > b$ or $b < a$ means that $a - b$ is positive. In particular, we may write

$$7 > 3, \quad 2 > -5, \quad -6 < -4, \quad \text{or} \quad \sqrt{2} < \pi.$$

If we refer to a coordinate line, then $a > b$ (or $b < a$) implies that the point with coordinate a lies to the right of the point with coordinate b. In addition, if $b > c$, then we have a situation similar to that shown in Figure 2.5, and evidently $a > c$. This provides a geometric demonstration of the following general rule for inequalities:

$$\text{If } a > b \text{ and } b > c, \quad \text{then} \quad a > c.$$

Figure 2.5

When working with inequalities, we often employ techniques which are similar to those used for solving equations. As we shall see, adding the same real number to both sides leaves the inequality sign invariant. To illustrate, if we add 2 to both sides of $7 > 3$, we obtain

$$7 + 2 > 3 + 2 \quad \text{or} \quad 9 > 5.$$

This fact is not surprising, since adding 2 to both sides of $7 > 3$ amounts to shifting the corresponding points on a coordinate line 2 units to the right, as illustrated in Figure 2.6. Similarly, adding -2 (or, equivalently, subtracting 2) would shift the points to the left. These are special cases of the following general rule:

$$\text{If } a > b, \quad \text{then} \quad a + c > b + c.$$

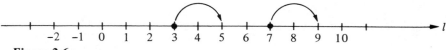

Figure 2.6

We can also multiply both sides of an inequality by the same positive real number. For example, multiplying both sides of $7 > 3$ by 2 gives us

$$7 \cdot 2 > 3 \cdot 2 \quad \text{or} \quad 14 > 6.$$

A major difference from our work with equations is that if we multiply by a negative real number, then the inequality sign is reversed. To illustrate, multiplying both sides of $7 > 3$ by -2 gives us

$$7(-2) < 3(-2) \quad \text{or} \quad -14 < -6.$$

The four fundamental properties of inequalities we have discussed are stated below, where a, b, and c denote real numbers.

Properties of Inequalities

(i)   If $a > b$ and $b > c$, then $a > c$.
(ii) If $a > b$, then $a + c > b + c$.
(iii) If $a > b$ and $c > 0$, then $ac > bc$.
(iv) If $a > b$ and $c < 0$, then $ac < bc$.

We can supply algebraic proofs for all four properties by making use of the fact that both the sum and product of any two positive real numbers are positive. Thus, to prove property (i), we first note that if $a > b$ and $b > c$, then $a - b$ and $b - c$ are both positive. Consequently, the sum $(a - b) + (b - c)$ is positive. Since the sum reduces to $a - c$, we see that $a - c$ is positive, which means that $a > c$.

To establish property (ii), we again note that if $a > b$, then $a - b$ is positive. Since $(a + c) - (b + c) = a - b$, it follows that $(a + c) - (b + c)$ is positive; that is, $a + c > b + c$.

For property (iii), if $a > b$ and $c > 0$, then $a - b$ and c are both positive and hence, so is the product $(a - b)c$. Consequently, $ac - bc$ is positive; that is, $ac > bc$.

Finally, to establish property (iv), we first note that if $c < 0$, then $0 - c$, or $-c$, is positive. In addition, if $a > b$, then $a - b$ is positive, and hence, the product $(a - b)(-c)$ is positive. However, $(a - b)(-c) = -ac + bc$, and hence, $bc - ac$ is positive. This means that $bc > ac$, or $ac < bc$.

The following analogous properties hold for $<$.

Properties of Inequalities

(i)   If $a < b$ and $b < c$, then $a < c$.
(ii) If $a < b$, then $a + c < b + c$.
(iii) If $a < b$ and $c > 0$, then $ac < bc$.
(iv) If $a < b$ and $c < 0$, then $ac > bc$.

We shall next consider inequalities which contain variables. The domains of the variables will be sets of real numbers. As a first illustration, let us consider

$$x^2 - 3 < 2x + 4.$$

If certain numbers such as 4 or 5 are substituted for x, we obtain the false statements $13 < 12$ or $22 < 14$, respectively. Other numbers such as 1 or 2 produce the true statements $-2 < 6$ or $1 < 8$, respectively. In general, if we are given an inequality in x, and if a true statement is obtained when x is replaced by a real number a, then a is called a **solution** of the inequality. Thus, 1 and 2 are solutions of the inequality $x^2 - 3 < 2x + 4$, whereas 4 and 5 are not solutions. To **solve** an inequality means to find all solutions. We say that two inequalities are **equivalent** if they have exactly the same solutions.

As with equations, a standard method for solving an inequality is to replace it with a chain of equivalent inequalities, terminating in one for which the solutions are obvious. For example, if x represents a real number, then adding the same expression in x to both sides leads to an equivalent inequality. We may multiply both sides of an inequality by an expression containing x if we are certain that the expression is positive for all values of x under consideration. To illustrate, multiplication by $x^4 + 3x^2 + 5$ would be permissible since this expression is always positive. If we multiply both sides of an inequality by an expression that is always negative, such as $-7 - x^2$, then the inequality sign is reversed.

The reader should supply reasons for the solutions of the following examples.

Example 1 Solve the inequality $-3x + 4 > 11$.

Solution The following inequalities are all equivalent.

$$-3x + 4 > 11$$
$$(-3x + 4) + (-4) > 11 + (-4)$$
$$-3x > 7$$
$$(-\tfrac{1}{3})(-3x) < (-\tfrac{1}{3})(7)$$
$$x < -\tfrac{7}{3}$$

Since the last inequality is equivalent to the first, the solutions of $-3x + 4 > 11$ consist of all real numbers x such that $x < -7/3$. ∎

Example 2 Solve the inequality $4x - 3 < 2x + 5$.

Solution The following is a list of equivalent inequalities:

$$4x - 3 < 2x + 5$$
$$(4x - 3) + 3 < (2x + 5) + 3$$
$$4x < 2x + 8$$
$$4x + (-2x) < (2x + 8) + (-2x)$$
$$2x < 8$$
$$\tfrac{1}{2}(2x) < \tfrac{1}{2}(8)$$
$$x < 4$$

Hence, the solutions of the given inequality consist of all real numbers x such that $x < 4$. ∎

It is convenient to give graphical interpretations for solutions of inequalities. By the **graph** of a set of real numbers we mean the collection of all points on a co-ordinate line l which correspond to the numbers. To **sketch a graph** we darken (or color) an appropriate portion of l. The graph corresponding to the solutions of Example 2 consists of all points to the left of the point with coordinate 4 and is sketched in Figure 2.7. The parenthesis in the figure indicates that the point corresponding to 4 is not part of the graph. If we wish to include such an **endpoint** of a graph, we use a bracket instead of a parenthesis. This will be illustrated in Example 4.

Figure 2.7

If $a < b$, the symbol (a, b) is sometimes used for all real numbers between a and b. This set is called an **open interval**.

Definition of Open Interval

$$(a, b) = \{x : a < x < b\}$$

The symbol to the right of the equals sign is translated "the set of all x such that $a < x < b$." The numbers a and b are called the **endpoints** of the interval. The graph of (a, b) consists of all points on a coordinate line which lie between the points corresponding to a and b. In Figure 2.8 we have sketched the graph of a general open interval (a, b) and also the special open intervals $(-1, 3)$ and $(2, 4)$. The parentheses in the figure indicate that the endpoints of the intervals are not to be included. For convenience, we shall use the terms *open interval* and *graph of an open interval* interchangeably.

Figure 2.8   *Open intervals* (a, b), $(-1, 3)$, *and* $(2, 4)$

The solutions of an inequality may often be described in terms of intervals, as illustrated in the next example.

Example 3   Solve the inequality $-6 < 2x - 4 < 2$ and represent the solutions graphically.

Solution   A real number x is a solution of the given inequality if and only if it is a solution of *both* of the inequalities

$$-6 < 2x - 4 \quad \text{and} \quad 2x - 4 < 2.$$

The first inequality is equivalent to each of the following:

$$-6 + 4 < (2x - 4) + 4$$
$$-2 < 2x$$
$$\tfrac{1}{2}(-2) < \tfrac{1}{2}(2x)$$
$$-1 < x.$$

The second inequality is equivalent to each of the following:

$$2x - 4 < 2$$
$$2x < 6$$
$$x < 3.$$

Thus, x is a solution of the given inequality if and only if *both*

$$-1 < x \quad \text{and} \quad x < 3,$$

that is,

$$-1 < x < 3.$$

Hence, the solutions are all numbers in the open interval $(-1, 3)$. The graph is sketched in Figure 2.9.

Figure 2.9

An alternate (and shorter) solution may be given by working with both inequalities simultaneously as follows:

$$-6 < 2x - 4 < 2$$
$$-6 + 4 < 2x < 2 + 4$$
$$-2 < 2x < 6$$
$$-1 < x < 3.$$

The last inequality agrees with the first solution.   ∎

For convenience, we refer to the interval $(-1, 3)$ as the **interval solution** of the inequality in Example 3.

If we wish to include an endpoint of an interval, a bracket is used instead of a parenthesis. If $a < b$, then **closed intervals**, denoted by $[a, b]$, and **half-open intervals**, denoted by $[a, b)$ or $(a, b]$, are defined as follows:

Definition of Closed and Half-open Intervals

$$[a, b] = \{x : a \leq x \leq b\}$$
$$[a, b) = \{x : a \leq x < b\}$$
$$(a, b] = \{x : a < x \leq b\}$$

Typical graphs are sketched in Figure 2.10, where a bracket indicates that the corresponding endpoint is part of the graph.

Figure 2.10

Example 4   Solve the inequality

$$-5 \leq \frac{4 - 3x}{2} < 1$$

and sketch the graph corresponding to the solutions.

Solution   A number x is a solution if and only if it satisfies both of the inequalities

$$-5 \le \frac{4 - 3x}{2} \quad \text{and} \quad \frac{4 - 3x}{2} < 1.$$

We could work with each of these separately or proceed as in the alternate solution of Example 3. Let us employ the latter technique as follows:

$$-5 \le \frac{4 - 3x}{2} < 1$$

$$-10 \le 4 - 3x < 2$$

$$-10 - 4 \le -3x < 2 - 4$$

$$-14 \le -3x < -2$$

$$\left(-\frac{1}{3}\right)(-14) \ge \left(-\frac{1}{3}\right)(-3x) > \left(-\frac{1}{3}\right)(-2)$$

$$\frac{14}{3} \ge x > \frac{2}{3}$$

$$\frac{2}{3} < x \le \frac{14}{3}.$$

Hence, the interval solution of the inequality is (2/3, 14/3]. The graph is sketched in Figure 2.11.

$$
\begin{array}{ccc}
0 & \frac{2}{3} & \frac{14}{3}
\end{array}
$$

Figure 2.11

The next example is an elementary illustration of an applied problem involving inequalities.

Example 5   The temperature readings on the Fahrenheit and Celsius scales are related by the equation $C = \frac{5}{9}(F - 32)$. What range of F corresponds to $30 \le C \le 40$?

Solution   The following inequalities are equivalent:

$$30 \le C \le 40$$

$$30 \le \tfrac{5}{9}(F - 32) \le 40$$

$$\tfrac{9}{5}(30) \le F - 32 \le \tfrac{9}{5}(40)$$

$$54 \le F - 32 \le 72$$

$$86 \le F \le 104.$$

Thus, a Celsius temperature range from $30°$ C to $40°$ C is the same as a Fahrenheit range from $86°$ F to $104°$ F. ∎

In order to describe solutions of inequalities such as $x < 4$, $x > -2$, $x \leq 7$, or $x \geq 3$, it is convenient to employ **infinite intervals**. In particular, if a is a real number, we define

$$(-\infty, a) = \{x : x < a\}.$$

That is, $(-\infty, a)$ denotes the set of all real numbers less than a. The symbol ∞ (**infinity**) is merely a notational device and is not to be interpreted as representing a real number. To illustrate, the interval solution in Example 1 is the infinite interval $(-\infty, -7/3)$, and that in Example 2 is $(-\infty, 4)$. The graph of $(-\infty, 4)$ is sketched in Figure 2.7. If we wish to include the point corresponding to a we write

$$(-\infty, a] = \{x : x \leq a\}.$$

Other types of infinite intervals are defined by

$$(a, \infty) = \{x : x > a\}.$$
$$[a, \infty) = \{x : x \geq a\}.$$

Example 6 Solve $\dfrac{1}{x-2} > 0$ and sketch the graph corresponding to the solutions.

Solution Since the numerator is positive, the given fraction is positive if and only if $x - 2 > 0$, or equivalently, $x > 2$. Hence, the interval solution is $(2, \infty)$. The graph is sketched in Figure 2.12.

$$\begin{array}{ccc} & & \\ 0 & 1 & 2 \end{array}$$

Figure 2.12 ∎

EXERCISES 2.5

1 Given $-6 < -4$, what inequality is obtained if:
(a) 5 is added to both sides?
(b) 5 is subtracted from both sides?
(c) both sides are multiplied by 1/2?
(d) both sides are multiplied by $-1/2$?

2 Given $3 > -3$, what inequality is obtained if:
(a) 3 is added to both sides?
(b) -3 is added to both sides?
(c) both sides are divided by 3?
(d) both sides are divided by -3?

In each of Exercises 3–12, express the inequality in interval notation and sketch a graph of the interval.

3 $2 < x < 5$

4 $-1 < x < 2$

5 $-1 < x \leq 3$

6 $-2 \geq x \geq -3$

7 $4 \geq x \geq 1$

8 $0 < x \leq 4$

9 $x > -1$

10 $x < 1$

11 $x \leq 2$

12 $x \geq -3$

Express each of the intervals in Exercises 13–18 as an inequality in the variable x.

13 $(-1, 7)$

14 $[5, 12]$

15 $(8, 9]$

16 $[-5, 0)$

17 $(5, \infty)$

18 $(-\infty, 7]$

Solve the inequalities in Exercises 19–42 and express the solutions in terms of intervals.

19 $5x - 6 > 11$

20 $3x - 5 < 10$

21 $2 - 7x \leq 16$

22 $7 - 2x \geq -3$

23 $3x + 1 < 5x - 4$

24 $6x - 5 > 9x + 1$

25 $4 - \frac{1}{2}x \geq -7 + \frac{1}{4}x$

26 $\frac{1}{3}x - 4 \leq \frac{1}{4}x - 6$

27 $-4 < 3x + 5 < 8$

28 $6 > 2x - 6 > 4$

29 $3 \geq \dfrac{7 - x}{2} \geq 1$

30 $-2 \leq \dfrac{5 - 3x}{4} \leq \dfrac{1}{2}$

31 $0 < 2 - \frac{3}{4}x \leq \frac{1}{2}$

32 $-3 < \frac{1}{2}x - 4 \leq 0$

33 $(3x + 1)(2x - 4) < (6x - 2)(x + 5)$

34 $(x + 2)(x - 5) < (x - 1)^2$

35 $(x + 2)^2 > x(x - 2)$

36 $2x(8x - 1) \geq (4x + 1)(4x - 3)$

37 $\dfrac{5}{3x + 7} > 0$

38 $\dfrac{1}{6 - 2x} < 0$

39 $(1 - 4x)^{-1} < 0$

40 $(x + 4)^{-1} > 0$

41 $\dfrac{1}{(x - 1)^2} > 0$

42 $\dfrac{1}{x^2 + 1} < 0$

43 The relationship between the Fahrenheit and Celsius temperature scales is given by $C = (5/9)(F - 32)$. If $60 \leq F \leq 80$, express the corresponding range for C in terms of an inequality.

44 In the study of electricity, Ohm's Law states that $R = E/I$ where E is measured in volts, I in amperes, and R in ohms. If $E = 110$, what values of R correspond to $I \leq 10$?

45 According to Hooke's Law, the force F (in pounds) required to stretch a certain spring x inches beyond its natural length is given by $F = (4.5)x$. If $10 \leq F \leq 18$, what is the corresponding range for x?

46 Boyle's Law for a certain gas states that $pv = 200$, where p denotes the pressure (lbs/in²) and v denotes the volume (in³). If $25 \leq v \leq 50$, what is the corresponding range for p?

47 If $0 < a < b$, prove that $(1/a) > (1/b)$. Why is the restriction $0 < a$ necessary?

48 If $0 < a < b$, prove that $a^2 < b^2$. Why is the restriction $0 < a$ necessary?

49 If $a < b$ and $c < d$, prove that $a + c < b + d$.

50 If $a < b$ and $c < d$, is it always true that $ac < bd$? Explain.

2.6 MORE ON INEQUALITIES

Among the most important inequalities which occur in advanced mathematics are those involving absolute values. From the discussion in Section 1.3, if a represents a real number, then $|a|$ is the distance on a coordinate line between the origin and the point corresponding to a. It follows that if we are given an inequality of the form $|a| < 5$, then the solutions consist of all real numbers a such that $-5 < a < 5$, or equivalently, the open interval $(-5, 5)$. In general, if b is any positive real number, the following can be proved.

Properties of Absolute Values

(i)	$\lvert a \rvert < b$ if and only if	$-b < a < b$
(ii)	$\lvert a \rvert > b$ if and only if	either $a > b$ or $a < -b$
(iii)	$\lvert a \rvert = b$ if and only if	$a = b$ or $a = -b$.

Properties (ii) and (iii) are also true if $b = 0$. The next three examples illustrate uses of these important properties.

Example 1   Solve each of the following:

(a) $|x| < 4$ (b) $|x| > 4$ (c) $|x| = 4.$

Solution

(a)   Using property (i) with $a = x$ and $b = 4$, we see that

$$|x| < 4 \quad \text{if and only if} \quad -4 < x < 4.$$

Hence, the interval solution is the open interval $(-4, 4)$.

(b)   By property (ii), $|x| > 4$ means that either $x > 4$ or $x < -4$. Thus, the solutions consist of all real numbers in the two infinite intervals $(-\infty, -4)$ and $(4, \infty)$.

(c)   By property (iii),

$$|x| = 4 \quad \text{if and only if} \quad x = 4 \text{ or } x = -4. \qquad \blacksquare$$

For solutions of the type obtained in (b) of Example 1 it is sometimes convenient to use the union symbol, $\cup$, and write

$$(-\infty, -4) \cup (4, \infty)$$

to denote all real numbers x such that either x is in $(-\infty, -4)$ or x is in $(4, \infty)$.

Example 2   Solve the inequality $|x - 3| < 0.1.$

Solution   By property (i), with $a = x - 3$ and $b = 0.1$, the given inequality is equivalent to

$$-0.1 < x - 3 < 0.1$$

and hence to $-0.1 + 3 < (x - 3) + 3 < 0.1 + 3$

or $2.9 < x < 3.1.$

Consequently, the interval solution of the given inequality is $(2.9, 3.1)$. $\qquad \blacksquare$

Example 3   Solve $|2x + 3| > 9$ and illustrate the solutions graphically.

Solution   By property (ii), with $a = 2x + 3$ and $b = 9$, the solutions of the given inequality are the solutions of the *two* inequalities

$$2x + 3 > 9 \quad \text{and} \quad 2x + 3 < -9.$$

The first inequality is equivalent to $2x > 6$, or $x > 3$. This gives us the infinite interval $(3, \infty)$. The second inequality is equivalent to $2x < -12$, or $x < -6$, which leads to the interval $(-\infty, -6)$. Consequently, the solutions of $|2x + 3| > 9$ consist of the numbers in the union $(-\infty, -6) \cup (3, \infty)$. The graph is sketched in Figure 2.13.

Figure 2.13   ■

Before proceeding, let us make an important remark about notation. The solutions in Example 3 consist of all real numbers x such that either $x > 3$ or $x < -6$. Some students employ the *incorrect* form $3 < x < -6$ to denote these inequalities. The notation $a < b < c$ should be used if and only if b satisfies *both* of the conditions $a < b$ and $b < c$. In particular, there is no real number x such that $3 < x < -6$, since it is impossible for x to satisfy $3 < x$ and $x < -6$ simultaneously.

Most of the inequalities considered up to this time were equivalent to inequalities containing only first-degree polynomials. To solve inequalities involving polynomials of higher degree we shall make use of an important result. This result, which will be discussed further in Section 4.2, implies that if real numbers c and d are *successive* solutions of a (polynomial) equation $a_nx^n + \cdots + a_1x + a_0 = 0$, that is, *if there are no other solutions between c and d, then either all values of the polynomial are positive, or all values are negative, when x is in the interval (c, d).* Thus, if we choose *any* number k, such that $c < k < d$, and if the value of the polynomial is positive for $x = k$, then the polynomial is positive for *every* x in (c, d). Similarly, if the polynomial is negative for $x = k$, then it is negative throughout (c, d). We shall call the value of the polynomial at $x = k$ a **test value** for the interval (c, d). Test values may also be used on infinite intervals of the form $(-\infty, a)$ or (a, ∞), provided the polynomial equation has no solutions on these intervals. The use of test values is demonstrated in the following examples.

Example 4   Solve $2x^2 - x - 3 < 0$.

Solution   The following inequalities are equivalent:

$$2x^2 - x - 3 < 0$$
$$(x + 1)(2x - 3) < 0.$$

We see from the factored form that the equation $2x^2 - x - 3 = 0$ has solutions -1 and $3/2$. For reference, let us plot the corresponding points on a real axis, as in Figure 2.14.

Figure 2.14

These points divide the axis into three parts and determine the following intervals:

$$(-\infty, -1), \quad (-1, 3/2), \quad (3/2, \infty).$$

According to the discussion preceding this example, the sign of the polynomial $2x^2 - x - 3$ in each interval can be determined by finding a suitable test value.

If we choose -2 in $(-\infty, -1)$, then the polynomial $2x^2 - x - 3$ has the value

$$2(-2)^2 - (-2) - 3 = 8 + 2 - 3 = 7.$$

Since 7 is positive, it follows that $2x^2 - x - 3 > 0$ for every x in $(-\infty, -1)$.

If we choose 0 in $(-1, 3/2)$, then the polynomial has the value

$$2(0)^2 - (0) - 3 = -3.$$

Since -3 is negative, $2x^2 - x - 3 < 0$ for every x in $(-1, 3/2)$.

Finally, choosing 2 in $(3/2, \infty)$, we obtain

$$2(2)^2 - 2 - 3 = 8 - 2 - 3 = 3,$$

and since 3 is positive, $2x^2 - x - 3 > 0$ throughout $(3/2, \infty)$. It is convenient to tabulate these facts as follows.

Interval	k	Test value of $2x^2 - x - 3$ at k	Sign of $2x^2 - x - 3$ in interval
$(-\infty, -1)$	-2	7	$+$
$(-1, \frac{3}{2})$	0	-3	$-$
$(\frac{3}{2}, \infty)$	2	3	$+$

Hence, the interval solution of the inequality $2x^2 - x - 3 < 0$ is $(-1, 3/2)$. ∎

Example 5 Solve the inequality $x^2 - 7x + 10 > 0$ and represent the solutions graphically.

Solution   The given inequality may be written in the form

$$(x - 2)(x - 5) > 0.$$

Points corresponding to the solutions 2 and 5 of $x^2 - 7x + 10 = 0$ are plotted in Figure 2.15.

Figure 2.15

Referring to the figure, we obtain the following intervals:

$$(-\infty, 2), \quad (2, 5), \quad (5, \infty).$$

We next use test values to determine the sign of $x^2 - 7x + 10$ in each interval. The following table summarizes results (*Check each entry!*).

Interval	k	Test value of $x^2 - 7x + 10$ at k	Sign of $x^2 - 7x + 10$ in interval
$(-\infty, 2)$	0	$0^2 - 7(0) + 10 = 10$	$+$
$(2, 5)$	3	$3^2 - 7(3) + 10 = -2$	$-$
$(5, \infty)$	6	$6^2 - 7(6) + 10 = 4$	$+$

Referring to the table, we see that $x^2 - 7x + 10 > 0$ if x is in either $(-\infty, 2)$ or $(5, \infty)$. Thus, the solutions of the inequality are given by $(-\infty, 2) \cup (5, \infty)$. The graph corresponding to the solutions is sketched in Figure 2.16.

Figure 2.16

Example 6 Solve $\dfrac{x + 1}{x + 3} \le 2$ and represent the solutions graphically.

Solution The following inequalities are equivalent:

$$\frac{x + 1}{x + 3} \le 2$$

$$\frac{x + 1}{x + 3} - 2 \le 0$$

$$\frac{x + 1 - 2x - 6}{x + 3} \le 0 \qquad (x+1) - 2(x+3)$$

$$\frac{-x - 5}{x + 3} \le 0$$

$$\frac{x + 5}{x + 3} \ge 0.$$

The numerator and denominator of $(x + 5)/(x + 3)$ equal zero at $x = -5$ and $x = -3$, respectively. For reference, we plot these points as in Figure 2.17.

Figure 2.17

Note that -5 is a solution of $(x + 5)/(x + 3) \geq 0$; however, -3 is *not* a solution. The points in the figure determine the following intervals.

$$(-\infty, -5), \quad (-5, -3), \quad (-3, \infty).$$

Since $(x + 5)/(x + 3)$ is a quotient of two polynomials, it is always positive or always negative throughout each interval. As before, we may use test values to determine the sign in each interval. The following table summarizes results.

Interval	k	Test value of $(x + 5)/(x + 3)$ at k	Sign of $(x + 5)/(x + 3)$ in interval
$(-\infty, -5)$	-6	$\frac{1}{3}$	$+$
$(-5, -3)$	-4	-1	$-$
$(-3, \infty)$	0	$\frac{5}{3}$	$+$

Referring to the table, we see that the solutions of $(x + 5)/(x + 3) > 0$ are given by $(-\infty, -5) \cup (-3, \infty)$. Hence, the solutions of $(x + 5)/(x + 3) \geq 0$ are given by $(-\infty, -5] \cup (-3, \infty)$. The graph is sketched in Figure 2.18.

Figure 2.18   ■

Example 7   Solve the inequality $(x + 2)(x - 1)(x - 5) > 0$ and represent the solutions graphically.

Solution   The expression $(x + 2)(x - 1)(x - 5)$ is zero at -2, 1, and 5. The corresponding points are plotted on a real axis in Figure 2.19.

Figure 2.19

These points determine the four intervals:

$$(-\infty, -2), \quad (-2, 1), \quad (1, 5), \quad (5, \infty).$$

We next use text values, as shown in the following table.

Interval	k	Test value of $(x + 2)(x - 1)(x - 5)$	Sign of $(x + 2)(x - 1)(x - 5)$
$(-\infty, -2)$	-3	$(-1)(-4)(-8) = -32$	$-$
$(-2, 1)$	0	$(2)(-1)(-5) = 10$	$+$
$(1, 5)$	2	$(4)(1)(-3) = -12$	$-$
$(5, \infty)$	6	$(8)(5)(1) = 40$	$+$

Referring to the fourth column of the table, we see that the solutions of $(x + 2)(x - 1)(x - 5) > 0$ are given by $(-2, 1) \cup (5, \infty)$. The graph is sketched in Figure 2.20.

Figure 2.20

EXERCISES 2.6

Solve the inequalities in Exercises 1–38 and express the solutions in terms of intervals.

1 $|x| < 2$

2 $|x| < 25$

3 $|x| > 6$

4 $|x| > 1/2$

5 $|-x| \le 10$

6 $|x| \ge 7$

7 $|x - 10| < 0.05$

8 $|x - 5| < 0.001$

9 $\left|\dfrac{x + 4}{3}\right| \le 2$

10 $\left|\dfrac{x + 1}{5}\right| \le 1$

11 $|5 - 3x| < 7$

12 $|7x + 4| < 10$

13 $|2x + 4| > 8$

14 $|6x - 7| \ge 4$

15 $\dfrac{3}{|5 - 2x|} \le 1$

16 $\left|\dfrac{4 - x}{2}\right| > 9$

17 $x^2 - x - 6 < 0$

18 $x^2 + 6x + 5 < 0$

19 $x(2x + 3) > 5$

20 $x(3x - 1) > 4$

21 $x^2 \le 10x$

22 $4x^2 \ge x$

23 $\dfrac{2x + 1}{10 - 3x} < 0$

24 $\dfrac{9 - 4x}{x + 1} \ge 0$

25 $\dfrac{6}{x^2 - 9} < 0$

26 $\dfrac{10}{25x^2 - 16} > 0$

27 $\dfrac{7x}{x^2 - 16} \ge 0$

28 $\dfrac{x - 2}{x^2 - 9} < 0$

29 $\dfrac{x^2 - 25}{4x^2 - 9} \le 0$

30 $\dfrac{10}{x^4 - 16} > 0$

31 $\dfrac{x + 4}{2x - 1} < 3$

32 $\dfrac{1}{x - 2} > \dfrac{3}{x + 1}$

33 $x^3 < x$

34 $x^4 + 3x^2 > 4$

35 $x^3 - x^2 - 4x + 4 \ge 0$

36 $(x^2 - 4x + 4)(3x - 7) < 0$

37 $\dfrac{x^2 - x - 2}{x^2 - 4x + 3} \ge 0$

38 $\dfrac{x^2 + 2x - 3}{x^2 - 4x} \le 0$

39 If a projectile is fired straight upward from level ground with an initial velocity of 72 ft/sec, its altitude s (in feet) after t seconds is given by $s = -16t^2 + 72t$. During what time interval will the projectile be at least 32 feet above the ground?

40 The period T (sec) of a simple pendulum of length l (cm) is given by $T = 2\pi\sqrt{l/g}$, where g is a physical constant. If, under certain conditions, $g = 980$ and $98 \le l \le 100$, what is the corresponding range for T?

2.7   REVIEW

Concepts

Define or discuss each of the following.

1   Solution of an equation

2   Root of an equation

3   Identity

4   Conditional equation

5   Equivalent equations

6   Linear equation

7   Quadratic equation

8   The Quadratic Formula

9   Discriminant of a quadratic equation

10   Extraneous solution

11   Properties of inequalities

12   Solution of an inequality

13   Equivalent inequalities

14   The graph of a set of real numbers

15   Open interval

16   Closed interval

17   Half-open interval

18   Infinite interval

Exercises

Solve the equations and inequalities in Exercises 1–30.

1   $\dfrac{3x+1}{5x+7} = \dfrac{6x+11}{10x-3}$

2   $2 - \dfrac{1}{x} = 1 + \dfrac{4}{x}$

3   $\dfrac{2}{x+5} - \dfrac{3}{2x+1} = \dfrac{5}{6x+3}$

4   $\dfrac{7}{x-2} - \dfrac{6}{x^2-4} = \dfrac{3}{2x+4}$

5   $\dfrac{1}{\sqrt{x}} - 2 = \dfrac{1-2\sqrt{x}}{\sqrt{x}}$

6   $2x^2 + 5x - 12 = 0$

7   $3x^2 + 4x - 5 = 0$

8   $\dfrac{x}{3x+1} = \dfrac{x-1}{2x+3}$

9   $(x-2)(x+1) = 3$

10   $4x^4 - 33x^2 + 50 = 0$

11   $x^{2/3} - 2x^{1/3} - 15 = 0$

12   $20x^3 + 8x^2 - 35x - 14 = 0$

13   $\dfrac{1}{x} + 6 = \dfrac{5}{\sqrt{x}}$

14   $\sqrt[3]{4x-5} - 2 = 0$

15   $\sqrt{7x+2} + x = 6$

16   $\sqrt{x+4} = \sqrt[4]{6x+19}$

17   $\sqrt{3x+1} - \sqrt{x+4} = 1$

18   $10 - 7x < 4 + 2x$

19   $-\dfrac{1}{2} < \dfrac{2x+3}{5} < \dfrac{3}{2}$

20   $(3x-1)(10x+4) \geq (6x-5)(5x-7)$

21   $\dfrac{6}{10x+3} < 0$

22   $|4x+7| < 21$

23   $|16-3x| \geq 5$

24   $2 < |x-6| < 4$

25   $10x^2 + 11x - 6 > 0$

26   $x^2 - 3x \leq 10$

27   $\dfrac{3}{2x+3} < \dfrac{1}{x-2}$

28   $\dfrac{x+1}{x^2-25} \leq 0$

29   $x^3 > x^2$

30   $(x^2-x)(x^2-5x+6) < 0$

In Exercises 31–36 solve for the indicated variable in terms of the remaining variables.

31   $S = \pi(r+R)s$ for R

32   $S = 2(ab + bc + ac)$ for a

33   $n = \dfrac{\pi P R^4}{8VL}$ for R

34   $V = \frac{4}{3}\pi r^3$ for r

35 $\dfrac{1}{R} = (n-1)\left(\dfrac{1}{R_1} + \dfrac{1}{R_2}\right)$ for R_2

36 $V = \frac{1}{3}\pi h(R_1^2 + R_2^2 + R_1 R_2)$ for R_1

37 The diagonal of a square is 50 cm long. What change in the length of a side will increase the area by 46 cm²?

38 The surface area S of a sphere of radius r is given by the formula $S = 4\pi r^2$. If $r = 6$ inches, what change in radius will increase the surface area by 36π square inches?

39 An airplane flew with the wind for 30 minutes and returned the same distance in 45 minutes. If the cruising speed of the airplane is 320 mph, find the speed of the wind.

40 A merchant wishes to mix peanuts costing $1.50 per pound with cashews costing $4.00 per pound, obtaining 50 pounds of a mixture costing $2.40 per pound. How many pounds of each should be used?

41 A chemist has 80 ml of a solution containing 25% acid. How many ml should be removed and replaced by pure acid in order to obtain a solution containing 40% acid?

42 A motorist averaged 45 mph driving outside the city limits and 25 mph within the city limits. If an 80-mile trip took the motorist two hours, how much time was spent driving within the city limits?

43 The width of a page in a book is 2 inches smaller than its length. The printed area is 72 square inches, with 1-inch margins at the top and bottom and $\frac{1}{2}$-inch margins at each side. Find the dimensions of the page.

44 A man puts a fence around a rectangular field and then subdivides the field into three smaller rectangular plots by placing two fences parallel to one of the sides. If the area of the field is 31,250 square yards, and 1,000 yards of fencing were used, find the dimensions of the field.

45 A North-South highway intersects an East-West highway at a point P. An automobile crosses P at 10:00 A.M., traveling east at a constant rate of 20 mph. At that same instant another automobile is two miles north of P, traveling south at 50 mph. Find a formula which expresses the distance d between the automobiles at time t (hours) after 10:00 A.M. At what time will the automobiles be 104 miles apart?

46 A sales representative for a company estimates that gasoline consumption for her automobile averages 28 miles per gallon on the highway and 22 miles per gallon in the city. On a recent trip she covered 627 miles and used 24 gallons of gasoline. How much of the trip was spent driving in the city?

3
FUNCTIONS

One of the most useful concepts in mathematics is that of function. Indeed, it is safe to say that, without the notion of function, little progress could be made in mathematics or in any area of science. In the first section of this chapter we consider coordinate systems in two dimensions. The remainder of the chapter contains a discussion of functions and graphs.

3.1 COORDINATE SYSTEMS IN TWO DIMENSIONS

In Section 1.3 we discussed how coordinates may be assigned to points on a straight line. Coordinate systems can also be introduced in planes by means of ordered pairs. The term **ordered pair** refers to two real numbers, where one is designated as the "first" number and the other as the "second." The symbol (a, b) is used to denote the ordered pair consisting of the real numbers a and b, where a is first and b is second. There are many uses for ordered pairs. They were used in Chapter 2 to denote open intervals. In this chapter we shall use ordered pairs to represent points in a plane. Although ordered pairs are employed in different situations, there is little chance for confusion, since it should always be clear from the discussion whether the symbol (a, b) represents an interval, a point, or some other mathematical object. We consider two ordered pairs (a, b) and (c, d) equal, and write

$$(a, b) = (c, d) \quad \text{if and only if} \quad a = c \text{ and } b = d.$$

This implies, in particular, that $(a, b) \neq (b, a)$ if $a \neq b$.

A **rectangular**, or **Cartesian**,* **coordinate system** may be introduced in a plane by considering two perpendicular coordinate lines in the plane which intersect in the origin O on each line. Unless specified otherwise, the same unit of length is chosen

---

* The term "Cartesian" is used in honor of the French mathematician and philosopher René Descartes (1596–1650), who was one of the first to employ such coordinate systems.

on each line. Usually one of the lines is horizontal with positive direction to the right, and the other line is vertical with positive direction upward, as indicated by the arrowheads in Figure 3.1. The two lines are called **coordinate axes**, and the point O is called the **origin**. The horizontal line is often referred to as the **x-axis** and the vertical line as the **y-axis**, and they are labeled x and y, respectively. The plane is then called a **coordinate plane** or, with the preceding notation for coordinate axes, an **xy-plane**. Although the symbols x and y are used to denote lines as well as numbers, there should be no misunderstanding as to what these letters represent when they appear alongside of coordinate lines as in Figure 3.1. In certain applications different labels, such as d, t, etc., are used for the coordinate lines. The coordinate axes divide the plane into four parts called the **first, second, third,** and **fourth quadrants** and labeled I, II, III, and IV, respectively, as shown in Figure 3.1.

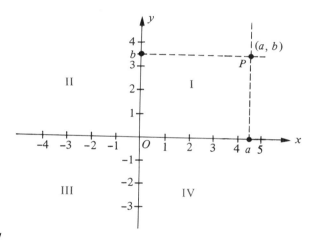

Figure 3.1

Each point P in an xy-plane may be assigned a unique ordered pair. If vertical and horizontal lines through P intersect the x- and y-axes at points with coordinates a and b, respectively (see Figure 3.1), then P is assigned the ordered pair (a, b). The number a is called the **x-coordinate** (or **abscissa**), of P, and b is called the **y-coordinate** (or **ordinate**), of P. We sometimes say that P *has coordinates* (a, b), or that P *is the point* (a, b).

Conversely, every ordered pair (a, b) determines a point P in the xy-plane with coordinates a and b. Specifically, P is the point of intersection of lines perpendicular to the x-axis and y-axis at the points having coordinates a and b, respectively. This establishes a one-to-one correspondence between the set of all points in the xy-plane and the set of all ordered pairs.

The symbol $P(a, b)$ will denote the point P with coordinates (a, b). To **plot a point** $P(a, b)$ means to represent P by a dot in the appropriate position, as illustrated in Figure 3.1. Note that abscissas are positive for points in quadrants I or IV and negative for points in quadrants II or III. Ordinates are positive for points in quadrants I or II and negative for points in quadrants III or IV. Some typical points in a coordinate plane are plotted in Figure 3.2.

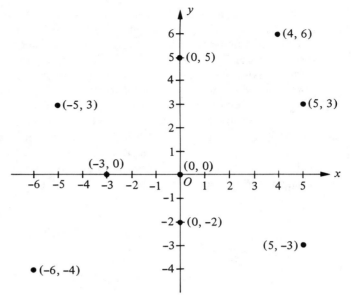

Figure 3.2

We shall next derive a formula for finding the distance between any two points in a coordinate plane. The distance between two points P and Q will be denoted by $d(P, Q)$. If $P = Q$, then we agree that $d(P, Q) = 0$, whereas, if $P \neq Q$, the distance is positive. Let us consider any two points $P_1(x_1, y_1)$ and $P_2(x_2, y_2)$ in the plane. If the points lie on the same horizontal line, then $y_1 = y_2$, and we may denote the points by $P_1(x_1, y_1)$ and $P_2(x_2, y_1)$. If lines through P_1 and P_2 parallel to the y-axis intersect the x-axis at $A_1(x_1, 0)$ and $A_2(x_2, 0)$, as shown in (i) of Figure 3.3, then we see that

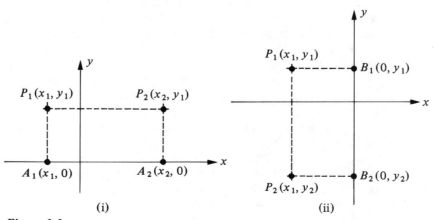

Figure 3.3

$d(P_1, P_2) = d(A_1, A_2)$. However, from Section 1.3, $d(A_1, A_2) = |x_2 - x_1|$, and hence,

$$d(P_1, P_2) = |x_2 - x_1|.$$

Since $|x_2 - x_1| = |x_1 - x_2|$, the last formula is valid whether P_1 lies to the left of P_2 or to the right of P_2. Moreover, the formula is independent of the quadrants in which the points lie.

In similar fashion, if P_1 and P_2 are on the same vertical line, then $x_1 = x_2$, and we may denote the points by $P_1(x_1, y_1)$ and $P_2(x_1, y_2)$. If we consider the points $B_1(0, y_1)$ and $B_2(0, y_2)$ on the y-axis, as shown in (ii) of Figure 3.3, then

$$d(P_1, P_2) = d(B_1, B_2) = |y_2 - y_1|.$$

Finally, let us consider the general case, in which the points $P_1(x_1, y_1)$ and $P_2(x_2, y_2)$ do not lie on the same horizontal or vertical line. The line through $P_1(x_1, y_1)$ parallel to the x-axis and the line through $P_2(x_2, y_2)$ parallel to the y-axis intersect at some point P_3. Since P_3 has the same y-coordinate as P_1 and the same x-coordinate as P_2, we can denote it by $P_3(x_2, y_1)$ (see Figure 3.4). From the previous discussion, $d(P_1, P_3) = |x_2 - x_1|$ and $d(P_3, P_2) = |y_2 - y_1|$. Since P_1, P_2, and P_3 form a right triangle with hypotenuse from P_1 to P_2, we have, by the Pythagorean Theorem,

$$[d(P_1, P_2)]^2 = [d(P_1, P_3)]^2 + [d(P_3, P_2)]^2$$

and hence,

$$[d(P_1, P_2)]^2 = |x_2 - x_1|^2 + |y_2 - y_1|^2.$$

Using the fact that $d(P_1, P_2)$ is nonnegative and that $|a|^2 = a^2$ for every real number a, we obtain the following important formula.

Distance Formula

> If $P_1(x_1, y_1)$ and $P_2(x_2, y_2)$ are points in a coordinate plane, then the **distance between** P_1 **and** P_2 is given by
>
> $$d(P_1, P_2) = \sqrt{(x_2 - x_1)^2 + (y_2 - y_1)^2}.$$

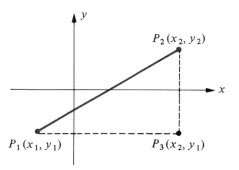

Figure 3.4

Although we referred to the special case indicated in Figure 3.4, the argument used in the proof of the Distance Formula is independent of the positions of P_1 and P_2.

Example 1   Plot the points $A(-1, -3)$, $B(6, 1)$, and $C(2, -5)$, and prove that the triangle with vertices A, B, and C is a right triangle.

Solution   The points and the triangle are shown in Figure 3.5. From plane geometry, a triangle is a right triangle if and only if the sum of the squares of two of its sides is equal to the square of the remaining side. Using the Distance Formula we obtain

$$d(A, B) = \sqrt{(-1 - 6)^2 + (-3 - 1)^2}$$
$$= \sqrt{49 + 16} = \sqrt{65}$$
$$d(B, C) = \sqrt{(6 - 2)^2 + (1 + 5)^2}$$
$$= \sqrt{16 + 36} = \sqrt{52}$$
$$d(A, C) = \sqrt{(-1 - 2)^2 + (-3 + 5)^2}$$
$$= \sqrt{9 + 4} = \sqrt{13}.$$

Since $[d(A, B)]^2 = [d(B, C)]^2 + [d(A, C)]^2$, the triangle is a right triangle with hypotenuse joining A to B.

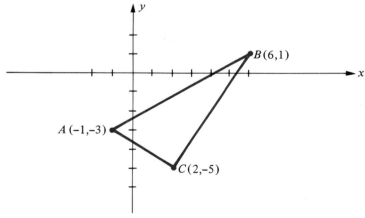

Figure 3.5   ■

It is easy to obtain a formula for the midpoint of a line segment. Let $P_1(x_1, y_1)$ and $P_2(x_2, y_2)$ be two points in a coordinate plane and let M be the midpoint of the segment $P_1 P_2$. The lines through P_1 and P_2 parallel to the y-axis intersect the x-axis at $A_1(x_1, 0)$ and $A_2(x_2, 0)$ and, from plane geometry, the line through M parallel to the y-axis bisects the segment $A_1 A_2$ (see Figure 3.6). If $x_1 < x_2$, then $x_2 - x_1 > 0$, and hence,

$$d(A_1, A_2) = x_2 - x_1.$$

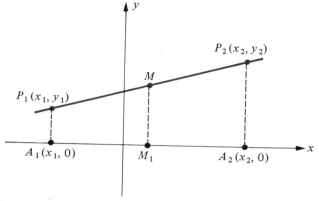

Figure 3.6

Since M_1 is halfway from A_1 to A_2, the abscissa of M_1 is

$$x_1 + \tfrac{1}{2}(x_2 - x_1) = x_1 + \tfrac{1}{2}x_2 - \tfrac{1}{2}x_1$$
$$= \tfrac{1}{2}x_1 + \tfrac{1}{2}x_2$$
$$= \frac{x_1 + x_2}{2}.$$

It follows that the abscissa of M is also $(x_1 + x_2)/2$. It can be shown in similar fashion that the ordinate of M is $(y_1 + y_2)/2$. Moreover, these formulas hold for all positions of P_1 and P_2. This gives us the following result.

Midpoint Formula

> The midpoint of the line segment from $P_1(x_1, y_1)$ to $P_2(x_2, y_2)$ is
> $$\left(\frac{x_1 + x_2}{2}, \frac{y_1 + y_2}{2}\right).$$

Example 2 Find the midpoint M of the line segment from $P_1(-2, 3)$ to $P_2(4, -2)$. Plot the points P_1, P_2, and M and verify that $d(P_1, M) = d(P_2, M)$.

Solution Applying the Midpoint Formula, the coordinates of M are

$$\left(\frac{-2 + 4}{2}, \frac{3 + (-2)}{2}\right) \quad \text{or} \quad \left(1, \frac{1}{2}\right).$$

The three points P_1, P_2, and M are plotted in Figure 3.7. Using the Distance Formula we obtain

$$d(P_1, M) = \sqrt{(-2 - 1)^2 + (3 - \tfrac{1}{2})^2} = \sqrt{9 + (\tfrac{25}{4})}$$
$$d(P_2, M) = \sqrt{(4 - 1)^2 + (-2 - \tfrac{1}{2})^2} = \sqrt{9 + (\tfrac{25}{4})}.$$

Hence, $d(P_1, M) = d(P_2, M)$.

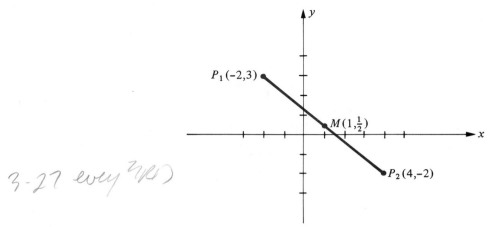

Figure 3.7

■

EXERCISES 3.1

1   Plot the following points on a rectangular coordinate system: $A(5, -2)$, $B(-5, -2)$, $C(5, 2)$, $D(-5, 2)$, $E(3, 0)$, $F(0, 3)$.

2   Plot the points $A(-3, 1)$, $B(3, 1)$, $C(-2, -3)$, $D(0, 3)$, and $E(2, -3)$ on a rectangular coordinate system and then draw the line segments AB, BC, CD, DE, and EA.

3   Plot $A(0, 0)$, $B(1, 1)$, $C(3, 3)$, $D(-1, -1)$, and $E(-2, -2)$. Describe the set of all points of the form (x, x) where x is a real number.

4   Plot $A(0, 0)$, $B(1, -1)$, $C(2, -2)$, $D(-1, 1)$, and $E(-3, 3)$. Describe the set of all points of the form $(a, -a)$ where a is a real number.

5   Describe the set of all points $P(x, y)$ in a coordinate plane such that:
   (a) $x = 3$ (b) $y = -1$
   (c) $x \geq 0$ (d) $xy > 0$
   (e) $y < 0$

6   Describe the set of all points $P(x, y)$ in a coordinate plane such that:
   (a) $y = 0$ (b) $x = -5$
   (c) $x/y < 0$ (d) $xy = 0$
   (e) $y > 1$

In Exercises 7–12, find (a) the distance $d(A, B)$ between the given points A and B; (b) the midpoint of the segment AB.

7 $A(4, -3), B(6, 2)$ **8** $A(-2, -5), B(4, 6)$

9 $A(-5, 0), B(-2, -2)$ **10** $A(6, 2), B(6, -2)$

11 $A(7, -3), B(3, -3)$ **12** $A(-4, 7), B(0, -8)$

In Exercises 13 and 14 prove that the triangle with the indicated vertices is a right triangle and find its area.

13 $A(8, 5), B(1, -2), C(-3, 2)$

14 $A(-6, 3), B(3, -5), C(-1, 5)$

15   Prove that the following points are vertices of a square: $A(-4, 2)$, $B(1, 4)$, $C(3, -1)$, $D(-2, -3)$.

16   Prove that the following points are vertices of a parallelogram: $A(-4, -1)$, $B(0, -2)$, $C(6, 1)$, $D(2, 2)$.

17   Given $A(-3, 8)$, find the coordinates of the point B such that $M(5, -10)$ is the midpoint of AB.

18   Given $A(5, -8)$ and $B(-6, 2)$, find the point on AB that is three-fourths of the way from A to B.

19   Given $A(-4, -3)$ and $B(6, 1)$, prove that $P(5, -11)$ is on the perpendicular bisector of AB.

20   Given $A(-4, -3)$ and $B(6, 1)$, find a formula which expresses the fact that $P(x, y)$ is on the perpendicular bisector of AB.

21 Find a formula which expresses the fact that $P(x, y)$ is a distance 5 from the origin. Describe the totality of all such points.

22 If r is a positive real number, find a formula that states that $P(x, y)$ is a distance r from a fixed point $C(h, k)$. Describe the totality of all such points.

23 Find all points on the y-axis that are 6 units from the point $(5, 3)$.

24 Find all points on the x-axis that are 5 units from $(-2, 4)$.

25 Let S denote the set of points of the form $(2x, x)$ where x is a real number. Find the point in S that is in the third quadrant and is a distance 5 from the point $(1, 3)$.

26 Let S denote the set of points of the form (x, x)

where x is a real number. Find all points in S that are a distance 3 from the point $(-2, 1)$.

27 For what values of a is the distance between $(a, 3)$ and $(5, 2a)$ greater than $\sqrt{26}$?

28 Given the points $A(-2, 0)$ and $B(2, 0)$, find a formula not containing radicals that expresses the fact that the sum of the distances from $P(x, y)$ to A and to B, respectively, is 5.

29 Prove that the midpoint of the hypotenuse of any right triangle is equidistant from the vertices. (*Hint*: Label the vertices of the triangle $O(0, 0)$, $A(a, 0)$, and $B(0, b)$.)

30 Prove that the diagonals of any parallelogram bisect each other. (*Hint*: Label three of the vertices of the parallelogram $O(0, 0)$, $A(a, b)$, and $C(0, c)$.)

3.2 GRAPHS

If W is a set of ordered pairs, we may consider the point $P(x, y)$ in a coordinate plane which corresponds to the ordered pair (x, y) in W. The **graph** of W is the set of all such points. The phrase "sketch the graph of W" means to illustrate the significant features of the graph geometrically on a coordinate plane, as illustrated in the next example.

Example 1 Sketch the graph of $W = \{(x, y): -1 < x \leq 4, 2 \leq y < 3\}$.

Solution The notation describing W may be translated "the set of all ordered pairs (x, y) such that $-1 < x \leq 4$ and $2 \leq y < 3$." Thus, a point $P(x, y)$ is in the graph of W if and only if the abscissa x is greater than -1 and less than or equal to 4, while the ordinate y is greater than or equal to 2 and less than 3. Hence, the graph of W consists of all points inside the shaded rectangle in Figure 3.8, together with the points on the lower and right-hand sides of the rectangle.

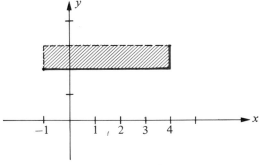

Figure 3.8 $\{(x, y): -1 < x \leq 4, 2 \leq y < 3\}$ ∎

Example 2   Sketch the graph of $W = \{(x,y): y = 2x - 1\}$.

Solution   We begin by finding points with coordinates of the form (x, y), where the ordered pair (x, y) is in W. It is convenient to list these coordinates in tabular form as shown below, where for each real number x the corresponding value for y is $2x - 1$.

x	-2	-1	0	1	2	3
y	-5	-3	-1	1	3	5

After plotting, it appears that the points with these coordinates all lie on a line and we sketch the graph accordingly (see Figure 3.9). Ordinarily, the few points we have plotted would not be enough to illustrate the graph; however, in this elementary case we can be reasonably sure that the graph is a line. It will be proved in Section 3.5 that our conjecture is correct.

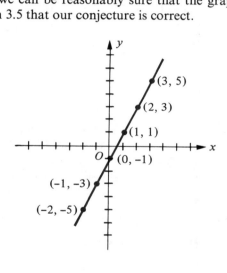

Figure 3.9   ■

It is impossible to sketch the entire graph in Example 2 since x may be assigned values which are numerically as large as desired. Nevertheless, we often call a drawing of the type given in Figure 3.9 *the graph of W* or *a sketch of the graph*, where it is understood that the drawing is only a device for visualizing the actual graph and the line does not terminate as shown in the figure. In general, when sketching a graph you should illustrate enough of the graph so that the remaining parts are evident.

Given an equation in x and y, we say that an ordered pair (a, b) is a **solution** of the equation if equality is obtained when a is substituted for x and b for y. For example, $(2, 3)$ is a solution of $y = 2x - 1$ since substitution of 2 for x and 3 for y leads to $3 = 4 - 1$, or $3 = 3$. Two equations in x and y are **equivalent** if they have exactly the same solutions. The solutions of an equation in x and y determine a set S of ordered pairs, and we define the **graph of the equation** as the graph of S. Note that the graph of the equation $y = 2x - 1$ is the same as the graph of the set W in Example 2 (see Figure 3.9).

For some equations encountered in this chapter, the technique used for sketching the graph will consist of plotting a sufficient number of points until some pattern emerges, and then sketching the graph accordingly. This is obviously a crude (and often inaccurate) way to arrive at the graph; however, it is a method often employed at the beginning of elementary courses. As we progress through this text, techniques will be introduced which will enable us to sketch a variety of graphs without plotting many points. In order to give accurate descriptions of graphs when complicated expressions are involved, it is usually necessary to employ more advanced mathematical tools of the types introduced in the study of calculus.

Example 3   Sketch the graph of the equation $y = x^2$.

Solution   The following table exhibits the coordinates of some points on the graph.

x	-3	$-\frac{5}{2}$	-2	$-\frac{3}{2}$	-1	$-\frac{1}{2}$	0	$\frac{1}{2}$	1	$\frac{3}{2}$	2	$\frac{5}{2}$	3
y	9	$\frac{25}{4}$	4	$\frac{9}{4}$	1	$\frac{1}{4}$	0	$\frac{1}{4}$	1	$\frac{9}{4}$	4	$\frac{25}{4}$	9

Larger numerical values of x produce even larger values of y. For example, the points $(4, 16)$, $(5, 25)$, and $(6, 36)$ are on the graph, as are $(-4, 16)$, $(-5, 25)$, and $(-6, 36)$. Plotting the points given in the table and drawing a smooth curve through these points, we obtain the sketch in Figure 3.10, where we have labeled several points.

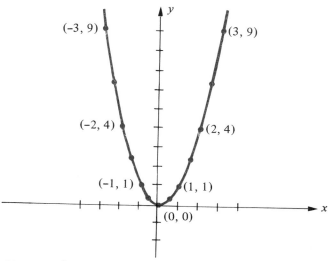

Figure 3.10 $y = x^2$ ∎

The graph in Example 3 is called a **parabola**. The lowest point $(0, 0)$ is called the **vertex** of the parabola and we say that the parabola **opens upward**. If the graph were inverted, as would be the case for $y = -x^2$, then the parabola **opens downward**. The y-axis is called the **axis of the parabola**.

If the coordinate plane in Figure 3.10 is folded along the y-axis, then the graph which lies in the left half of the plane coincides with that in the right half. We say that **the graph is symmetric with respect to the y-axis**. As in (i) of Figure 3.11, a graph is symmetric with respect to the y-axis provided that the point $(-x, y)$ is on the graph whenever (x, y) is on the graph. Similarly, as in (ii) of Figure 3.11, **a graph is symmetric with respect to the x-axis** if, whenever a point (x, y) is on the graph, then $(x, -y)$ is also on the graph. In this case if we fold the coordinate plane along the x-axis, the part of the graph which lies above the x-axis will coincide with the part which lies below. Another type of symmetry which certain graphs possess is called **symmetry with respect to the origin**. In this situation, whenever a point (x, y) is on the graph, then $(-x, -y)$ is also on the graph, as illustrated in (iii) of Figure 3.11.

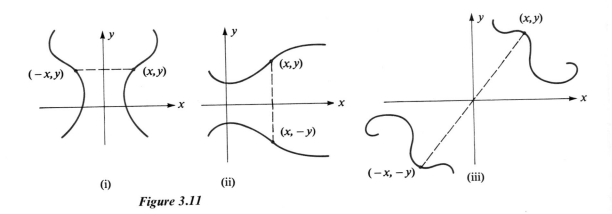

(i)   (ii)   (iii)

Figure 3.11

The following tests are useful for investigating these three types of symmetry.

Tests for Symmetry

> (i)   The graph of an equation is symmetric with respect to the y-axis if substitution of $-x$ for x leads to an equivalent equation.
>
> (ii) The graph of an equation is symmetric with respect to the x-axis if substitution of $-y$ for y leads to an equivalent equation.
>
> (iii) The graph of an equation is symmetric with respect to the origin if the simultaneous substitution of $-x$ for x and $-y$ for y leads to an equivalent equation.

If, in the equation of Example 3, we substitute $-x$ for x, we obtain $y = (-x)^2$, which is equivalent to $y = x^2$. Hence, by Test (i), the graph is symmetric with respect to the y-axis.

If symmetry with respect to an axis exists, then it is sufficient to determine the graph in half of the coordinate plane, since the remainder of the graph is a mirror image, or reflection, of that half.

Example 4 Sketch the graph of $y^2 = x$.

Solution   Since substitution of $-y$ for y does not change the equation, the graph is symmetric with respect to the x-axis. (See Symmetry Test (ii).) It is sufficient, therefore, to plot points with nonnegative ordinates and then reflect through the x-axis. Since $y^2 = x$, the ordinates of points above the x-axis are given by $y = \sqrt{x}$. Coordinates of some points on the graph are tabulated below. A portion of the graph is sketched in Figure 3.12. The graph is a parabola with vertex at the origin and which opens to the right. In this case the x-axis is the axis of the parabola.

x	0	1	2	3	4	9
y	0	1	$\sqrt{2}$	$\sqrt{3}$	2	3

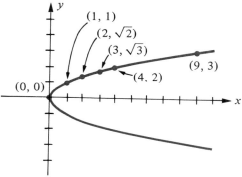

Figure 3.12 $y^2 = x$ ■

Example 5 Sketch the graph of the equation $y = x^3$.

Solution   If we substitute $-x$ for x and $-y$ for y, then

$$-y = (-x)^3 = -x^3.$$

Multiplying both sides by -1, we see that the last equation has the same solutions as the given equation $y = x^3$. Hence, from Symmetry Test (iii), the graph is symmetric with respect to the origin. The following table lists some points on the graph.

x	0	$\frac{1}{4}$	$\frac{1}{2}$	$\frac{3}{4}$	1	$\frac{3}{2}$	2
y	0	$\frac{1}{64}$	$\frac{1}{8}$	$\frac{27}{64}$	1	$\frac{27}{8}$	8

By symmetry (or substitution) we see that the points $(-\frac{1}{4}, -\frac{1}{64})$, $(-\frac{1}{2}, -\frac{1}{8})$, etc., are on the graph. Plotting points leads to the graph in Figure 3.13.

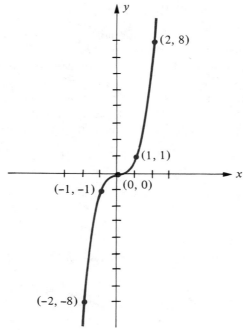

Figure 3.13 $y = x^3$ ■

Given a geometric figure in a coordinate plane, it is sometimes possible to find its equation, in the sense that the figure is the graph of the equation. We shall demonstrate how this can be accomplished for circles.

If $C(h, k)$ is a point in a coordinate plane, then a circle in the plane with center C and radius r may be defined as the collection of all points in the plane that are r units from C. If $P(x, y)$ is an arbitrary point in the plane, then, as illustrated in Figure 3.14, P is on the circle if and only if $d(C, P) = r$.

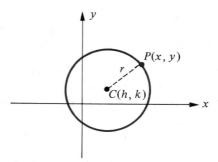

Figure 3.14 $(x - h)^2 + (y - k)^2 = r^2$

Using the Distance Formula gives us

$$\sqrt{(x - h)^2 + (y - k)^2} = r.$$

The equivalent equation

Circle Equation

$$(x - h)^2 + (y - k)^2 = r^2, \qquad r > 0$$

is called the **equation of a circle of radius *r* and center (*h, k*)**, or simply the **circle equation**. If $h = 0$ and $k = 0$, this reduces to

$$x^2 + y^2 = r^2$$

which is an equation of a circle of radius *r* with center at the origin (see Figure 3.15). If we let $r = 1$, the graph is a **unit circle** with center at the origin. A point $P(x, y)$ is on this unit circle if and only if $x^2 + y^2 = 1$.

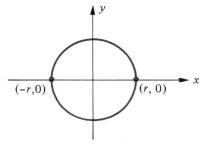

Figure 3.15 $x^2 + y^2 = r^2$

Example 6 Find an equation of the circle having center $C(-2, 3)$ and containing the point $D(4, 5)$.

Solution The circle is illustrated in Figure 3.16. Since D is on the circle, the radius r is the distance from the center to D. By the Distance Formula,

$$r = d(C, D) = \sqrt{(-2 - 4)^2 + (3 - 5)^2} = \sqrt{36 + 4} = \sqrt{40}.$$

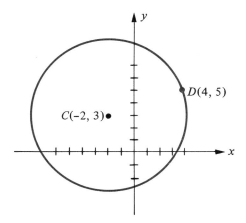

Figure 3.16

Using the circle equation, with $h = -2$ and $k = 3$, we obtain

$$(x + 2)^2 + (y - 3)^2 = 40.$$ ∎

Squaring terms in the circle equation and simplifying, we obtain an equation of the form

$$x^2 + y^2 + ax + by + c = 0$$

where a, b, and c are real numbers. Conversely, if we begin with such an equation, it is always possible, by **completing the squares** in x and y, to obtain an equation of the form

$$(x - h)^2 + (y - k)^2 = d.$$

The method will be illustrated in Example 7. If $d > 0$, the graph is a circle with center (h, k) and radius $r = \sqrt{d}$. If $d = 0$, then, since $(x - h)^2 \geq 0$ and $(y - k)^2 \geq 0$, the only solution of the equation is (h, k), and hence, the graph consists of only one point. Finally, if $d < 0$, the equation has no real solutions and there is no graph.

Example 7   Find the center and radius of the circle with equation

$$x^2 + y^2 - 4x + 6y - 3 = 0.$$

Solution   We begin by arranging the equation as follows:

$$(x^2 - 4x) + (y^2 + 6y) = 3.$$

Next we complete the squares by adding appropriate numbers within the parentheses. Of course, to obtain equivalent equations we must add the numbers to *both* sides of the equation. In order to complete the square for an expression of the form $x^2 + ax$, we add the square of half the coefficient of x, that is, $(a/2)^2$, to both sides of the equation. Similarly, for $y^2 + by$, we add $(b/2)^2$ to both sides. In this example $a = -4$, $b = 6$, $(a/2)^2 = (-2)^2 = 4$, and $(b/2)^2 = 3^2 = 9$. This leads to

$$(x^2 - 4x + 4) + (y^2 + 6y + 9) = 3 + 4 + 9$$

or $$(x - 2)^2 + (y + 3)^2 = 16.$$

Hence, by the preceding discussion, the center is $(2, -3)$ and the radius is 4.   ∎

EXERCISES 3.2

In each of Exercises 1–8 sketch the graph of the given set W of ordered pairs.

1   $W = \{(x, y) : x = 4\}$

2   $W = \{(x, y) : y = -3\}$

3   $W = \{(x, y) : xy < 0\}$

4   $W = \{(x, y) : xy = 0\}$

5   $W = \{(x, y) : |x| < 2, |y| > 1\}$

6   $W = \{(x, y) : |x| > 1, |y| \leq 2\}$

7   $W = \{(x, y) : |x - 2| < 3, |y| \leq 1\}$

8   $W = \{(x, y) : |x - 1| < 5, |y + 1| < 2\}$

In each of Exercises 9–28 sketch the graph of the equation after plotting a sufficient number of points.

9 $y = 3x + 1$

10 $y = 4x - 3$

11 $y = -2x + 3$

12 $y = 2 - 3x$

13 $2y + 3x + 6 = 0$

14 $3y - 2x = 6$

15 $y = 2x^2$

16 $y = -x^2$

17 $y = 2x^2 - 1$

18 $y = -x^2 + 2$

19 $4y = x^2$

20 $3y + x^2 = 0$

21 $y = -\frac{1}{2}x^3$

22 $y = \frac{1}{2}x^3$

23 $y = x^3 - 2$

24 $y = 2 - x^3$

25 $y = \sqrt{x}$

26 $y = \sqrt{x} - 1$

27 $y = \sqrt{-x}$

28 $y = \sqrt{x-1}$

Describe the graphs in Exercises 29–36.

29 $x^2 + y^2 = 16$

30 $x^2 + y^2 = 25$

31 $9x^2 + 9y^2 = 1$

32 $2x^2 + 2y^2 = 1$

33 $(x - 2)^2 + (y + 1)^2 = 4$

34 $(x + 3)^2 + (y - 4)^2 = 1$

35 $x^2 + (y - 3)^2 = 9$

36 $(x + 5)^2 + y^2 = 16$

In each of Exercises 37–46 find an equation of a circle satisfying the stated conditions.

37 Center $C(3, -2)$, radius 4

38 Center $C(-5, 2)$, radius 5

39 Center $C(\frac{1}{2}, -\frac{3}{2})$, radius 2

40 Center $C(\frac{1}{3}, 0)$, radius $\sqrt{3}$

41 Center at the origin, passing through $P(-3, 5)$

42 Center $C(-4, 6)$, passing through $P(1, 2)$

43 Center $C(-4, 2)$, tangent to the x-axis

44 Center $C(3, -5)$, tangent to the y-axis

45 Endpoints of a diameter $A(4, -3)$ and $B(-2, 7)$

46 Tangent to both axes, center in the first quadrant, radius 2

In each of Exercises 47–56 find the center and radius of the circle with the given equation.

47 $x^2 + y^2 + 2x - 10y + 10 = 0$

48 $x^2 + y^2 - 8x + 4y + 15 = 0$

49 $x^2 + y^2 - 6y + 5 = 0$

50 $x^2 + y^2 + 14x + 46 = 0$

51 $x^2 + y^2 - 10x + 8y = 0$

52 $x^2 + y^2 + 2y = 0$

53 $4x^2 + 4y^2 + 8x - 8y + 7 = 0$

54 $2x^2 + 2y^2 + 8x + 7 = 0$

55 $3x^2 + 3y^2 - 3x + 2y + 1 = 0$

56 $9x^2 + 9y^2 + 12x - 6y + 4 = 0$

CALCULATOR EXERCISES 3.2

Sketch the graphs of the following equations, after approximating the coordinates of a sufficient number of points.

1 $y = x^{2/3}$

2 $y = x^{3/2}$

3 $y = x^{5/2}$

4 $y = 2x^{3/5} - 1$

5 $y = \sqrt[5]{x} - 1$

6 $y = \sqrt[4]{x+1}$

7 $x^{2/3} + y^{2/3} = 1$

8 $y = x/(x^2 + 1)$

3.3 DEFINITION OF FUNCTION

The notion of **correspondence** is encountered frequently in everyday life. For example, to each book in a library there corresponds the number of pages in the book. As another example, to each human being there corresponds a birth date. To cite a third example, if the temperature of the air is recorded throughout a day, then, at each instant of time, there is a corresponding temperature.

The examples of correspondences we have given involve two sets X and Y. In our first example, X denotes the set of books in a library and Y the set of positive integers. For each book x in X there corresponds a positive integer y, namely the number of pages in the book.

We sometimes represent correspondences by diagrams of the type shown in Figure 3.17, where the sets X and Y are represented by points within regions in a plane. The curved arrow indicates that the element y of Y corresponds to the element x of X. We have pictured X and Y as different sets. However, X and Y may have elements in common. As a matter of fact, we often have $X = Y$.

Figure 3.17

Our examples indicate that to each x in X there corresponds *one and only one y* in Y; that is, *y is unique* for a given x. However, the same element of Y may correspond to different elements of X. For example, two different books may have the same number of pages, two different people may have the same birthday, and so on.

In most of our work X and Y will be sets of numbers. To illustrate, let X and Y both denote the set $\mathbb{R}$ of real numbers, and to each real number x let us assign its square x^2. Thus, to 3 we assign 9, to -5 we assign 25, and to $\sqrt{2}$ the number 2. This gives us a correspondence from $\mathbb{R}$ to $\mathbb{R}$.

All the examples of correspondences we have given are *functions*, as defined below.

Definition

> A **function** f from a set X to a set Y is a correspondence that assigns to each element x of X a unique element y of Y. The element y is called the **image** of x under f and is denoted by $f(x)$. The set X is called the **domain** of the function. The **range** of the function consists of all images of elements of X.

The notation $f(x)$ for the element of Y which corresponds to x is read "f of x." We also call $f(x)$ the **value** of f at x. In terms of the pictorial representation given earlier, we may now sketch a diagram as in Figure 3.18. The curved arrows indicate that the elements $f(x)$, $f(w)$, $f(z)$, and $f(a)$ of Y correspond to the elements x, w, z, and a of X. Let us repeat the important fact that to each x in X there is assigned precisely one image $f(x)$ in Y; however, different elements of X, such as w and z in Figure 3.18, may have the same image in Y.

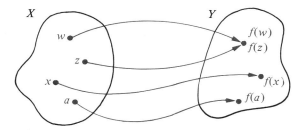

Figure 3.18

Beginning students are sometimes confused by the symbols f and $f(x)$. Remember that f is used to represent the function. It is neither in X nor in Y. However, $f(x)$ is an element of Y, namely the element which f assigns to x.

Two functions f and g from X to Y are said to be **equal**, and we write

$$f = g \quad \text{if and only if} \quad f(x) = g(x)$$

for every x in X.

Example 1   Let f be the function with domain $\mathbb{R}$ such that $f(x) = x^2$ for every x in $\mathbb{R}$. Find $f(-6)$, $f(\sqrt{3})$, and $f(a)$, where a is any real number. What is the range of f?

Solution   Values of f (or images under f) may be found by substituting for x in the equation $f(x) = x^2$. Thus,

$$f(-6) = (-6)^2 = 36, \quad f(\sqrt{3}) = (\sqrt{3})^2 = 3, \quad \text{and} \quad f(a) = a^2.$$

If T denotes the range of f, then by definition T consists of all numbers of the form $f(a)$ where a is in $\mathbb{R}$. Hence, T is the set of all squares a^2, where a is a real number. Since the square of any real number is nonnegative, T is contained in the set of all nonnegative real numbers. Moreover, every nonnegative real number c is an image under f, since $f(\sqrt{c}) = (\sqrt{c})^2 = c$. Hence, the range of f is the set of all nonnegative real numbers.   ∎

If a function is defined as in the preceding example, the symbol used for the variable is immaterial; that is, expressions such as $f(x) = x^2$, $f(s) = s^2$, $f(t) = t^2$, and so on, all define the same function f. This is true because if a is any number in the domain of f, then the same image a^2 is obtained no matter which expression is employed.

Example 2   Let X denote the set of nonnegative real numbers and let f be the function from X to $\mathbb{R}$ defined by $f(x) = \sqrt{x} + 1$ for every x in X. Find $f(4)$ and $f(\pi)$. If b and c are in X find $f(b + c)$ and $f(b) + f(c)$.

Solution   As in Example 1, finding images under f is simply a matter of substituting the appropriate number for x in the expression for $f(x)$. Thus,

$$f(4) = \sqrt{4} + 1 = 2 + 1 = 3$$
$$f(\pi) = \sqrt{\pi} + 1$$
$$f(b + c) = \sqrt{b + c} + 1$$
$$f(b) + f(c) = (\sqrt{b} + 1) + (\sqrt{c} + 1) = \sqrt{b} + \sqrt{c} + 2. \qquad \blacksquare$$

In certain branches of mathematics, such as calculus, it is important to carry out manipulations of the type given in the next example.

Example 3   Suppose $f(x) = x^2 + 3x - 2$ for every real number x. If a and t are real numbers, and $t \neq 0$, find

$$\frac{f(a + t) - f(a)}{t}.$$

Solution   Since

$$f(a + t) = (a + t)^2 + 3(a + t) - 2 = (a^2 + 2at + t^2) + (3a + 3t) - 2$$

and

$$f(a) = a^2 + 3a - 2$$

we see that

$$\frac{f(a + t) - f(a)}{t} = \frac{(a^2 + 2at + t^2) + (3a + 3t) - 2 - (a^2 + 3a - 2)}{t}$$

$$= \frac{2at + t^2 + 3t}{t}$$

$$= 2a + t + 3. \qquad \blacksquare$$

Occasionally one of the notations

$$X \xrightarrow{f} Y, \quad f : X \to Y, \quad \text{or} \quad f : x \to f(x)$$

is used to signify that f is a function from X to Y. It is not unusual in this event to say f ***maps*** X ***into*** Y or f ***maps*** x ***into*** $f(x)$. If f is the function in Example 1, then f maps x into x^2 and we may write $f : x \to x^2$.

Many formulas which occur in mathematics and the sciences determine functions. As an illustration, the formula $A = \pi r^2$ for the area A of a circle of radius r assigns to each positive real number r a unique value of A. This determines a function f, where $f(r) = \pi r^2$, and we may write $A = f(r)$. The letter r, which represents an arbitrary number from the domain of f, is often called an **independent variable**. The letter A, which represents a number from the range of f, is called a **dependent**

variable, since its value depends on the number assigned to r. If two variables r and A are related in this manner, it is customary to use the phrase A *is a function of* r. To cite another example, if an automobile travels at a uniform rate of 50 miles per hour, then the distance d (miles) traveled in time t (hours) is given by $d = 50t$, and hence, the distance d is a function of time t.

We have seen that different elements in the domain of a function may have the same image. If images are always different, then, as in the next definition, the function is called *one-to-one*.

Definition

> A function f from X to Y is a **one-to-one function** if, whenever $a \neq b$ in X, then $f(a) \neq f(b)$ in Y.

If f is one-to-one, then each $f(x)$ in the range is the image of *precisely one* x in X. The function illustrated in Figure 3.18 is not one-to-one since two different elements w and z of X have the same image in Y. A one-to-one function is often called a **one-to-one correspondence**. To illustrate, the association between real numbers and points on a coordinate line is an example of a one-to-one correspondence.

Example 4

(a)   If $f(x) = 3x + 2$, where x is real, prove that f is one-to-one.

(b)   If $g(x) = x^2 + 5$, where x is real, prove that g is not one-to-one.

Solution

(a)   If $a \neq b$, then $3a \neq 3b$ and hence $3a + 2 \neq 3b + 2$, or $f(a) \neq f(b)$. Hence, f is one-to-one.

(b)   The function g is not one-to-one since different numbers in the domain may have the same image. For example, although $-1 \neq 1$, both $g(-1)$ and $g(1)$ are equal to 6. ∎

Definition

> A function f from X to Y is a **constant function** if there is some (fixed) element c in Y such that $f(x) = c$ for every x in X.

The diagram in Figure 3.19 illustrates the fact that if f is constant function, then every arrow from X terminates at the same element in Y.

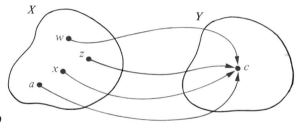

Figure 3.19

In the remainder of our work, unless specified otherwise, the phrase *f is a function* will mean that the domain and range are sets of real numbers. If a function is defined by means of some expression as in Examples 1–3, and the domain X is not stated explicitly, then X is considered to be the totality of real numbers for which the given expression is meaningful. To illustrate, if $f(x) = \sqrt{x}/(x - 1)$, then the domain is assumed to be the set of nonnegative real numbers different from 1. If x is in the domain we sometimes say that f **is defined at** x, or that $f(x)$ **exists**. If a subset S is contained in the domain we often say that f **is defined on** S. The terminology f **is undefined at** x means that x is not in the domain of f.

The types of functions described in the next definition occur frequently.

Definition

> A function f with domain X is
>
> (i)   **even** if $f(-x) = f(x)$ for every x in X.
> (ii) **odd** if $f(-x) = -f(x)$ for every x in X.

Example 5

(a)   If $f(x) = 3x^4 - 2x^2 + 5$, show that f is an even function.

(b)   If $g(x) = 2x^5 - 7x^3 + 4x$, show that f is an odd function.

Solution   If x is any real number, then

(a) $f(-x) = 3(-x)^4 - 2(-x)^2 + 5$
$\qquad\qquad = 3x^4 - 2x^2 + 5 = f(x)$

and hence f is even.

(b) $g(-x) = 2(-x)^5 - 7(-x)^3 + 4(-x)$
$\qquad\qquad = -2x^5 + 7x^3 - 4x = -g(x)$

that is, g is odd.   ∎

The results of Example 5 can be generalized. Specifically, if $f(x)$ is a polynomial, and only even powers of x have nonzero coefficients, then f is an even function (see Exercise 45). Similarly, if only odd powers of x have nonzero coefficients, then f is odd (see Exercise 46).

The concept of ordered pair can be used to obtain an alternate approach to functions. We first observe that a function f from X to Y determines the following set W of ordered pairs:

$$W = \{(x, f(x)) : x \text{ is in } X\}.$$

Thus, W is the totality of ordered pairs for which the first number is in X and the second number is the image of the first. In Example 2, where $f(x) = \sqrt{x} + 1$, W consists of all pairs of the form $(x, \sqrt{x} + 1)$, where x is a nonnegative real number. It is important to note that, for each x, there is exactly one ordered pair (x, y) in W having x in the first position.

Conversely, if we begin with a set W of ordered pairs such that each x in X appears exactly once in the first position of an ordered pair, and numbers from Y appear in the second position, then W determines a function from X to Y. Specifically, for any x in X there is a unique pair (x, y) in W, and by letting y correspond to x, we obtain a function from X to Y.

It follows from the preceding discussion that the statement given below could also be used as a definition of function. We prefer, however, to think of it as an alternate approach to this concept.

Alternate Definition of a Function

> A function with domain X is a set W of ordered pairs such that, for each x in X, there is exactly one ordered pair (x, y) in W having x in the first position.

In terms of the preceding definition the function f of Example 1, where $f(x) = x^2$, is the set of all ordered pairs of the form (x, x^2). Similarly, the ordered pairs $(x, x^2 + 3x - 2)$ determine the function of Example 3, where we had $f(x) = x^2 + 3x - 2$.

EXERCISES 3.3

3–42 EV 3RD

1 If $f(x) = 2x^2 - 3x + 4$, find $f(1)$, $f(-1)$, $f(0)$, and $f(2)$.

2 If $f(x) = x^3 + 5x^2 - 1$, find $f(2)$, $f(-2)$, $f(0)$, and $f(-1)$.

3 If $f(x) = \sqrt{x - 1} + 2x$, find $f(1)$, $f(3)$, $f(5)$, and $f(10)$.

4 If $f(x) = \dfrac{x}{x - 2}$, find $f(1)$, $f(3)$, $f(-2)$, and $f(0)$.

In Exercises 5–8 find each of the following, where a, b, and h are real numbers:
(a) $f(a)$ (b) $f(-a)$
(c) $-f(a)$ (d) $f(a + h)$
(e) $f(a) + f(h)$

(f) $\dfrac{f(a + h) - f(a)}{h}$ provided $h \neq 0$

5 $f(x) = 5x - 2$ **6** $f(x) = 3 - 4x$

7 $f(x) = 2x^2 - x + 3$ **8** $f(x) = x^3 - 2x$

In Exercises 9–12 find the following:
(a) $g(1/a)$ (b) $1/g(a)$
(c) $g(a^2)$ (d) $(g(a))^2$
(e) $g(\sqrt{a})$ (f) $\sqrt{g(a)}$

9 $g(x) = 3x^2$ **10** $g(x) = 3x - 8$

11 $g(x) = \dfrac{2x}{x^2 + 1}$ **12** $g(x) = \dfrac{x^2}{x + 1}$

In each of Exercises 13–20 find the largest subset of $\mathbb{R}$ that can serve as the domain of the function f.

13 $f(x) = \sqrt{3x - 5}$ **14** $f(x) = \sqrt{7 - 2x}$

15 $f(x) = \sqrt{4 - x^2}$ **16** $f(x) = \sqrt{x^2 - 9}$

17 $f(x) = \dfrac{x + 1}{x^3 - 9x}$

18 $f(x) = \dfrac{4x + 7}{6x^2 + 13x - 5}$

19 $f(x) = \dfrac{\sqrt{x}}{2x^2 - 11x + 12}$

20 $f(x) = \dfrac{x^3 - 1}{x^2 - 1}$

In each of Exercises 21–26 find the number that maps into 4. If $a > 0$, what number maps into a? Find the range of f.

21 $f(x) = 7x - 5$ **22** $f(x) = 3x$

23 $f(x) = \sqrt{x - 3}$ **24** $f(x) = 1/x$

25 $f(x) = x^3$ **26** $f(x) = \sqrt[3]{x - 4}$

In each of Exercises 27–34 determine if the function f is one-to-one.

27 $f(x) = 2x + 9$ **28** $f(x) = 1/(7x + 9)$

29 $f(x) = 5 - 3x^2$ **30** $f(x) = 2x^2 - x - 3$

31 $f(x) = \sqrt{x}$ **32** $f(x) = x^3$

33 $f(x) = |x|$ **34** $f(x) = 4$

In each of Exercises 35–44 determine whether f is even, odd, or neither even nor odd.

35 $f(x) = 3x^3 - 4x$ **36** $f(x) = 7x^6 - x^4 + 7$

37 $f(x) = 9 - 5x^2$ **38** $f(x) = 2x^5 - 4x^3$

39 $f(x) = 2$ **40** $f(x) = 2x^3 + x^2$

41 $f(x) = 2x^2 - 3x + 4$ **42** $f(x) = \sqrt{x^2 + 1}$

43 $f(x) = \sqrt[3]{x^3 - 4}$ **44** $f(x) = |x| + 5$

45 If $f(x)$ is a polynomial, and if the coefficients of all odd powers of x are 0, show that f is an even function.

46 If $f(x)$ is a polynomial, and if the coefficients of all even powers of x are 0, show that f is an odd function.

47 Find a formula which expresses the radius r of a circle as a function of its circumference C. If the circumference of *any* circle is increased by 12 inches, determine how much the radius increases.

48 Find a formula which expresses the volume of a cube as a function of its surface area. Find the volume if the surface area is 36 square inches.

49 An open box is to be made from a rectangular piece of cardboard having dimensions 20 inches by 30 inches by cutting out identical squares of area x^2 from each corner and turning up the sides. Express the volume V of the box as a function of x.

50 Find a formula which expresses the area A of an equilateral triangle as a function of the length s of a side.

51 Express the perimeter P of a square as a function of its area A.

52 Express the surface area S of a sphere as a function of its volume V.

53 A weather balloon is released at 1:00 P.M. and rises vertically at a rate of 2 meters per second. An observer is situated 100 meters from a point on the ground directly below the balloon. If t denotes the time (in seconds) after 1:00 P.M., express the distance d between the balloon and the observer as a function of t.

54 Two ships leave port at 9:00 A.M., one sailing south at a rate of 16 mph and the other west at a rate of 20 mph. If t denotes the time (in hours) after 9:00 A.M., express the distance d between the ships as a function of t.

55 A manufacturer sells a certain article to dealers at a rate of $20 each if less than 50 are ordered. If 50 or more are ordered (up to 600), the price per article is reduced at a rate of 2 cents times the number ordered. Let A denote the amount of money received when x articles are ordered. Express A as a function of x.

56 A travel agency offers a tour at a cost of $30 per person if not more than 60 take the tour. If more than 60 take the tour the cost is to be reduced 10 cents for each person in excess of 60. Let A denote the total cost if x people go on the tour. Express A as a function of x.

In each of Exercises 57–64 determine whether the set W of ordered pairs is a function in the sense of the alternate definition of function.

57 $W = \{(x, y) : 2y = x^2 + 5\}$

58 $W = \{(x, y) : x = 3y + 2\}$

59 $W = \{(x, y) : x^2 + y^2 = 4\}$

60 $W = \{(x, y) : y^2 - x^2 = 1\}$

61 $W = \{(x, y) : y = 3\}$

62 $W = \{(x, y) : x = y\}$

63 $W = \{(x, y) : xy = 0\}$

64 $W = \{(x, y) : x + y = 0\}$

3.4 GRAPHS OF FUNCTIONS

Graphs, or more precisely, *sketches* of graphs, are often used to describe the variation of physical quantities. For example, a scientist may use Figure 3.20 to indicate the temperature T of a certain solution at various times t during an experiment. The sketch shows that the temperature increased gradually from time $t = 0$ to time $t = 5$, did not change between $t = 5$ and $t = 8$, and then decreased rapidly from $t = 8$ to $t = 9$. A visual aid of this type reveals the behavior of T more clearly than a long table of numerical values.

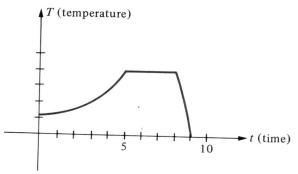

Figure 3.20

If f is a function, we may use a graph to exhibit the behavior of $f(x)$ as x varies through the domain of f. By definition, the **graph of a function** f is the set of all points $(x, f(x))$ in a coordinate plane, where x is in the domain of f. Thus, the graph of f is the same as the graph of the equation $y = f(x)$, and if $P(a, b)$ is on the graph, then the ordinate b is the functional value $f(a)$, as illustrated in Figure 3.21. It is important to note that, since there is a unique $f(a)$ for each a in the domain, there is only *one* point on the graph with abscissa a. Thus, every vertical line intersects the graph of a function in at most one point. Consequently, for graphs of functions it is impossible to obtain a sketch such as that shown in Figure 3.12 where some vertical lines intersect the graph in more than one point.

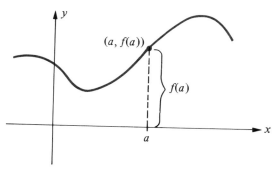

Figure 3.21

Example 1   Sketch the graph of f if $f(x) = 2x + 3$.

Solution   The graph consists of all points $(x, f(x))$, where $f(x) = 2x + 3$. The following table lists coordinates of several points on the graph, where $y = f(x)$.

x	-3	-2	-1	0	1	2
y	-3	-1	1	3	5	7

Plotting, it appears that the points lie on a straight line, and we sketch the graph as in Figure 3.22. (In the next section we shall *prove* that the graph is a straight line.) The graph of f is the same as the graph of the equation $y = 2x + 3$.

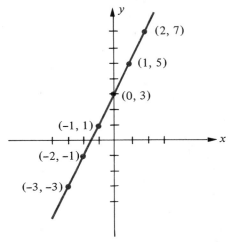

Figure 3.22 $f(x) = 2x + 3$ ∎

It is useful to determine the points at which the graph of a function f intersects the x-axis. The abscissas of these points are called the **x-intercepts** of the graph and are found by locating all points $(x, f(x))$ such that $f(x) = 0$. In Example 1 the x-intercept is $-3/2$, since that number is the solution of $2x + 3 = 0$. A number a such that $f(a) = 0$ is also called a **zero of the function** f.

If the number 0 is in the domain of f, then $f(0)$ is called the **y-intercept** of the graph of f. It is the ordinate of the point at which the graph intersects the y-axis. The graph of a function can have at most one y-intercept (Why?). The y-intercept in Example 1 is 3.

If f is the function in Example 1, and if $x_1 < x_2$, then $2x_1 + 3 < 2x_2 + 3$; that is, $f(x_1) < f(x_2)$. This means that, as abscissas of points increase, ordinates also increase. The function f is then said to be *increasing*. If f is increasing, then the graph rises as x increases. For certain functions, we have $f(x_1) > f(x_2)$ whenever $x_1 < x_2$. In this case the graph of f falls as x increases and the function is called a *decreasing* function. In general we shall speak of functions which increase or decrease on certain subsets of their domains, as in the following definition. Functions which neither increase nor decrease are called *constant*.

Definition

> If S is a subset of the domain of a function f, then
>
> (i) f is **increasing** on S if $f(x_1) < f(x_2)$ whenever $x_1 < x_2$ in S.
> (ii) f is **decreasing** on S if $f(x_1) > f(x_2)$ whenever $x_1 < x_2$ in S.
> (iii) f is **constant** on S if $f(x_1) = f(x_2)$ for every x_1, x_2 in S.

If we regard Figure 3.20 as the graph of a function f, then f is increasing on the closed interval $[0, 5]$, is constant on the closed interval $[5, 8]$, and is decreasing on $[8, 9]$.

Example 2   Sketch the graph of f if $f(x) = x^2 - 3$.

Solution   We list coordinates $(x, f(x))$ of some points on the graph of f in tabular form, as follows:

x	-3	-2	-1	0	1	2	3
$f(x)$	6	1	-2	-3	-2	1	6

The x-intercepts are the solutions of the equation $f(x) = 0$, that is, of $x^2 - 3 = 0$. These are $\pm\sqrt{3}$. The y-intercept is $f(0) = -3$. Plotting the points given by the table and using the x-intercepts leads to the sketch in Figure 3.23.

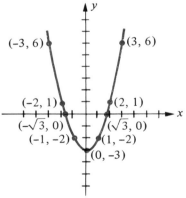

Figure 3.23 $f(x) = x^2 - 3$ ■

In arriving at the sketch in Figure 3.23 we assumed that f is decreasing on the interval $(-\infty, 0]$ and increasing on the interval $[0, \infty)$. These facts can be proved using properties of inequalities; however, for more complicated functions it is often necessary to use techniques studied in calculus. In this text we shall usually rely on sufficient point plotting to determine where functions increase or decrease.

Referring to Figure 3.23 we see that $f(x) = x^2 - 3$ takes on its least value at $x = 0$. This smallest value, -3, is called the **minimum value** of f. The corresponding point $(0, -3)$ is the lowest point on the graph. Clearly, $f(x)$ does not attain a **maximum value**, that is, a *largest* value.

The solution to Example 2 could have been shortened by observing that, since $(-x)^2 - 3 = x^2 - 3$, the graph of $y = x^2 - 3$ is symmetric with respect to the y-axis. This fact also follows from (i) of the next theorem.

Theorem on Symmetry

(i) The graph of an even function is symmetric with respect to the y-axis.

(ii) The graph of an odd function is symmetric with respect to the origin.

Proof   If f is even, then $f(-x) = f(x)$, and hence the equation $y = f(x)$ is not changed if $-x$ is substituted for x. Statement (i) now follows from Symmetry Test (i) of Section 3.2. The proof of (ii) is left to the reader.   ∎

Example 3   Sketch the graph of f if $f(x) = |x|$.

Solution   If $x \geq 0$, then $f(x) = x$ and hence the points (x, x) in the first quadrant are on the graph of f. Some special cases are $(0,0)$, $(1, 1)$, $(2, 2)$, $(3, 3)$, and $(4, 4)$. Since $|-x| = |x|$ we see that f is an even function and hence, by the preceding theorem, the graph is symmetric with respect to the y-axis. Plotting points and using symmetry leads to the sketch in Figure 3.24. As in Example 2, this function decreases on $(-\infty, 0]$ and increases on $[0, \infty)$.

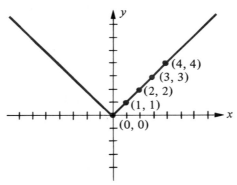

Figure 3.24 $f(x) = |x|$ ∎

Example 4   Sketch the graph of f if $f(x) = \sqrt{x - 1}$.

Solution   The domain of f does not include values of x such that $x - 1 < 0$, since $f(x)$ is not a real number in this case. Consequently there are no points (x, y) on the graph with $x < 1$. Moreover, no part of the graph lies below the x-axis (Why?).

The following table lists some points $(x, f(x))$ on the graph.

x	1	2	3	4	5	6
$f(x)$	0	1	$\sqrt{2}$	$\sqrt{3}$	2	$\sqrt{5}$

Plotting points leads to the sketch shown in Figure 3.25. The function is increasing throughout its domain. The x-intercept is 1 and there is no y-intercept.

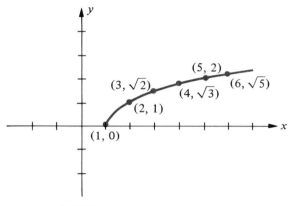

Figure 3.25 $f(x) = \sqrt{x-1}$ ∎

Example 5 Sketch the graph of a constant function with domain $\mathbb{R}$.

Solution By definition f is a constant function if there is some (fixed) real number c such that $f(x) = c$ for every x. The graph of f consists of all points with coordinates (x, c), where x is any real number. In particular this includes $(-1, c)$, $(0, c)$, and $(2, c)$. Since all the ordinates equal c, the graph is a line parallel to the x-axis with y-intercept c. A sketch for the case $c > 0$ is shown in Figure 3.26.

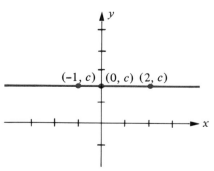

Figure 3.26 $f(x) = c$ ∎

Example 6   Given $f(x) = x^2 + c$, sketch the graph of f if (a) $c = 4$; (b) $c = -2$.

Solution   We shall sketch both graphs on the same coordinate axes. The graph of $y = x^2$ was sketched in Figure 3.10, and for reference it is represented by dashes in Figure 3.27. To find the graph of $f(x) = x^2 + 4$ we may simply increase the ordinate of each point on the graph of $y = x^2$ by 4, as shown in the figure (Why?). This amounts to *shifting* the graph of $y = x^2$ *upward* 4 units. For $c = -2$ we decrease ordinates by 2 and hence this graph may be obtained by shifting the graph of $y = x^2$ downward 2 units. Each graph is a parabola symmetric with respect to the y-axis. In order to get the correct positions it is advisable to plot several points on each graph.

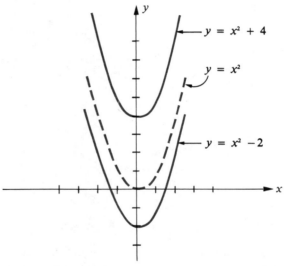

Figure 3.27 $f(x) = x^2 + c$ ■

The graphs in the preceding example illustrate **vertical shifts** of the graph of $y = x^2$, and are special cases of the following general rules.

Vertical Shifts of Graphs (c > 0)

To obtain the graph of:	shift the graph of $y = f(x)$:
$y = f(x) - c$	c units downward
$y = f(x) + c$	c units upward

Similar rules can be stated for **horizontal shifts**. Specifically, if $c > 0$, consider the graphs of $y = f(x)$ and $y = f(x - c)$ sketched on the same coordinate axes, as illustrated in Figure 3.28.

Since $f(a) = f(a + c - c)$ we see that the point with abscissa a on the graph of $y = f(x)$ has the same y-coordinate as the point with abscissa $a + c$ on the graph of

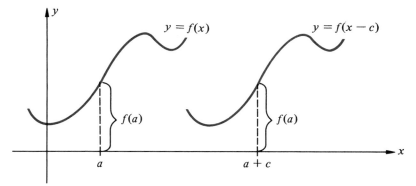

Figure 3.28

$y = f(x - c)$. This implies that the latter graph can be obtained by shifting the graph of $y = f(x)$ to the right c units. Similarly, the graph of $y = f(x + c)$ can be obtained by shifting the graph of f to the left c units. These rules are listed for reference in the next box.

Horizontal Shifts of Graphs ($c > 0$)

To obtain the graph of:	shift the graph $y = f(x)$:
$y = f(x - c)$ $y = f(x + c)$	c units to the right c units to the left

Example 7   Sketch the graph of f if:

(a) $f(x) = (x - 4)^2$ (b) $f(x) = (x + 2)^2$

Solution   The graph of $y = x^2$ is sketched, with dashes, in Figure 3.29. According to the last two rules, shifting this graph to the right 4 units gives us the graph of

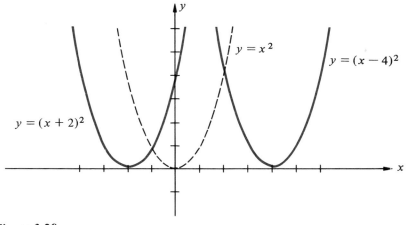

Figure 3.29

$y = (x - 4)^2$, whereas shifting to the left 2 units leads to the graph of $y = (x + 2)^2$. Students who are not convinced of this technique are urged to plot several points on each graph.   ∎

To obtain the graph of $y = cf(x)$ we may *multiply* the ordinates of points on the graph of $y = f(x)$ by c. For example, if $y = 2f(x)$, we double ordinates, or if $y = \frac{1}{2}f(x)$, we multiply each ordinate by $1/2$. If $c > 0$ (and $c \neq 1$) we shall refer to this procedure as **stretching** the graph of $y = f(x)$.

Example 8   Sketch the graph of $y = cx^2$ if   (a) $c = 4$;   (b) $c = 1/4$.

Solution   To obtain the graph of $y = 4x^2$ we may refer to the graph of $y = x^2$ (see dashes in Figure 3.30) and multiply the ordinate of each point by 4. This gives us a narrower parabola which is sharper at the vertex, as illustrated in (i) of the figure. In order to get the correct position, several points such as $(0, 0)$, $(1/2, 1)$, and $(1, 4)$ should be plotted.

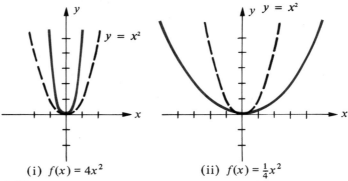

(i) $f(x) = 4x^2$ (ii) $f(x) = \frac{1}{4}x^2$

Figure 3.30

Similarly, for the graph of $y = (1/4)x^2$, we multiply ordinates of points on the graph of $y = x^2$ by $1/4$. This gives us a wider parabola which is flatter at the vertex, as shown in (ii) of Figure 3.30.   ∎

The graph of $y = -f(x)$ may be obtained by multiplying the y-coordinate of each point on the graph of $y = f(x)$ by -1. In particular, every point (a, b) on the graph of $y = f(x)$ which lies above the x-axis determines a point $(a, -b)$ below the x-axis. Similarly, if (c, d) lies below the x-axis (that is, $d < 0$), then $(c, -d)$ lies above the x-axis. We call the graph of $y = -f(x)$ a **reflection** of the graph of $y = f(x)$ through the x-axis.

Example 9   Sketch the graph of $y = -x^2$.

Solution   The graph may be obtained by plotting points; however, since the graph of $y = x^2$ is well-known, we sketch it with dashes as in Figure 3.31, and then multiply ordinates of points by -1. This gives us the reflection through the x-axis indicated in the figure.

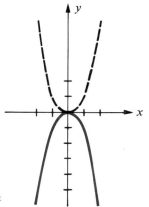

Figure 3.31 $f(x) = -x^2$ ■

Sometimes functions are described in terms of more than one expression, as in the next examples.

Example 10 Sketch the graph of the function f which is defined as follows:

$$f(x) = \begin{cases} 2x + 3 & \text{if } x < 0 \\ x^2 & \text{if } 0 \le x < 2 \\ 1 & \text{if } x \ge 2. \end{cases}$$

Solution If $x < 0$, then $f(x) = 2x + 3$. This means that, when x is negative, the expression $2x + 3$ should be used to find functional values. Consequently, if $x < 0$, then the graph of f coincides with the graph in Figure 3.22 and we sketch that portion of the graph to the left of the y-axis as indicated in Figure 3.32.

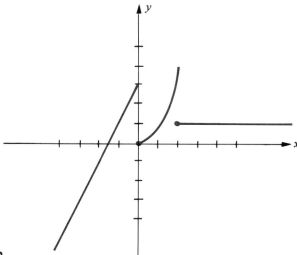

Figure 3.32

If $0 \leq x < 2$, we use x^2 to find functional values of f, and therefore, this part of the graph of f coincides with the graph of the equation $y = x^2$. We then sketch the part of the graph of f between $x = 0$ and $x = 2$, as indicated in Figure 3.32.

Finally, if $x \geq 2$, the graph of f coincides with the constant function having values equal to 1. That part of the graph is the horizontal half-line illustrated in Figure 3.32. ∎

Example 11 If x is any real number, then there exist consecutive integers n and $n + 1$ such that $n \leq x < n + 1$. Let f be the function defined as follows: If $n \leq x < n + 1$, then $f(x) = n$. Sketch the graph of f.

Solution   Abscissas and ordinates of points on the graph may be listed as follows:

Values of x	$f(x)$
. . .	. . .
$-2 \leq x < -1$	-2
$-1 \leq x < 0$	-1
$0 \leq x < 1$	0
$1 \leq x < 2$	1
$2 \leq x < 3$	2
. . .	. . .

Since f behaves in the same manner as a constant function when x is between integral values, the corresponding part of the graph is a segment of a horizontal line. Part of the graph is sketched in Figure 3.33.

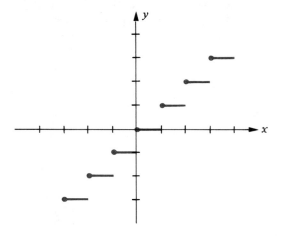

Figure 3.33   ∎

The symbol $[x]$ is often used to denote the greatest integer z such that $z \leq x$. For example, $[1.6] = 1$, $[\sqrt{5}] = 2$, $[\pi] = 3$, and $[-3.5] = -4$. Using this notation, the function f of Example 9 may be defined by $f(x) = [x]$. It is customary to refer to f as the **greatest integer function**.

EXERCISES 3.4

In Exercises 1–20 sketch the graph of f and state where f is increasing, is decreasing, or is constant.

1 $f(x) = 5x$ **2** $f(x) = -3x$

3 $f(x) = -4x + 2$ **4** $f(x) = 4x - 2$

5 $f(x) = 3 - x^2$ **6** $f(x) = 4x^2 + 1$

7 $f(x) = 2x^2 - 4$ **8** $f(x) = \frac{1}{100}x^2$

9 $f(x) = \sqrt{x + 4}$ **10** $f(x) = \sqrt{4 - x}$

11 $f(x) = \sqrt{x} + 2$ **12** $f(x) = 2 - \sqrt{x}$

13 $f(x) = 4/x$ **14** $f(x) = 1/x^2$

15 $f(x) = |x - 2|$ **16** $f(x) = |x + 2|$

17 $f(x) = |x| - 2$ **18** $f(x) = |x| + 2$

19 $f(x) = \dfrac{x}{|x|}$ **20** $f(x) = x + |x|$

In Exercises 21–30 sketch, on the same coordinate axes, the graphs of f for the stated values of c. (Make use of vertical shifts, horizontal shifts, stretching, or reflecting).

21 $f(x) = 3x + c$; $c = 0, c = 2, c = -1$

22 $f(x) = -2x + c$; $c = 0, c = 1, c = -3$

23 $f(x) = x^3 + c$; $c = 0, c = 1, c = -2$

24 $f(x) = -x^3 + c$; $c = 0, c = 2, c = -1$

25 $f(x) = \sqrt{4 - x^2} + c$; $c = 0, c = 4, c = -3$

26 $f(x) = c - |x|$; $c = 0, c = 5, c = -2$

27 $f(x) = 3(x - c)$; $c = 0, c = 2, c = 3$

28 $f(x) = -2(x - c)^2$; $c = 0, c = 3, c = 1$

29 $f(x) = (x + c)^3$; $c = 0, c = 2, c = -2$

30 $f(x) = c\sqrt{9 - x^2}$; $c = 0, c = 2, c = 3$

31 If (i) of Figure 3.34 is the graph of a function f with domain $0 \le x \le 4$, sketch the graph of each of the following.
  (a) $y = f(x + 2)$ (b) $y = f(x - 2)$
  (c) $y = f(x) + 2$ (d) $y = f(x) - 2$
  (e) $y = 2f(x)$ (f) $y = \frac{1}{2}f(x)$
  (g) $y = -2f(x)$ (h) $y = f(x - 3) + 1$

(i)

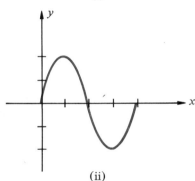

(ii)

Figure 3.34

32 If (ii) of Figure 3.34 is the graph of a function f with domain $0 \le x \le 4$, sketch the graph of each of the following.
  (a) $y = f(x - 1)$ (b) $y = f(x + 3)$
  (c) $y = f(x) - 2$ (d) $y = f(x) + 1$
  (e) $y = 3f(x)$ (f) $y = \frac{1}{2}f(x)$
  (g) $y = f(x + 2) - 2$ (h) $y = f(2x)$

In each of Exercises 33–38 sketch the graph of f.

33 $f(x) = \begin{cases} 2 & \text{if } x < 0 \\ -1 & \text{if } x \ge 0 \end{cases}$

34 $f(x) = \begin{cases} -1 & \text{if } x \text{ is an integer} \\ 1 & \text{if } x \text{ is not an integer} \end{cases}$

35 $f(x) = \begin{cases} 3 & \text{if } x < -3 \\ -x & \text{if } -3 \le x \le 3 \\ -3 & \text{if } x > 3 \end{cases}$

36 $f(x) = \begin{cases} x & \text{if } x < 0 \\ -2 & \text{if } 0 \le x < 1 \\ x^2 & \text{if } x \ge 1 \end{cases}$

37 $f(x) = \begin{cases} x^2 & \text{if } x \le -1 \\ x^3 & \text{if } |x| < 1 \\ 2x & \text{if } x \ge 1 \end{cases}$

38 $f(x) = \begin{cases} x & \text{if } x \le 1 \\ -x^2 & \text{if } 1 < x < 2 \\ x & \text{if } x \ge 2 \end{cases}$

39 If $[x]$ denotes values of the greatest integer function (see Example 11), sketch the graph of f in each of the following:
(a) $f(x) = [x - 3]$ (b) $f(x) = 2[x]$
(c) $f(x) = -[x]$

40 Explain why the graph of the equation $x^2 + y^2 = 1$ is not the graph of a function.

41 Sketch the graph of f if $f(x) = x$ for every x.

42 Define a function f whose graph is the upper half of a circle with center at the origin and radius 1.

43 Define a function f whose graph is the lower half of a circle with center at the origin and radius 1.

44 If a function f is increasing throughout its domain, prove that f is one-to-one.

45 If a function f is decreasing throughout its domain, prove that f is one-to-one.

46 Prove that a function f is one-to-one if and only if every horizontal line pierces the graph of f in at most one point.

3.5   LINEAR FUNCTIONS

The following type of function is very important in mathematics and its applications.

Definition

A function f is a **linear function** if

$$f(x) = ax + b$$

where a and b are real numbers and $a \ne 0$.

The reason for the term "linear" is that the graph of f is a line, as we shall see later in this section. Let us first introduce several fundamental concepts pertaining to lines. All lines referred to are considered to be in a coordinate plane.

Definition

If l is a line which is not parallel to the y-axis, and if $P_1(x_1, y_1)$ and $P_2(x_2, y_2)$ are distinct points on l, then the **slope m** of l is given by

$$m = \frac{y_2 - y_1}{x_2 - x_1}.$$

If l is parallel to the y-axis, then the slope is not defined.

The numerator $y_2 - y_1$ in the formula for m is sometimes called the **rise** from P_1 to P_2. It measures the vertical change in direction in proceeding from P_1 to P_2 and may be positive, negative, or zero. The denominator $x_2 - x_1$ is called the **run** from P_1 to P_2. It measures the amount of horizontal change in going from P_1 to P_2. The run may be positive or negative but is never zero because l is not parallel to the y-axis. Using this terminology,

$$\text{slope of } l = \frac{\text{rise from } P_1 \text{ to } P_2}{\text{run from } P_1 \text{ to } P_2}.$$

In finding the slope of a line it is immaterial which point is labeled P_1 and which is labeled P_2, since

$$\frac{y_2 - y_1}{x_2 - x_1} = \frac{y_1 - y_2}{x_1 - x_2}.$$

Consequently, we may as well assume that the points are labeled so that $x_1 < x_2$, as in Figure 3.35. In this event $x_2 - x_1 > 0$, and hence the slope is positive, negative, or zero, depending on whether $y_2 > y_1$, $y_2 < y_1$, or $y_2 = y_1$. The slope of the line shown in (i) of Figure 3.35 is positive, whereas the slope of the line shown in (ii) of the figure is negative.

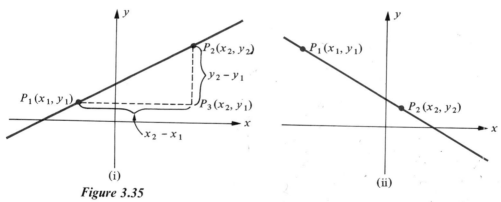

(i)

(ii)

Figure 3.35

A **horizontal line** is a line that is parallel to the x-axis. Note that a line is horizontal if and only if its slope is 0. A **vertical line** is a line that is parallel to the y-axis. The slope of a vertical line is undefined.

Example 1 Sketch the lines through the following pairs of points and find their slopes.

(a) $A(-1, 4)$ and $B(3, 2)$

(b) $A(2, 5)$ and $B(-2, -1)$

(c) $A(4, 3)$ and $B(-2, 3)$

(d) $A(4, -1)$ and $B(4, 4)$.

Solution   The lines are sketched in Figure 3.36. Using the definition of slope gives
us

(a)
$$m = \frac{2-4}{3-(-1)} = \frac{-2}{4} = -\frac{1}{2}$$

(b)
$$m = \frac{5-(-1)}{2-(-2)} = \frac{6}{4} = \frac{3}{2}$$

(c)
$$m = \frac{3-3}{-2-4} = \frac{0}{-6} = 0$$

(d) The slope is undefined since the line is vertical. This is also seen by noting
that, if the formula for m is used, the denominator is zero.

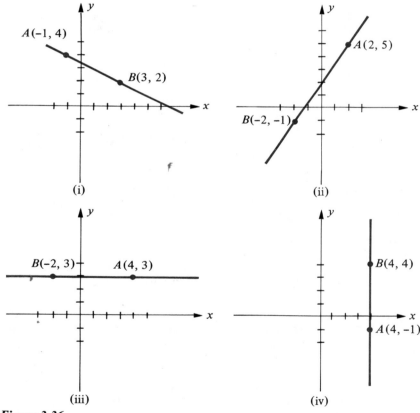

Figure 3.36

It is important to note that the definition of slope is independent of the two points
that are chosen on l, for if other points $P'_1(x'_1, y'_1)$ and $P'_2(x'_2, y'_2)$ are used, then, as in
Figure 3.37 the triangle with vertices P'_1, P'_2, and $P'_3(x'_2, y'_1)$ is similar to the triangle
with vertices P_1, P_2, and $P_3(x_2, y_1)$. Since the ratios of corresponding sides are equal,

it follows that

$$\frac{y_2 - y_1}{x_2 - x_1} = \frac{y'_2 - y'_1}{x'_2 - x'_1}.$$

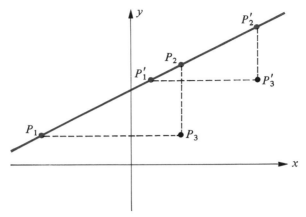

Figure 3.37 ∎

Example 2 Construct a line through $P(2, 1)$ that has slope   (a) $5/3$;   (b) $-5/3$.

Solution   If the slope of a line is a/b, where b is positive, then for every change of b units in the horizontal direction, the line rises or falls $|a|$ units, depending on whether a is positive or negative, respectively. If $P(2, 1)$ is on the line and $m = 5/3$, we can obtain another point on the line by starting at P and moving 3 units to the right and 5 units upward. This gives us the point $Q(5, 6)$ and the line is determined (see (i) of Figure 3.38). Similarly, if $m = -5/3$, we move 3 units to the right and 5 units downward, obtaining $Q(5, -4)$ as in (ii) of Figure 3.38.

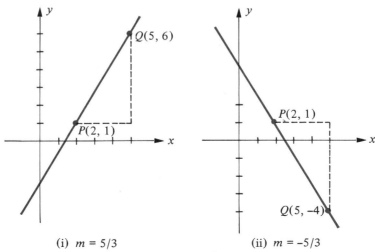

(i) $m = 5/3$ (ii) $m = -5/3$

Figure 3.38 ∎

Theorem

> (i)   The graph of the equation $x = a$ is a vertical line with x-intercept a.
> (ii) The graph of the equation $y = b$ is a horizontal line with y-intercept b.

Proof

The equation $x = a$, where a is a real number, may be considered as an equation in two variables x and y, since we can write it in the form

$$x + (0y) = a.$$

Some typical solutions of the equation are $(a, -2)$, $(a, 1)$, and $(a, 3)$. Evidently, all solutions of the equation consist of pairs of the form (a, y), where y may have any value and a is fixed. It follows that the graph of $x = a$ is a line parallel to the y-axis with x-intercept a, as illustrated in Figure 3.39. This proves (i).

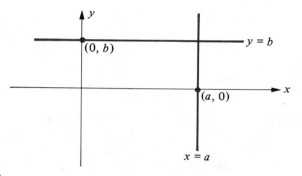

Figure 3.39

Part (ii) follows from the fact that the graph of $y = b$ is the same as the graph of the constant function f, where $f(x) = b$ (see Figure 3.26). The graph is sketched in Figure 3.39. ∎

Let us now find an equation of a line l through a point $P_1(x_1, y_1)$ with slope m (only one such line exists). If $P(x, y)$ is any point with $x \neq x_1$ (see Figure 3.40), then

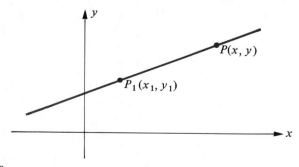

Figure 3.40

P is on l if and only if the slope of the line through P_1 and P is m, that is, if and only if

$$\frac{y - y_1}{x - x_1} = m.$$

This equation may be written in the form

$$y - y_1 = m(x - x_1).$$

Note that (x_1, y_1) is also a solution of the last equation, and hence the points on l are precisely the points which correspond to the solutions. This equation for l is referred to as the **Point-Slope Form**. Our discussion may be summarized as follows:

Point-Slope Form for the Equation of a Line

> An equation for the line through the point $P(x_1, y_1)$ with slope m is
> $$y - y_1 = m(x - x_1).$$

Example 3 Find an equation of the line through the points $A(1, 7)$ and $B(-3, 2)$.

Solution The slope m of the line is

$$m = \frac{7 - 2}{1 - (-3)} = \frac{5}{4}.$$

Using the coordinates of A in the Point-Slope Form gives us

$$y - 7 = \frac{5}{4}(x - 1)$$

which is equivalent to

$$4y - 28 = 5x - 5 \quad \text{or} \quad 5x - 4y + 23 = 0.$$

The same equation would have been obtained if the coordinates of point B had been substituted in the Point-Slope Form. ∎

The Point-Slope Form may be rewritten as $y = mx - mx_1 + y_1$, which is of the form

$$y = mx + b$$

where $b = -mx_1 + y_1$. The real number b is the y-intercept of the graph, as may be seen by setting $x = 0$. Since the equation $y = mx + b$ displays the slope m and y-intercept b of l, it is called the **Slope-Intercept Form** for the equation of a line. Conversely, if we start with $y = mx + b$, we may write

$$y - b = m(x - 0).$$

Comparing with the Point-Slope Form, we see that the graph is a line with slope m and passing through the point $(0, b)$. This gives us the next result.

**Slope-
Intercept
Form for the
Equation of
a Line**

> The graph of the equation $y = mx + b$ is a line having slope m and y-intercept b.

The work we have done shows that every line is the graph of an equation of the form

$$ax + by + c = 0$$

where a, b, and c are real numbers, and a and b are not both zero. We call such an equation a **linear equation** in x and y. Let us show, conversely, that the graph of $ax + by + c = 0$, where a and b are not both zero is always a line. On the one hand, if $b \neq 0$, we may solve for y, obtaining,

$$y = (-a/b)x + (-c/b)$$

which, by the Slope-Intercept Form, is an equation of a line with slope $-a/b$ and y-intercept $-c/b$. On the other hand, if $b = 0$ but $a \neq 0$, then we may solve for x, obtaining $x = -c/a$, which is the equation of a vertical line with x-intercept $-c/a$. This establishes the following important theorem.

Theorem

> The graph of a linear equation $ax + by + c = 0$ is a line and, conversely, every line is the graph of a linear equation.

For simplicity, we shall often use the terminology *the line* $ax + by + c = 0$ instead of the more accurate phrase *the line with equation* $ax + by + c = 0$.

Example 4   Sketch the graph of $2x - 5y = 8$.

Solution   From the previous theorem the graph is a line, and hence it is sufficient to find two points on the graph. Let us find the x- and y-intercepts. Substituting $y = 0$ in the given equation, we obtain the x-intercept 4. Substituting $x = 0$, we see that the y-intercept is $-8/5$. This leads to the graph in Figure 3.41. Another method of solution is to express the given equation in Slope-Intercept Form. To do this we begin by isolating the term involving y on one side of the equals sign, obtaining

$$5y = 2x - 8.$$

Next, dividing both sides by 5 gives us

$$y = \frac{2}{5}x + \left(\frac{-8}{5}\right)$$

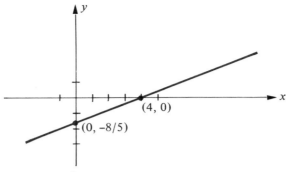

Figure 3.41 $2x - 5y = 8$

which is of the form $y = mx + b$. Hence, the slope is $m = 2/5$, and the y-intercept is $b = -8/5$. We may then sketch a line through the point $(0, -8/5)$ with slope $2/5$. ∎

The following can be proved geometrically.

Slopes of Parallel Lines

> Two nonvertical lines are parallel if and only if they have the same slope.

We shall use this fact in the next example.

Example 5 Find an equation of a line through the point $(5, -7)$ which is parallel to the line $6x + 3y - 4 = 0$.

Solution Let us express the given equation in Slope-Intercept Form. We begin by writing

$$3y = -6x + 4$$

and then divide both sides by 3, obtaining

$$y = -2x + \frac{4}{3}.$$

The last equation is in Slope-Intercept Form with $m = -2$, and hence the slope is -2. Since parallel lines have the same slope, the required line also has slope -2. Applying the Point-Slope Form gives us

$$y + 7 = -2(x - 5).$$

This is equivalent to

$$y + 7 = -2x + 10 \quad \text{or} \quad 2x + y - 3 = 0.$$ ∎

The next result specifies conditions for perpendicular lines.

Slopes of Perpendicular Lines

> Two lines with slopes m_1 and m_2 are perpendicular if and only if $m_1 m_2 = -1$.

Proof

For simplicity, let us consider the special case where the lines intersect at the origin O, as illustrated in Figure 3.42. In this case equations of the lines are $y = m_1 x$ and $y = m_2 x$.

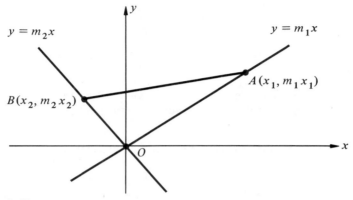

Figure 3.42

If, as in the figure, we choose points $A(x_1, m_1 x_1)$ and $B(x_2, m_2 x_2)$ different from O on the lines, then the lines are perpendicular if and only if angle AOB is a right angle. Applying the Pythagorean Theorem to triangle AOB, this is equivalent to the condition

$$[d(A, B)]^2 = [d(O, B)]^2 + [d(O, A)]^2$$

that is,

$$(m_2 x_2 - m_1 x_1)^2 + (x_2 - x_1)^2 = (m_2 x_2)^2 + x_2^2 + (m_1 x_1)^2 + x_1^2.$$

Squaring the indicated terms and simplifying gives us

$$-2m_1 m_2 x_1 x_2 - 2x_1 x_2 = 0.$$

Dividing both sides by $-2x_1 x_2$ we see that the lines are perpendicular if and only if $m_1 m_2 + 1 = 0$, or $m_1 m_2 = -1$.

The same type of proof may be given if the lines intersect at *any* point (a, b), the only difference being that the algebraic manipulations are more complicated (see Exercise 56). ∎

A convenient way to remember the conditions for perpendicularity is to note that m_1 and m_2 must be *negative reciprocals* of one another, that is, $m_1 = -1/m_2$ and $m_2 = -1/m_1$.

Example 6 Find an equation of a line through the point $(5, -7)$ which is perpendicular to the line $6x + 3y - 4 = 0$.

Solution The slope m of the given line is -2 (see Example 5). Hence, the slope m' of the required line is the negative reciprocal of -2, that is, $m' = -[1/(-2)] = 1/2$. Thus, the equation is

$$y + 7 = \frac{1}{2}(x - 5).$$

This simplifies to

$$2y + 14 = x - 5, \quad \text{or} \quad x - 2y - 19 = 0. \qquad \blacksquare$$

We shall conclude this section with the following important theorem.

Theorem

The graph of a linear function is a line.

Proof If f is linear we may write

$$f(x) = mx + b$$

where $m \neq 0$. Since the graph of f is the same as the graph of the equation $y = mx + b$, the conclusion follows at once from the Slope-Intercept Form. $\blacksquare$

#3,6,
9-42 every
3RD

EXERCISES 3.5

In each of Exercises 1–6 plot the points A and B and find the slope of the line through A and B.

1 $A(-3, 8), B(-2, 16)$ **2** $A(7, -1), B(-4, 9)$

3 $A(\frac{1}{3}, \frac{2}{3}), B(-\frac{3}{2}, \frac{1}{2})$ **4** $A(-5, -2), B(6, 4)$

5 $A(5, 0), B(-3, 2)$ **6** $A(0, -4), B(8, 3)$

7 Use slopes to show that $A(-4, -1)$, $B(2, 3)$, $C(1, -3)$, and $D(7, 1)$ are vertices of a parallelogram.

8 Use slopes to show that $A(-3, -9)$, $B(1, -1)$, and $C(4, 5)$ lie on a straight line.

In each of Exercises 9–22 find an equation for the line satisfying the given conditions.

9 Through $A(5, -2)$; slope $1/3$

10 Through $B(-3, 7)$; slope -2

11 Through $A(-8, -3)$ and $B(2, -5)$

12 Through $P(4, -1)$ and $Q(-6, 3)$

13 Slope -4, y-intercept 7

14 Slope $3/4$, x-intercept 5

15 x-intercept -3, y-intercept 7

16 Through $A(-7, -8)$; y-intercept 0

17 Through $A(7, -3)$, parallel to the (a) x-axis (b) y-axis

18 Through $A(-8, 2)$, perpendicular to the (a) y-axis (b) x-axis

19 Bisecting the second and fourth quadrants

20 Coinciding with the y-axis

21 Through $A(-4, 10)$, parallel to the line through $B(0, 5)$ and $C(-8, -8)$.

22 Through $P(3/2, -1/4)$, parallel to the line with equation $2x + 4y = 5$.

23 Find equations for the medians of the triangle with vertices $A(-4, -3)$, $B(2, 3)$, and $C(5, -1)$.

24 If $A(x_1, y_1)$, $B(x_2, y_2)$, $C(x_3, y_3)$, and $D(x_4, y_4)$ are vertices of an arbitrary quadrilateral, prove that the line segments joining the midpoints of adjacent sides form a parallelogram.

In each of Exercises 25–34 use the Slope-Intercept Form to find the slope and y-intercept of the line with the given equation and sketch the graph.

25 $2x - 5y + 10 = 0$ **26** $2y - 3x = 4$

27 $3x + 5y = 0$ **28** $6x = 1 - 2y$

29 $y + 2 = 0$ **30** $3y = 8$

31 $5x = 20 - 6y$ **32** $9x - 4y = 0$

33 $y = 0$ **34** $x = (2/3)y + 4$

35 Find a real number k such that the point $P(-1, 2)$ is on the line $kx + 2y - 7 = 0$.

36 Find a real number k such that the line $5x + ky - 3 = 0$ has y-intercept -5.

37 Sketch the graph of f if $f(x) = -5x - 2$.

38 Sketch the graph of f if $f(x) = 4x - 3$.

39 Given $f(x) = ax + 3$, sketch the graph of f if:
(a) $a = 0$ (b) $a = 2$
(c) $a = 5$ (d) $a = -3$

40 Given $f(x) = -5x + b$, sketch the graph of f if:
(a) $b = 0$ (b) $b = 2$
(c) $b = 5$ (d) $b = -3$

41 If $f(x) = 3x + 7$ and $g(x) = -2x + 5$, find the abscissa of a point which is on the graphs of both f and g.

42 If the graph of a linear function f contains the points $(2, 7)$ and $(0, 1)$, find $f(x)$.

43 If f is a linear function such that $f(0) = -2$ and $f(-1) = 4$, find $f(x)$.

44 If the graph of a linear function f is parallel to the line $3x + 5y - 2 = 0$ and has x-intercept -4, find $f(x)$.

45 If $a > 0$, prove that the linear function f defined by $f(x) = ax + b$ is an increasing function throughout its domain. If $a < 0$, prove that f is decreasing throughout its domain.

46 Prove that the graph of the equation $ax + by = 0$, where a and b are not both zero, is a line passing through the origin.

47 If a line l has nonzero x- and y-intercepts a and b, respectively, prove that an equation for l is

$$\frac{x}{a} + \frac{y}{b} = 1.$$

(This is called the **intercept form** for the equation of a line.) Express the equation $4x - 2y = 6$ in intercept form.

48 Prove that an equation of the line through $P_1(x_1, y_1)$ and $P_2(x_2, y_2)$ is

$$(y - y_1)(x_2 - x_1) = (y_2 - y_1)(x - x_1).$$

(This is called the **two-point form** for the equation of a line.) Use the two-point form to find an equation of the line through $A(7, -1)$ and $B(4, 6)$.

49 Find all values of r such that the slope of the line through the points $(r, 4)$ and $(1, 3 - 2r)$ is less than 5.

50 Find all values of t such that the slope of the line through $(t, 3t + 1)$ and $(1 - 2t)$ is greater than 4.

51 Find an equation of the line through the point $(7, -3)$ which is perpendicular to the line with equation $2x - 5y = 8$.

52 Find an equation of the line through the point $(-4, 8)$ which is perpendicular to the line through $(5, -1)$ and $(-2, -3)$.

53 Find an equation for the perpendicular bisector of the line segment from $A(3, -1)$ to $B(-2, 6)$. (*Hint*: Find the midpoint of segment AB.)

54 Find an equation for the perpendicular bisector of the line segment from the origin to $P(-5, 6)$.

55 Find equations for the altitudes of the triangle with vertices $A(-3, 2)$, $B(5, 4)$, and $C(3, -8)$.

56 Prove the result on slopes of perpendicular lines if the lines intersect at any point (a, b).

3.6 COMPOSITE AND INVERSE FUNCTIONS

Suppose X, Y, and Z are sets of real numbers, and let f be a function from X to Y and g a function from Y to Z. A function from X to Z may be defined in a natural way. For every x in X, the number $f(x)$ is in Y. Since the domain of g is Y, we may then find the image of $f(x)$ under g. Of course, this element of Z is written as $g(f(x))$. By associating $g(f(x))$ with x, we obtain a function from X to Z called the *composite function* of g by f. This is illustrated pictorially in Figure 3.43, where the dashes indicate the correspondence we have defined from X to Z. We sometimes use an operational symbol $\circ$ and denote this function by $g \circ f$. The following definition summarizes our discussion.

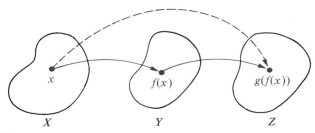

Figure 3.43

Definition

> If f is a function from X to Y and g is a function from Y to Z, then the **composite function** $g \circ f$ is the function from X to Z defined by
>
> $$(g \circ f)(x) = g(f(x))$$
>
> for every x in X.

Example 1 If $f(x) = x^3$ and $g(x) = 5x^2 + 2x + 1$, find $(g \circ f)(x)$.

Solution By definition,

$$(g \circ f)(x) = g(f(x)) = g(x^3).$$

Since $g(x^3)$ means that x^3 should be substituted for x in the expression for $g(x)$, we have

$$g(x^3) = 5(x^3)^2 + 2(x^3) + 1.$$

Consequently,

$$(g \circ f)(x) = 5x^6 + 2x^3 + 1.$$ ∎

In applying the definition of $g \circ f$, it is not essential that the domain of g be all of Y, but merely that it *contain* the range of f. In certain cases we may wish to restrict x to some subset of X so that $f(x)$ is in the domain of g as illustrated in the next example.

Example 2   If the functions f and g are defined by $f(x) = x - 2$ and $g(x) = 5x + \sqrt{x}$, find $(g \circ f)(x)$.

Solution   Formal substitutions give us the following:

$$
\begin{aligned}
(g \circ f)(x) &= g(f(x)) & \text{(definition of } g \circ f) \\
&= g(x - 2) & \text{(definition of } f) \\
&= 5(x - 2) + \sqrt{x - 2} & \text{(definition of } g) \\
&= 5x - 10 + \sqrt{x - 2} & \text{(simplifying)}.
\end{aligned}
$$

The domain X of f is the set of all real numbers; however, the last equality implies that $(g \circ f)(x)$ is a real number only if $x \geq 2$. Thus, when working with the composite function $g \circ f$ we must restrict x to the interval $[2, \infty)$.   ∎

Given f and g, it may also be possible to find $f(g(x))$. In this case we first obtain the image of x under g and then apply f to $g(x)$. This gives us a function called the *composite function* of f by g and denoted by $f \circ g$. Thus, by definition,

Composite Function $f \circ g$

$$
(f \circ g)(x) = f(g(x)).
$$

Example 3   If $f(x) = x^2 - 1$ and $g(x) = 3x + 5$, find $(f \circ g)(x)$ and $(g \circ f)(x)$.

Solution   We have the following identities:

$$
\begin{aligned}
(f \circ g)(x) &= f(g(x)) & \text{(definition of } f \circ g) \\
&= f(3x + 5) & \text{(definition of } g) \\
&= (3x + 5)^2 - 1 & \text{(definition of } f) \\
&= 9x^2 + 30x + 24 & \text{(simplifying)}.
\end{aligned}
$$

Similarly,

$$
\begin{aligned}
(g \circ f) &= g(f(x)) & \text{(definition of } g \circ f) \\
&= g(x^2 - 1) & \text{(definition of } f) \\
&= 3(x^2 - 1) + 5 & \text{(definition of } g) \\
&= 3x^2 + 2 & \text{(simplifying)}.
\end{aligned}
$$
∎

We see from Example 3 that $f(g(x))$ and $g(f(x))$ are not always the same, that is, $f \circ g \neq g \circ f$. In certain cases it may happen that equality *does* occur. Of major importance is the case in which $f(g(x))$ and $g(f(x))$ are not only identical, but both are equal to x. Needless to say, f and g must be very special functions in order for this to happen. In the following discussion we indicate the manner in which they will be restricted.

Suppose f is a *one-to-one function* with domain X and range Y. As mentioned in Section 3.3 this implies that each element of Y is the image of precisely one element of X. Another way of phrasing this is to say that *each element of Y can be written in one and only one way in the form $f(x)$, where x is in X*. We may then define a function g from Y to X by demanding that

$$g(f(x)) = x \quad \text{for every } x \text{ in } X.$$

This amounts to *reversing* the correspondence given by f. If f is represented geometrically by drawing arrows, as in (i) of Figure 3.44, then g can be represented by simply *reversing* these arrows, as illustrated in (ii) of the figure. It follows that g is a one-to-one function with domain Y and range X. (See Exercise 38.)

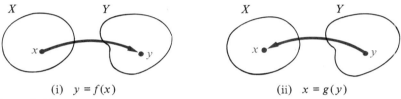

(i) $y = f(x)$ (ii) $x = g(y)$

Figure 3.44

As illustrated in Figure 3.44, if $f(x) = y$, then $x = g(y)$. This means that

$$f(g(y)) = y \quad \text{for every } y \text{ in } Y.$$

Since the notation used for the variable is immaterial, we may write

$$f(g(x)) = x \quad \text{for every } x \text{ in } Y.$$

The functions f and g are called *inverse functions* of one another, according to the following definition.

Definition

> If f is a one-to-one function with domain X and range Y, then a function g with domain Y and range X is called the **inverse function of f** if
>
> $$f(g(x)) = x \quad \text{for every } x \text{ in } Y \text{ and}$$
> $$g(f(x)) = x \quad \text{for every } x \text{ in } X.$$

The symbol f^{-1} is often used to denote the inverse function of f. Employing this notation,

> $$f^{-1}(f(x)) = x \quad \text{for every } x \text{ in } X$$
> $$f(f^{-1}(x)) = x \quad \text{for every } x \text{ in } Y.$$

The symbol -1 used here should not be mistaken for an exponent; that is, $f^{-1}(x)$ does not mean $1/f(x)$. The reciprocal $1/f(x)$ may be denoted by $[f(x)]^{-1}$.

Inverse functions are very important in the study of trigonometry. In Chapter Five we shall discuss two other special types of inverse functions. It is important to remember that in order to define the inverse of a function f *it is absolutely essential that f be one-to-one*. The most common examples of one-to-one functions are those which are either increasing or decreasing on their domains, for, in this case, if $a \neq b$ in the domain, then $f(a) \neq f(b)$ in the range.

An algebraic method can sometimes be used to find the inverse of a one-to-one function f with domain X and range Y. Given any x in X, its image y in Y may be found by means of the equation $y = f(x)$. In order to determine the inverse function f^{-1} we wish to *reverse* this procedure, in the sense that, given y, the element x may be found. Since x and y are related by means of $y = f(x)$, it follows that, if the latter equation can be solved for x in terms of y, we may arrive at the inverse function f^{-1}. This technique is illustrated in the following examples.

Example 4   If $f(x) = 3x - 5$ for every real number x, find the inverse function of f.

Solution   The graph of the given linear function f is a line of slope 3 (Why?) and hence f is increasing for all x. It follows that f is a one-to-one function with domain and range $\mathbb{R}$, and hence the inverse function g exists. If we let

$$y = 3x - 5$$

and then solve for x in terms of y, we get

$$x = \frac{y + 5}{3}.$$

The last equation enables us to find x when given y. Letting

$$g(y) = \frac{y + 5}{3}$$

gives us a function g from Y to X that reverses the correspondence determined by f. Since the symbol used for the independent variable is immaterial, we may replace y by x in the expression for g, obtaining

$$g(x) = \frac{x + 5}{3}.$$

To verify that g is actually the inverse function of f, we must verify that the two conditions $f(g(x)) = x$ and $g(f(x)) = x$ are fulfilled. Thus,

$$f(g(x)) = f\left(\frac{x + 5}{3}\right) \qquad \text{(definition of } g\text{)}$$

$$= 3\left(\frac{x + 5}{3}\right) - 5 \qquad \text{(definition of } f\text{)}$$

$$= x \qquad \text{(simplifying).}$$

Also,

$$g(f(x)) = g(3x - 5) \qquad \text{(definition of } f)$$

$$= \frac{(3x - 5) + 5}{3} \qquad \text{(definition of } g)$$

$$= x \qquad \text{(simplifying)}.$$

This proves that g is the inverse function of f. Using the f^{-1} notation,

$$f^{-1}(x) = \frac{x + 5}{3}. \qquad \blacksquare$$

Example 5 Find the inverse function of f if the domain X is the interval $[0, \infty)$ and $f(x) = x^2 - 3$ for all x in X.

Solution   The domain has been restricted so that f is increasing and hence is one-to-one. The range of f is the interval $[-3, \infty)$. As in Example 4 we begin by considering the equation

$$y = x^2 - 3.$$

Solving for x gives us

$$x = \pm\sqrt{y + 3}.$$

Since x is nonnegative we reject $x = -\sqrt{y + 3}$ and, as in the preceding example, we let

$$g(y) = \sqrt{y + 3} \quad \text{or equivalently,} \quad g(x) = \sqrt{x + 3}.$$

Since

$$f(g(x)) = f(\sqrt{x + 3}) = (\sqrt{x + 3})^2 - 3 = (x + 3) - 3 = x$$

and

$$g(f(x)) = g(x^2 - 3) = \sqrt{(x^2 - 3) + 3} = x$$

we see that

$$f^{-1}(x) = \sqrt{x + 3} \quad \text{where } x \geq -3. \qquad \blacksquare$$

There is an interesting relationship between the graphs of a function f and its inverse function f^{-1}. We first note that f maps a into b if and only if f^{-1} maps b into a; that is, $b = f(a)$ means the same thing as $a = f^{-1}(b)$. These equations imply that the point (a, b) is on the graph of f if and only if the point (b, a) is on the graph of f^{-1}. As an illustration, in Example 5 we found that the functions f and f^{-1} given by

$$f(x) = x^2 - 3 \quad \text{and} \quad f^{-1}(x) = \sqrt{x + 3}$$

are inverse functions of one another, provided x is suitably restricted. Some points on the graph of f are $(0, -3)$, $(1, -2)$, $(2, 1)$, and $(3, 6)$. Corresponding points on the graph of f^{-1} are $(-3, 0)$, $(-2, 1)$, $(1, 2)$, and $(6, 3)$. The graphs of f and f^{-1} are sketched on the same coordinate axes in Figure 3.45. If the page is folded along the line l which bisects quadrants I and III (as indicated by the dashes in the figure), then the graphs of f and f^{-1} coincide. The two graphs are said to be *reflections* of one another through the line l (or *symmetric* with respect to l). This is typical of the graph of every function f that has an inverse function f^{-1}.

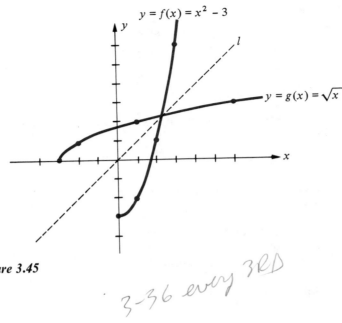

Figure 3.45

3-36 every 3RD

EXERCISES 3.6

In Exercises 1–18 find $(f \circ g)(x)$ and $(g \circ f)(x)$.

1 $f(x) = 3x + 2, g(x) = 2x - 1$

2 $f(x) = x + 4, g(x) = 5x - 3$

3 $f(x) = 4x^2 - 5, g(x) = 3x$

4 $f(x) = 7x + 1, g(x) = 2x^2$

5 $f(x) = 3x^2 + 2x, g(x) = 2x - 1$

6 $f(x) = 5x + 7, g(x) = 4 - x^2$

7 $f(x) = x - 1, g(x) = x^3$

8 $f(x) = x^3 + 5x, g(x) = 4x$

9 $f(x) = x^2 + 9x, g(x) = \sqrt{x + 9}$

10 $f(x) = \sqrt[3]{x^2 + 1}, g(x) = x^3 + 1$

11 $f(x) = \dfrac{1}{2x - 5}, g(x) = 1/x^2$

12 $f(x) = \dfrac{x}{x + 1}, g(x) = x - 1$

13 $f(x) = |x|, g(x) = -5$

14 $f(x) = 7, g(x) = 10$

15 $f(x) = x^2, g(x) = 1/x^2$

16 $f(x) = \dfrac{1}{x + 1}, g(x) = x + 1$

17 $f(x) = 2x - 3, g(x) = \dfrac{x + 3}{2}$

18 $f(x) = x^3 - 1, g(x) = \sqrt[3]{x + 1}$

In each of Exercises 19–22 prove that f and g are inverse functions of one another and sketch the graphs of f and g on the same coordinate plane.

19 $f(x) = 7x + 5; g(x) = (x - 5)/7$

20 $f(x) = x^2 - 1, x \geq 0; g(x) = \sqrt{x + 1}, x \geq -1$

21 $f(x) = \sqrt{2x - 4}, x \geq 2; g(x) = (x^2 + 4)/2, x \geq 0$

22 $f(x) = x^3 + 1; g(x) = \sqrt[3]{x - 1}$

In Exercises 23–36 find the inverse function of f.

23 $f(x) = 4x - 3$ **24** $f(x) = 9 - 7x$

25 $f(x) = \dfrac{1}{2x + 5}, x > -5/2$

26 $f(x) = \dfrac{1}{3x - 1}, x > 1/3$

27 $f(x) = 9 - x^2, x \geq 0$

28 $f(x) = 4x^2 + 1, x \geq 0$

29 $f(x) = 5x^3 - 2$ **30** $f(x) = 7 - 2x^3$

31 $f(x) = \sqrt{3x - 5}, x \geq 5/3$

32 $f(x) = \sqrt{4 - x^2}, 0 \leq x \leq 2$

33 $f(x) = \sqrt[3]{x} + 8$ **34** $f(x) = (x^3 + 1)^5$

35 $f(x) = x$ **36** $f(x) = -x$

37 (a) Prove that the function f defined by $f(x) = ax + b$, where $a \neq 0$, has an inverse function, and find $f^{-1}(x)$.

(b) Does a constant function have an inverse? Explain.

38 If f is a one-to-one function with domain X and range Y, prove that f^{-1} is a one-to-one function with domain Y and range X.

39 Prove that a one-to-one function can have at most one inverse function.

40 If $f(x) = ax + b$ and $g(x) = cx + d$, find conditions on c and d in terms of a and b which will guarantee that $f \circ g = g \circ f$. Discuss the case where $a = b = 1$.

3.7 VARIATION

We shall next introduce terminology which is used in science to describe relationships among variable quantities. In the following definition the letters u, v and w are variables. The domains and ranges of the variables will not be specified. In any particular problem they should be evident.

Definition

   (i) The phrase **u varies directly as v**, or **u is directly proportional to v**, means that $u = kv$ for some real number k.

   (ii) If n is a positive real number, then the phrase **u varies directly as the nth power of v**, or **u is directly proportional to the nth power of v**, means that $u = kv^n$ for some real number k.

   (iii) The phrase **u varies inversely as v**, or **u is inversely proportional to v**, means that $u = k/v$, where k is a real number. If $u = k/v^n$ for some positive real number n, then **u varies inversely as the nth power of v**.

   (iv) The phrase **w varies jointly as u and v** means $w = kuv$ for some real number k. If $w = ku^n v^m$ for positive real numbers n and m, then **w varies jointly as the nth power of u and the mth power of v**.

The number k is sometimes called the **constant of variation** or the **constant of proportionality**. For example, if an automobile is moving at a rate of 50 miles per hours, then the distance d it travels in t hours is given by $d = 50t$. Hence, as in (i), the distance d is directly proportional to the time t and the constant of proportionality is 50.

To illustrate (ii) of the definition, the formula $A = \pi r^2$ for the area of a circle states that the area A varies directly as the square of the radius r. The constant of proportionality is π. As another illustration, the formula $V = \frac{4}{3}\pi r^3$ for the volume of a sphere of radius r states that the volume V is directly proportional to the cube of the radius. The constant of proportionality in this case is $\frac{4}{3}\pi$.

Note that in direct variation (i) the dependent variable increases numerically as the independent variable increases, whereas in inverse variation (iii) the absolute value of the dependent variable decreases as the independent variable increases.

In many applied problems the constant of proportionality can be determined by examining experimental facts, as illustrated in the following example.

Example 1   If the temperature remains constant, then the pressure of an enclosed gas is inversely proportional to the volume. The pressure of a certain gas within a spherical balloon of radius 9 inches is 20 pounds per square inch. If the radius of the balloon increases to 12 inches, find the new pressure of the gas.

Solution   The original volume is $\frac{4}{3}\pi(9)^3 = 972\pi$ cubic inches. If we denote the pressure by P and the volume by V, then by (iii) of the definition,

$$P = \frac{k}{V}$$

for some real number k. Since $P = 20$ when $V = 972\pi$,

$$20 = \frac{k}{972\pi}$$

and hence, $k = 20(972\pi) = 19440\pi$. Consequently, a formula for P is

$$P = \frac{19440\pi}{V}.$$

If the radius is 12 inches, then $V = \frac{4}{3}(12)^3\pi = 2304\pi$ cubic inches. Substituting this number for V in the previous equation gives us

$$P = \frac{19440\pi}{2304\pi} = \frac{135}{16} = 8.4375.$$

Thus, the pressure is 8.4375 pounds per square inch when the radius is 12 inches.

∎

Combinations of the types of variation may occur. It is inconvenient to discuss all possible situations, but perhaps the following illustration will suffice. If a variable s varies jointly as u and the cube of v and inversely as the square of w, then

$$s = k\,\frac{uv^3}{w^2}$$

where k is some real number.

Example 2   The weight that can be safely supported by a beam with a rectangular cross section varies jointly as the width and square of the depth of the cross section and inversely as the length of the beam. If a 2-inch by 4-inch beam which is 8 feet long safely supports a load of 500 pounds, what weight can be safely supported by a 2-inch by 8-inch beam which is 10 feet long? (Assume that the width is the *shorter* dimension of the cross section.)

Solution   If the width, depth, length, and weight are denoted by w, d, l, and W, respectively, then

$$W = k\,\frac{wd^2}{l}.$$

According to the given data,

$$500 = k\,\frac{2(4^2)}{8}.$$

Solving for k we obtain $k = 125$, and hence the formula for W is

$$W = 125\left(\frac{wd^2}{l}\right).$$

To answer the question we substitute $w = 2$, $d = 8$, and $l = 10$, obtaining

$$W = 125\left(\frac{2 \cdot 8^2}{10}\right) = 1600 \text{ pounds.} \qquad \blacksquare$$

EXERCISES 3.7

In Exercises 1–10 express each statement as a formula and determine the constant of proportionality from the given conditions.

1 a is directly proportional to v. If $v = 30$, then $a = 12$.

2 s varies directly as t. If $t = 10$, then $s = 18$.

3 r varies directly as s and inversely as t. If $s = -2$ and $t = 4$, then $r = 7$.

4 w varies directly as z and inversely as the square root of u. If $z = 2$ and $u = 9$, then $w = 6$.

5 y is directly proportional to the square of x and inversely proportional to the cube of z. If $x = 5$ and $z = 3$, then $y = 25$.

6 q is inversely proportional to the sum of x and y. If $x = 0.5$ and $y = 0.7$, then $q = 1.4$.

7 c varies jointly as the square of a and the cube of b. If $a = 7$ and $b = -2$, then $c = 16$.

8 r varies jointly as s and v and inversely as the cube of p. If $s = 2$, $v = 3$, and $p = 5$, then $r = 40$.

9 F varies jointly as m_1 and m_2 and inversely as the square of d. If $m_1 = 100$, $m_2 = 50$, and $d = 25$, then $F = 225$.

10 z is directly proportional to the sum of the squares of x and y. If $x = 4$ and $y = 1$, then $z = 50$.

11 The pressure acting at a point in a liquid is directly proportional to the distance from the surface of the liquid to the point. In a certain oil tank the pressure at a depth of 2 feet is 118 pounds per square foot. Find the pressure at a depth of 5 feet.

12 The volume V of a gas varies directly as the temperature T and inversely as the pressure P. The temperature, pressure, and volume of a certain gas are 240°, 20 pounds per square inch, and 30 cubic feet, respectively. If the temperature drops 40° and the pressure increases by 2 pounds per square inch, find the new volume.

13 Hooke's Law states that the force required to stretch a spring x units beyond its natural length is directly proportional to x. If a weight of 4 pounds stretches a spring from its natural length of 10 inches to a length of 10.3 inches, what weight will stretch it to a length of 11.5 inches?

14 If an object is dropped from a position near the surface of the earth, then the distance s that it falls in t seconds varies as the square of t. If $s = 36$ when $t = 1.5$, find a formula for s as a function of t. If an object is dropped from a height of 1,000 feet, how long does it take to reach the ground?

15 The volume V of a gas varies directly as the temperature T and inversely as the pressure P. Find a formula for V in terms of T and P, if $V = 50$ when $P = 40$ and $T = 300$. What is V when $T = 320$ and $P = 30$?

16 The kinetic energy K of an object in motion varies jointly as its mass m and the square of its velocity v. If an object weighing 25 pounds and moving with a velocity of 90 feet per second has a kinetic energy of 400 foot-pounds, find its kinetic energy when the velocity is 130 feet per second.

17 The electrical resistance of a wire varies directly as its length and inversely as the square of its diameter. If a wire 100 feet long of diameter 0.01 inches has a resistance of 25 ohms, find the resistance in a wire made of the same material which has a diameter of 0.015 inches and is 50 feet long.

18 The intensity of illumination I from a source of light varies inversely as the square of the distance d from the source. If a searchlight has an intensity of 1,000,000 candlepower at 50 feet, what is the intensity at a distance of 1 mile?

19 The period of a simple pendulum, that is, the time required for one complete oscillation, varies directly as the square root of its length. If a pendulum 2 feet long has a period of 1.5 seconds, find the period of a pendulum 6 feet long.

20 The distance a ball rolls down an inclined plane varies as the square of the time. If a ball rolls 5 feet in the first second, how far will it roll in 4 seconds?

3.8 REVIEW

Concepts

Define or discuss each of the following.

1 Ordered pair

2 Rectangular coordinate system in a plane

3 Coordinate axes

4 Quadrants

5 The abscissa and ordinate of a point

6 The distance formula

7 The midpoint formula

8 The graph of an equation in x and y

9 Tests for symmetry

10 The standard equation of a circle

11 Unit circle

12 Function

13 The domain and range of a function

14 The graph of a function

15 Increasing function

16 Decreasing function

17 One-to-one function

18 Identity function

19 Constant function

20 Even function

21 Odd function

22 Vertical shifts of graphs

23 Horizontal shifts of graphs

24 Stretching of graphs

25 Reflections of graphs

26 Linear function

27 Slope of a line

28 The Point-Slope Form

29 The Slope-Intercept Form

30 Linear equation in x and y

31 The composite function of two functions

32 The inverse of a function

33 Direct variation

34 Inverse variation

Exercises

1 Plot the points $A(3, 1)$, $B(-5, -3)$, and $C(4, -1)$ and prove that they are vertices of a right triangle. What is the area of the triangle?

2 Given the points $P(-5, 9)$ and $Q(-8, -7)$, (a) find the midpoint of the segment PQ (b) find a point T such that Q is the midpoint of PT.

3 Describe the set of all points (x, y) in a coordinate plane such that $xy < 0$.

4 Find the slope of the line through $C(11, -5)$ and $D(-8, 6)$.

5 Prove that the points $A(-3, 1)$, $B(1, -1)$, $C(4, 1)$, and $D(3, 5)$ are vertices of a trapezoid.

6 Find an equation of the circle with center $C(7, -4)$, and passing through the point $Q(-3, 3)$.

7 Find an equation of the circle with center $C(-5, -1)$ which is tangent to the line $x = 4$.

8 Express the equation $8x + 3y - 24 = 0$ in Slope-Intercept Form.

9 Find an equation of the line through $A(1/2, -1/3)$ that is (a) parallel to the line $6x + 2y + 5 = 0$ and (b) perpendicular to the line $6x + 2y + 5 = 0$.

10 Find an equation of the line having x-intercept -3 and passing through the center of the circle $(x - 2)^2 + (y + 5)^2 = 3$.

Sketch the graphs of the equations in Exercises 11–19.

11 $2y + 5x - 8 = 0$

12 $x = 3y + 4$

13 $x + 5 = 0$

14 $2y - 7 = 0$

15 $y = \sqrt{1 - x}$

16 $y - x^2 = 1$

17 $y^2 - x = 1$

18 $y^2 = 16 - x^2$

19 $x^2 + y^2 + 4x - 16y + 64 = 0$

In each of Exercises 20–26 sketch the graph of f and determine the intervals in which f is increasing or decreasing.

20 $f(x) = |x + 3|$

21 $f(x) = \dfrac{1 - 3x}{2}$

22 $f(x) = \sqrt{2 - x}$

23 $f(x) = 1 - \sqrt{x + 1}$

24 $f(x) = 9 - x^2$

25 $f(x) = 1000$

26 $f(x) = \begin{cases} x^2 & \text{if } x < 0 \\ 3x & \text{if } 0 \le x < 2 \\ 6 & \text{if } x \ge 2 \end{cases}$

27 If $f(x) = x/\sqrt{x + 3}$ find:

(a) $f(1)$ (b) $f(-1)$

(c) $f(0)$ (d) $f(-x)$

(e) $-f(x)$ (f) $f(x^2)$

(g) $(f(x))^2$

28 Find the domain and range of f if:

(a) $f(x) = \sqrt{3x - 4}$

(b) $f(x) = 1/(x + 3)^2$

29 Sketch the graphs of the following, using results on shifting, stretching, or relecting.

(a) $y = \sqrt{x}$ (b) $y = \sqrt{x + 4}$

(c) $y = \sqrt{x} + 4$ (d) $y = 4\sqrt{x}$

(e) $y = \frac{1}{4}\sqrt{x}$ (f) $y = -\sqrt{x}$

30 Determine whether f is even, odd, or neither even nor odd.

(a) $f(x) = \sqrt[3]{x^3 + 4x}$

(b) $f(x) = \sqrt[3]{3x^2 - x^3}$

(c) $f(x) = \sqrt[3]{x^4 + 3x^2 + 5}$

In Exercises 31 and 32 find $(f \circ g)(x)$ and $(g \circ f)(x)$.

31 $f(x) = 2x^2 - 5x + 1, g(x) = 3x + 2$

32 $f(x) = \sqrt{3x + 2}, g(x) = 1/x^2$

In Exercises 33 and 34 find $f^{-1}(x)$ and sketch the graphs of f and f^{-1} on the same coordinate plane.

33 $f(x) = 10 - 15x$

34 $f(x) = 9 - 2x^2, \quad x \leq 0$

35 If the altitude and radius of a right circular cylinder are equal, express the volume V as a function of the circumference C of the base.

36 Find a formula which expresses the fact that w varies jointly as x and the square of y, and inversely as z, if $w = 30$ when $x = 3$, $y = 2$, and $z = 5$.

POLYNOMIAL FUNCTIONS, RATIONAL FUNCTIONS, AND CONIC SECTIONS

Polynomial functions are the most basic functions in algebra, since they are defined only in terms of additions and multiplications of real numbers and variables. Techniques for sketching their graphs are discussed in the first two sections. Next we consider quotients of polynomial functions, that is, rational functions. The chapter concludes with a brief survey of graphs of conic sections.

4.1 QUADRATIC FUNCTIONS

Among the most important functions in mathematics are those defined as follows.

Definition

A function f is a **polynomial function** if

$$f(x) = a_n x^n + a_{n-1} x^{n-1} + \cdots + a_1 x + a_0$$

where the coefficients $a_0, a_1, \ldots, a_n$ are real numbers and the exponents are nonnegative integers.

According to our work in Section 1.7, f is a polynomial function if $f(x)$ is a polynomial, and we say that f has **degree n** if $f(x)$ has degree n. Note that a polynomial function of degree 1 is a linear function. If the degree is 2, then, as in the next definition, f is called a *quadratic function*.

Definition

A function f is a **quadratic function** if

$$f(x) = ax^2 + bx + c$$

where a, b, and c are real numbers and $a \neq 0$.

If $b = c = 0$, then $f(x) = ax^2$ and the graph is a parabola with vertex at the origin, opening upward if $a > 0$ or downward if $a < 0$ (see Figures 3.30 and 3.31). If $b = 0$ and $c \neq 0$, then

$$f(x) = ax^2 + c$$

and from the discussion of vertical shifts in Section 3.4, the graph is a parabola with vertex at the point $(0, c)$ on the y-axis (see for example, Figure 3.27).

Example 1   Sketch the graph of f if $f(x) = -\frac{1}{2}x^2 + 4$.

Solution   The graph $f(x) = -\frac{1}{2}x^2$ is similar to the graph of $f(x) = -x^2$ sketched in Figure 3.31, but is somewhat wider. The graph of $f(x) = -\frac{1}{2}x^2 + 4$ may be found by shifting the graph of $f(x) = -\frac{1}{2}x^2$ four units upward. Coordinates of several points on the graph are listed in the following table.

x	0	1	2	$\sqrt{8}$	3
$f(x)$	4	$\frac{7}{2}$	2	0	$-\frac{1}{2}$

Plotting and using symmetry with respect to the y-axis gives us the sketch in Figure 4.1.

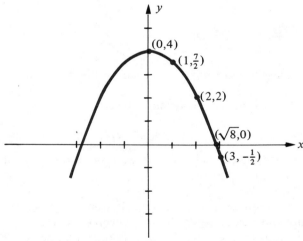

Figure 4.1 $f(x) = -\frac{1}{2}x^2 + 4$

If $f(x) = ax^2 + bx + c$ and $b \neq 0$, then the technique of completing the square can be used to change the form of $f(x)$ to

$$f(x) = a(x - h)^2 + k$$

where h and k are real numbers. As we shall see, the graph will be similar to that of $f(x) = ax^2$.

Example 2   Express $f(x)$ in the form $a(x - h)^2 + k$ if:

(a) $f(x) = x^2 - 6x - 5$ (b) $f(x) = 3x^2 + 24x + 50$

Solution

(a)   We begin by writing

$$f(x) = (x^2 - 6x \quad) - 5.$$

We next complete the square by determining the number to add to the expression within the parentheses so that it takes on the form $(x - h)^2$ for some h. *This number is always the square of one-half the coefficient of x.* In this example the coefficient of x is -6, and consequently the square of one-half of -6 is $(-3)^2$, or 9. Adding 9 to $x^2 - 6x$ gives $x^2 - 6x + 9$, or $(x - 3)^2$. Of course, if we add 9 to $f(x)$, then we must also subtract 9, so that $f(x)$ does not change. Thus,

$$\begin{aligned} f(x) &= (x^2 - 6x \quad) - 5 \\ &= (x^2 - 6x + 9) - 5 - 9 \\ &= (x - 3)^2 - 14 \end{aligned}$$

which has the desired form with $a = 1$, $h = 3$, and $k = -14$.

(b)   Given $f(x) = 3x^2 + 24x + 50$, we begin by removing a factor 3 from the first two terms as follows:

$$f(x) = 3(x^2 + 8x \quad) + 50.$$

To complete the square for the expression in parentheses we add the square of one-half the coefficient of x, that is, $(8/2)^2$, or 16. However, if we add 16 to the expression within parentheses, then because of the factor 3 we are actually adding 48 to $f(x)$. Hence, we must compensate by subtracting 48 as follows:

$$\begin{aligned} f(x) &= 3(x^2 + 8x \quad) + 50 \\ &= 3(x^2 + 8x + 16) + 50 - 48 \\ &= 3(x + 4)^2 + 2 \end{aligned}$$

which has proper form with $a = 3$, $h = -4$, and $k = 2$.   ∎

From the discussion in Section 3.4, the graph of $y = a(x - h)^2$ can be obtained by shifting the graph of $y = ax^2$ to the right or left, depending on whether h is positive or negative, respectively. Thus, the graph of $y = a(x - h)^2$ is a parabola with vertex $(h, 0)$ and axis parallel to the y-axis (see, for example, Figure 3.28). Since the graph of

$$y = a(x - h)^2 + k$$

can be obtained by shifting the graph of $y = a(x - h)^2$ upward or downward $|k|$ units, according as k is positive or negative, the graph is a parabola with vertex (h, k), axis parallel to the y-axis, and opening upward if $a > 0$, or downward if $a < 0$. This gives us the following:

**Standard
Equation of
a Parabola
(Vertical
Axis)**

$$y - k = a(x - h)^2.$$

Since the vertex (h, k) is the highest or lowest point of the parabola, it follows that $f(x) = a(x - h)^2 + k$ has its *maximum or minimum value at h*. This value is $f(h) = k$.

Example 3   Sketch the graph of $f(x) = 2x^2 - 6x + 4$ and find the minimum value of $f(x)$.

Solution   The graph of f is the same as the graph of the equation $y = 2x^2 - 6x + 4$. From the preceding discussion we know that the graph is a parabola, and we begin by completing the square as follows:

$$\begin{aligned} y &= 2x^2 - 6x + 4 \\ &= 2(x^2 - 3x \quad\;\;) + 4 \\ &= 2\left(x^2 - 3x + \frac{9}{4}\right) + \left(4 - \frac{9}{2}\right) \\ &= 2\left(x - \frac{3}{2}\right)^2 - \frac{1}{2}. \end{aligned}$$

If we write the last equation as

$$y + \frac{1}{2} = 2\left(x - \frac{3}{2}\right)^2$$

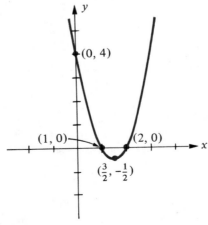

Figure 4.2 $f(x) = 2x^2 - 6x + 4$

$$y = 2(x^2 - 3x) + y$$

$$y = \frac{1}{2} \qquad y + \frac{9}{2} + y$$

and compare with the standard equation of a parabola we see that $h = 3/2$ and $k = -1/2$. Consequently, the vertex (h, k) of the parabola is $(3/2, -1/2)$. Since $a = 2 > 0$, the parabola opens upward. The y-intercept is $f(0) = 4$. To find the x-intercepts we solve $2x^2 - 6x + 4 = 0$, or the equivalent equation $(2x - 2)(x - 2) = 0$, obtaining $x = 2$ and $x = 1$. The vertex and the x- and y-intercepts are sufficient for a reasonably accurate sketch (see Figure 4.2). The minimum value of f occurs at the vertex, and hence is $f(3/2) = -1/2$. ■

Example 4 Sketch the graph of f if $f(x) = 8 - 2x - x^2$, and find the maximum value of $f(x)$.

Solution We know from the discussion in this section that the graph is a parabola. Let us set $y = f(x)$ and find the vertex by completing the square as follows:

$$\begin{aligned} y &= -x^2 - 2x + 8 \\ &= -(x^2 + 2x) + 8 \\ &= -(x^2 + 2x + 1) + 8 + 1 \\ &= -(x + 1)^2 + 9. \end{aligned}$$

If we now write

$$y - 9 = -(x + 1)^2$$

and compare with the standard equation we see that $h = -1$, $k = 9$, and hence the vertex is $(-1, 9)$. Since $a = -1 < 0$, the parabola opens downward. To find the x-intercepts we solve the equation $8 - 2x - x^2 = 0$. Either by factoring or by using the quadratic formula we obtain $x = -4$ and $x = 2$. The y-intercept is 8. Using this information gives us the sketch in Figure 4.3. The maximum value of f occurs at the vertex and hence is $f(-1) = 9$.

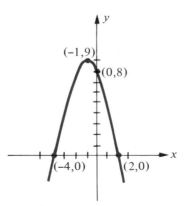

Figure 4.3 $y = 8 - 2x - x^2$ ■

Certain applied problems may be solved by finding maximum or minimum values of quadratic functions. The next example gives one illustration.

Example 5   A long rectangular sheet of metal, 12 inches wide, is to be made into a rain gutter by turning up two sides so that they are perpendicular to the sheet. How many inches should be turned up to give the gutter its greatest capacity?

Solution   The gutter is illustrated in Figure 4.4, where, if x denotes the number of inches turned up on each side, then the width of the base of the gutter is $12 - 2x$ inches (Why?). The capacity will be greatest when the area of the rectangle with sides of length x and $12 - 2x$ has its greatest value. Letting $f(x)$ denote this area we have

$$f(x) = x(12 - 2x) = 12x - 2x^2 = -2x^2 + 12x.$$

We may find the maximum value of $f(x)$ by completing the square as follows:

$$\begin{aligned} f(x) &= -2(x^2 - 6x) \\ &= -2(x^2 - 6x + 9) + 18 \\ &= -2(x - 3)^2 + 18. \end{aligned}$$

The maximum value of $f(x)$ occurs at $x = 3$ (Why?). Hence, 3 inches should be turned up to achieve maximum capacity.

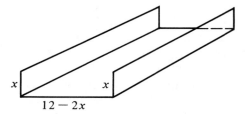

Figure 4.4 $12 - 2x$ ■

Parabolas (and hence quadratic functions) are very useful in applications of mathematics to the physical world. It can be shown that if a projectile is fired and is acted upon only by the force of gravity—that is, if air resistance and other outside factors are inconsequential—then the path of the projectile is a parabola. Properties of parabolas are used in the design of mirrors for telescopes and searchlights. They are also employed in the design of field microphones used in television broadcasts of football games. These are only a few of many physical applications.

EXERCISES 4.1

1 Given $f(x) = ax^2 + 2$, sketch the graph of f if:

(a) $a = 2$ (b) $a = 5$
(c) $a = \frac{1}{2}$ (d) $a = -3$

2 Given $f(x) = 4x^2 + c$, sketch the graph of f if:

(a) $c = 2$ (b) $c = 5$
(c) $c = \frac{1}{2}$ (d) $c = -3$

In each of Exercises 3–6 sketch the graph of f and find the vertex.

3 $f(x) = 4x^2 - 9$ **4** $f(x) = 2x^2 + 3$

5 $f(x) = 9 - 4x^2$ **6** $f(x) = 16 - 9x^2$

In Exercises 7–14 use the Quadratic Formula to find all real numbers a such that $f(a) = 0$.

7 $f(x) = 4x^2 - 11x - 3$

8 $f(x) = 25x^2 + 10x + 1$

9 $f(x) = 9x^2 - 12x + 4$

10 $f(x) = 10x^2 + x - 21$

11 $f(x) = 2x^2 + x - 5$

12 $f(x) = 2x^2 + 4x - 3$

13 $f(x) = \dfrac{2x+1}{x-3} - \dfrac{x-1}{x+2}$

14 $f(x) = 9x^2 + 13x$

In Exercises 15–20 use the technique of completing the square to express $f(x)$ in the form $a(x - h)^2 + k$ and find the maximum or minimum value of $f(x)$.

15 $f(x) = 2x^2 - 16x + 23$

16 $f(x) = 3x^2 - 12x + 7$

17 $f(x) = -5x^2 - 10x + 3$

18 $f(x) = -2x^2 - 12x - 12$

19 $f(x) = 4x^2 - 4x + 2$

20 $f(x) = 9x^2 + 12x - 8$

In each of Exercises 21–30 sketch the graph of f and find the maximum or minimum value of $f(x)$.

21 $f(x) = x^2 + 5x + 4$ **22** $f(x) = x^2 - 6x$

23 $f(x) = 8x - 12 - x^2$

24 $f(x) = 10 + 3x - x^2$

25 $f(x) = x^2 + x + 3$ **26** $f(x) = x^2 + 2x + 5$

27 $f(x) = 3x^2 - 12x + 16$

28 $f(x) = 2x^2 + 12x + 13$

29 $f(x) = -5x^2 - 10x - 4$

30 $f(x) = -3x^2 + 24x - 42$

31 If $f(x) = x^2 + kx - 5 + 3k$, find two different values for k such that f has only one zero.

32 If $f(x) = 3x^2 + kx - 4k$, find all values of k such that f has no real zeros.

33 If one zero of $f(x) = kx^2 + 3x + k$ is -2, find the other zero.

34 Find a real number k such that the graph of $y = x^2 + kx + 3$ contains the point $P(-1, -2)$.

35 Find two real numbers whose difference is 40 and whose product is a minimum. (*Hint*: Let x and $x - 40$ denote the numbers.)

36 Find two positive real numbers whose sum is 40 and whose product is a maximum.

37 A piece of wire 24 inches long is bent into the shape of a rectangle. Prove that the enclosed area will be greatest if the rectangle is a square.

38 A person standing on the top of a building projects an object directly upward with a velocity of 144 ft/sec. Its height $s(t)$ in feet above the ground after t seconds is given by $s(t) = -16t^2 + 144t + 100$. What is its maximum height? What is the height of the building?

39 A man wishes to put a fence around a rectangular field and then subdivide this field into three smaller rectangular plots by placing two fences parallel to one of the sides. If he can afford only 1000 yd of fencing, what dimensions will give him the maximum area?

40 A manufacturer sells a certain article to dealers at a rate of $20 each if less than 50 are ordered. If 50 or more are ordered (up to 600) the price per article is reduced at a rate of 2 cents times the number ordered. What size order will produce the maximum amount of money for the manufacturer?

4.2 GRAPHS OF POLYNOMIAL FUNCTIONS OF DEGREE GREATER THAN 2

Suppose f is a polynomial function of degree n, that is,

$$f(x) = a_n x^n + a_{n-1} x^{n-1} + \cdots + a_1 x + a_0$$

where $a_n \neq 0$. If all the coefficients except a_n are zero, then we may write

$$f(x) = ax^n, \quad \text{where } a = a_n \neq 0.$$

In this case if $n = 1$, the graph of f is a line passing through the origin, whereas, if $n = 2$, the graph is a parabola with vertex at the origin. The special case in which $a = 1$ and $n = 3$, that is, $f(x) = x^3$, is sketched in Figure 3.13. Several other illustrations with $n = 3$ are given in the next example.

Example 1   Sketch the graph of f if   (a) $f(x) = \frac{1}{2}x^3$ (b) $f(x) = -\frac{1}{2}x^3$.

Solution

(a)   The table below gives several points on the graph.

x	0	$\frac{1}{2}$	1	$\frac{3}{2}$	2	$\frac{5}{2}$
$f(x)$	0	$\frac{1}{16}$	$\frac{1}{2}$	$\frac{27}{16}$	4	$\frac{125}{16} \approx 7.8$

Since f is an odd function, the graph of f is symmetric with respect to the origin (see Section 3.4) and hence the points $(-\frac{1}{2}, -\frac{1}{16})$, $(-1, -\frac{1}{2})$, etc., are also on the graph. The graph is sketched in (i) of Figure 4.5.

(b)   If $f(x) = -\frac{1}{2}x^3$, the graph can be obtained from that in part (a) by multiplying all ordinates by -1. This amounts to reflecting the graph through the x-axis, as shown in (ii) of Figure 4.5.

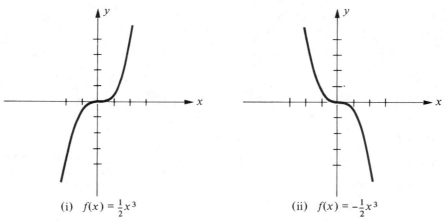

(i) $f(x) = \frac{1}{2}x^3$ (ii) $f(x) = -\frac{1}{2}x^3$

Figure 4.5   ∎

In general, if $f(x) = ax^3$, then increasing the numerical value of the coefficient a causes the graph to rise or fall more sharply. For example, if $f(x) = 10x^3$, then $f(1) = 10$, $f(2) = 80$, and $f(-2) = -80$. The effect of using a larger exponent and holding a fixed as, for example, in $f(x) = \frac{1}{2}x^5$, also leads to a graph which rises or falls more rapidly.

If n is an *even* integer, then the graph of $f(x) = ax^n$ is symmetric with respect to the y-axis. (Why?) Two illustrations (with $a = 1$) are given in (i) and (ii) of Figure 4.6. Note that as the exponent increases, the graph becomes flatter at the origin. It also rises (or falls) more rapidly if we let x increase through positive (or negative) values. Using a negative coefficient such as -1 inverts the graphs.

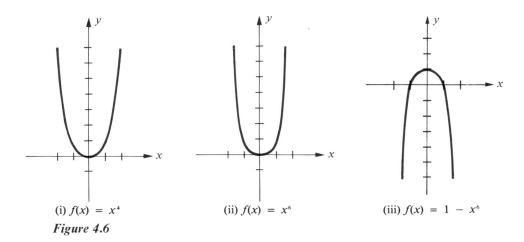

(i) $f(x) = x^4$ (ii) $f(x) = x^6$ (iii) $f(x) = 1 - x^6$

Figure 4.6

As usual, the graph of $f(x) = ax^n + c$, where $c \neq 0$, can be found by raising or lowering the graph of $f(x) = ax^n$. For example, if $f(x) = 1 - x^6$, we obtain the sketch in (iii) of Figure 4.6.

A complete analysis of graphs of polynomial functions of degree greater than 2 requires methods studied in calculus. As the degree increases, the graphs usually become more complicated. However, they always have a smooth appearance with a number of "hills" and "valleys," as illustrated in Figure 4.7. This number may be large if the degree is large, but it may also be small, as illustrated in Figures 4.5 and 4.6.

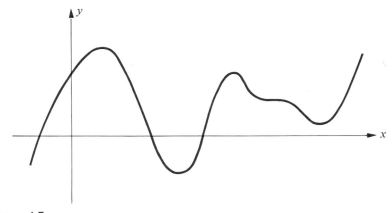

Figure 4.7

A crude method for obtaining a rough sketch is to plot many points and fit a curve to the resulting configuration. This is usually an extremely tedious procedure. In arriving at graphs in this way we use a property of polynomial functions called **continuity**. Continuity, which is studied extensively in calculus, implies that a small change in x produces a small change in the functional value $f(x)$. The next theorem specifies another important property of polynomial functions. The proof requires advanced mathematical methods.

The Intermediate Value Theorem for Polynomial Functions

> If f is a polynomial function and $f(a) \neq f(b)$, where $a < b$, then f takes on every value between $f(a)$ and $f(b)$ in the interval $[a, b]$.

The Intermediate Value Theorem states that if w is any number between $f(a)$ and $f(b)$, then there is a number c between a and b such that $f(c) = w$. If the graph of the polynomial function f is regarded as extending in an unbroken manner from the point $(a, f(a))$ to the point $(b, f(b))$, as illustrated in Figure 4.8, then for any number w between $f(a)$ and $f(b)$ it appears that a horizontal line with y-intercept w should intersect the graph in at least one point P. The abscissa c of P is a number such that $f(c) = w$.

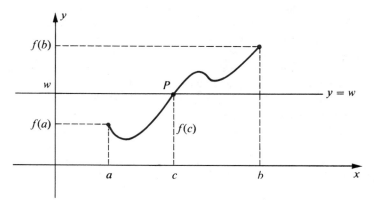

Figure 4.8

A corollary to the Intermediate Value Theorem is that if $f(a)$ and $f(b)$ have opposite signs, then there is a number c between a and b such that $f(c) = 0$; that is, f has a zero at c. Geometrically, this implies that if the point $(a, f(a))$ on the graph of a polynomial function lies below the x-axis, and the point $(b, f(b))$ lies above the x-axis, or vice versa, then the graph crosses the x-axis at least once between the points $(a, 0)$ and $(b, 0)$. A by-product of this fact is that if c and d are *successive* zeros of $f(x)$, that is, there are no other zeros between c and d, then $f(x)$ *does not change sign on the interval* (c, d). Thus, if we choose any number k such that $c < k < d$, and if $f(k)$ is positive, then $f(x)$ is positive throughout (c, d). Similarly, if $f(k)$ is negative, then $f(x)$ is negative throughout (c, d). We shall call $f(k)$ a **test value** for $f(x)$ on the interval (c, d). Test values may also be used on infinite intervals of the form $(-\infty, a)$ or (a, ∞), provided $f(x)$ has no zeros on these intervals. The use of test values in graphing is similar to that used for inequalities in Section 2.6.

Example 2 If $f(x) = x^3 + x^2 - 4x - 4$, determine all values of x such that $f(x) > 0$, and all values of x such that $f(x) < 0$. Use this information as an aid in sketching the graph of f.

Solution   The geometric significance of the stated inequalities is that the graph lies above the x-axis for values of x such that $f(x) > 0$, whereas the graph lies below the x-axis if $f(x) < 0$. We may factor $f(x)$ by grouping terms as follows:

$$\begin{aligned}
f(x) &= (x^3 + x^2) - (4x + 4) \\
&= x^2(x + 1) - 4(x + 1) \\
&= (x^2 - 4)(x + 1) \\
&= (x + 2)(x - 2)(x + 1).
\end{aligned}$$

It follows that the zeros of $f(x)$ are -2, -1, and 2. The corresponding points on the graph (see Figure 4.9) divide the x-axis into four parts, and we consider the intervals

$$(-\infty, -2), \quad (-2, -1), \quad (-1, 2), \quad \text{and} \quad (2, \infty).$$

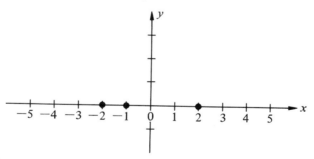

Figure 4.9

The sign of $f(x)$ in each of these intervals can be determined by finding a suitable test value. Thus, if we choose -3 in $(-\infty, -2)$, then

$$\begin{aligned}
f(-3) &= (-3)^3 + (-3)^2 - 4(-3) - 4 \\
&= -27 + 9 + 12 - 4 = -10.
\end{aligned}$$

Since the test value $f(-3)$ is negative, $f(x)$ is negative throughout the interval $(-\infty, -2)$.

If we choose $-3/2$ in the interval $(-2, -1)$, then the corresponding test value is

$$f\left(-\frac{3}{2}\right) = \left(-\frac{3}{2}\right)^3 + \left(-\frac{3}{2}\right)^2 - 4\left(-\frac{3}{2}\right) - 4$$

$$= -\frac{27}{8} + \frac{9}{4} + 6 - 4 = \frac{7}{8}.$$

Since $f(-3/2)$ is positive, $f(x)$ is positive throughout $(-2, -1)$. The following table summarizes these facts and lists suitable test values for the remaining two intervals.

Interval	Test value	Sign of $f(x)$	Position of graph
$(-\infty, -2)$	$f(-3) = -10$	$-$	Below x-axis
$(-2, -1)$	$f(-\frac{3}{2}) = \frac{7}{8}$	$+$	Above x-axis
$(-1, 2)$	$f(0) = -4$	$-$	Below x-axis
$(2, \infty)$	$f(3) = 20$	$+$	Above x-axis

Using this information and plotting several points gives us the graph in Figure 4.10. In order to find the high and low points it is necessary to use methods developed in calculus.

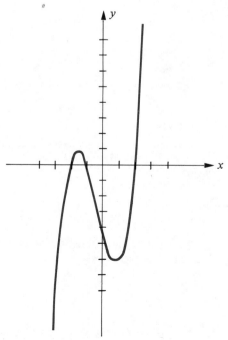

Figure 4.10 $f(x) = x^3 + x^2 - 4x - 4$

The graph of every polynomial function of degree 3 has an S-shaped appearance similar to that shown in Figure 4.10, or an inverted version of that graph if the co-efficient of x^3 is negative. However, sometimes there may be only one x-intercept, or the S may be elongated, as in Figure 4.5.

Example 3 If $f(x) = x^4 - 4x^3 + 3x^2$, determine all values of x such that $f(x) > 0$, and all x such that $f(x) < 0$. Sketch the graph of f.

Solution We begin by factoring $f(x)$. Thus,

$$f(x) = x^2(x^2 - 4x + 3)$$
$$= x^2(x - 1)(x - 3).$$

The zeros of $f(x)$, that is, the x-intercepts of the graph, are, in increasing order, 0, 1, and 3. The corresponding points on the graph are shown in Figure 4.11.

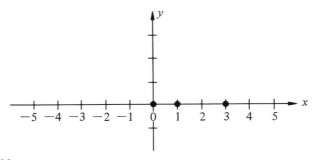

Figure 4.11

We next determine the sign of $f(x)$ in each of the intervals $(-\infty, 0)$, $(0, 1)$, $(1, 3)$, and $(3, \infty)$. This may be accomplished by finding suitable test values, as in the preceding example. The following table summarizes results. The reader should carefully check each entry.

Interval	Test value	Sign of $f(x)$	Position of graph
$(-\infty, 0)$	$f(-1) = 8$	$+$	Above x-axis
$(0, 1)$	$f(\frac{1}{2}) = \frac{5}{16}$	$+$	Above x-axis
$(1, 3)$	$f(2) = -4$	$-$	Below x-axis
$(3, \infty)$	$f(4) = 48$	$+$	Above x-axis

Making use of this information and plotting several points gives us the sketch in Figure 4.12.

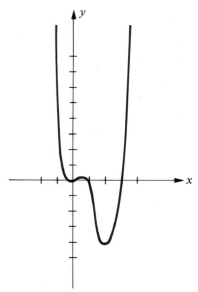

Figure 4.12 $f(x) = x^4 - 4x^3 + 3x^2$ ■

Example 4   Show that $f(x) = x^5 + 2x^4 - 6x^3 + 2x - 3$ has a zero between 1 and 2.

Solution   Substitution for x gives us

$$f(1) = 1 + 2 - 6 + 2 - 3 = -4$$
$$f(2) = 32 + 32 - 48 + 4 - 3 = 17.$$

Since $f(1)$ and $f(2)$ have opposite signs, it follows from the Intermediate Value Theorem that $f(c) = 0$ for some real number c between 1 and 2.   ■

Example 4 illustrates a scheme for locating zeros of polynomials. By using a method of *successive approximation*, each such zero can be approximated to any degree of accuracy (see, for example, Calculator Exercises 1–10). Additional techniques for locating zeros of polynomials will be introduced in Chapter 10.

EXERCISES 4.2

1   If $f(x) = ax^3 + 2$, sketch the graph of f if:

   (a) $a = 2$ (b) $a = 4$
   (c) $a = \frac{1}{4}$ (d) $a = -2$

2   If $f(x) = 2x^3 + c$, sketch the graph of f if:
   (a) $c = 2$ (b) $c = 4$
   (c) $c = \frac{1}{4}$ (d) $c = -2$

In each of Exercises 3–18 determine all x such that $f(x) > 0$, and all x such that $f(x) < 0$. Sketch the graph of f.

3   $f(x) = \frac{1}{2}x^3 - 4$ 4 $f(x) = -\frac{1}{4}x^3 - 16$

5   $f(x) = \frac{1}{8}x^4 + 2$ 6 $f(x) = 1 - x^5$

7   $f(x) = x^3 - 9x$ 8 $f(x) = 16x - x^3$

9 $f(x) = -x^3 - x^2 + 2x$

10 $f(x) = x^3 + x^2 - 12x$

11 $f(x) = (x + 4)(x - 1)(x - 5)$

12 $f(x) = (x + 2)(x - 3)(x - 4)$

13 $f(x) = x^4 - 16$ **14** $f(x) = 16 - x^4$

15 $f(x) = -x^4 - 3x^2 + 4$

16 $f(x) = x^4 - 7x^2 - 18$

17 $f(x) = x(x - 2)(x + 1)(x + 3)$

18 $f(x) = x(x + 1)^2(x - 3)(x - 5)$

19 Find a number k such that the graph of $f(x) = 3x^3 - kx^2 + x - 5k$ contains the point $(-1, 4)$.

20 If one zero of $f(x) = x^3 - 2x^2 - 16x + 16k$ is 2, find two other zeros.

In Exercises 21–26 show that f has a zero between a and b.

21 $f(x) = x^3 - 4x^2 + 3x - 2$; $a = 3, b = 4$

22 $f(x) = 2x^3 + 5x^2 - 3$; $a = -3, b = -2$

23 $f(x) = -x^4 + 3x^3 - 2x + 1$; $a = 2, b = 3$

24 $f(x) = 2x^4 + 3x - 2$; $a = \frac{1}{2}, b = \frac{3}{4}$

25 $f(x) = x^5 + x^3 + x^2 + x + 1$; $a = -\frac{3}{2}$, $b = -1$

26 $f(x) = x^5 - 3x^4 - 2x^3 + 3x^2 - 9x - 6$; $a = 3, b = 4$

CALCULATOR EXERCISES 4.2

If a zero of a polynomial lies between two integers, then a method of *successive approximation* may be used to approximate the zero to any degree of accuracy. One such method is illustrated in Exercise 1.

1 If $f(x) = x^3 - 3x + 1$:
   (a) show that f has a zero between 1 and 2.
   (b) by increasing values of x by tenths, show that f has a zero between 1.5 and 1.6.
   (c) by increasing values of x by hundredths, show that f has a zero between 1.53 and 1.54.

In (c) of the preceding exercise, a zero of $x^3 - 3x + 1$ is said to be *isolated between successive hundredths* in the interval $[1, 2]$. In each of Exercises 2–6 isolate a zero of $f(x)$ between successive hundredths in the interval $[a, b]$.

2 $f(x) = x^3 + 5x - 3$; $a = 0, b = 1$

3 $f(x) = 2x^3 - 4x^2 - 3x + 1$; $a = 2, b = 3$

4 $f(x) = x^4 - 4x^3 + 3x^2 - 8x + 2$; $a = 3, b = 4$

5 $f(x) = x^5 - 2x^2 + 4$; $a = -2, b = -1$

6 $f(x) = x^4 - 2x^3 + 10x - 25$; $a = 2, b = 3$

In Exercises 7–10 isolate a zero of $f(x)$ between successive thousandths in the interval $[a, b]$.

7 $f(x) = x^4 + 2x^3 - 5x^2 + 1$; $a = 1, b = 2$

8 $f(x) = x^4 - 5x^2 + 2x - 5$; $a = 2, b = 3$

9 $f(x) = x^5 + x^2 - 9x - 3$; $a = -2, b = -1$

10 $f(x) = x^5 + 3x^4 - x^3 + 2x^2 + 6x - 2$; $a = -4, b = -3$

4.3 RATIONAL FUNCTIONS

A **rational function** is a quotient of two polynomial functions. Thus, f is rational if, for all x in its domain,

$$f(x) = \frac{g(x)}{h(x)}$$

where $g(x)$ and $h(x)$ are polynomials. Of major importance are the zeros of the numerator and denominator (we shall assume that $g(x)$ and $h(x)$ have no *common* zeros). Note that $f(c) = 0$ if and only if $g(c) = 0$, and hence the zeros of the numerator $g(x)$ are the zeros of $f(x)$. However, if c is a zero of the denominator $h(x)$, then $f(c)$ is undefined, and the behavior of $f(x)$ when x is near c requires special attention.

Example 1   Sketch the graph of f if $f(x) = \dfrac{1}{x-2}$.

Solution   Since the numerator 1 is never zero, $f(x)$ has no zeros. Hence, the graph has no x-intercepts. Since the denominator $x - 2$ has the zero 2, $f(2)$ is undefined. If x is very close to 2, and $x > 2$, then $f(x)$ is very large. For example,

$$f(2.1) = \frac{1}{2.1 - 2} = \frac{1}{0.1} = 10$$

$$f(2.01) = \frac{1}{2.01 - 2} = \frac{1}{0.01} = 100$$

$$f(2.001) = \frac{1}{2.001 - 2} = \frac{1}{0.001} = 1,000.$$

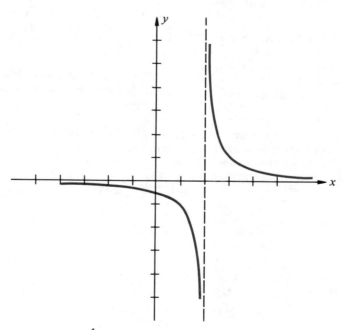

Figure 4.13 $f(x) = \dfrac{1}{x-2}$

If x is close to 2, and $x < 2$, then $f(x)$ is *numerically* large, but negative. Thus,

$$f(1.9) = \frac{1}{1.9 - 2} = \frac{1}{-0.1} = -10$$

$$f(1.99) = \frac{1}{1.99 - 2} = \frac{1}{-0.01} = -100$$

$$f(1.999) = \frac{1}{1.999 - 2} = \frac{1}{-0.001} = -1000.$$

Several other values of $f(x)$ are displayed in the following table.

x	-8	-1	0	1	3	4	12
$f(x)$	$-\frac{1}{10}$	$-\frac{1}{3}$	$-\frac{1}{2}$	-1	1	$\frac{1}{2}$	$\frac{1}{10}$

Observe that, as $|x|$ increases, $f(x)$ approaches zero. Plotting points and taking into account what happens near $x = 2$ gives us the sketch in Figure 4.13. ■

In the preceding example, $f(x)$ can be made as large as desired by choosing x sufficiently close to 2 (and $x > 2$). This will be denoted symbolically by

$$f(x) \to \infty \quad \text{as} \quad x \to 2^+$$

which may be read $f(x)$ *increases without bound as x approaches* 2 *from the right.* The phrase "$f(x)$ becomes positively infinite" is sometimes used in place of "$f(x)$ increases without bound." It is important to remember that the symbol ∞ (read "infinity") does not represent a real number, but is used merely as an abbreviation for certain types of functional behavior.

In like manner, we write

$$f(x) \to -\infty \quad \text{as} \quad x \to 2^-$$

which is read $f(x)$ *decreases without bound as x approaches* 2 *from the left.* The phrase "$f(x)$ becomes negatively infinite" is sometimes used instead of "$f(x)$ decreases without bound."

Let us extend the previous discussion to arbitrary functions and to the case where x approaches *any* real number a. In particular, the symbol $x \to a^+$ will signify that x approaches a from the *right*, that is, through values *greater* than a. The symbol $x \to a^-$ will mean that x approaches a from the *left*, that is, through values *less* than a. Some illustrations of the manner in which a function may increase or decrease without bound, together with the notation used, are shown in Figure 4.14, where a is pictured as positive. Of course, we can also have $a < 0$.

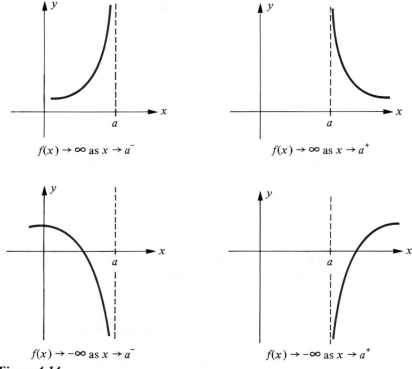

$f(x) \to \infty$ as $x \to a^-$ $f(x) \to \infty$ as $x \to a^+$

$f(x) \to -\infty$ as $x \to a^-$ $f(x) \to -\infty$ as $x \to a^+$

Figure 4.14

Definition

> The line $x = a$ is a **vertical asymptote** for the graph of a function f if
>
> $$f(x) \to \infty \quad \text{or} \quad f(x) \to -\infty$$
>
> as x approaches a, either from the left or right.

In Figure 4.14, the dashed lines represent vertical asymptotes. In Example 1 the line $x = 2$ is a vertical asymptote for the graph of $f(x) = 1/(x - 2)$.

Vertical asymptotes are common characteristics of graphs of rational functions. Indeed, *if a is a zero of the denominator $h(x)$, then the graph of $f(x) = g(x)/h(x)$ has the vertical asymptote $x = a$.*

We shall also be interested in values of $f(x)$ when $|x|$ is large. As an illustration, if, in Example 1, we assign very large values to x, then $f(x)$ is close to 0. Thus,

$$f(1{,}002) = \frac{1}{1{,}000} = 0.001 \quad \text{and} \quad f(1{,}000{,}002) = \frac{1}{1{,}000{,}000} = 0.000001.$$

Moreover, we can make $f(x)$ as close to 0 as desired by choosing x sufficiently large. This is expressed symbolically by

$$f(x) \to 0 \quad \text{as} \quad x \to \infty$$

which is read $f(x)$ *approaches* 0 *as x increases without bound* (*or as x becomes positively infinite*).

Similarly, in Example 1, we write

$$f(x) \to 0 \quad \text{as} \quad x \to -\infty$$

which is read $f(x)$ *approaches* 0 *as x decreases without bound* (*or as x becomes negatively infinite*).

In Example 1 the line $y = 0$, that is, the x-axis, is called a *horizontal asymptote* for the graph. In general, we have the following definition, where the notation should be self-evident.

Definition

> The line $y = b$ is a **horizontal asymptote** for the graph of a function f if
>
> $$f(x) \to b \quad \text{as} \quad x \to \infty \quad \text{or as} \quad x \to -\infty.$$

Graphs of rational functions often have horizontal asymptotes. Some typical cases (for $x \to \infty$) are illustrated in Figure 4.15. The manner in which the graph "approaches" the line $y = b$ may vary, depending on the nature of the function. Similar sketches may be made for the case $x \to -\infty$. Note, as in the third sketch, that the graph of f may cross a horizontal asymptote.

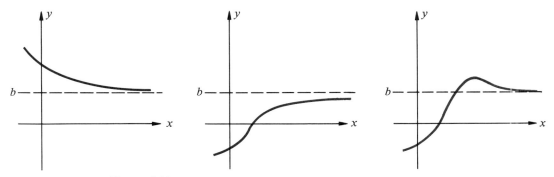

Figure 4.15

We shall next list some guidelines for sketching the graph of a rational function. Their use will be illustrated in Examples 2, 3, and 4.

> **Guidelines for sketching the graph of** $f(x) = \dfrac{g(x)}{h(x)}$
>
> **where** $g(x)$ **and** $h(x)$ **are polynomials**

Step 1: Find the real zeros of the numerator $g(x)$, and use them to plot the points corresponding to the x-intercepts.

Step 2: Find the real zeros of the denominator $h(x)$. For each zero a, the line $x = a$ is a vertical asymptote. Represent $x = a$ with dashes.

Step 3: Find the sign of $f(x)$ in each of the intervals determined by the zeros of $g(x)$ and $h(x)$. Use these signs to determine whether the graph lies above, or below, the x-axis in each interval.

Step 4: If $x = a$ is a vertical asymptote, use the information in Step 3 to determine whether $f(x) \to \infty$ or $f(x) \to -\infty$ as:

$$\text{(i)} \quad x \to a^-; \qquad \text{(ii)} \quad x \to a^+.$$

Make note of this by sketching a portion of the graph on each side of $x = a$.

Step 5: Use the information in Step 3 to determine the manner in which the graph intersects the x-axis.

Step 6: Determine the behavior of $f(x)$ as $x \to \infty$ or $x \to -\infty$. If $f(x) \to b$, then the line $y = b$ is a horizontal asymptote. If $b \neq 0$, represent $y = b$ with dashes.

Step 7: Sketch the graph, plotting points wherever necessary.

Example 2 Sketch the graph of f if $f(x) = \dfrac{x - 1}{x^2 - x - 6}$.

Solution It is convenient to factor the denominator as follows:

$$f(x) = \frac{x - 1}{(x + 2)(x - 3)}.$$

We shall obtain the graph by following the steps listed in the guidelines.

Step 1: The numerator $x - 1$ has the zero 1, and hence we plot the point $(1, 0)$ on the graph, as shown in Figure 4.16.

Step 2: The denominator has zeros -2 and 3. Hence, the lines $x = -2$ and $x = 3$ are vertical asymptotes, and we represent them with dashes, as in Figure 4.16.

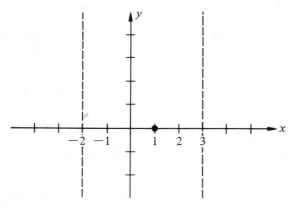

Figure 4.16

Step 3: The zeros -2, 1, and 3 of the numerator and denominator of $f(x)$ determine the following intervals:

$$(-\infty, -2), \quad (-2, 1), \quad (1, 3), \quad \text{and} \quad (3, \infty).$$

Since $f(x)$ is a quotient of two polynomials, it follows from our work in Section 4.2 that $f(x)$ is always positive or always negative throughout each of these intervals. Using test values to determine the sign of $f(x)$ we arrive at the following table.

Interval	Test value	Sign of $f(x)$	Position of graph
$(-\infty, -2)$	$f(-3) = -\frac{2}{3}$	$-$	Below x-axis
$(-2, 1)$	$f(0) = \frac{1}{6}$	$+$	Above x-axis
$(1, 3)$	$f(2) = -\frac{1}{4}$	$-$	Below x-axis
$(3, \infty)$	$f(4) = \frac{1}{2}$	$+$	Above x-axis

Step 4: We shall use the fourth column of the table in Step 3 to investigate the behavior of $f(x)$ near each vertical asymptote.

(a) Consider the vertical asymptote $x = -2$. Since the graph lies *below* the x-axis throughout the interval $(-\infty, -2)$ it follows that

$$f(x) \to -\infty \quad \text{as} \quad x \to -2^-.$$

Since the graph lies *above* the x-axis throughout the interval $(-2, 1)$ it follows that

$$f(x) \to \infty \quad \text{as} \quad x \to -2^+.$$

We note these facts in Figure 4.17 by sketching portions of the graph on each side of the line $x = -2$.

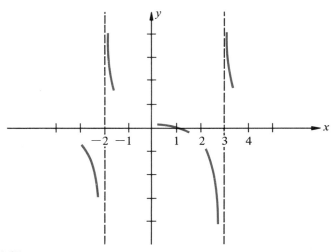

Figure 4.17

(b) Consider the vertical asymptote $x = 3$. The graph lies *below* the x-axis throughout the interval $(1, 3)$ and hence

$$f(x) \to -\infty \quad \text{as} \quad x \to 3^-.$$

The graph lies *above* the x-axis throughout $(3, \infty)$ and hence

$$f(x) \to \infty \quad \text{as} \quad x \to 3^+.$$

We note these facts in Figure 4.17 by sketching portions of the graph on each side of $x = 3$.

Step 5: Referring to the fourth column of the table in Step 3, we see that the graph crosses the x-axis at $(1, 0)$ in a manner similar to that illustrated in Figure 4.17.

Step 6: To determine what is true if $x \to \infty$ or $x \to -\infty$, we divide numerator and denominator of $f(x)$ by x^2, obtaining

$$f(x) = \frac{(1/x) - (1/x^2)}{1 - (1/x) - (6/x^2)}.$$

If x is numerically very large, then the expressions in parentheses are close to 0, and hence

$$f(x) \approx \frac{0 - 0}{1 - 0 - 0} = \frac{0}{1} = 0.$$

This indicates that

$$f(x) \to 0 \quad \text{as} \quad x \to \infty \qquad \text{and} \qquad f(x) \to 0 \quad \text{as} \quad x \to -\infty.$$

Thus, the line $y = 0$ (the x-axis) is a horizontal asymptote for the graph.

Step 7: Using the information in Steps 4, 5, and 6, and plotting several points gives us the sketch in Figure 4.18.

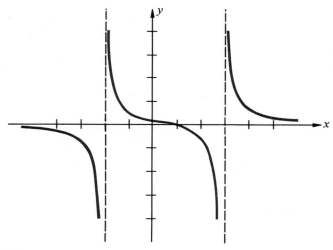

Figure 4.18

■

The technique used in Step 6 of Example 2 can be generalized. Specifically, if we wish to investigate $f(x) = g(x)/h(x)$ as $x \to \infty$ or as $x \to -\infty$, and if the degree of $g(x)$ is not greater than the degree of $h(x)$, then we divide numerator and denominator by x^k, where k is the degree of $h(x)$. If the degree of $g(x)$ is greater than the degree of $h(x)$, it can be shown that $f(x)$ either increases or decreases without bound as $x \to \infty$ or as $x \to -\infty$.

Example 3   Sketch the graph of f if $f(x) = \dfrac{x^2}{x^2 - x - 2}$.

Solution   Factoring the denominator gives us

$$f(x) = \frac{x^2}{(x + 1)(x - 2)}.$$

We shall again follow the guidelines listed earlier.

Step 1:   The numerator x^2 has 0 as a zero and hence the graph intersects the x-axis at $(0,0)$, as shown in Figure 4.19.

Step 2:   Since the denominator has zeros -1 and 2, the lines $x = -1$ and $x = 2$ are vertical asymptotes, and we represent them with dashes, as in Figure 4.19.

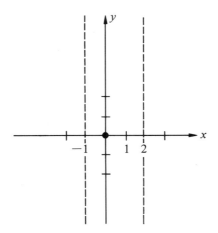

Figure 4.19

Step 3:   The intervals determined by the zeros in Steps 1 and 2 are

$$(-\infty, -1), \quad (-1, 0), \quad (0, 2), \quad \text{and} \quad (2, \infty).$$

Following the procedure used in Step 3 of Example 2 we arrive at the following table.

Interval	Test value	Sign of $f(x)$	Position of graph
$(-\infty, -1)$	$f(-2) = 1$	$+$	Above x-axis
$(-1, 0)$	$f(-\tfrac{1}{2}) = -\tfrac{1}{5}$	$-$	Below x-axis
$(0, 2)$	$f(1) = -\tfrac{1}{2}$	$-$	Below x-axis
$(2, \infty)$	$f(3) = \tfrac{9}{4}$	$+$	Above x-axis

Step 4:   We refer to the fourth column of the table in Step 3 and proceed as follows:

(a)   Consider the vertical asymptote $x = -1$. Since the graph is above the x-axis in $(-\infty, -1)$ it follows that

$$f(x) \to \infty \quad \text{as} \quad x \to -1^-.$$

Since the graph is below the x-axis in $(-1, 0)$ we have

$$f(x) \to -\infty \quad \text{as} \quad x \to -1^+.$$

(b)   Consider the vertical asymptote $x = 2$. The graph is below the x-axis in the interval $(0, 2)$ and hence

$$f(x) \to -\infty \quad \text{as} \quad x \to 2^-.$$

The graph is above the x-axis in $(2, \infty)$ and hence

$$f(x) \to \infty \quad \text{as} \quad x \to 2^+.$$

These facts are noted in Figure 4.20.

Figure 4.20

Step 5:   Referring to the table in Step 3 we see that the graph lies below the x-axis in both of the intervals $(-1, 0)$ and $(0, 2)$. Consequently, the graph intersects, but does not cross, the x-axis at $(0, 0)$.

Step 6: In order to determine what happens as $x \to \infty$ or as $x \to -\infty$ we use the method discussed prior to this example. Since the degree of the denominator $x^2 - x - 2$ is 2, we divide numerator and denominator of $f(x)$ by x^2, obtaining

$$f(x) = \frac{x^2/x^2}{(x^2 - x - 2)/x^2} = \frac{1}{1 - (1/x) - (2/x^2)}.$$

If we assign very large numerical values to x, the expressions $1/x$ and $2/x^2$ are very close to 0, and hence

$$f(x) \approx \frac{1}{1 - 0 - 0} = 1.$$

This indicates that

$$f(x) \to 1 \quad \text{as} \quad x \to \infty \qquad \text{and} \qquad f(x) \to 1 \quad \text{as} \quad x \to -\infty.$$

Hence, the line $y = 1$ is a horizontal asymptote for the graph, and we sketch it with dashes in Figure 4.21.

Step 7: Using the information gathered in Steps 4, 5, and 6, and plotting several points, we obtain the graph in Figure 4.21. The graph intersects the horizontal asymptote at $x = -2$, as may be verified by solving the equation $x^2/(x^2 - x - 2) = 1$. The fact that the graph lies below the horizontal asymptote if $x < -2$ and above it if $-2 < x < -1$ may be verified by plotting points.

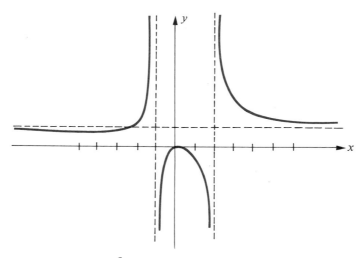

Figure 4.21 $f(x) = \dfrac{x^2}{x^2 - x - 2}$ ■

Example 4 Sketch the graph of f if $f(x) = \dfrac{2x^4}{x^4 + 1}$.

Solution In this solution we shall not formally write down each step in the guidelines. Note that, since $f(-x) = f(x)$, the function is even, and hence the graph is symmetric with respect to the y-axis.

The graph intersects the x-axis at $(0,0)$. Since the denominator of $f(x)$ has no real zeros, there are no vertical asymptotes. In order to examine what happens to $f(x)$ as $x \to \infty$ or $x \to -\infty$, we divide numerator and denominator by x^4, obtaining

$$f(x) = \frac{2}{1 + (1/x^4)}.$$

Since

$$\frac{1}{x^4} \to 0 \quad \text{as} \quad x \to \infty$$

it follows that

$$f(x) \to \frac{2}{1 + 0} = 2 \quad \text{as} \quad x \to \infty.$$

Hence, the line $y = 2$ is a horizontal asymptote. Using this fact, plotting several points, and making use of the symmetry with respect to the y-axis, leads to the graph in Figure 4.22.

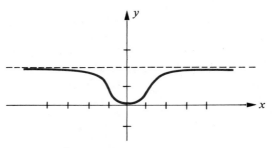

Figure 4.22 $f(x) = \dfrac{2x^4}{x^4 + 1}$

■

Graphs of rational functions may become very complicated as the degrees of the polynomials in the numerator and denominator increase. For a thorough treatment it is necessary to employ techniques developed in calculus.

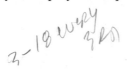

EXERCISES 4.3

Sketch the graph of f in each of Exercises 1–20.

1 $f(x) = \dfrac{1}{x + 2}$

2 $f(x) = \dfrac{1}{x + 3}$

3 $f(x) = \dfrac{-2}{x + 4}$

4 $f(x) = \dfrac{-3}{x - 1}$

5 $f(x) = \dfrac{x}{x - 5}$

6 $f(x) = \dfrac{x}{3x + 2}$

7 $f(x) = \dfrac{4}{(x - 1)^2}$

8 $f(x) = \dfrac{-1}{(x + 2)^2}$

9 $f(x) = \dfrac{1}{x^2 - 4}$

10 $f(x) = \dfrac{2}{x^2 + x - 2}$

11 $f(x) = \dfrac{5x}{4 - x^2}$

12 $f(x) = \dfrac{x^2}{x^2 - 4}$

13 $f(x) = \dfrac{x^2}{x^2 - 7x + 10}$ **14** $f(x) = \dfrac{x}{x^2 - x - 6}$ **17** $f(x) = \dfrac{4}{x^2 + 4}$ **18** $f(x) = \dfrac{3x}{x^2 + 1}$

15 $f(x) = \dfrac{3x + 2}{x}$ **16** $f(x) = \dfrac{x^2 - 4}{x^2}$ **19** $f(x) = \dfrac{1}{x^3 + x^2 - 6x}$ **20** $f(x) = \dfrac{x^2 - x}{16 - x^2}$

4.4 PARTIAL FRACTIONS

It is easy to verify that

$$\frac{2}{x^2 - 1} = \frac{1}{x - 1} + \frac{-1}{x + 1}.$$

The expression on the right side of this equation is called *the partial fraction decomposition* of $2/(x^2 - 1)$. In general, if f is *any* rational function, it is theoretically possible to express $f(x)$ as a sum of quotients of polynomials whose denominators involve powers of polynomials of degree not greater than two. This fact has important applications in the study of calculus.

If $g(x)$ and $h(x)$ are polynomials *and the degree of $g(x)$ is less than the degree of $h(x)$*, then it follows from a theorem in algebra that

$$\frac{g(x)}{h(x)} = F_1 + F_2 + \cdots + F_k$$

where each F_i has one of the forms

$$\frac{A}{(px + q)^m} \quad \text{or} \quad \frac{Cx + D}{(ax^2 + bx + c)^n}$$

for some nonnegative integers m and n, and where $ax^2 + bx + c$ is **irreducible**, in the sense that this quadratic polynomial has no real zeros, that is, $b^2 - 4ac < 0$. The sum $F_1 + F_2 + \cdots + F_k$ is called the **partial fraction decomposition** of $g(x)/h(x)$ and each F_i is called a **partial fraction**. We shall not prove this algebraic result but will, instead, give rules for obtaining the decomposition.

To find the partial fraction decomposition of $g(x)/h(x)$ it is essential that $g(x)$ have lower degree than $h(x)$. If this is not the case, then long division should be employed to arrive at such an expression. For example, given

$$\frac{x^3 - 6x^2 + 5x - 3}{x^2 - 1}$$

we obtain, by long division,

$$\frac{x^3 - 6x^2 + 5x - 3}{x^2 - 1} = x - 6 + \frac{6x - 9}{x^2 - 1}.$$

The partial fraction decomposition is then found for $(6x - 9)/(x^2 - 1)$.

In order to obtain the decomposition of $g(x)/h(x)$, we begin by expressing the denominator $h(x)$ as a product of factors $px + q$, or irreducible quadratic factors $ax^2 + bx + c$. Repeated factors are then collected so that $h(x)$ is a product of *different* factors of the form $(px + q)^m$ or $(ax^2 + bx + c)^n$, where m and n are nonnegative integers and $ax^2 + bx + c$ is irreducible. We then apply the following rules.

Rule 1.   For each factor of $h(x)$ of the form $(px + q)^m$ where $m \geq 1$, the partial fraction decomposition of $g(x)/h(x)$ contains a sum of m partial fractions of the form

$$\frac{A_1}{px + q} + \frac{A_2}{(px + q)^2} + \cdots + \frac{A_m}{(px + q)^m}$$

where each A_i is a real number.

Rule 2.   For each factor of $h(x)$ of the form $(ax^2 + bx + c)^n$ where $n \geq 1$ and $b^2 - 4ac < 0$, the partial fraction decomposition of $g(x)/h(x)$ contains a sum of n partial fractions of the form

$$\frac{A_1 x + B_1}{ax^2 + bx + c} + \frac{A_2 x + B_2}{(ax^2 + bx + c)^2} + \cdots + \frac{A_n x + B_n}{(ax^2 + bx + c)^n}$$

where, for each i, A_i and B_i are real numbers.

Example 1   Find the partial fraction decomposition of

$$\frac{4x^2 + 13x - 9}{x^3 + 2x^2 - 3x}.$$

Solution   The denominator of the quotient has the factored form $x(x + 3)(x - 1)$. Each of the linear factors is handled according to Rule 1 with $m = 1$. Thus, for the factor x there corresponds a partial fraction of the form A/x. Similarly, for the factors $x + 3$ and $x - 1$, there correspond partial fractions $B/(x + 3)$ and $C/(x - 1)$, respectively. The partial fraction decomposition has the form

$$\frac{4x^2 + 13x - 9}{x(x + 3)(x - 1)} = \frac{A}{x} + \frac{B}{x + 3} + \frac{C}{x - 1}.$$

Multiplying by the lowest common denominator $x(x + 3)(x - 1)$ gives us

$$4x^2 + 13x - 9 = A(x + 3)(x - 1) + Bx(x - 1) + Cx(x + 3).$$

In a case such as this, in which the factors are all linear and nonrepeated, the values for A, B, and C can be found by substituting values for x which make the various factors zero. If we let $x = 0$ in the last equation, then $-9 = -3A$, or $A = 3$. Letting $x = 1$ gives us $8 = 4C$, or $C = 2$. Finally, if $x = -3$, then $-12 =$

$12B$, or $B = -1$. The partial fraction decomposition is, therefore,

$$\frac{4x^2 + 13x - 9}{x(x + 3)(x - 1)} = \frac{3}{x} + \frac{-1}{x + 3} + \frac{2}{x - 1}.$$ ■

Example 2 Find the partial fraction decomposition of

$$\frac{x^2 + 10x - 36}{x(x - 3)^2}.$$

Solution   By Rule 1, with $m = 1$, there is a partial fraction A/x corresponding to the factor x. Next, applying Rule 1, with $m = 2$, the squared factor $(x - 3)^2$ gives rise to a sum of two partial fractions $B/(x - 3) + C/(x - 3)^2$. Thus, the partial fraction decomposition has the form

$$\frac{x^2 + 10x - 36}{x(x - 3)^2} = \frac{A}{x} + \frac{B}{x - 3} + \frac{C}{(x - 3)^2}.$$

Multiplying both sides by $x(x - 3)^2$ gives us

(∗) $x^2 + 10x - 36 = A(x - 3)^2 + Bx(x - 3) + Cx$

where the symbol (∗) is used for later identification of this equation. Two of the unknown constants may be determined easily. Letting $x = 3$ in (∗) gives us

$$9 + 30 - 36 = A(0) + B(0) + 3C.$$

This reduces to $3 = 3C$ and hence $C = 1$. In like manner, letting $x = 0$, we get

$$0 + 0 - 36 = A(-3)^2 + B(0) + C(0).$$

Hence, $-36 = 9A$, or $A = -4$. The remaining constant can be found by comparing coefficients. If the right side of (∗) is expanded and like powers of x collected we obtain

$$x^2 + 10x - 36 = (A + B)x^2 + (-3A - 3B + C)x + 9A.$$

The coefficient $A + B$ of x^2 on the right must equal the coefficient of x^2 on the left, that is, $A + B = 1$. Since $A = -4$, it follows that $B = 1 - A = 1 - (-4) = 5$. The partial fraction decomposition is, therefore,

$$\frac{x^2 + 10x - 36}{x(x - 3)^2} = \frac{-4}{x} + \frac{5}{x - 3} + \frac{1}{(x - 3)^2}.$$ ■

Example 3 Find the partial fraction decomposition of

$$\frac{3x^3 - 18x^2 + 29x - 4}{(x + 1)(x - 2)^3}.$$

Solution   By Rule 1, there is a partial fraction of the form $A/(x + 1)$ corresponding to the factor $x + 1$ in the denominator. For the factor $(x - 2)^3$ we apply Rule 1 (with $m = 3$), obtaining a sum of three partial fractions $B/(x - 2)$, $C/(x - 2)^2$, and $D/(x - 2)^3$. Consequently, the decomposition has the form

$$\frac{3x^3 - 18x^2 + 29x - 4}{(x + 1)(x - 2)^3} = \frac{A}{x + 1} + \frac{B}{x - 2} + \frac{C}{(x - 2)^2} + \frac{D}{(x - 2)^3}.$$

Multiplying both sides by $(x + 1)(x - 2)^3$ gives us

(*)
$$3x^3 - 18x^2 + 29x - 4$$
$$= A(x - 2)^3 + B(x + 1)(x - 2)^2 + C(x + 1)(x - 2) + D(x + 1).$$

Two of the unknown constants may be determined easily. If we let $x = 2$ in (*) then

$$24 - 72 + 58 - 4 = 3D, \quad 6 = 3D, \quad \text{and} \quad D = 2.$$

Similarly, letting $x = -1$ in (*) gives us

$$-3 - 18 - 29 - 4 = -27A, \quad -54 = -27A, \quad \text{and} \quad A = 2.$$

The remaining constants may be found by comparing coefficients. If the right side of (*) is expanded and like powers of x collected, we see that the coefficient of x^3 is $A + B$. This must be equal to the coefficient of x^3 on the left, that is,

$$A + B = 3.$$

Since $A = 2$, it follows that $B = 3 - A = 3 - 2 = 1$. Finally, we compare the constant terms in (*) by letting $x = 0$. This gives us

$$-4 = -8A + 4B - 2C + D.$$

Substituting the values we have found for A, B, and D leads to

$$-4 = -16 + 4 - 2C + 2$$

which has the solution $C = -3$. The partial fraction decomposition is, therefore,

$$\frac{3x^3 - 18x^2 + 29x - 4}{(x + 1)(x - 2)^3} = \frac{2}{x + 1} + \frac{1}{x - 2} + \frac{-3}{(x - 2)^2} + \frac{2}{(x - 2)^3}. \qquad \blacksquare$$

Example 4   Find the partial fraction decomposition of

$$\frac{x^2 - x - 21}{2x^3 - x^2 + 8x - 4}.$$

Solution   The denominator may be factored by grouping as follows:

$$2x^3 - x^2 + 8x - 4 = x^2(2x - 1) + 4(2x - 1) = (x^2 + 4)(2x - 1).$$

Applying Rule 2 to the irreducible quadratic factor $x^2 + 4$, we see that one of the partial fractions has the form $(Ax + B)/(x^2 + 4)$. By Rule 1, there is also a partial fraction $C/(2x - 1)$ corresponding to the factor $2x - 1$. Consequently,

$$\frac{x^2 - x - 21}{2x^3 - x^2 + 8x - 4} = \frac{Ax + B}{x^2 + 4} + \frac{C}{2x - 1}.$$

As in previous examples, this leads to

$(*)$ $$x^2 - x - 21 = (Ax + B)(2x - 1) + C(x^2 + 4).$$

Substituting $x = 1/2$, we obtain $(1/4) - (1/2) - 21 = (17/4)C$, which has the solution $C = -5$. The remaining constants may be found by comparing coefficients. Rearranging the right side of $(*)$ gives us

$$x^2 - x - 21 = (2A + C)x^2 + (-A + 2B)x - B + 4C.$$

Comparing the coefficients of x^2, we see that $2A + C = 1$. Since $C = -5$ it follows that $2A = 6$ or $A = 3$. Similarly, comparing the constant terms, $-B + 4C = -21$ and hence $-B - 20 = -21$ or $B = 1$. Thus, the partial fraction decomposition is

$$\frac{x^2 - x - 21}{2x^3 - x^2 + 8x - 4} = \frac{3x + 1}{x^2 + 4} + \frac{-5}{2x - 1}. \qquad \blacksquare$$

Example 5   Find the partial fraction decomposition of

$$\frac{5x^3 - 3x^2 + 7x - 3}{(x^2 + 1)^2}.$$

Solution   Applying Rule 2, with $n = 2$,

$$\frac{5x^3 - 3x^2 + 7x - 3}{(x^2 + 1)^2} = \frac{Ax + B}{x^2 + 1} + \frac{Cx + D}{(x^2 + 1)^2}.$$

Multiplying both sides by $(x^2 + 1)^2$ gives us

$$5x^3 - 3x^2 + 7x - 3 = (Ax + B)(x^2 + 1) + Cx + D$$

or $$5x^3 - 3x^2 + 7x - 3 = Ax^3 + Bx^2 + (A + C)x + (B + D).$$

Comparing the coefficients of x^3 and x^2, we obtain $A = 5$ and $B = -3$. From the coefficients of x we see that $A + C = 7$ or $C = 7 - A = 7 - 5 = 2$. Finally, the constant terms give us $B + D = -3$ or $D = -3 - B = -3 - (-3) = 0$. Therefore,

$$\frac{5x^3 - 3x^2 + 7x - 3}{(x^2 + 1)^2} = \frac{5x - 3}{x^2 + 1} + \frac{2x}{(x^2 + 1)^2}. \qquad \blacksquare$$

EXERCISES 4.4

Find the partial fraction decompositions in Exercises 1–24.

1 $\dfrac{8x - 1}{(x - 2)(x + 3)}$

2 $\dfrac{x - 29}{(x - 4)(x + 1)}$

3 $\dfrac{x + 34}{x^2 - 4x - 12}$

4 $\dfrac{5x - 12}{x^2 - 4x}$

5 $\dfrac{4x^2 - 15x - 1}{(x - 1)(x + 2)(x - 3)}$

6 $\dfrac{x^2 + 19x + 20}{x(x + 2)(x - 5)}$

7 $\dfrac{4x^2 - 5x - 15}{x^3 - 4x^2 - 5x}$

8 $\dfrac{37 - 11x}{(x + 1)(x^2 - 5x + 6)}$

9 $\dfrac{2x + 3}{(x - 1)^2}$

10 $\dfrac{5x^2 - 4}{x^2(x + 2)}$

11 $\dfrac{19x^2 + 50x - 25}{3x^3 - 5x^2}$

12 $\dfrac{10 - x}{x^2 + 10x + 25}$

13 $\dfrac{x^2 - 6}{(x + 2)^2(2x - 1)}$

14 $\dfrac{2x^2 + x}{(x - 1)^2(x + 1)^2}$

15 $\dfrac{3x^3 + 11x^2 + 16x + 5}{x(x + 1)^3}$

16 $\dfrac{4x^3 + 3x^2 + 5x - 2}{x^3(x + 2)}$

17 $\dfrac{x^2 + x - 6}{(x^2 + 1)(x - 1)}$

18 $\dfrac{x^2 - x - 21}{(x^2 + 4)(2x - 1)}$

19 $\dfrac{9x^2 - 3x + 8}{x^3 + 2x}$

20 $\dfrac{2x^3 + 2x^2 + 4x - 3}{x^4 + x^2}$

21 $\dfrac{4x^3 - x^2 + 4x + 2}{(x^2 + 1)^2}$

22 $\dfrac{3x^3 + 13x - 1}{(x^2 + 4)^2}$

23 $\dfrac{2x^4 - 2x^3 + 6x^2 - 5x + 1}{x^3 - x^2 + x - 1}$

24 $\dfrac{x^3}{x^3 - 3x^2 + 9x - 27}$

4.5 CONIC SECTIONS

Each of the geometric figures to be discussed in this section can be obtained by the intersection of a double-napped right circular cone and a plane, as illustrated in Figure 4.23. For this reason they are called **conic sections**, or simply **conics**. If, as in (i) of the figure, the plane cuts through one nappe and is perpendicular to the axis of the cone, then the curve of intersection is a circle. Equations of circles were considered in Section 3.2. If the plane cuts entirely across one nappe and is not perpendicular to the axis, as in (ii) of the figure, then an *ellipse* is obtained. If, as in (iii), the plane does not cut completely across one nappe and does not intersect both nappes, then the curve of intersection is a *parabola*. Finally, if the plane cuts through both nappes, as in (iv), then the resulting figure is called a *hyperbola*.

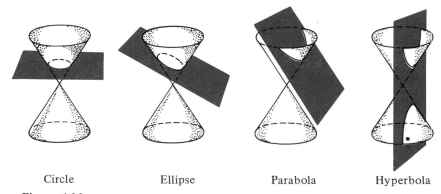

Circle   Ellipse   Parabola   Hyperbola

Figure 4.23

The conic sections were studied extensively by the early Greek mathematicians who used the methods of Euclidean geometry. A remarkable fact about conic sections is that, although they were studied thousands of years ago, they are far from obsolete. Indeed, they are important tools for current investigations in outer space and for the study of the behavior of atomic particles. Several applications of parabolas were mentioned in Section 4.1. There are also numerous applications involving ellipses and hyperbolas. For example, orbits of planets are ellipses. If the ellipse is very flat, the curve resembles the path of a comet. Elliptic gears or cams are sometimes used in machines. The hyperbola is useful for describing the path of an alpha particle in the electric field of the nucleus of an atom. The interested person can find many other applications of conic sections.

We know from our work in Section 4.1 that if $a \neq 0$, then the graph of the equation $y = ax^2 + bx + c$ is a parabola which opens upward if $a > 0$ or downward if $a < 0$. By interchanging the variables x and y we obtain the equation

$$x = ay^2 + by + c.$$

If $b = c = 0$, then $x = ay^2$ and the graph is a parabola with vertex at the origin, symmetric with respect to the x-axis, and opening to the right if $a > 0$ or to the left if $a < 0$, as illustrated in Figure 4.24.

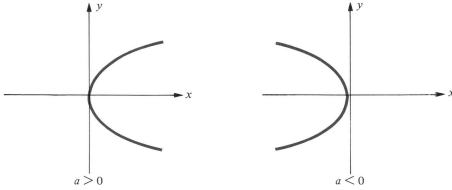

$a > 0$ $a < 0$

Figure 4.24 $x = ay^2$

If $b = 0$ and $c \neq 0$, then $x = ay^2 + c$, and the graph may be obtained by shifting one of the graphs in Figure 4.24 to the right or left, depending on the sign of c.

If $b \neq 0$ we may complete the square in y, obtaining the following:

**Standard
Equation of
a Parabola
(Horizontal
Axis)**

$$x - h = a(y - k)^2.$$

The parabola has vertex (h, k) and opens to the right if $a > 0$ or to the left if $a < 0$.

Example 1   Sketch the graph of the equation $x = 3y^2 + 8y - 3$.

Solution   Since $a = 3 > 0$, the graph is a parabola which opens to the right. We complete the square in y as follows:

$$x = 3y^2 + 8y - 3$$
$$= 3\left(y^2 + \frac{8}{3}y\right) - 3$$
$$= 3\left(y^2 + \frac{8}{3}y + \frac{16}{9}\right) - 3 - \frac{16}{3}$$
$$= 3\left(y + \frac{4}{3}\right)^2 - \frac{25}{3}.$$

If we write the last equation as

$$x + \frac{25}{3} = 3\left(y + \frac{4}{3}\right)^2$$

and compare with the standard equation we see that $h = -25/3$ and $k = -4/3$. Hence, the vertex is $(-25/3, -4/3)$. To find the y-intercepts of the graph we let $x = 0$ in the given equation and obtain

$$0 = 3y^2 + 8y - 3 = (3y - 1)(y + 3).$$

Figure 4.25 $x = 3y^2 + 8y - 3$

Consequently, the y-intercepts are $1/3$ and -3. The x-intercept -3 is found by setting $y = 0$ in the equation of the parabola. Plotting the vertex and the points corresponding to the intercepts leads to the sketch in Figure 4.25. ■

The graph of the equation

Equation of an Ellipse

$$\frac{x^2}{a^2} + \frac{y^2}{b^2} = 1$$

where a and b are positive real numbers, and $a \neq b$, is an **ellipse** with center at the origin. (If $a = b$ the graph is a circle.) To find the x-intercepts of the graph we let $y = 0$, obtaining $\pm a$. Similarly, letting $x = 0$ gives us the y-intercepts $\pm b$. We may solve the equation for y in terms of x as follows:

$$\frac{y^2}{b^2} = 1 - \frac{x^2}{a^2} = \frac{a^2 - x^2}{a^2}$$

$$y^2 = \frac{b^2}{a^2}(a^2 - x^2)$$

$$y = \pm \frac{b}{a}\sqrt{a^2 - x^2}.$$

In order to obtain points on the graph, the radicand $a^2 - x^2$ must be nonnegative. This will be true if $-a \leq x \leq a$. Consequently, the entire graph lies between the vertical lines $x = -a$ and $x = a$.

Note that for each permissible value of x there correspond two values for y. Let us consider the nonnegative values given by

$$y = \frac{b}{a}\sqrt{a^2 - x^2}.$$

If we let x vary from $-a$ to 0, we see that y increases from 0 to b. As x varies from 0 to a, y decreases from b to 0. This gives us the upper half of the graph in Figure 4.26. The lower half is the graph of $y = (-b/a)\sqrt{a^2 - x^2}$.

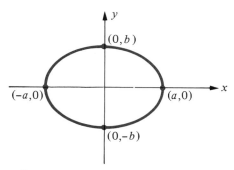

Figure 4.26 $\dfrac{x^2}{a^2} + \dfrac{y^2}{b^2} = 1$

Multiplying both sides of $(x^2/a^2) + (y^2/b^2) = 1$ by a^2b^2 we obtain

$$b^2x^2 + a^2y^2 = a^2b^2$$

which may be written in the form

$$Ax^2 + By^2 = C$$

where A, B, and C are positive real numbers. Conversely, an equation of this type has, for its graph, an ellipse with center at the origin. Knowing this, the graph may be readily sketched by using the x- and y-intercepts and plotting several points.

Example 2   Sketch the graph of $9x^2 + 16y^2 = 144$.

Solution   From the preceding discussion we know that the graph is an ellipse with center at the origin. The x-intercepts (found by letting $y = 0$) are ± 4, and the y-intercepts (found by letting $x = 0$) are ± 3. Let us next locate the points on the graph with abscissa 2. Substituting $x = 2$ in the given equation we obtain

$$9(2)^2 + 16y^2 = 144 \quad \text{or} \quad y^2 = \tfrac{108}{16} = \tfrac{27}{4}$$

and hence

$$y = \pm\frac{\sqrt{27}}{2} \approx 2.6.$$

Consequently, the points $(2, \pm\sqrt{27}/2)$ are on the graph. Similarly, $(-2, \pm\sqrt{27}/2)$ are solutions of the given equation. The graph is sketched in Figure 4.27.

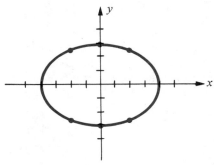

Figure 4.27 $9x^2 + 16y^2 = 144$ ∎

The horizontal and vertical line segments which join the x- and y-intercepts in Figure 4.27 are called the **axes** of the ellipse. The longer axis is called the **major axis**, and the shorter axis is called the **minor axis** of the ellipse. The major axis may be on the y-axis, as in the graph of $16x^2 + 9y^2 = 144$ sketched in Figure 4.28.

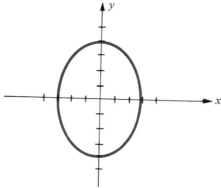

Figure 4.28 $16x^2 + 9y^2 = 144$

The graph of

**Equation of
a Hyperbola**

$$\frac{x^2}{a^2} - \frac{y^2}{b^2} = 1$$

is a **hyperbola** with center at the origin. The graph has the general appearance illustrated in Figure 4.29. (The significance of the dashes will be pointed out later). Note that the x-intercepts are $\pm a$. The corresponding points $V(a, 0)$ and $V'(-a, 0)$ are called the **vertices**, and the line segment $V'V$ is known as the **transverse axis** of the hyperbola. There are no y-intercepts, since the equation $-y^2/b^2 = 1$ has no solution.

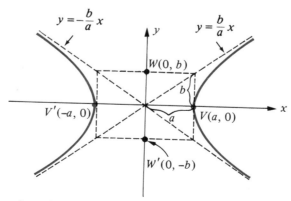

Figure 4.29 $\dfrac{x^2}{a^2} - \dfrac{y^2}{b^2} = 1$

If the equation $x^2/a^2 - y^2/b^2 = 1$ is solved for y, we obtain

$$y = \pm \frac{b}{a} \sqrt{x^2 - a^2}.$$

Hence, there are no points (x, y) on the graph if $x^2 - a^2 < 0$, that is, if $-a < x < a$. However, there *are* points $P(x, y)$ on the graph if $x \geq a$ or $x \leq -a$. If $x \geq a$, we may write the last equation in the form

$$y = \pm \frac{b}{a} x \sqrt{1 - \frac{a^2}{x^2}}.$$

It follows that if x is large (in comparison to a), the radicand is close to 1, and hence the ordinate y of the point $P(x, y)$ on the hyperbola is close to either $(b/a)x$ or $-(b/a)x$. This means that the point $P(x, y)$ is close to the line with equation $y = (b/a)x$ when y is positive, or the line with equation $y = -(b/a)x$ when y is negative. As x increases (or decreases), we say that the point $P(x, y)$ *approaches* one of these lines. A corresponding situation exists when $x \leq -a$. The lines with equations

$$y = \pm \frac{b}{a} x$$

are called the **asymptotes** of the hyperbola under discussion. The asymptotes serve as excellent guides for sketching the graph. This is illustrated in Figure 4.29, where we have represented the asymptotes by dashed lines and indicated the manner in which the points on the hyperbola approach the asymptotes as x increases or decreases. The two curves that make up the hyperbola are called the **branches** of the hyperbola.

A convenient way to sketch the asymptotes is first to plot the vertices $V(a, 0)$ and $V'(-a, 0)$ and the points $W(0, b)$ and $W'(0, -b)$ (see Figure 4.29). The line segment $W'W$ of length $2b$ is called the **conjugate axis** of the hyperbola. If horizontal and vertical lines are drawn through the endpoints of the conjugate and transverse axes, respectively, then the diagonals of the resulting rectangle have slopes b/a and $-b/a$. Consequently, by extending these diagonals, we obtain the asymptotes. The hyperbola is then sketched using the asymptotes as a guide.

Example 3   Discuss and sketch the graph of the equation

$$9x^2 - 4y^2 = 36.$$

Solution   Dividing both sides by 36, we obtain

$$\frac{x^2}{4} - \frac{y^2}{9} = 1$$

which is of the form shown in Figure 4.29 with $a^2 = 4$ and $b^2 = 9$. Hence, $a = 2$ and $b = 3$. The vertices $(\pm 2, 0)$ and the endpoints $(0, \pm 3)$ of the conjugate axis determine a rectangle whose diagonals (extended) give us the asymptotes. The graph is sketched in Figure 4.30.

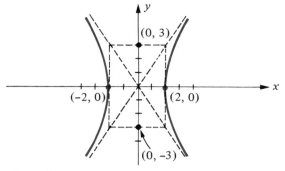

Figure 4.30 $\dfrac{x^2}{4} - \dfrac{y^2}{9} = 1$ ∎

If we interchange x and y in our discussion we obtain

$$\frac{y^2}{a^2} - \frac{x^2}{b^2} = 1.$$

In this case the graph is a hyperbola having vertices $(0, \pm a)$ and endpoints of conjugate axes $(\pm b, 0)$. A typical graph is illustrated in the next example.

Example 4 Discuss and sketch the graph of the equation

$$4y^2 - 2x^2 = 1.$$

Solution We may rewrite the equation as

$$\frac{y^2}{(1/4)} - \frac{x^2}{(1/2)} = 1$$

which is in the preceding form with $a^2 = 1/4$, $b^2 = 1/2$, and therefore, $a = 1/2$, $b = \sqrt{2}/2 \approx 0.7$. Thus, the vertices are $(0, \pm 1/2)$ and the endpoints of the conjugate axes are $(\pm \sqrt{2}/2, 0)$. As in Example 3 we use these four points to construct a rectangle and then obtain the asymptotes from the diagonals. This leads to the sketch in Figure 4.31.

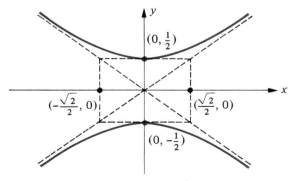

Figure 4.31 $4y^2 - 2x^2 = 1$ ∎

Our discussion of ellipses and hyperbolas can be extended to the case where the center is any point (h, k) and the axes are parallel to the x- and y-axes. Two typical graphs are illustrated in Figure 4.32. Note that their equations may be obtained from those discussed earlier by replacing x by $x - h$ and y by $y - k$. We shall not prove these facts.

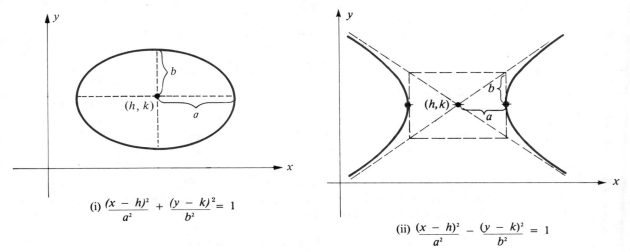

(i) $\dfrac{(x - h)^2}{a^2} + \dfrac{(y - k)^2}{b^2} = 1$

(ii) $\dfrac{(x - h)^2}{a^2} - \dfrac{(y - k)^2}{b^2} = 1$

Figure 4.32

If $a > b$, as in (i) of Figure 4.32, the major axis of the ellipse is parallel to the x-axis; however, if $a < b$, the major axis is parallel to the y-axis. If, in (ii) of the figure, we change the equation to $(y - k)^2/a^2 - (x - h)^2/b^2 = 1$, we obtain a hyperbola with transverse axis parallel to the y-axis.

Squaring the terms $(x - h)^2$ and $(y - k)^2$ of the equations in Figure 4.32 and then simplifying leads to an equation of the form

$$Ax^2 + By^2 + Cx + Dy + F = 0.$$

For ellipses, A and B have the same sign, whereas for hyperbolas, A and B have opposite signs. Conversely, given such an equation, we may use the technique of completing squares to obtain the center (h, k) of the conic (whenever the graph exists). The procedure is similar to that used for circles in Section 3.2, and is illustrated next, in Examples 5 and 6.

Example 5   Sketch the graph of the equation

$$16x^2 + 9y^2 + 64x - 18y - 71 = 0$$

Solution   We begin by writing the given equation in the form

$$16(x^2 + 4x) + 9(y^2 - 2y) = 71.$$

Next we complete the squares for the expressions within parentheses, obtaining

$$16(x^2 + 4x + 4) + 9(y^2 - 2y + 1) = 71 + 64 + 9.$$

Note that, by adding 4 to the expression within the first parentheses, we have added 64 to the left side of the equation and hence must compensate by adding 64 to the right side. Similarly, by adding 1 to the expression within the second parentheses, 9 is added to the left side and consequently 9 must also be added to the right side. The last equation may be written

$$16(x + 2)^2 + 9(y - 1)^2 = 144.$$

Dividing by 144 we obtain

$$\frac{(x + 2)^2}{9} + \frac{(y - 1)^2}{16} = 1$$

which has the form shown in (i) of Figure 4.32 with $h = -2$ and $k = 1$. Thus, the graph is an ellipse with center $(-2, 1)$. Since $16 > 9$, the major axis is parallel to the y-axis. The lengths of *half* the major and minor axes are 4 and 3, respectively. We may, therefore, obtain the vertices as indicated in Figure 4.33. This gives us a rough sketch of the graph. If more accuracy is desired we could find the x- and y-intercepts, or plot several additional points.

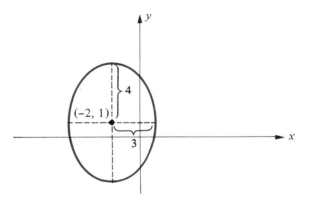

Figure 4.33 ■

Example 6 Discuss and sketch the graph of the equation

$$9x^2 - 4y^2 - 54x - 16y + 29 = 0.$$

Solution As in Example 5, we arrange our work as follows:

$$9(x^2 - 6x) - 4(y^2 + 4y) = -29$$
$$9(x^2 - 6x + 9) - 4(y^2 + 4y + 4) = -29 + 81 - 16.$$

Note that, because of the -4 outside the second parentheses, adding 4 within that parentheses results in *subtracting* 16 from the left side of the equation. Hence, we must compensate by subtracting 16 from the right side. Next we write

$$9(x - 3)^2 - 4(y + 2)^2 = 36$$

$$\frac{(x - 3)^2}{4} - \frac{(y + 2)^2}{9} = 1$$

which has the form shown in (ii) of Figure 4.32 with $h = 3$, $k = -2$, $a = 2$, and $b = 3$. Thus, the graph is a hyperbola with center $(3, -2)$ and transverse axis parallel to the x-axis. Since $a = 2$ we may find the vertices by proceeding 2 units to the left and right of the center. The endpoints of the conjugate axis are found in like manner, using $b = 3$. These four points may be used to construct a rectangle whose diagonals determine the asymptotes. We then sketch the graph as in Figure 4.34 (compare with Figure 4.30).

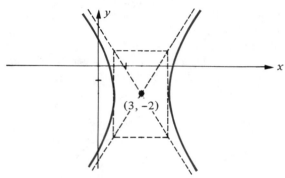

Figure 4.34 $\dfrac{(x - 3)^2}{4} - \dfrac{(y + 2)^2}{9} = 1$

EXERCISES 4.5

Sketch the graphs of the equations in Exercises 1–32.

1   $x = 4y^2$

2   $x = -2y^2$

3   $x = 4y^2 - 3$

4   $x = 8 - 2y^2$

5   $x = 9 - y^2$

6   $x = y^2 + 4$

7   $x = y^2 + 4y + 1$

8   $x = y^2 + 6y + 5$

9   $x = -2y^2 + 12y - 13$

10   $x = 4y - y^2$

11   $y = x^2 - 4x - 6$

12   $y = 3x^2 - 30x + 75$

13   $x^2 + 5y^2 = 25$

14   $4x^2 + 25y^2 = 100$

15   $9x^2 + 4y^2 = 36$

16   $9x^2 + y^2 = 81$

17   $5x^2 + 3y^2 = 15$

18   $4x^2 + 3y^2 = 24$

19   $25x^2 + 9y^2 = 16$

20   $x^2 + 8y^2 = 4$

21   $\dfrac{x^2}{4} + \dfrac{y^2}{9} = 2$

22   $\dfrac{x^2}{4} + \dfrac{y^2}{4} = 1$

23   $9x^2 - 25y^2 = 225$

24   $16x^2 - 9y^2 = 144$

25   $25y^2 - 9x^2 = 225$

26   $16y^2 - 9x^2 = 144$

27 $y^2 - 36x^2 = 36$ **28** $x^2 - 5y^2 = 10$

29 $x^2 - y^2 = 1$ **30** $y^2 - x^2 = 4$

31 $4x^2 - 9y^2 = -1$ **32** $y^2 - 25x^2 = 1$

33 $\dfrac{(x+5)^2}{16} + \dfrac{(y-4)^2}{9} = 1$

34 $8(x+3)^2 + (y-6)^2 = 32$

35 $\dfrac{(x-2)^2}{36} - \dfrac{(y+7)^2}{49} = 1$

36 $\dfrac{(y-2)^2}{9} - (x+4)^2 = 1$

37 $4(x+2)^2 + (y-2)^2 = 1$

38 $(x+1)^2 + 9y^2 = 36$

39 $16y^2 - 100(x-3)^2 = 1600$

40 $(x-7)^2 - (y+5)^2 = 1$

41 $4x^2 + y^2 + 24x - 10y + 45 = 0$

42 $9x^2 + 16y^2 + 36x + 96y + 36 = 0$

43 $9x^2 + y^2 - 108x - 4y + 319 = 0$

44 $x^2 + 4y^2 - 2x = 0$

45 $y^2 - 4x^2 + 6y - 40x - 107 = 0$

46 $25y^2 - 9x^2 - 100y - 54x + 10 = 0$

47 $9x^2 - y^2 - 36x + 12y - 9 = 0$

48 $4y^2 - x^2 + 32y - 8x + 49 = 0$

In Exercises 49 and 50 find an equation of an ellipse with center at the origin which has the given intercepts.

49 x-intercepts ± 10, y-intercepts ± 5

50 x-intercepts ± 8, y-intercepts $\pm 1/2$

51 An arch of a bridge is semielliptical with its longer axis horizontal. The base of the arch is 30 feet across and the highest part of the arch is 10 feet above the horizontal roadway. Find the height of the arch 6 feet from the center of the base.

52 Determine A so that the point $(2, -3)$ is on the conic $Ax^2 + 2y^2 = 4$. Is the conic an ellipse or a hyperbola?

53 If a square with sides parallel to the coordinate axes is inscribed in the ellipse with equation $x^2/a^2 + y^2/b^2 = 1$, express the area A of the square in terms of a and b.

54 The **eccentricity** e of the ellipse having equation $(x^2/a^2) + (y^2/b^2) = 1$ is defined as the ratio $\sqrt{a^2 - b^2}/a$. Prove that $0 < e < 1$. If a is fixed and b varies, describe the general shape of the ellipse when the eccentricity is close to 1 and when it is close to 0.

55 The graphs of the equations

$$\frac{x^2}{a^2} - \frac{y^2}{b^2} = 1 \quad \text{and} \quad \frac{x^2}{a^2} - \frac{y^2}{b^2} = -1$$

are called **conjugate hyperbolas**. Sketch the graphs of both equations on the same coordinate plane if $a = 5$ and $b = 3$. Describe the relationship between the two graphs.

56 Find an equation of a hyperbola having x-intercepts ± 2 and asymptotes $y = \pm 3x$.

4.6 REVIEW

Concepts

Define or discuss each of the following.

 1 Polynomial function

 2 Quadratic function

 3 Graph of a quadratic function

 4 Maximum or minimum values of quadratic functions

 5 Intermediate Value Theorem

 6 Graphs of polynomial functions of degree greater than 2

7 Rational function

8 Vertical asymptote

9 Horizontal asymptote

10 Partial fractions

11 Equations of parabolas, ellipses, and hyperbolas

12 Asymptotes of a hyperbola

13 Conic sections

Exercises

Sketch the graphs of the equations in Exercises 1–10.

1 $y^2 = 1 - x$ **2** $4x^2 + y = 10$

3 $y = x^2 + 6x + 16$ **4** $y^2 - 16y - x = 15$

5 $4y = (x + 2)(x - 1)^2(3 - x)$

6 $y = -4x^2 + 8x - 6$

7 $x^2 + 4y^2 = 64$ **8** $4x^2 + y^2 = 64$

9 $x^2 - 4y^2 = 64$ **10** $4y^2 - x^2 = 64$

In each of Exercises 11–20 sketch the graph of f.

11 $f(x) = (x - 4)^2$

12 $f(x) = 12x^2 + 5x - 3$

13 $f(x) = \dfrac{-2}{(x + 1)^2}$

14 $f(x) = \dfrac{1}{(x - 1)^3}$

15 $f(x) = \dfrac{3x^2}{16 - x^2}$

16 $f(x) = \dfrac{x}{(x + 5)(x^2 - 5x + 4)}$

17 $f(x) = (x + 2)^3$

18 $f(x) = 2x^2 + x^3 - x^4$

19 $f(x) = x^3 + 2x^2 - 8x$

20 $f(x) = x^6 - 32$

In Exercises 21 and 22 find the maximum or minimum value of $f(x)$ by completing the square.

21 $f(x) = 5x^2 + 30x + 49$

22 $f(x) = -3x^2 + 30x - 82$

23 The interior of a half-mile race track consists of a rectangle with semicircles at two opposite ends. Find the dimensions which will maximize the area of the rectangle.

24 At 1:00 P.M. ship A is 30 miles due south of ship B and is sailing north at a rate of 15 miles per hour. If ship B is sailing west at a rate of 10 miles per hour, at what time will the distance between the ships be minimal?

Find the partial fraction decompositions of the following:

25 $\dfrac{4x^2 + 54x + 134}{(x + 3)(x^2 + 4x - 5)}$

26 $\dfrac{x^2 + 14x - 13}{x^3 + 5x^2 + 4x + 9}$

Sketch the graph of each of the following equations:

27 $4x^2 + 9y^2 + 24x - 36y + 36 = 0$

28 $4x^2 + y^2 - 24x + 4y + 36 = 0$

29 $y^2 - 2x^2 + 6y + 8x - 3 = 0$

30 $4x^2 - y^2 - 40x - 8y + 88 = 0$

5

EXPONENTIAL AND LOGARITHMIC FUNCTIONS

Exponential and logarithmic functions are very important in mathematics and the sciences. In this chapter we shall define these functions and study some of their properties.

5.1 EXPONENTIAL FUNCTIONS

Throughout this section the letter a will denote a positive real number. In Section 1.6 we defined a^r, where r is any rational number. It is also possible to define a unique real number a^x for every real number x (rational or irrational) in such a way that the Laws of Exponents remain valid. To illustrate, given a number such as a^π, we may use the nonterminating decimal representation 3.14159 . . . for π and consider the numbers a^3, $a^{3.1}$, $a^{3.14}$, $a^{3.141}$, $a^{3.1415}$, ... We might expect that each successive power gets closer to a^π. This is precisely what happens if a^x is properly defined. However, the definition requires deeper concepts than are available to us and, consequently, it is better to leave it for a more advanced mathematics course such as calculus.

Although definitions and proofs are omitted, we shall assume, henceforth, that the Laws of Exponents (see Section 1.4) are valid for all *real* exponents.

Before continuing the discussion of real exponents, we shall establish several results about rational exponents. Let us first show that if $a > 1$, then $a^r > 1$ for every positive rational number r. If $a > 1$, then multiplying both sides by a we obtain $a^2 > a$, and hence $a^2 > a > 1$. Multiplying by a again we obtain $a^3 > a^2 > a > 1$. Continuing, we see that $a^n > 1$ for every positive integer n. (A complete proof requires the method of mathematical induction discussed in Section 11.1.) Similarly, if $0 < a < 1$, it follows that $a^n < 1$ for every positive integer n. Next suppose that $a > 1$ and $r = p/q$, where p and q are positive integers. If it were true that $a^{p/q} \leq 1$, then from the previous discussion, $(a^{p/q})^q \leq 1$, or $a^p \leq 1$, which contradicts the fact that $a^n > 1$ for every positive integer n. Consequently, $a^r > 1$ for every positive rational number r. We may use this fact to prove the following theorem.

Theorem

> If $a > 1$ and if r and s are rational numbers such that $r < s$, then $a^r < a^s$.

Proof

If $r < s$, then $s - r$ is a positive rational number, and from the previous discussion, $a^{s-r} > 1$. Multiplying both sides by a^r we obtain

$$(a^{s-r})a^r > 1 \cdot a^r \quad \text{or} \quad a^s > a^r$$

which is what we wished to prove.   ∎

It is possible to extend the preceding theorem to *real* exponents r and s.

Since to each real number x there corresponds a unique real number a^x, we can define a function as follows.

Definition

> If $a > 0$, then the **exponential function f with base a** is defined by
>
> $$f(x) = a^x$$
>
> where x is any real number.

If $a > 1$, and x_1 and x_2 are real numbers such that $x_1 < x_2$, then $a^{x_1} < a^{x_2}$, that is, $f(x_1) < f(x_2)$. This means that if $a > 1$, then the exponential function f with base a is increasing throughout its domain. It can also be shown that if $0 < a < 1$, then f is decreasing.

Example 1   Sketch the graph of f if $f(x) = 2^x$.

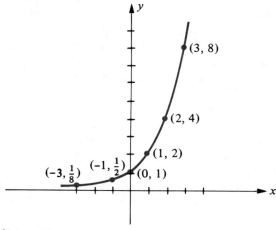

Figure 5.1 $f(x) = 2^x$ ∎

Solution   Coordinates of some points on the graph are listed in the following table.

x	-3	-2	-1	0	1	2	3	4
2^x	$\frac{1}{8}$	$\frac{1}{4}$	$\frac{1}{2}$	1	2	4	8	16

Plotting points and using the fact that f is increasing, gives us the sketch in Figure 5.1.   ∎

The graph in Figure 5.1 is typical of the exponential function with base a if $a > 1$. Since $a^0 = 1$, the y-intercept is always 1. Observe that, as x decreases through negative values, the graph approaches the x-axis but never intersects it, since $a^x > 0$ for all x. This means that the x-axis is a **horizontal asymptote** for the graph. As x increases through positive values the graph rises very rapidly. Indeed, if we begin with $x = 0$ and consider successive unit changes in x, then the corresponding changes in y are 1, 2, 4, 8, 16, 32, 64, and so on. This type of variation is very common in nature and is characteristic of the **exponential law of growth**. In this case f is called a **growth function**. At the end of this section we shall give several practical applications.

Example 2   If $f(x) = (3/2)^x$ and $g(x) = 3^x$, sketch the graphs of f and g on the same coordinate plane.

Solution   The following table displays coordinates of several points on the graphs.

x	-2	-1	0	1	2	3	4
$(\frac{3}{2})^x$	$\frac{4}{9}$	$\frac{2}{3}$	1	$\frac{3}{2}$	$\frac{9}{4}$	$\frac{27}{8}$	$\frac{81}{16}$
3^x	$\frac{1}{9}$	$\frac{1}{3}$	1	3	9	27	81

With the aid of these points we obtain Figure 5.2, where dashes have been used for the graph of f in order to distinguish it from the graph of g.

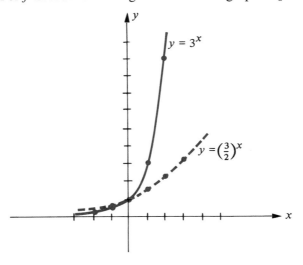

Figure 5.2   ∎

Example 2 brings out the fact that if $1 < a < b$, then $a^x < b^x$ for positive values of x and $b^x < a^x$ for negative values of x. In particular, since $\frac{3}{2} < 2 < 3$, this tells us that the graph of $y = 2^x$ in Example 1 lies between the graphs in Example 2.

Example 3   Sketch the graph of the equation $y = (\frac{1}{2})^x$.

Solution   Some points on the graph may be obtained from the following table.

x	-3	-2	-1	0	1	2	3
$(\frac{1}{2})^x$	8	4	2	1	$\frac{1}{2}$	$\frac{1}{4}$	$\frac{1}{8}$

The graph is sketched in Figure 5.3. Since $(\frac{1}{2})^x = 2^{-x}$, the graph is the same as the graph of the equation $y = 2^{-x}$.

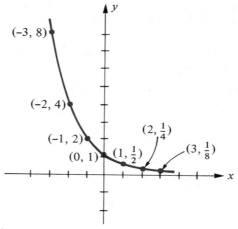

Figure 5.3 $y = (\frac{1}{2})^x = 2^{-x}$

In advanced mathematics and applications it is often necessary to consider a function f such that $f(x) = a^p$, where p is some expression in x. We do not intend to study such functions in detail; however, let us consider one example.

Example 4   Sketch the graph of f if $f(x) = 2^{-x^2}$.

Solution   Since $f(x) = 1/2^{x^2}$, it follows that if x increases numerically, the corresponding point $(x, f(x))$ on the graph approaches the x-axis. Thus, the x-axis is a horizontal asymptote for the graph. The maximum value of $f(x)$ occurs at $x = 0$. Since f is an even function, the graph is symmetric with respect to the y-axis (see Section 3.4). Several points on the graph are $(0, 1)$, $(1, 1/2)$, and $(2, 1/16)$. Plotting and using symmetry gives us the sketch in Figure 5.4. Functions of this type arise in the study of the branch of mathematics called *probability*.

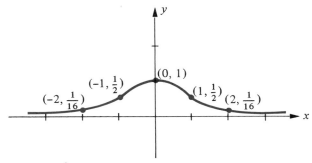

Figure 5.4 $y = 2^{-x^2}$ ∎

Application (Biology)

The variation of many physical entities can be described by means of exponential functions. One of the most common examples occurs in the growth of certain populations. As an illustration, it might be observed experimentally that the number of bacteria in a culture doubles every hour. If 1,000 bacteria are present at the start of the experiment, then the experimenter would obtain the readings listed below, where t is the time in hours and $f(t)$ is the bacteria count at time t.

t	0	1	2	3	4
$f(t)$	1,000	2,000	4,000	8,000	16,000

It appears that $f(t) = (1,000)2^t$. This formula makes it possible to predict the number of bacteria present at any time t. For example, at $t = 1.5$ we have

$$f(t) = (1,000)2^{3/2} = 1,000\sqrt{2^3} = 1,000\sqrt{8} \approx 2,828.$$

Exponential growth patterns of this type may also be observed in some human and animal populations.

Application (Compound Interest)

Compound interest provides another illustration of exponential growth. If a sum of money P (called the **principal**) is invested at a simple interest rate of i percent, then the interest at the end of one interest period is Pi. For example, if $P = \$100$ and i is 8% per year, then the interest at the end of one year is $\$100(0.08)$, or $\$8$. If the interest is reinvested at the end of this period then the new principal is

$$P + Pi, \quad \text{or} \quad P(1 + i).$$

Note that, to find the new principal, we multiply the original principal by $(1 + i)$. In the preceding illustration the new principal is $\$100(1.08)$, or $\$108$.

If another time period elapses, then the new principal may be found by multiplying $P(1 + i)$ by $(1 + i)$. Thus the principal after two time periods is $P(1 + i)^2$. If we reinvest and continue this process, the principal after three periods is $P(1 + i)^3$;

after four it is $P(1 + i)^4$; and in general, after n time periods the principal P_n is given by

$$P_n = P(1 + i)^n.$$

Interest accumulated in this way is called **compound interest**. We see that the principal is given in terms of an exponential function whose base is $1 + i$ and exponent is n. The time period may vary, being measured in years, months, weeks, days, or any other suitable unit of time. When this formula for P_n is employed, it must be remembered that i is the interest rate per time period. For example, if the rate is stated as 9% *per year compounded monthly*, then the rate per month is $\frac{9}{12}\%$, or equivalently, $\frac{3}{4}\%$. In this case, $i = 0.75\% = 0.0075$ and n is the number of months. The sketch in Figure 5.5 illustrates the growth of $100 invested at this rate over a period of 15 years. We have connected the principal amounts by a curve in order to indicate the growth during this time. Due to the complexity of the base, the use of the formula for P_n is extremely tedious unless tables or calculators are available. For those with calculators, several problems on compound interest have been included in the exercises.

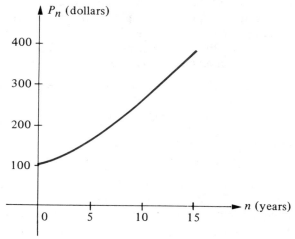

Figure 5.5   *Compound interest:* $P_n = 100(1.075)^{12n}$

Application (Radioactive Decay)

Certain physical quantities may *decrease* exponentially. In such cases the base a of the exponential function is between 0 and 1. One of the most common examples is the decay of a radioactive substance. As an illustration, the polonium isotope ^{210}Po has a half-life of approximately 140 days; that is, given any amount, one-half of it will disintegrate in 140 days. If there is initially 20 mg present, then the following table indicates the amount remaining after various intervals of time.

t (days)	0	140	280	420	560
Amount remaining (mg)	20	10	5	2.5	1.25

The sketch in Figure 5.6 illustrates the exponential nature of the disintegration.

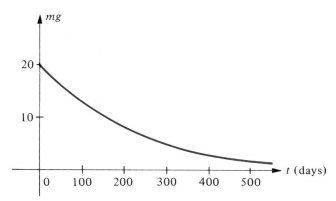

Figure 5.6 *Decay of polonium*

The behavior of an electrical condenser can be used to illustrate exponential decay. If the condenser is allowed to discharge, the initial rate of discharge is relatively high, but it then tapers off as in the preceding example of radioactive decay.

In calculus and its applications a certain irrational number, denoted by e, is often used as the base of an exponential function. To five decimal places,

$$e \approx 2.71828.$$

The base e arises naturally in the discussion of many physical phenomena. For this reason the function f defined by $f(x) = e^x$ is called the **natural exponential function**. Since $2 < e < 3$, the graph of f lies "between" the graphs of the equations $y = 2^x$ and $y = 3^x$. Many hand-held calculators have an e^x key which can be used to approximate values of the natural exponential function. There may also be a key marked y^x that can be used for any positive base y. For those who have access to calculators, we have included some exercises involving e^x and y^x.

7 – 36   3/2/()

EXERCISES 5.1

In each of Exercises 1–24 sketch the graph of the function f.

1 $f(x) = 4^x$

2 $f(x) = 5^x$

3 $f(x) = 10^x$

4 $f(x) = 8^x$

5 $f(x) = 3^{-x}$

6 $f(x) = 4^{-x}$

7 $f(x) = -2^x$

8 $f(x) = -3^x$

9 $f(x) = 4 - 2^{-x}$

10 $f(x) = 2 + 3^{-x}$

11 $f(x) = (2/3)^x$

12 $f(x) = (3/4)^{-x}$

13 $f(x) = (5/2)^{-x}$

14 $f(x) = (4/3)^x$

15 $f(x) = 2^{|x|}$

16 $f(x) = 2^{-|x|}$

17 $f(x) = 2^{x+3}$

18 $f(x) = 3^{x+2}$

19 $f(x) = 2^{3-x}$

20 $f(x) = 3^{-2-x}$

21 $f(x) = 3^{1-x^2}$

22 $f(x) = 2^{-(x+1)^2}$

23 $f(x) = 3^x + 3^{-x}$

24 $f(x) = 3^x - 3^{-x}$

25 If $f(x) = 2^x$ and $g(x) = x^2$, illustrate the difference in the rates of growth of f and g for $x \geq 0$ by sketching the graphs of both functions on the same coordinate plane.

26 Repeat Exercise 25 if $f(x) = 2^{-x}$ and $g(x) = x^{-2}$.

27 The number of bacteria in a certain culture increased from 600 to 1800 between 7:00 A.M. and 9:00 A.M. Assuming exponential growth, it can be shown, using methods of calculus, that the number $f(t)$ of bacteria t hours after 7:00 A.M. was given by $f(t) = 600(3)^{t/2}$.
(a) Estimate the number of bacteria in the culture at 8:00 A.M.; 10:00 A.M.; 11:00 A.M.
(b) Sketch the graph of f from $t = 0$ to $t = 4$.

28 According to Newton's Law of Cooling, the rate at which an object cools is directly proportional to the difference in temperature between the object and the surrounding medium. If a certain object cools from $125°$ to $100°$ in 30 minutes when surrounded by air at a temperature of $75°$, then it can be shown that its temperature $f(t)$ after t hours of cooling is given by $f(t) = 50(2)^{-2t} + 75$.
(a) If $t = 0$ corresponds to 1:00 P.M. approximate, to the nearest tenth of a degree, the temperature at 2:00 P.M.; 3:30 P.M.; 4:00 P.M.
(b) Sketch the graph of f from $t = 0$ to $t = 4$.

29 The radioactive isotope ^{210}Bi has a half-life of 5 days, that is, the number of radioactive particles will decrease to one-half the number in 5 days. If there are 100 mg of ^{210}Bi present at $t = 0$, then the amount $f(t)$ remaining after t days is given by $f(t) = 100(2)^{-t/5}$.
(a) How much remains after 5 days? 10 days? 12.5 days?
(b) Sketch the graph of f from $t = 0$ to $t = 30$.

30 The number of bacteria in a certain culture at time t is given by $Q(t) = 2(3^t)$, where t is measured in hours and $Q(t)$ in thousands. What is the initial number of bacteria? What is the number after 10 minutes? After 30 minutes? After 1 hour?

31 The half-life of radium is 1,600 years; that is, given any quantity, one-half of it will disintegrate in 1,600 years. If the initial amount is q_0 milligrams, then it can be shown that the quantity $q(t)$ remaining after t years is given by $q(t) = q_0 2^{kt}$. Find k.

32 If 10 grams of salt are added to a certain quantity of water, then the amount $q(t)$ which is undissolved after t minutes is given by $q(t) = 10(\frac{4}{5})^t$. Sketch a graph which shows the value $q(t)$ at any time from $t = 0$ to $t = 10$.

33 If $1,000 is invested at a rate of 12% per year compounded monthly, what is the principal after 1 month? 2 months? 6 months? 1 year?

34 If a savings fund pays interest at a rate of 10% compounded semiannually, how much money invested now will amount to $5,000 after one year?

35 If a certain make of automobile is purchased for C dollars, then its trade-in value $v(t)$ at the end of t years is given by $v(t) = 0.78C(0.85)^{t-1}$. If the original cost is $10,000, calculate, to the nearest dollar, the value after (a) 1 year; (b) 4 years; (c) 7 years.

36 If the value of real estate increases at a rate of 10% per year, then after t years the value V of a house purchased for P dollars is given by $V = P(1.1)^t$. If a house was purchased for $80,000 in 1980, what will it be worth in 1984?

37 Why was $a < 0$ ruled out in the discussion of a^x?

38 Prove that if $0 < a < 1$ and r and s are rational numbers such that $r < s$, then $a^r > a^s$.

39 How does the graph of $y = a^x$ compare with the graph of $y = -a^x$?

40 If $a > 1$, how does the graph of $y = a^x$ compare with the graph of $y = a^{-x}$?

CALCULATOR EXERCISES 5.1

1 Approximate $3^{\sqrt{2}}$ and compare this number with the successive approximations

$$3^{1.4}, \quad 3^{1.41}, \quad 3^{1.414}, \quad 3^{1.4142}.$$

2 Approximate 2^{π} and compare this number with the successive approximations

$$2^{3.1}, \quad 2^{3.14}, \quad 2^{3.141}, \quad 2^{3.1415}.$$

In Exercises 3 and 4 use decimal approximations to replace $\Box$ by $<$, $>$, or $=$.

3 $(\sqrt{2})^{\sqrt{3}} \Box (\sqrt{3})^{\sqrt{2}}$ **4** $\pi^{22/7} \Box (22/7)^{\pi}$

In Exercises 5 and 6 demonstrate that the given equality is true by expressing each side as a decimal (do not use laws of exponents).

5 (a) $5^{\sqrt{2}}5^{\sqrt{3}} = 5^{\sqrt{2}+\sqrt{3}}$
  (b) $5^{\sqrt{3}}/5^{\sqrt{2}} = 5^{\sqrt{3}-\sqrt{2}}$

6 $(5^{\sqrt{3}})^{\sqrt{2}} = 5^{\sqrt{6}}$

In Exercises 7 and 8 sketch the graph corresponding to $-3 \le x \le 3$ by choosing values of x at intervals of length 0.5 as follows: $x = -3, -2.5, -2, -1.5$, etc.

7 $y = (1.8)^x$ **8** $y = (2.3)^{-x^2}$

Solve Exercises 9–12 by using the compound interest formula $P_n = P(1 + i)^n$.

9 If \$1,000 is invested at an interest rate of 6% per year compounded quarterly, find the principal at the end of
  (a) one year.   (b) two years.
  (c) five years.   (d) ten years.
  (*Hint*: $i = \frac{1}{4}(6\%) = 1.5\%$ and, at the end of one year, $n = 4$.)

10 Rework Exercise 9 if the interest rate is 6% per year compounded monthly.

11 If \$10,000 is invested at a rate of 9% per year compounded semi-annually, how long will it take for the principal to exceed
  (a) \$15,000?   (b) \$20,000?
  (c) \$30,000?

12 A certain department store requires its credit card customers to pay interest at the rate of 18% per year compounded monthly on any unpaid bills. If a man buys a \$500 television set on credit and then makes no payments for one year, how much does he owe?

13 Sketch the graph of f if $f(x)$ is given by:
  (a) e^x (b) e^{-x}

  (c) $\dfrac{e^x + e^{-x}}{2}$ (d) $\dfrac{2}{e^x + e^{-x}}$

  (e) $\dfrac{e^x - e^{-x}}{2}$ (f) $\dfrac{2}{e^x - e^{-x}}$

14 Sketch the graph of the equation $y = e^{-x^2/2}$.

15 Sketch the graph of $y = x(2^x)$.

16 If $f(x) = x^2(5^{\sqrt{x}})$, find $f(1.01) - f(1)$.

17 If $f(x) = (1 + x)^{1/x}$, approximate the following to four decimal places:

$$f(1), \quad f(0.1), \quad f(0.01), \quad f(0.001),$$
$$f(0.0001), \quad f(0.00001).$$

Compare your answers with the value of $e \approx 2.71828$ and arrive at a conjecture.

18 The population of a certain city is increasing at the rate of 5% per year and the present population is 500,000. Using methods developed in calculus, it is estimated that the population after t more years will be approximately $500,000e^{0.05t}$. Estimate the population after (a) 10 years; (b) 20 years; (c) 40 years.

19 Under certain conditions the atmospheric pressure p (in inches) at altitude h ft is given by

$$p = 29e^{-0.000034h}.$$

What is the pressure at an altitude of 40,000 ft?

20 Starting with c milligrams of the polonium isotope ^{210}Po, the amount remaining after t days may be approximated by $A = ce^{-0.00495t}$. If the initial amount is 50 milligrams, find, to the nearest hundredth, the amount remaining after (a) 30 days; (b) 180 days; (c) 365 days.

21 Growths of physical quantities are often more stable than that given by $f(t) = ke^{at}$, where t is time. In biology the formula

$$G(t) = ke^{-(Ae^{-Bt})}$$

where A, B, and k are positive constants, is used to estimate certain populations. The graph of G is called a **Gompertz growth curve**. Sketch the graph of G for the special case $k = 10$, $A = 1/2$, $B = 1$, and $t \ge 0$, and describe what happens as t increases without bound.

22 The graph of

$$f(x) = a + b(1 - e^{-cx})$$

where a, b, and c are positive constants may be used to describe certain learning processes. To illustrate, suppose a manufacturer estimates that a new employee can produce 3 items the first day on the job. As the employee becomes more proficient, more items per day can be produced until a certain maximum production is reached. Suppose that after n days on the job, the number $f(n)$ of items produced is approximated by

$$f(n) = 3 + 20(1 - e^{-0.1n}).$$

(a) Estimate the number of items produced on the fifth day; the ninth day; the twenty-fourth day; the thirtieth day.

(b) Sketch the graph of f from $n = 0$ to $n = 30$. (Graphs of this type are called **learning curves**, and are used frequently in education and psychology).

(c) What happens as n increases without bound?

5.2   LOGARITHMS

Throughout this section and the next, it is assumed that a is a positive real number different from the number 1. Let us begin by examining the graph of f, where $f(x) = a^x$ (see Figure 5.7 for the case $a > 1$). It appears that every positive real number u is the ordinate of some point on the graph; that is, there is a number v such that $u = a^v$. Indeed, v is the abscissa of the point where the line $y = u$ intersects the graph. Moreover, since f is increasing throughout its domain, the number u can occur as an ordinate only once. A similar situation exists if $0 < a < 1$. This makes the following theorem plausible.

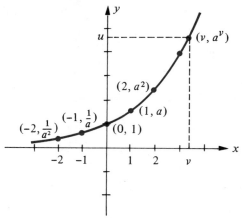

Figure 5.7 $f(x) = a^x, a > 1$

Theorem

For each positive real numbers u there is a unique real number v such that

$$a^v = u.$$

A rigorous proof of this theorem requires concepts studied in calculus.

Definition

> If u is any positive real number, then the (unique) exponent v such that $a^v = u$ is called the **logarithm of u with base a** and denoted by $\log_a u$.

A convenient way to memorize this definition is by means of the following statement:

$$v = \log_a u \quad \text{if and only if} \quad a^v = u.$$

As illustrations we may write

$$3 = \log_2 8 \quad \text{since } 2^3 = 8$$
$$-2 = \log_5 \tfrac{1}{25} \quad \text{since } 5^{-2} = \tfrac{1}{25}$$
$$4 = \log_{10} 10{,}000 \quad \text{since } 10^4 = 10{,}000.$$

The fact that $\log_a u$ is the exponent v such that $a^v = u$ gives us the following important identity:

$$a^{\log_a u} = u.$$

Since $a^r = a^r$ for every real number r, we may use the definition of logarithm with $v = r$ and $u = a^r$ to obtain

$$r = \log_a a^r.$$

In particular, letting $r = 1$ we see that $\log_a a = 1$.

Since $a^0 = 1$, it follows from the definition of logarithm that

$$\log_a 1 = 0.$$

Example 1 In each of the following find s.

    (a) $s = \log_4 2$ (b) $\log_5 s = 2$ (c) $\log_s 8 = 3$

Solutions

    (a)   If $s = \log_4 2$, then $4^s = 2$ and hence $s = \tfrac{1}{2}$.

    (b)   If $\log_5 s = 2$, then $5^2 = s$ and hence $s = 25$.

    (c)   If $\log_s 8 = 3$, then $s^3 = 8$ and hence $s = \sqrt[3]{8} = 2$.   ∎

Example 2 Solve the equation $\log_4 (5 + x) = 3$.

Solution If $\log_4 (5 + x) = 3$, then

$$5 + x = 4^3 \quad \text{or} \quad 5 + x = 64.$$

Hence, the solution is $x = 59$.   ∎

The following laws are fundamental for all work with logarithms, where it is assumed that u and w are positive real numbers.

**Laws of
Logarithms**

> (i) $\log_a(uw) = \log_a u + \log_a w$
>
> (ii) $\log_a(u/w) = \log_a u - \log_a w$
>
> (iii) $\log_a(u^c) = c\log_a u$, for every real number c

Proof

To prove (i), we begin by letting

$$r = \log_a u \quad \text{and} \quad s = \log_a w.$$

By the definition of logarithm, this implies that $a^r = u$ and $a^s = w$. Consequently,

$$a^r a^s = uw$$

and hence, by a law of exponents,

$$a^{r+s} = uw.$$

Again using the definition of logarithm, we see that the last equation is equivalent to

$$r + s = \log_a(uw).$$

Substituting for r and s from the first step in the proof gives us

$$\log_a u + \log_a w = \log_a uw.$$

This completes the proof of (i).

To prove (ii) we begin as in the proof of (i), but this time we divide a^r by a^s, obtaining

$$\frac{a^r}{a^s} = \frac{u}{w}, \quad \text{or} \quad a^{r-s} = \frac{u}{w}.$$

The last equation is equivalent to

$$r - s = \log_a(u/w).$$

Substituting for r and s gives us (ii).

Finally, if c is any real number, then using the same notation as above,

$$(a^r)^c = u^c, \quad \text{or} \quad a^{cr} = u^c.$$

According to the definition of logarithm, the last equality implies that

$$cr = \log_a u^c.$$

Substituting for r, we obtain

$$c\log_a u = \log_a u^c.$$

This proves Law (iii).   ∎

The following examples illustrate uses of the Laws of Logarithms.

Example 3   If $\log_a 3 = 0.4771$ and $\log_a 2 = 0.3010$, find each of the following.

(a) $\log_a 6$ (b) $\log_a (\tfrac{3}{2})$ (c) $\log_a \sqrt{2}$ (d) $\dfrac{\log_a 3}{\log_a 2}$

Solutions

(a)   Since $6 = 2 \cdot 3$, we may use Law (i) to obtain

$$\log_a 6 = \log_a (2 \cdot 3) = \log_a 2 + \log_a 3$$
$$= 0.4771 + 0.3010 = 0.7781.$$

(b)   By Law (ii),
$$\log_a (\tfrac{3}{2}) = \log_a 3 - \log_a 2$$
$$= 0.4771 - 0.3010 = 0.1761.$$

(c)   Using Law (iii),
$$\log_a \sqrt{2} = \log_a 2^{1/2} = (\tfrac{1}{2}) \log_a 2$$
$$= (\tfrac{1}{2})(0.3010) = 0.1505.$$

(d)   There is no law of logarithms which allows us to simplify $(\log_a 3)/(\log_a 2)$. Consequently, we *divide* 0.4771 by 0.3010, obtaining the approximation 1.585. It is important to notice the difference between this problem and the one stated in part (b).   ∎

Example 4   Solve each of the following equations.

(a) $\log_5 (2x + 3) = \log_5 11 + \log_5 3$

(b) $\log_4 (x + 6) - \log_4 10 = \log_4 (x - 1) - \log_4 2$

Solutions

(a)   Using Law (i), the given equation may be written as

$$\log_5 (2x + 3) = \log_5 (11 \cdot 3) = \log_5 33.$$

Consequently, $2x + 3 = 33$, or $2x = 30$, and therefore, the solution is $x = 15$.

(b)   The given equation is equivalent to

$$\log_4 (x + 6) - \log_4 (x - 1) = \log_4 10 - \log_4 2.$$

Applying Law (ii) gives us

$$\log_4 \left(\frac{x + 6}{x - 1} \right) = \log_4 \frac{10}{2} = \log_4 5$$

and hence,

$$\frac{x + 6}{x - 1} = 5.$$

The last equation implies that

$$x + 6 = 5x - 5 \quad \text{or} \quad 4x = 11.$$

Thus, the solution is $x = 11/14$.   ∎

Extraneous solutions sometimes occur in the process of solving logarithmic equations, as illustrated in the next example.

Example 5 Solve the equation $2\log_7 x = \log_7 36$.

Solution   Applying Law (iii), we obtain $2\log_7 x = \log_7 x^2$, and substitution in the given equation leads to

$$\log_7 x^2 = \log_7 36.$$

Consequently, $x^2 = 36$ and hence either $x = 6$ or $x = -6$. However, $x = -6$ is not a solution of the original equation since x must be positive in order for $\log_7 x$ to exist. Thus, there is only one solution, $x = 6$.   ■

Sometimes it is necessary to *change the base* of a logarithm by expressing $\log_b u$ in terms of $\log_a u$, for some positive real number b different from 1. This can be accomplished as follows. We begin with the equivalent equations

$$v = \log_b u \quad \text{and} \quad b^v = u.$$

Taking the logarithm, base a, of both sides of the second equation gives us

$$\log_a b^v = \log_a u.$$

Applying Law (iii), we get

$$v \log_a b = \log_a u.$$

Solving for v (that is, $\log_b u$), we obtain formula (i) in the next box.

Change of Base Formulas

$$\text{(i)} \quad \log_b u = \frac{\log_a u}{\log_a b} \qquad \text{(ii)} \quad \log_b a = \frac{1}{\log_a b}$$

To obtain formula (ii) we let $u = a$ in (i) and use the fact that $\log_a a = 1$.
The Laws of Logarithms are often used as in the next example.

Example 6 Express $\log_a \dfrac{x^3\sqrt{y}}{z^2}$ in terms of the logarithms of x, y, and z.

Solution   Using the three Laws of Logarithms,

$$\log_a \frac{x^3\sqrt{y}}{z^2} = \log_a (x^3\sqrt{y}) - \log_a z^2$$

$$= \log_a x^3 + \log_a \sqrt{y} - \log_a z^2$$

$$= 3\log_a x + (\tfrac{1}{2})\log_a y - 2\log_a z.   ■$$

The procedure in Example 6 can also be reversed; that is, beginning with the final equation, we can retrace our steps to obtain the original expression (see Exercises 67–70).

As a final remark, note that there is no general law for expressing $\log_a (u + w)$ in terms of simpler logarithms. It is evident that it does not always equal $\log_a u + \log_a w$, since the latter equals $\log_a (uw)$.

EXERCISES 5.2

Change the equations in Exercises 1–8 to logarithmic form.

1 $4^3 = 64$ **2** $3^5 = 243$

3 $2^7 = 128$ **4** $5^3 = 125$

5 $10^{-3} = 0.001$ **6** $10^{-2} = 0.01$

7 $t^r = s$ **8** $v^w = u$

Change the equations in Exercises 9–16 to exponential form.

9 $\log_{10} 1000 = 3$ **10** $\log_3 81 = 4$

11 $\log_3 \dfrac{1}{243} = -5$ **12** $\log_4 \dfrac{1}{64} = -3$

13 $\log_7 1 = 0$ **14** $\log_9 1 = 0$

15 $\log_t r = p$ **16** $\log_v w = q$

Find the numbers in Exercises 17–30.

17 $\log_4 (1/16)$ **18** $\log_2 32$

19 $\log_{10} 100$ **20** $\log_8 64$

21 $10^{\log_{10} 5}$ **22** $\log_{10} 0.0001$

23 $\log_7 \sqrt[3]{7}$ **24** $10^{2\log_{10} 3}$

25 $\log_{10} (1/10)$ **26** $\log_{10} 100,000$

27 $\log_{1/2} 8$ **28** $\log_3 (1/81)$

29 $3^{4\log_3 2}$ **30** $\log_6 \sqrt[5]{6}$

Given $\log_a 7 = 0.8$ and $\log_a 3 = 0.5$, change the expressions in Exercises 31–40 to decimal form.

31 $\log_a (7/3)$ **32** $\log_a (3/7)$

33 $\log_a 7^3$ **34** $\log_a (3^7)$

35 $\log_a (1/\sqrt{3})$ **36** $\log_a \sqrt[4]{3}$

37 $\log_a 63$ **38** $(\log_a 7)/(\log_a 3)$

39 $\log_7 a$ **40** $(\log_3 a)(\log_a 3)$

Find the solutions of the equations in Exercises 41–58.

41 $\log_3 (x - 4) = 2$ **42** $\log_2 (x - 5) = 4$

43 $\log_9 x = 3/2$ **44** $\log_4 x = -3/2$

45 $\log_5 x^2 = -2$ **46** $\log_{10} x^2 = -4$

47 $\log_2 (x^2 - 5x + 14) = 3$

48 $\log_2 (x^2 - 10x + 18) = 1$

49 $\log_x 7 = 3$ **50** $5^{\log_5 x} = 10$

51 $\log_6 (2x - 3) = \log_6 12 - \log_6 3$

52 $2\log_3 x = 3\log_3 5$

53 $\log_2 x - \log_2 (x + 1) = 3\log_2 4$

54 $\log_5 x + \log_5 (x + 6) = \frac{1}{2}\log_5 9$

55 $\log_{10} x^2 = \log_{10} x$

56 $\log_x 10 = 10$

57 $\frac{1}{2}\log_5 (x - 2) = 3\log_5 2 - \frac{3}{2}\log_5 (x - 2)$

58 $\log_{10} 5^x = \log_3 1$

In each of Exercises 59–66, express the logarithm in terms of logarithms of x, y, and z.

59 $\log_a \dfrac{x^2 y}{z^3}$ **60** $\log_a \dfrac{x^3 y^2}{z^5}$

61 $\log_a \dfrac{\sqrt{x}z^2}{y^4}$ **62** $\log_a x \sqrt[3]{\dfrac{y^2}{z^4}}$

63 $\log_a \sqrt[3]{\dfrac{x^2}{yz^5}}$ **64** $\log_a \dfrac{\sqrt{x}y^6}{\sqrt[3]{z^2}}$

65 $\log_a\sqrt{x\sqrt{yz^3}}$ **66** $\log_a\sqrt[3]{x^2 y\sqrt{z}}$

In each of Exercises 67–70 write the expression as one logarithm.

67 $2\log_a x + \frac{1}{3}\log_a (x - 2) - 5\log_a (2x + 3)$

68 $5\log_a x - \frac{1}{2}\log_a (3x - 4) + 3\log_a (5x + 1)$

69 $\log_a (y^2 x^3) - 2\log_a x\sqrt[3]{y} + 3\log_a \left(\dfrac{x}{y}\right)$

70 $2\log_a \dfrac{y^3}{x} - 3\log_a y + \frac{1}{2}\log_a x^4 y^2$

CALCULATOR EXERCISES 5.2

The LOG key refers to logarithms with base 10. In Exercises 1–6 use this key to demonstrate that the given equality is true by expressing each side as a decimal (do not use laws of logarithms).

1 $\log_{10}(2 \cdot 5) = \log_{10} 2 + \log_{10} 5$

2 $\log_{10}(5/2) = \log_{10} 5 - \log_{10} 2$

3 $\log_{10}(2^5) = 5 \log_{10} 2$

4 $\log_{10}(2 + 5) \neq \log_{10} 2 + \log_{10} 5$

5 $10^{\log_{10} 5} = 5$

6 $\log_{10} 1 = 0$

If $a = 10$ in the Change of Base Formula (i), then

$$\log_b u = (\log_{10} u)/(\log_{10} b).$$

Use this formula to find the following to four decimal place accuracy.

7 $\log_3 8$

8 $\log_9 7$

9 $\log_{17} 412$

10 $\log_\pi \sqrt{5}$

5.3 LOGARITHMIC FUNCTIONS

We next use the concept of logarithm to introduce a new function whose domain is the set of positive real numbers.

Definition

> The function f defined by
>
> $$f(x) = \log_a x$$
>
> for all positive real numbers x is called the **logarithmic function with base** a.

The graph of f is the same as the graph of the equation $y = \log_a x$ which, by the definition of logarithm, is equivalent to

$$x = a^y.$$

In order to find some pairs which are solutions of the preceding equation, we may substitute for y and find the corresponding values of x, as illustrated in the following table.

y	-3	-2	-1	0	1	2	3
x	$\dfrac{1}{a^3}$	$\dfrac{1}{a^2}$	$\dfrac{1}{a}$	1	a	a^2	a^3

If $a > 1$, we obtain the sketch in Figure 5.8. In this case f is an increasing function throughout its domain. If $0 < a < 1$, then the graph has the general shape shown in Figure 5.9, and hence f is a decreasing function. Note that for every a under consideration, the region to the left of the y-axis is excluded. There is no y-intercept and the x-intercept is 1.

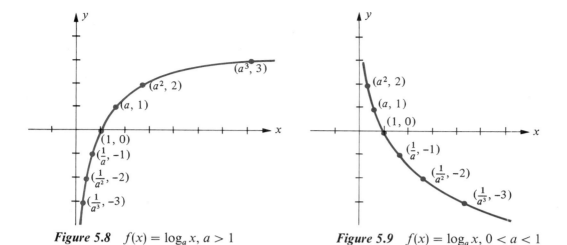

Figure 5.8 $f(x) = \log_a x, a > 1$ **Figure 5.9** $f(x) = \log_a x, 0 < a < 1$

Functions defined by expressions of the form $\log_a p$, where p is some expression in x, often occur in mathematics and its applications. Functions of this type are classified as members of the logarithmic family; however, the graphs may differ from those sketched in Figures 5.8 and 5.9, as is illustrated in the following examples.

Example 1   Sketch the graph of f if $f(x) = \log_3(-x)$, $x < 0$.

Solution   If $x < 0$, then $-x > 0$, and hence $\log_3(-x)$ is defined. We wish to sketch the graph of the equation $y = \log_3(-x)$, or equivalently, $3^y = -x$. The following table displays coordinates of some points on the graph, which is sketched in Figure 5.10.

y	-2	-1	0	1	2
x	$-\frac{1}{9}$	$-\frac{1}{3}$	-1	-3	-9

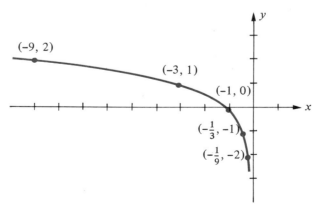

Figure 5.10 $f(x) = \log_3(-x)$ ■

Example 2   Sketch the graph of the equation $y = \log_3 |x|$, $x \neq 0$.

Solution   Since $|x| > 0$ for all $x \neq 0$, there are points on the graph corresponding to negative values of x as well as to positive values. If $x > 0$, then $|x| = x$ and hence to the right of the y-axis the graph coincides with the graph of $y = \log_3 x$, or equivalently, $x = 3^y$. If $x < 0$, then $|x| = -x$ and the graph is the same as that of $y = \log_3(-x)$ (see Example 1). The graph is sketched in Figure 5.11. Note that the graph is symmetric with respect to the y-axis.

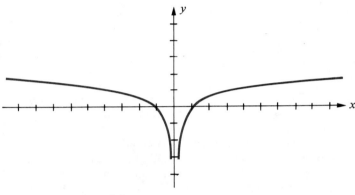

Figure 5.11 $f(x) = \log_3 |x|$ ■

There is a close relationship between exponential and logarithmic functions. This is to be expected since the logarithmic function was defined in terms of the exponential function. A precise description of the relationship is stated in the following theorem.

Theorem

> The exponential and logarithmic functions with base a are inverse functions of one another.

Proof   If $f(x) = a^x$ and $g(x) = \log_a x$, then, according to the definition of inverse function (see Section 2.4), we must show that

$$f(g(x)) = x \quad \text{and} \quad g(f(x)) = x.$$

For every positive real number x,

$$
\begin{aligned}
f(g(x)) &= a^{g(x)} \\
&= a^{\log_a x} \\
&= x.
\end{aligned}
$$

For every real number x,

$$
\begin{aligned}
g(f(x)) &= \log_a f(x) \\
&= \log_a a^x \\
&= x.
\end{aligned}
$$

This proves the theorem.   ■

Logarithmic functions occur frequently in applications. Indeed, if two variables u and v are related so that u is an exponential function of v, then v is a logarithmic function of u. As a specific example, if \$1.00 is invested at a rate of 9% per year compounded monthly, then from the discussion in Section 5.1, the principal P after n interest periods is

$$P = (1.0075)^n.$$

The logarithmic form of this equation is

$$n = \log_{1.0075} P.$$

Thus, the number of periods n required for \$1.00 to grow to an amount P is related logarithmically to P.

Another example we considered in Section 5.1 was that of population growth. In particular, the equation for the number N of bacteria in a certain culture after t hours was

$$N = (1000)2^t \quad \text{or} \quad 2^t = N/1000.$$

Changing to logarithmic form, we obtain

$$t = \log_2 (N/1000).$$

Hence, the time t is a logarithmic function of N.

In Section 5.1 we defined the natural exponential function f by means of the equation $f(x) = e^x$. The logarithmic function with base e, is called the **natural logarithmic function**. We use **ln x** as an abbreviation for $\log_e x$ and refer to it as the **natural logarithm of x**. Since $e \approx 3$, the graph of $y = \ln x$ is similar in appearance to the graph of $y = \log_3 x$. For natural logarithms the Laws of Logarithms are written

$$\ln (uv) = \ln u + \ln v, \qquad \ln \frac{u}{v} = \ln u - \ln v, \qquad \ln u^c = c \ln u.$$

Example 3 According to Newton's Law of Cooling, the rate at which an object cools is directly proportional to the difference in temperature between the object and the surrounding medium. Newton's Law can be used to show that under certain conditions the temperature T of an object is given by

$$T = 75e^{-2t}$$

where t is time. Express t as a function of T.

Solution The given equation may be written

$$e^{-2t} = T/75.$$

Using logarithms with base e yields

$$-2t = \log_e (T/75) = \ln (T/75).$$

Consequently,

$$t = -\tfrac{1}{2}\ln\left(T/75\right) \quad \text{or} \quad t = -\tfrac{1}{2}\left[\ln T - \ln 75\right]. \qquad \blacksquare$$

We shall not have many occasions to work with natural logarithms in this text. For computational purposes the base 10 is used. Base 10 gives us the system of common logarithms discussed in subsequent sections.

Many hand-held calculators have a key labeled $\ln x$ which can be used to approximate values of the natural logarithmic function. Several exercises are included at the end of this section for those who have a calculator of that type.

EXERCISES 5.3

Sketch the graph of f in each of Exercises 1–10.

1 (a) $f(x) = \log_2 x$ (b) $f(x) = \log_4 x$

2 (a) $f(x) = \log_5 x$ (b) $f(x) = \log_{10} x$

3 (a) $f(x) = \log_3 x$
  (b) $f(x) = \log_3(x - 2)$

4 (a) $f(x) = \log_2(-x)$
  (b) $f(x) = \log_2(3 - x)$

5 (a) $f(x) = \log_3(3x)$ (b) $f(x) = 3\log_3 x$

6 (a) $f(x) = \log_2(x^2)$ (b) $f(x) = \log_2(x^3)$

7 (a) $f(x) = \log_2\sqrt{x}$ (b) $f(x) = \log_2\sqrt[3]{x}$

8 (a) $f(x) = \log_2|x|$ (b) $f(x) = |\log_2 x|$

9 (a) $f(x) = \log_3(1/x)$
  (b) $f(x) = 1/(\log_3 x)$

10 (a) $f(x) = \log_3(2 + x)$
  (b) $f(x) = 2 + \log_3 x$

11 What is the geometric relationship between the graphs of $y = \log_a x$ and $y = a^x$? Why does this relationship exist?

12 Describe the relationship of the graph of $y = \ln x$ to the graphs of $y = \log_2 x$ and $y = \log_3 x$.

13 If one begins with q_0 milligrams of pure radium, then the amount q remaining after t years is given by the formula

$$q = q_0(2)^{-t/1600}.$$

Use logarithms with base 2 to solve for t in terms of q and q_0.

14 The number of bacteria in a certain culture at time t is given by

$$N = 10^4(3)^t.$$

Use logarithms with base 3 to solve for t in terms of N.

15 An electrical condenser with initial charge Q_0 is allowed to discharge. After t seconds the charge Q is given by

$$Q = Q_0 e^{kt}$$

where k is a constant of proportionality. Use natural logarithms to solve for t in terms of Q_0, Q, and k.

16 Under certain conditions the atmospheric pressure p at altitude h is given by

$$p = 29e^{-0.000034h}$$

Use natural logarithms to solve for h as a function of p.

17 The loudness of sound, as experienced by the human ear, is based upon intensity levels. A formula used for finding the intensity level α which corresponds to a sound intensity I is

$$\alpha = 10\log_{10}(I/I_0) \text{ decibels}$$

where I_0 is a special value of I agreed to be the weakest sound that can be detected by the ear under certain conditions. Find α if:

(a) I is 10 times as great as I_0.

(b) I is 1,000 times as great as I_0.

(c) I is 10,000 times as great as I_0. (This is the intensity level of the average voice.)

18 The current I at time t in a certain electrical circuit is given $I = 20e^{-Rt/L}$, where R and L denote the resistance and inductance, respectively. Use natural logarithms to solve for t in terms of the remaining variables.

19 Using the Richter scale, the magnitude R of an earthquake of intensity I may be found by means of the formula $R = \log_{10}(I/I_0)$, where I_0 is a certain minimum intensity. Find R if an earthquake has intensity:

(a) 100 times that of I_0.

(b) 10,000 times that of I_0.

(c) 100,000 times that of I_0.

20 Refer to Exercise 19. The magnitude of the San Francisco earthquake of 1906 was approximately 8 on the Richter scale. What is the corresponding intensity in terms of I_0?

CALCULATOR EXERCISES 5.3

1 Sketch the graph of

(a) $y = \ln x$ (b) $y = \ln |x|$

(c) $y = |\ln x|$

2 Sketch the graph of f if $f(x) = x^2 \ln x$.

3 Demonstrate that $\ln e^x = x$ and $e^{\ln x} = x$ for the following values of x:

(a) 1   (b) 3.4

(c) 0.056   (d) 8.143

4 Refer to Calculator Exercise 18 of Section 5.1 For what value of t will the population of the city be 1,000,000?

5 Letting $a = e$ in the change of base formula

$\log_b u = (\log_a u)/(\log_a b)$ leads to

$$\log_b u = \frac{\ln u}{\ln b}.$$

Use this formula to approximate

(a) $\log_3 4$ (b) $\log_5 (1.64)$

(c) $\log_7 \sqrt{2}$ (d) $\log_{10} 382$

(e) $\log_{10}(47/62)$

6 The LOG key on a calculator refers to the logarithmic function with base 10. Use a calculator as an aid to sketching the graph of f if $f(x) = \log_{10} x$.

5.4 COMMON LOGARITHMS

Logarithms with base 10 are useful for certain numerical computations. It is customary to refer to such logarithms as **common logarithms** and to use the symbol **log** x as an abbreviation for $\log_{10} x$.

Base 10 is used in computational work because every positive real number can be written in the scientific form $c \cdot 10^k$, where $1 \le c < 10$ and k is an integer (see Section 1.6). For example,

$$513 = (5.13)10^2, \quad 2375 = (2.375)10^3, \quad 720000 = (7.2)10^5,$$

$$0.641 = (6.41)10^{-1}, \quad 0.00000438 = (4.38)10^{-6}, \quad 4.601 = (4.601)10^0.$$

If x is any positive real number and we write

$$x = c \cdot 10^k$$

where $1 \leq c < 10$ and k is an integer, then applying (iii) of the Laws of Logarithms gives us

$$\log x = \log c + \log 10^k.$$

Since $\log 10^k = k$ (Why?) we see that

$$\log x = \log c + k.$$

The last equation tells us that to find $\log x$ for any positive real number x it is sufficient to know the logarithms of numbers between 1 and 10. The number $\log c$, where $1 \leq c < 10$, is called the **mantissa**, and the integer k is called the **characteristic** of $\log x$.

If $1 \leq c < 10$, then, since $\log x$ increases as x increases,

$$\log 1 \leq \log c < \log 10,$$

or equivalently,

$$0 \leq \log c < 1.$$

Hence, the mantissa of a logarithm is a number between 0 and 1. When working numerical problems, it is usually necessary to approximate logarithms. For example, it can be shown that

$$\log 2 = 0.3010299957 \ldots$$

where the decimal is nonrepeating and nonterminating. We shall round off such logarithms to four decimal places and write

$$\log 2 \approx 0.3010.$$

If a number between 0 and 1 is written as a finite decimal, it is sometimes referred to as a **decimal fraction**. Thus, the equation $\log x = \log c + k$ implies that if x is any positive real number, then $\log x$ *may be approximated by the sum of a positive decimal fraction (the mantissa) and an integer k (the characteristic)*. We shall refer to this representation as the **standard form** for $\log x$.

Logarithms of many of the numbers between 1 and 10 have been calculated. Table 1 contains four-decimal-place approximations for logarithms of numbers between 1.00 and 9.99 at intervals of 0.01. This table can be used to find the logarithm of any three-digit number to four-decimal-place accuracy. There exist far more extensive tables which provide logarithms of many additional numbers to much greater accuracy than four decimal places. The use of Table 1 is illustrated in the following examples.

Example 1   Approximate each of the following.

   (a)   log 43.6   (b)   log 43,600   (c)   log 0.0436

Solutions

(a) Since $43.6 = (4.36)10^1$, the characteristic of $\log 43.6$ is 1. Referring to Table 1, we find that the mantissa of $\log 4.36$ may be approximated by 0.6395. Hence, as in the preceding discussion,

$$\log 43.6 \approx 0.6395 + 1 = 1.6395.$$

(b) Since $43{,}600 = (4.36)10^4$, the mantissa is the same as in part (a); however, the characteristic is 4. Consequently,

$$\log 43{,}600 \approx 0.6395 + 4 = 4.6395.$$

(c) If we write $0.0436 = (4.36)10^{-2}$, then

$$\log 0.0436 = \log 4.36 + (-2).$$

Hence, $\qquad\qquad \log 0.0436 \approx 0.6395 + (-2).$

We could subtract 2 from 0.6395 and obtain

$$\log 0.0436 \approx -1.3605$$

but this is not standard form, since $-1.3605 = -0.3605 + (-1)$, a number in which the decimal fraction is *negative*. A common error is to write $0.6395 + (-2)$ as -2.6395. This is incorrect since $-2.6395 = -0.6395 + (-2)$, which is not the same as $0.6395 + (-2)$. ∎

If a logarithm has a negative characteristic, it is customary either to leave it in standard form or to rewrite the logarithm, keeping the decimal part positive. To illustrate the latter technique, let us add and subtract 8 on the right side of the equation

$$\log 0.0436 \approx 0.6395 + (-2).$$

This gives us

$$\log 0.0436 \approx 0.6395 + (8 - 8) + (-2)$$

or $\qquad\qquad \log 0.0436 \approx 8.6395 - 10.$

We could also write

$$\log 0.0436 \approx 18.6395 - 20 = 43.6395 - 45$$

and so on, as long as the *integral part* of the logarithm is -2.

Example 2 Approximate each of the following.

(a) $\log (0.00652)^2$ (b) $\log (0.00652)^{-2}$ (c) $\log (0.00652)^{1/2}$

Solutions

(a)   By (iii) of the Laws of Logarithms,

$$\log (0.00652)^2 = 2 \log 0.00652.$$

Since $0.00652 = (6.52)10^{-3}$,

$$\log 0.00652 = \log 6.52 + (-3).$$

Referring to Table 1, we see that $\log 6.52$ is approximately 0.8142 and, therefore,

$$\log 0.00652 \approx 0.8142 + (-3).$$

Hence, $\log (0.00652)^2 = 2 \log 0.00652$

$$\approx 2[0.8142 + (-3)]$$
$$= 1.6284 + (-6).$$

The standard form is $0.6284 + (-5)$.

(b)   Again using Law (iii) and the value for $\log 0.00652$ found in part (a),

$$\log (0.00652)^{-2} = -2 \log 0.00652$$
$$\approx -2[0.8142 + (-3)]$$
$$= -1.6284 + 6.$$

It is important to note that -1.6284 means $-0.6284 + (-1)$ and, consequently, the decimal part is negative. To obtain the standard form, we may write

$$-1.6284 + 6 = 6.0000 - 1.6284$$
$$= 4.3716.$$

This shows that the mantissa is 0.3716 and the characteristic is 4.

(c)   By Law (iii),

$$\log (0.00652)^{1/2} = (\tfrac{1}{2}) \log 0.00652$$
$$\approx \tfrac{1}{2}[0.8142 + (-3)].$$

If we multiply by 1/2, the standard form is not obtained since neither number in the resulting sum is the characteristic. In order to avoid this, we may adjust the expression within brackets by adding and subtracting a suitable number. If we use 1 in this way, we obtain

$$\log (0.00652)^{1/2} \approx \tfrac{1}{2}[1.8142 + (-4)]$$
$$= 0.9071 + (-2)$$

which is in standard form. We could also have added and subtracted a number other than 1. For example,

$$\tfrac{1}{2}[0.8142 + (-3)] = \tfrac{1}{2}[17.8142 + (-20)]$$
$$= 8.9071 + (-10).$$ ∎

Table 1 can be used to find an approximation to x if $\log x$ is given, as illustrated in the following example.

Example 3 Find a decimal approximation to x if $\log x$ is as follows.

(a) $\log x = 1.7959$ (b) $\log x = -3.5918$

Solutions

(a) The mantissa 0.7959 determines the sequence of digits in x and the characteristic determines the position of the decimal point. Referring to the *body* of Table 1, we see that the mantissa 0.7959 is the logarithm of 6.25. Since the characteristic is 1, x lies between 10 and 100. Consequently, $x \approx 62.5$.

(b) In order to find x from Table 1, $\log x$ must be written in standard form. In order to change $\log x = -3.5918$ to standard form, we may add and subtract 4, obtaining

$$\log x = (4 - 3.5918) - 4$$
$$= 0.4082 - 4.$$

Referring to Table 1, we see that the mantissa 0.4082 is the logarithm of 2.56. Since the characteristic of $\log x$ is -4, it follows that $x \approx 0.000256$. ∎

In each of the previous examples the mantissas appeared in Table 1. If this is not the case, then x can be approximated by the method of interpolation discussed in Section 5.6.

If a hand-held calculator with a log key is used to determine common logarithms, then the standard form for $\log x$ is obtained only if $x \geq 1$. For example, to find $\log 463$ on a typical calculator, we enter 463 and press the log key, obtaining the standard form

However, if we find $\log 0.0463$ in similar fashion, then the following number appears on the display panel:

This is not the standard form for the logarithm, but is similar to that which occurred in the solutions to Example 1(c) and Example 3(b). To find the standard form we could add 2 to the logarithm (using a calculator) obtaining 0.665580991 and then append -2. Thus,

$$\log 0.0463 \approx 0.665580991 - 2.$$

EXERCISES 5.4

In Exercises 1–16 use Table 1 and laws of logarithms to approximate the common logarithms of the given numbers. If a calculator with a log key is available, compare your answers with those obtained by means of the calculator.

1 347; 0.00347; 3.47

2 86.2; 8,620; 0.862

3 0.54; 540; 540,000

4 208; 2.08; 20,800

5 60.2; 0.0000602; 602

6 5; 0.5; 0.0005

7 $(44.9)^2$; $(44.9)^{1/2}$; $(44.9)^{-2}$

8 $(1810)^4$; $(1810)^{40}$; $(1810)^{1/4}$

9 $(0.943)^3$; $(0.943)^{-3}$; $(0.943)^{1/3}$

10 $(0.017)^{10}$; $10^{0.017}$; $10^{1.43}$

11 $(638)(17.3)$

12 $\dfrac{(2.73)(78.5)}{621}$

13 $\dfrac{(47.4)^3}{(29.5)^2}$

14 $\dfrac{(897)^4}{\sqrt{17.8}}$

15 $\sqrt[3]{20.6}(371)^3$

16 $\dfrac{(0.0048)^{10}}{\sqrt{0.29}}$

In Exercises 17–30 use Table 1 to find a decimal approximation to x. If a suitable calculator is available, compare your approximations with those obtained by means of the calculator.

17 $\log x = 3.6274$

18 $\log x = 1.8965$

19 $\log x = 0.9469$

20 $\log x = 0.5729$

21 $\log x = 5.2095$

22 $\log x = 6.7300 - 10$

23 $\log x = 9.7348 - 10$

24 $\log x = 7.6739 - 10$

25 $\log x = 8.8306 - 10$

26 $\log x = 4.9680$

27 $\log x = 2.2765$

28 $\log x = 3.0043$

29 $\log x = -1.6253$

30 $\log x = -2.2118$

31 Chemists use a number denoted by pH to describe quantitatively the acidity or basicity of solutions. By definition,

$$pH = -\log[H^+]$$

where $[H^+]$ is the hydrogen ion concentration in moles per liter. Given the following, with the indicated $[H^+]$, approximate the pH of each substance.
(a) vinegar: $[H^+] \approx 6.3 \times 10^{-3}$
(b) carrots: $[H^+] \approx 1.0 \times 10^{-5}$
(c) sea water: $[H^+] \approx 5.0 \times 10^{-9}$

32 Approximate the hydrogen ion concentration $[H^+]$ in each of the following substances. (Refer to Exercise 31 and use the closest entry in Table 1 to approximate $[H^+]$.)
(a) apples: pH ≈ 3.0 (b) beer: pH ≈ 4.2
(c) milk: pH ≈ 6.6

33 A solution is considered acidic if $[H^+] > 10^{-7}$ or basic if $[H^+] < 10^{-7}$. What are the corresponding inequalities involving pH?

34 Many solutions have a pH between 1 and 14. What is the corresponding range of the hydrogen ion content $[H^+]$?

35 The sound intensity level formula considered in Exercise 17 of Section 5.3 may be written

$$\alpha = 10 \log(I/I_0).$$

Find α if I is 285,000 times as great as I_0.

36 A sound intensity level of 140 decibels produces pain in the average human ear. Approximately how many times greater than I_0 must I be in order for α to reach this level? (Refer to Exercise 35 and use the nearest entry in Table 1 to find the approximation.)

37 The Richter scale formula given in Exercise 19 of Section 5.3 may be written $R = \log(I/I_0)$. An earthquake of magnitude 4.5 on this scale may cause damage to buildings. Find the intensity (in terms of I_0) of such an earthquake.

38 Refer to Exercise 37. The magnitudes of the largest recorded earthquakes have been between 8 and 9 on the Richter scale. Find the corresponding intensities in terms of I_0

5.5 EXPONENTIAL AND LOGARITHMIC EQUATIONS

The variables in certain equations appear as exponents or logarithms, as illustrated in the following examples.

Example 1 Solve the equation $3^x = 21$.

Solution   Taking the common logarithm of both sides and using (iii) of the Laws of Logarithms, we obtain

$$\log(3^x) = \log 21$$
$$x \log 3 = \log 21$$
$$x = \frac{\log 21}{\log 3}.$$

If an approximation is desired, we may use Table 1 to obtain

$$x \approx \frac{1.3222}{0.4771} \approx 2.77$$

where the last number is obtained by dividing 1.3222 by 0.4771. A partial check on the solution is to note that, since $3^2 = 9$ and $3^3 = 27$, the number x such that $3^x = 21$ should lie between 2 and 3, somewhat closer to 3 than to 2. ∎

Example 2 Solve $5^{2x+1} = 6^{x-2}$.

Solution   Taking the common logarithm of both sides and using (iii) of the Laws of Logarithms, we obtain

$$(2x + 1)\log 5 = (x - 2)\log 6.$$

We may now solve for x as follows:

$$2x \log 5 + \log 5 = x \log 6 - 2 \log 6$$
$$2x \log 5 - x \log 6 = -\log 5 - 2 \log 6$$
$$x(2 \log 5 - \log 6) = -(\log 5 + \log 6^2)$$
$$x = \frac{-(\log 5 + \log 36)}{2 \log 5 - \log 6}$$
$$= \frac{-\log(5 \cdot 36)}{\log 5^2 - \log 6}$$
$$x = \frac{-\log 180}{\log(25/6)}.$$

If an approximation to the solution is desired, we may proceed as in Example 1. ∎

Example 3   Solve the equation $\log(5x - 1) - \log(x - 3) = 2$.

Solution   The given equation may be written as

$$\log \frac{5x - 1}{x - 3} = 2.$$

Using the definition of logarithm with $a = 10$ gives us

$$\frac{5x - 1}{x - 3} = 10^2.$$

Consequently,

$$5x - 1 = 10^2(x - 3) = 100x - 300, \quad \text{or} \quad 299 = 95x.$$

Hence,

$$x = \frac{299}{95}.$$ ∎

Example 4   Solve the equation $\dfrac{5^x - 5^{-x}}{2} = 3$.

Solution   Multiplying both sides of the given equation first by 2 and then by 5^x gives us

$$5^x - 5^{-x} = 6$$
$$5^{2x} - 1 = 6(5^x)$$

which may be written $(5^x)^2 - 6(5^x) - 1 = 0$.

Letting $u = 5^x$ gives us the quadratic equation

$$u^2 - 6u - 1 = 0$$

in the variable u. Applying the Quadratic Formula,

$$u = \frac{6 \pm \sqrt{36 + 4}}{2} = 3 \pm \sqrt{10},$$

that is, $5^x = 3 \pm \sqrt{10}$. Since 5^x is never negative, the number $3 - \sqrt{10}$ must be discarded; therefore,

$$5^x = 3 + \sqrt{10}.$$

Taking the common logarithm of both sides and using (iii) of the Laws of Logarithms,

$$x \log 5 = \log(3 + \sqrt{10}) \quad \text{or} \quad x = \frac{\log(3 + \sqrt{10})}{\log 5}.$$

To obtain an approximate solution we may write $3 + \sqrt{10} \approx 6.16$ and use Table 1. This gives us

$$x \approx \frac{\log 6.16}{\log 5} \approx \frac{0.7896}{0.6990} \approx 1.13.$$ ∎

$3-30\ 3(R)$

EXERCISES 5.5

Find the solutions of the equations in Exercises 1–20.

1 $10^x = 7$ **2** $5^x = 8$

3 $4^x = 3$ **4** $10^x = 6$

5 $3^{4-x} = 5$ **6** $(1/3)^x = 100$

7 $3^{x+4} = 2^{1-3x}$ **8** $4^{2x+3} = 5^{x-2}$

9 $2^{-x} = 8$ **10** $2^{-x^2} = 5$

11 $\log x = 1 - \log(x - 3)$

12 $\log(5x + 1) = 2 + \log(2x - 3)$

13 $\log(x^2 + 4) - \log(x + 2) = 3 + \log(x - 2)$

14 $\log(x - 4) - \log(3x - 10) = \log(1/x)$

15 $\log(x^2) = (\log x)^2$ **16** $\log \sqrt{x} = \sqrt{\log x}$

17 $\log(\log x) = 2$ **18** $\log \sqrt{x^3 - 9} = 2$

19 $x^{\sqrt{\log x}} = 10^8$ **20** $\log(x^3) = (\log x)^3$

In Exercises 21–24 solve for x in terms of y.

21 $y = \dfrac{10^x + 10^{-x}}{2}$ **22** $y = \dfrac{10^x - 10^{-x}}{2}$

23 $y = \dfrac{10^x - 10^{-x}}{10^x + 10^{-x}}$ **24** $y = \dfrac{1}{10^x - 10^{-x}}$

In Exercises 25–28 use natural logarithms to solve for x in terms of y.

25 $y = \dfrac{e^x - e^{-x}}{2}$ **26** $y = \dfrac{e^x + e^{-x}}{2}$

27 $y = \dfrac{e^x - e^{-x}}{e^x + e^{-x}}$ **28** $y = \dfrac{e^x + e^{-x}}{e^x - e^{-x}}$

29 The current i in a certain electrical circuit is given by

$$i = \frac{E}{R}(1 - e^{-Rt/L}).$$

Use natural logarithms to solve for t in terms of the remaining symbols.

30 If a sum of money P_0 is invested at an interest rate of $100r$ percent per year, compounded m times per year, then the principal P at the end of t years is given by

$$P = P_0 \left(1 + \frac{r}{m}\right)^{mt}.$$

Solve for t in terms of the other symbols.

31 In Exercise 17 of Section 5.3 we considered the sound intensity level formula

$$\alpha = 10 \log(I/I_0).$$

(a) Solve for I in terms of α and I_0.
(b) Show that a one-decibel rise in the intensity level corresponds to a 26% increase in the intensity.

32 The chemical formula $\text{pH} = -\log[\text{H}^+]$ was introduced in Exercise 31 of Section 5.4. Solve for $[\text{H}^+]$ in terms of pH.

CALCULATOR EXERCISES 5.5

Use natural logarithms to approximate the solutions of the equations in Exercises 1–4.

1 $5^x = 7$ (*Hint*: $\ln 5^x = \ln 7$, or $x \ln 5 = \ln 7$)

2 $4^{-x} = 10$ **3** $e^{-x^2} = 0.163$

4 $e^{-2x+1} = 5^x$

5 How long will it take money to double if it is invested at a rate of 6% per year compounded monthly?

6 If \$10,000 is invested at an interest rate of 8% per year compounded quarterly, when will the principal exceed \$20,000? (*Hint*: Use Exercise 30.)

7 A certain radioactive substance decays according to the formula $q(t) = q_0 e^{-0.0063t}$, where q_0 is the initial amount of the substance and t is the time in days. Use natural logarithms to approximate its half life, that is, the number of days it takes for half of the substance to decay.

8 The air pressure $p(h)$, in lb/in.^{2}, at an altitude of h feet above sea level may be approximated by

$$p(h) = 14.7e^{-0.0000385h}.$$

At approximately what altitude h is the air pressure
(a) 10 lb/in.^{2}?
(b) one-half its value at sea level?

9 The graph of $G(t) = 10e^{-(1/2)e^{-t}}$ is a Gompertz growth curve (see Calculator Exercise 21 of Section 5.1). Approximate, to the nearest thousandth, the value of t such that $G(t) = 8$.

10 The formula $f(n) = 3 + 20(1 - e^{-0.1n})$ may be used to estimate certain learning processes (see Calculator Exercise 22 of Section 5.1). Approximate, to the nearest hundredth, the value of n such that $f(n) = 13$.

5.6 LINEAR INTERPOLATION

The only logarithms that can be found *directly* from Table 1 are logarithms of numbers that contain at most three nonzero digits. If *four* nonzero digits are involved, then it is possible to obtain an approximation by using the method of linear interpolation described in this section. The terminology *linear interpolation* is used because, as we shall see, the method is based upon approximating portions of the graph of $y = \log x$ by line segments. It should be pointed out that there are quicker, more mechanical, ways to find logarithms accurately. Indeed, if a hand-held calculator is available, it is usually only necessary to press a few keys to obtain accuracy to many decimal places. Nonetheless, linear interpolation is a very useful skill, since it can be applied to any tabular data in which entries vary in a manner similar to that of a linear function, at least over small portions of the table. Consequently, as you proceed through this section, keep in mind that you are learning a process that can be used with many tables other than those for logarithms.

In order to illustrate the process of linear interpolation, and at the same time give some justification for it, let us consider the specific example log 12.64. Since the logarithmic function with base 10 is increasing, this number lies between log 12.60 ≈ 1.1004 and log 12.70 ≈ 1.1038. Examining the graph of $y = \log x$, we have the situa-

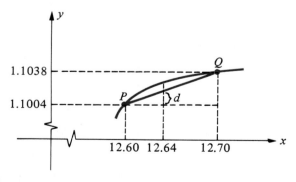

Figure 5.12

tion shown in Figure 5.12, where we have distorted the units on the x- and y-axes and also the portion of the graph shown. A more accurate drawing would indicate that the graph of $y = \log x$ is much closer to the line segment joining $P(12.60, 1.1004)$ to $Q(12.70, 1.1038)$ than is shown in the figure. Since $\log 12.64$ is the ordinate of the point on the graph having abscissa 12.64, it can be approximated by the ordinate of the point with abscissa 12.64 on the *line segment PQ*. Referring to Figure 5.12, we see that the latter ordinate is $1.1004 + d$. The number d can be approximated by using similar triangles. Referring to Figure 5.13, where the graph of $y = \log x$ has been deleted, we may form the following proportion:

$$\frac{d}{0.0034} = \frac{0.04}{0.1}.$$

Hence, $$d = \frac{(0.04)(0.0034)}{0.1} = 0.00136.$$

When using this technique, we always round off decimals to the same number of places as appear in the body of the table. Consequently, $d \approx 0.0014$ and

$$\log 12.64 \approx 1.1004 + 0.0014 = 1.1018.$$

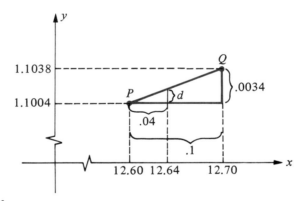

Figure 5.13

Hereafter we shall not sketch a graph when interpolating. Instead we shall use the scheme illustrated in the next example.

Example 1   Approximate $\log 572.6$.

Solution   It is convenient to arrange our work as follows:

$$1.0\left\{0.6\left\{\begin{matrix} \log 572.0 \approx 2.7574 \\ \log 572.6 = ? \\ \log 573.0 \approx 2.7582 \end{matrix}\right\} d\right\}0.0008$$

where we have indicated differences by appropriate symbols alongside of the braces. This leads to the proportion

$$\frac{d}{0.0008} = \frac{0.6}{1.0} = \frac{6}{10} \quad \text{or} \quad d = \left(\frac{6}{10}\right)(0.0008) = 0.00048 \approx 0.0005.$$

Hence, $\log 572.6 \approx 2.7574 + 0.0005 = 2.7579.$

Another way of working this type of problem is to reason that since 572.6 is $\frac{6}{10}$ of the way from 572.0 to 573.0, then $\log 572.6$ is (approximately) $\frac{6}{10}$ of the way from 2.7574 to 2.7582. Hence,

$$\log 572.6 \approx 2.7574 + (\tfrac{6}{10})(0.0008) \approx 2.7574 + 0.0005 = 2.7579. \quad ∎$$

Example 2   Approximate $\log 0.003678$.

Solution   We begin by arranging our work as in the solution of Example 1. Thus,

$$10\left\{8\left\{\begin{matrix} \log 0.003670 \approx 0.5647 + (-3) \\ \log 0.003678 = ? \\ \log 0.003680 \approx 0.5658 + (-3) \end{matrix}\right\}d\right\}0.0011$$

Since we are only interested in ratios, we have used the numbers 8 and 10 on the left side because their ratio is the same as the ratio of 0.000008 to 0.000010. This leads to the proportion

$$\frac{d}{0.0011} = \frac{8}{10} = 0.8 \quad \text{or} \quad d = (0.0011)(0.8) = 0.00088 \approx 0.0009.$$

Hence,
$$\log 0.003678 \approx [0.5647 + (-3)] + 0.0009$$
$$= 0.5656 + (-3). \quad ∎$$

If a number x is written in the scientific form $x = c \cdot 10^k$, where $1 \le c < 10$, then before using Table 1 to find $\log x$ by interpolation, c should be rounded off to three decimal places. Another way of saying this is that x should be rounded off to four **significant figures**. Some examples will help to clarify the procedure. If $x = 36.4635$, we round off to 36.46 before approximating $\log x$. The number 684,279 should be rounded off to 684,300. For a decimal such as 0.096202 we use 0.09620. The reason for doing this is that Table 1 does not guarantee more than four-digit accuracy, since the mantissas which appear in it are approximations. This means that if *more* than four-digit accuracy is required in a problem, then Table 1 cannot be used. If, in more extensive tables, the logarithm of a number containing n digits can be found directly, then interpolation is allowed for numbers involving $n + 1$ digits and numbers should be rounded off accordingly.

The method of interpolation can also be used to find x when we are given $\log x$. If we use Table 1, then x may be found to four significant figures. In this case we are given the *ordinate* of a point on the graph of $y = \log x$ and are asked to find the *abscissa.* A geometric argument similar to the one given earlier can be used to justify the procedure illustrated in the next example.

Example 3 Find x to four significant figures if $\log x = 1.7949$.

Solution The mantissa 0.7949 does not appear in Table 1, but it can be isolated between adjacent entries, namely the mantissas corresponding to 6.230 and 6.240. We shall arrange our work as follows:

$$0.1\left\{r\left\{\begin{matrix}\log 62.30 \approx 1.7945\\ \log x \quad\;\; = 1.7949\end{matrix}\right\}0.0004\\ \log 62.40 \approx 1.7952\end{matrix}\right\}0.0007.$$

This leads to the proportion

$$\frac{r}{0.1} = \frac{0.0004}{0.0007} = \frac{4}{7} \quad \text{or} \quad r = (0.1)\left(\frac{4}{7}\right) \approx 0.06.$$

Hence, $x \approx 62.30 + 0.06 = 62.36.$ ∎

If we are given $\log x$, then the number x is called the **antilogarithm** of $\log x$. In Example 3 the antilogarithm of $\log x = 1.7949$ is $x \approx 62.36$. Sometimes the notation antilog $(1.7949) \approx 62.36$ is used.

EXERCISES 5.6 3 – 18 3RD)

Use linear interpolation to approximate the common logarithms of the numbers in Exercises 1–20. If a calculator with a log key is available, compare your approximation with those obtained by means of the calculator.

1	25.48	**2**	421.6
3	5363	**4**	0.3817
5	0.001259	**6**	69,450
7	123,400	**8**	0.0212
9	0.7786	**10**	1.203
11	384.7	**12**	54.44
13	0.9462	**14**	7259
15	66,590	**16**	0.001428
17	0.04321	**18**	300,100
19	3.003	**20**	1.236

In Exercises 21–40 use linear interpolation to approximate x. If a suitable calculator is available, compare your approximations with those obtained by means of the calculator.

21	$\log x = 1.4437$	**22**	$\log x = 3.7455$
23	$\log x = 4.6931$	**24**	$\log x = 0.5883$
25	$\log x = 9.1664 - 10$	**26**	$\log x = 8.3902 - 10$
27	$\log x = 3.8153 - 6$	**28**	$\log x = 5.9306 - 9$
29	$\log x = 2.3705$	**30**	$\log x = 4.2867$
31	$\log x = 0.1358$	**32**	$\log x = 0.0194$
33	$\log x = 8.9752 - 10$	**34**	$\log x = 2.4979 - 5$
35	$\log x = 5.0409$	**36**	$\log x = 1.3796$
37	$\log x = -2.8712$	**38**	$\log x = -1.8164$
39	$\log x = -0.6123$	**40**	$\log x = -3.1426$

5.7   COMPUTATIONS WITH LOGARITHMS

The importance of logarithms for numerical computations has diminished in recent years because of the development of computers and hand-held calculators. However, since mechanical devices are not always available, it is worthwhile to have some familiarity with the use of logarithms for solving arithmetic problems. At the same time, practice in working numerical problems leads to a deeper understanding of logarithms and logarithmic functions. The following examples illustrate some computational techniques.

Example 1   Approximate $N = \dfrac{(59700)(0.0163)}{41.7}$.

Solution   Using laws of logarithms and Table 1 leads to

$$\begin{aligned}
\log N &= \log 59700 + \log 0.0163 - \log 41.7 \\
&\approx 4.7760 + (0.2122 - 2) - (1.6201) \\
&= 4.9882 - 3.6201 \\
&= 1.3681.
\end{aligned}$$

Referring to Table 1 for the antilogarithm we have, to three significant figures,

$$N \approx 23.3. \qquad\qquad \blacksquare$$

Example 2   Approximate $N = \sqrt[3]{56.11}$ to four significant figures.

Solution   Writing $N = (56.11)^{1/3}$ and using (iii) of the Laws of Logarithms gives us

$$\log N = \tfrac{1}{3}\log 56.11.$$

To find $\log 56.11$ we interpolate from Table 1 as follows:

$$10\left\{1\left\{\begin{aligned}
\log 56.10 &\approx 1.7490 \\
\log 56.11 &= ? \\
\log 56.20 &\approx 1.7497
\end{aligned}\right\}d\right\}0.0007$$

$$\frac{d}{0.0007} = \frac{1}{10}$$

$$d = 0.00007 \approx 0.0001.$$

Hence, $\log 56.11 \approx 1.7490 + 0.0001 = 1.7491$

and therefore, $\log N \approx \tfrac{1}{3}(1.7491) \approx 0.5830.$

The antilogarithm may be found by interpolation from Table 1 as follows

$$0.01 \left\{ r \left\{ \begin{array}{l} \log 3.820 \approx 0.5821 \\ \log N \quad\ \approx 0.5830 \end{array} \right\} 9 \\ \log 3.830 \approx 0.5832 \end{array} \right\} 11$$

$$\frac{r}{0.01} = \frac{9}{11}$$

$$r = \frac{9}{11}(0.01) \approx 0.008.$$

Consequently, $N \approx 3.820 + 0.008 = 3.828.$ ∎

If we were interested in only *three* significant figures, then the interpolations in Example 2 could have been avoided. In the remaining examples we shall, for simplicity, work only with three-digit numbers.

Example 3   Approximate $N = \dfrac{(1.32)^{10}}{\sqrt[5]{0.0268}}$.

Solution   Using laws of logarithms and Table 1,

$$\begin{aligned} \log N &= 10 \log 1.32 - \tfrac{1}{5} \log 0.0268 \\ &\approx 10(0.1206) - \tfrac{1}{5}(3.4281 - 5) \\ &= 1.206 - 0.6856 + 1 \\ &= 1.5204. \end{aligned}$$

Finding the antilogarithm (to three significant figures), we obtain

$$N \approx 33.1.$$ ∎

Example 4   Find x if $x^{2.1} = 6.5$.

Solution   Taking the common logarithm of both sides and using (iii) of the Laws of Logarithms gives us

$$(2.1) \log x = \log 6.5.$$

Hence, $$\log x = \frac{\log 6.5}{2.1}$$

and by the definition of logarithm,

$$x = 10^{(\log 6.5)/(2.1)}.$$

If an approximation to x is desired, then from the second equation above we have

$$\log x = \frac{\log 6.5}{2.1} \approx \frac{0.8129}{2.1} \approx 0.3871.$$

Using Table 1 the antilogarithm (to three significant figures) is

$$x \approx 2.44.$$ ∎

Example 5 Approximate $N = \dfrac{69.3 + \sqrt[3]{56.1}}{\log 807}$.

Solution Since we have no formula for the logarithm of a sum, the two terms in the numerator must be *added* before the logarithm can be found. From Example 2, $\sqrt[3]{56.1} \approx 3.8$ and hence the numerator is approximately 73.1. Using Table 1 we see that

$$\log 807 \approx 2.9069 \approx 2.91.$$

Hence, the given expression may be approximated by

$$N \approx \frac{73.1}{2.91}.$$

It is now an easy matter to find N, either by using logarithms or by long division. It is left to the reader to verify that $N \approx 25.1$. ∎

EXERCISES 5.7

$3-18$

Use logarithms to approximate the numbers in Exercises 1–20 to three significant figures. If a calculator is available, compare your answers with those obtained by means of the calculator.

1 $(638)(57.2)$

2 $(0.0178)(0.00729)$

3 $\dfrac{35{,}900}{8{,}430}$

4 $\dfrac{3.14}{63.2}$

5 $(4.21)^{10}$

6 $(0.712)^6$

7 $\sqrt[5]{0.517}$

8 $\sqrt[4]{2.23}$

9 $\dfrac{(26.7)^3(1.48)}{(67.4)^2}$

10 $\dfrac{(3.04)^3}{\sqrt{1.12}(86.6)}$

11 $\sqrt{1.65}\sqrt[3]{(70.8)^2}$

12 $\left[\dfrac{(11.1)^3}{\sqrt[5]{4.17}}\right]^{-1/2}$

13 $\sqrt{\dfrac{0.563}{(0.105)^3}}$

14 $\sqrt[5]{\dfrac{(124)^2}{(9.83)^3}}$

15 $10^{-3.14}$

16 $(100)^{0.523}$

17 $(4.12)^{0.220}$

18 $\sqrt{8.46\sqrt{3.07}}$

19 $\dfrac{\log 37.4}{\log 6.19}$

20 $\dfrac{56.8 + \log(7.13)}{\sqrt[10]{4.42}}$

Use interpolation in Table 1 to approximate the numbers in Exercises 21–26 to four significant figures.

21 $(2.461)^5$

22 $1/(33.89)^4$

23 $\sqrt[10]{0.5138}$

24 $\sqrt[3]{(17.04)^2}$

25 $(5.375)^{2/3}$

26 $(1776)^{11}$

27 The area A of a triangle with sides a, b, and c may be calculated from the formula $A = \sqrt{s(s-a)(s-b)(s-c)}$, where s is one-half the perimeter. Use logarithms to approximate the area of a triangle with sides 12.6, 18.2, and 14.1.

28 The volume V of a right circular cone of altitude h and radius of base r is $V = \frac{1}{3}\pi r^2 h$. Use logarithms to approximate the volume of a cone of radius 2.43 cm and altitude 7.28 cm.

29 The formula used in physics to approximate the period T (seconds) of a simple pendulum of length L (feet) is $T = 2\pi\sqrt{L/(32.2)}$. Approximate the period of a pendulum 33 inches long.

30 The pressure p (pounds per cubic foot) and volume v (cubic feet) of a certain gas are related by the formula $pv^{1.4} = 600$. Approximate the pressure if $v = 8.22$ cubic feet.

5.8 REVIEW

Concepts

Define or discuss the following.

1 The exponential function with base a

2 The natural exponential function

3 The logarithm of u with base a

4 The natural logarithmic function

5 The laws of logarithms

6 The logarithmic function with base a

7 Common logarithms

8 Mantissa

9 Characteristic

10 Linear interpolation

Exercises

Find the numbers in Exercises 1–6.

1 $\log_2 (1/16)$

2 $\log_5 \sqrt[3]{5}$

3 $6^{\log_6 4}$

4 $10^{3 \log 2}$

5 $\log 1{,}000{,}000$

6 $\ln e$

In Exercises 7–16 sketch the graph of f.

7 $f(x) = 3^{x+2}$

8 $f(x) = (3/5)^x$

9 $f(x) = (3/2)^{-x}$

10 $f(x) = 3^{-2x}$

11 $f(x) = 3^{-x^2}$

12 $f(x) = 1 - 3^{-x}$

13 $f(x) = \log_6 x$

14 $f(x) = \log_3 (x^2)$

15 $f(x) = 2 \log_3 x$

16 $f(x) = \log_2 (x + 4)$

Find the solutions of the equations in Exercises 17–24.

17 $\log_8 (x - 5) = 2/3$

18 $\log_4 (x + 1) = 2 + \log_4 (3x - 2)$

19 $2 \log_3 (x + 3) - \log_3 (x + 1) = 3 \log_3 2$

20 $\log \sqrt[4]{x + 1} = 1/2$

21 $2^{5-x} = 6$

22 $3^{x^2} = 7$

23 $2^{5x+3} = 3^{2x+1}$

24 $5^{\log_5 (x+1)} = 3$

25 Express $\log x^4 \sqrt[3]{y^2/z}$ in terms of logarithms of x, y, and z.

26 Express $\log (x^2/y^3) + 4 \log y - 6 \log \sqrt{xy}$ as one logarithm.

Solve the equations in Exercises 27 and 28 for x in terms of y.

27 $y = \dfrac{10^x + 10^{-x}}{10^x - 10^{-x}}$

28 $y = \dfrac{1}{10^x + 10^{-x}}$

Use linear interpolation to approximate the logarithms in Exercises 29–32.

29 $\log 47.82$

30 $\log 0.001347$

31 $\log 300{,}600$

32 $\log 0.2143$

Use linear interpolation to approximate x in Exercises 33–36.

33 $\log x = 2.4995$

34 $\log x = 1.5045$

35 $\log x = 8.7970 - 10$

36 $\log x = -1.3146$

Use logarithms to approximate the numbers in Exercises 37–40.

37 $\dfrac{(38.2)^3}{\sqrt{4.67}}$

38 $\sqrt[5]{\dfrac{21.8}{62.2}}$

39 $(5.16)^{2.1}$

40 $(1.01)^{44}$

6

THE TRIGONOMETRIC FUNCTIONS

Trigonometry originated over 2,000 years ago, when the Greeks developed precise methods for measuring angles and sides of triangles. Although such measurements are still important in certain branches of science and engineering, there are many modern applications of trigonometry in which the notion of angle either is secondary or does not enter into the picture. Of primary importance are properties of the so-called trigonometric functions, whose domains are sets of real numbers. For this reason, these functions are introduced in Section 6.2 without reference to angles. However, since the classical techniques are also important, our modern definition is followed almost immediately, in Section 6.5, by a more traditional approach involving angles and ratios. As a consequence, we have available, very early in our work, two different, but equivalent, formulations of the trigonometric functions. Students should strive to become comfortable using both approaches, since each has certain advantages. Graphs involving the trigonometric functions are considered in Sections 6.6–6.8. The chapter concludes with some modern and ancient applications of trigonometry to problems involving right triangles.

6.1 THE UNIT CIRCLE *U*

Let U be a unit circle, that is, a circle of radius 1, with center at the origin of a rectangular coordinate system. Thus, U is the graph of the equation $x^2 + y^2 = 1$. If A is the point with coordinates $(1, 0)$, and P is any other point on U, then proceeding along U in a *counterclockwise direction* from A (see Figure 6.1), there is a unique positive real number t called the **length of the arc** $\overset{\frown}{AP}$. Since the circumference of U is 2π, we have $0 < t < 2\pi$. If we let $t = 0$ when $P = A$, then with each point P on U there is associated precisely one real number in the half-open interval $[0, 2\pi)$.

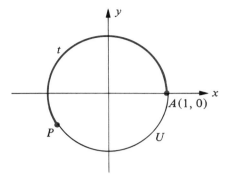

Figure 6.1

Conversely, it can be shown that, for each real number t in the half-open interval $[0, 2\pi)$, there is one and only one point P on U such that the length of $\overset{\frown}{AP}$, obtained by proceeding in the counterclockwise direction from A to P, is t. This establishes a one-to-one correspondence between the real numbers in the interval $[0, 2\pi)$ and the points on U. Since the position of P depends on t, we shall use functional notation and denote it by $P(t)$. In certain applications a (variable) point is regarded as moving along U, and we refer to the point as *traveling*, or *traversing*, *a distance* t from A to $P(t)$.

For most values of t it is very difficult to determine the rectangular coordinates of the point $P(t)$ on U. Three simple cases are illustrated in Figure 6.2. In Case (i) we note that, since the circumference of the unit circle is 2π, the number $\pi/2$ is one-fourth the circumference, and hence the point $P(\pi/2)$ has coordinates $(0, 1)$. Similarly, in Case (ii), π is one-half the circumference, and hence $P(\pi)$ may be found by starting at A and traveling half-way around U to the point $(-1, 0)$. Finally, in Case (iii), $P(3\pi/2)$ can be located by traversing three-fourths of the circumference (in the counterclockwise direction) to the point $(0, -1)$.

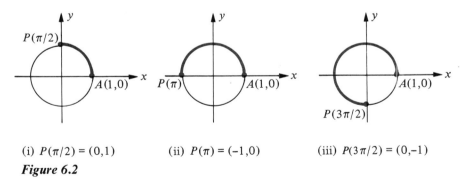

(i) $P(\pi/2) = (0,1)$ (ii) $P(\pi) = (-1,0)$ (iii) $P(3\pi/2) = (0,-1)$

Figure 6.2

Example 1   Find the rectangular coordinates of $P(\pi/4)$, $P(3\pi/4)$, $P(5\pi/4)$, and $P(7\pi/4)$.

Solution   The rectangular coordinates of the point $P(\pi/4)$ can be found by noting that $\pi/4$ is one-half of $\pi/2$, and consequently, $P(\pi/4)$ is in the first quadrant and

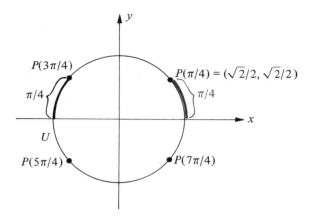

Figure 6.3

bisects the circular arc from $(1, 0)$ to $(0, 1)$. It follows that $P(\pi/4)$ lies on the line $y = x$ and hence has coordinates (c, c) for some positive real number c (see Figure 6.3). Since (c, c) is a point on the unit circle $x^2 + y^2 = 1$, we must have $c^2 + c^2 = 1$, or $2c^2 = 1$. Thus, $c = \sqrt{1/2} = \sqrt{2}/2$ and

$$P(\pi/4) = (\sqrt{2}/2, \sqrt{2}/2).$$

Since $3\pi/4 = \pi - (\pi/4)$, we see that $P(3\pi/4)$ and $P(\pi/4)$ are symmetric with respect to the y-axis, as illustrated in Figure 6.3. It follows that

$$P(3\pi/4) = (-\sqrt{2}/2, \sqrt{2}/2).$$

Again using symmetry, it is evident from the figure that

$$P\left(\frac{5\pi}{4}\right) = \left(-\frac{\sqrt{2}}{2}, -\frac{\sqrt{2}}{2}\right) \quad \text{and} \quad P\left(\frac{7\pi}{4}\right) = \left(\frac{\sqrt{2}}{2}, -\frac{\sqrt{2}}{2}\right). \qquad ■$$

Example 2   Find the rectangular coordinates of   (a) $P(\pi/6)$;   (b) $P(\pi/3)$.

Solutions

(a)   Since $\pi/6 = (1/3)(\pi/2)$, the point $P(\pi/6)$ on U is in the first quadrant, one-third the way from $A(1, 0)$ to $B(0, 1)$. If P has coordinates (c, d), then, as illustrated in (i) of Figure 6.4, the point $P'(c, -d)$ is also on U. Since the lengths of arcs $\overset{\frown}{P'P}$ and $\overset{\frown}{PB}$ are equal (both are $\pi/3$), it follows that the length $2d$ of chord $P'P$ is the same as the length of chord PB. Employing the Distance Formula,

$$2d = \sqrt{(c - 0)^2 + (d - 1)^2}.$$

Squaring both sides of this equation we obtain

$$4d^2 = c^2 + d^2 - 2d + 1.$$

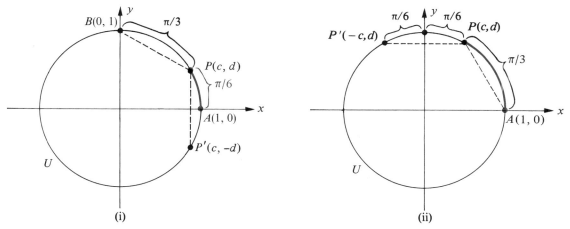

Figure 6.4

Since (c, d) is a point on U we must have $c^2 + d^2 = 1$, and hence,

$$4d^2 = 1 - 2d + 1, \quad \text{or} \quad 4d^2 + 2d - 2 = 0.$$

Factoring gives us

$$2(2d - 1)(d + 1) = 0.$$

Since d is positive the last equation implies that $d = 1/2$. Consequently, since $c^2 + d^2 = 1$,

$$c = \sqrt{1 - d^2} = \sqrt{1 - (1/4)} = \sqrt{3/4} = \sqrt{3}/2$$

and therefore,

$$P\left(\frac{\pi}{6}\right) = \left(\frac{\sqrt{3}}{2}, \frac{1}{2}\right).$$

(b) Since $\pi/3 = (2/3)(\pi/2)$, the point $P(\pi/3)$ on U is in the first quadrant, two-thirds of the way from $A(1, 0)$ to $B(0, 1)$. If P has coordinates (c, d), then by symmetry, the point $P'(-c, d)$ is also on U. Moreover, referring to (ii) of Figure 6.4, the lengths of arcs $\overparen{PP'}$ and $\overparen{AP}$ both equal $\pi/3$, and hence the length $2c$ of chord PP' is the same as the length of chord AP. Applying the Distance Formula,

$$2c = \sqrt{(c - 1)^2 + d^2}$$

or equivalently,

$$4c^2 = c^2 - 2c + 1 + d^2.$$

Proceeding as in part (a) we obtain $c = 1/2$ and $d = \sqrt{3}/2$ (Verify this fact!). Hence,

$$P(\pi/3) = (1/2, \sqrt{3}/2). \qquad \blacksquare$$

Using Example 2 and the symmetry of the unit circle U, we can find the rectangular coordinates of $P(t)$, where t is any integral multiple of $\pi/6$ that lies between

0 and 2π. The positions of these points are illustrated in Figure 6.5 and may be verified by noting that $2\pi/3 = \pi - \pi/3$, $5\pi/6 = \pi - \pi/6$, etc.

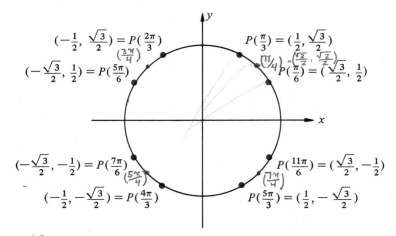

Figure 6.5

Although geometric arguments can be used to find the rectangular coordinates of special points on U, such as $P(\pi/4)$, $P(\pi/6)$, and $P(\pi/3)$, it is impossible for us to cope with values of t such as $2/3$, $\sqrt{5}$, 4.627, or for that matter, integer values such as 1, 2, 3, and 4. The reason for the difficulty is that techniques studied in calculus are required to handle all possibilities. Fortunately, in our future work it will be unnecessary to calculate the exact coordinates (x, y) of $P(t)$ for every value of t.

The preceding discussion can be extended so that to *every* real number there corresponds exactly one point on U. Specifically, to associate a point $P(t)$ with any *positive* real number t, we proceed along U in a counterclockwise direction from A, *making several revolutions*, if necessary. For example, if $t = 5\pi/2$, we first note that

$$\frac{5\pi}{2} = \frac{\pi}{2} + 2\pi.$$

Thus, the point $P(5\pi/2)$ may be found by first traversing the distance $\pi/2$ from A to the point $(0, 1)$, and then making one complete counterclockwise revolution through the distance 2π. Consequently, $P(5\pi/2) = (0, 1)$. Similarly, to locate $P(9\pi/2)$, we note that

$$\frac{9\pi}{2} = \frac{\pi}{2} + 4\pi$$

and hence *two* complete revolutions should be made after reaching the point $(0, 1)$. This gives us $P(9\pi/2) = (0, 1)$. In general, if t_1 and t_2 are positive real numbers such that $t_2 = t_1 + 2\pi n$ for some positive integer n, then $P(t_2) = P(t_1)$, since the circumference of U is 2π.

To associate a point on U with each *negative* real number t, we start at A and proceed a distance $|t|$ in a *clockwise* direction along U. The three special cases

$t = -\pi/2$, $t = -\pi$, and $t = -7\pi/4$ are illustrated in Figure 6.6. The reader should compare the third case with Example 1. Of course, for real numbers less than -2π, it is necessary to make more than one revolution (in the clockwise direction) along U.

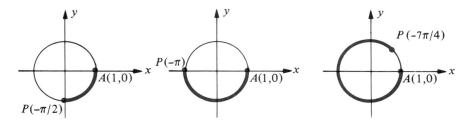

(i) $P(-\pi/2) = (0, -1)$ (ii) $P(-\pi) = (-1, 0)$ (iii) $P(-7\pi/4) = (\sqrt{2}/2, \sqrt{2}/2)$

Figure 6.6

We have demonstrated that, to each real number t, there is associated a unique point $P(t)$ on the unit circle U. We shall refer to $P(t)$ as **the point on U that corresponds to t**. Since the circumference of U is 2π, it follows that two real numbers t_1 and t_2 are associated with the same point if and only if $t_2 = t_1 + 2\pi n$ for some integer n. This gives us the important formula

$$P(t) = P(t + 2\pi n)$$

for every real number t and every integer n. Several uses of this formula are given in the next two examples.

Example 3   Find the rectangular coordinates of $P(27\pi/4)$.

Solution   Since $27/4 = 6 + (3/4)$ we see that

$$\frac{27\pi}{4} = 6\pi + \frac{3\pi}{4}.$$

Using the formula $P(t) = P(t + 2\pi n)$, with $n = 3$,

$$P\left(\frac{27\pi}{4}\right) = P\left(\frac{3\pi}{4} + 6\pi\right) = P\left(\frac{3\pi}{4}\right).$$

Finally, referring to Example 1, we obtain $P(27\pi/4) = (-\sqrt{2}/2, \sqrt{2}/2)$.   ■

Example 4   Find the quadrant which contains $P(21)$.

Solution   From the discussion in this section, $P(21) = P(21 + 2\pi n)$ for every integer n. By choosing n negative we can find a number of the form $21 + 2\pi n$ which is in the interval $[0, 2\pi)$. The quadrant may then be readily determined. In particular, if we let $n = -3$, then

$$21 - 6\pi \approx 21 - 18.85 = 2.15.$$

Since $\pi/2 < 2.15 < \pi$, we see that

$$\frac{\pi}{2} < 21 - 6\pi < \pi.$$

It follows that $P(21)$ is in the second quadrant. ∎

EXERCISES 6.1

In Exercises 1–10 find the rectangular coordinates of the given points.

1 $P(5\pi)$ **2** $P(-3\pi)$

3 $P(7\pi/2)$ **4** $P(5\pi/2)$

5 $P(-6\pi)$ **6** $P(11\pi/2)$

7 $P(37\pi)$ **8** $P(99\pi)$

9 $P(-9\pi/2)$ **10** $P(1000000\pi)$

Find the rectangular coordinates of the points in Exercises 11–22 (*Hint*: See Figures 6.3 and 6.5).

11 $P(-3\pi/4)$ **12** $P(-\pi/4)$

13 $P(-9\pi/4)$ **14** $P(-5\pi/4)$

15 $P(11\pi/4)$ **16** $P(9\pi/4)$

17 $P(-\pi/3)$ **18** $P(-\pi/6)$

19 $P(-5\pi/6)$ **20** $P(-4\pi/3)$

21 $P(8\pi/3)$ **22** $P(17\pi/6)$

If $P(t)$ has the rectangular coordinates given in each of Exercises 23–30, find (a) $P(t + \pi)$, (b) $P(t - \pi)$, (c) $P(-t)$, (d) $P(-t - \pi)$.

23 $(3/5, 4/5)$ **24** $(4/5, -3/5)$

25 $(-8/17, 15/17)$ **26** $(-15/17, 8/17)$

27 $(-1, 0)$ **28** $(0, -1)$

29 $(a/\sqrt{a^2 + b^2}, b/\sqrt{a^2 + b^2})$ where $a^2 + b^2 \neq 0$

30 $(a, \sqrt{1 - a^2})$

In each of Exercises 31–34, approximate the y-coordinate of $P(t)$ if $P(t)$ is in the indicated quadrant and has the given x-coordinate.

31 I, $x = 1/2$ **32** II, $x = -\sqrt{3}/2$

33 III, $x = -2/3$ **34** IV, $x = 0.1$

In each of Exercises 35–38, approximate the x-coordinate of $P(t)$ if $P(t)$ is in the indicated quadrant and has the given y-coordinate.

35 II, $y = \sqrt{3}/2)$ **36** III, $y = -1/2$

37 IV, $y = -0.01$ **38** II, $y = 0.9$

CALCULATOR EXERCISES 6.1

Find the quadrants containing the points in Exercises 1–8.

1 $P(43)$ **2** $P(23.64)$

3 $P(-28)$ **4** $P(3.141591)$

5 $P(1776)$ **6** $P(-2001)$

7 $P(\sqrt[4]{782\pi})$ **8** $P(6\sqrt{2} + 5\sqrt{7})$

9 Approximate the y-coordinate of $P(t)$ if $P(t)$ is in the fourth quadrant and has x-coordinate 0.9703.

10 Approximate the x-coordinate of $P(t)$ if $P(t)$ is in the third quadrant and has y-coordinate -0.6561.

6.2 THE TRIGONOMETRIC FUNCTIONS

In the preceding section we introduced a technique for associating, with each real number t, a unique point $P(t)$ on the unit circle U. The rectangular coordinates (x, y) of $P(t)$ can be used to define the six **trigonometric** or **(circular) functions**. These functions are referred to as the **sine, cosine, tangent, cotangent, secant,** and **cosecant functions**, and are designated by the symbols **sin, cos, tan, cot, sec,** and **csc,** respectively. If t is a real number, then the real number which the sine function associates with t will be denoted by either $\sin(t)$ or $\sin t$, and similarly for the other five functions.

Definition of the Trigonometric Functions

For any real number t, let $P(t)$ be the point on the unit circle U that corresponds to t. If the rectangular coordinates of $P(t)$ are (x, y), then

$$\sin t = y \qquad\qquad \csc t = \frac{1}{y} \quad (\text{if } y \neq 0)$$

$$\cos t = x \qquad\qquad \sec t = \frac{1}{x} \quad (\text{if } x \neq 0)$$

$$\tan t = \frac{y}{x} \quad (\text{if } x \neq 0) \qquad \cot t = \frac{x}{y} \quad (\text{if } y \neq 0).$$

In order to use this definition to find the values of trigonometric functions, it is necessary to determine the rectangular coordinates (x, y) of the point $P(t)$ on U and then substitute for x and y in the appropriate formulas. In later sections we shall introduce other techniques for finding functional values. The domains and ranges of the trigonometric functions will be discussed in Section 6.3.

Example 1 Find the values of the trigonometric functions corresponding to the number $t = \pi/4$.

Solution Referring to Figure 6.3 we see that $P(\pi/4) = (\sqrt{2}/2, \sqrt{2}/2)$. Letting $x = \sqrt{2}/2$ and $y = \sqrt{2}/2$ in the last definition gives us

$$\sin \frac{\pi}{4} = \frac{\sqrt{2}}{2} \qquad\qquad \csc \frac{\pi}{4} = \frac{2}{\sqrt{2}} = \sqrt{2}$$

$$\cos \frac{\pi}{4} = \frac{\sqrt{2}}{2} \qquad\qquad \sec \frac{\pi}{4} = \frac{2}{\sqrt{2}} = \sqrt{2}$$

$$\tan \frac{\pi}{4} = \frac{\sqrt{2}/2}{\sqrt{2}/2} = 1 \qquad \cot \frac{\pi}{4} = \frac{\sqrt{2}/2}{\sqrt{2}/2} = 1.$$

Example 2 Find the values of the trigonometric functions at $t = \pi/6$.

Solution   From Figure 6.5, $P(\pi/6) = (\sqrt{3}/2, 1/2)$. Using the definition with $x = \sqrt{3}/2$ and $y = 1/2$ we obtain

$$\sin \frac{\pi}{6} = \frac{1}{2} \qquad\qquad \csc \frac{\pi}{6} = \frac{1}{(1/2)} = 2$$

$$\cos \frac{\pi}{6} = \frac{\sqrt{3}}{2} \qquad\qquad \sec \frac{\pi}{6} = \frac{1}{(\sqrt{3}/2)} = \frac{2}{\sqrt{3}} = \frac{2\sqrt{3}}{3}$$

$$\tan \frac{\pi}{6} = \frac{(1/2)}{(\sqrt{3}/2)} = \frac{1}{\sqrt{3}} = \frac{\sqrt{3}}{3} \qquad \cot \frac{\pi}{6} = \frac{(\sqrt{3}/2)}{(1/2)} = \sqrt{3}. \qquad ■$$

Before considering additional functional values, we shall discuss several important relationships which exist among the trigonometric functions. The formulas listed in the next box are, without doubt, the most important identities in trigonometry, because they may be used to simplify and unify many different aspects of the subject. Since the formulas are true for every allowable value of t, and are part of the foundation for work in trigonometry, they are called the *Fundamental Identities*. These identities should be memorized before proceeding to the next section of this text.

Three of the Fundamental Identities involve squares such as $(\sin t)^2$ and $(\cos t)^2$. In general, if n is an integer different from -1, then powers such as $(\cos t)^n$ are written in the form $\cos^n t$. The symbols $\sin^{-1} t$ and $\cos^{-1} t$ are reserved for inverse trigonometric functions, to be discussed in the next chapter. With this agreement on notation we have, for example,

$$\cos^2 t = (\cos t)^2 = (\cos t)(\cos t)$$
$$\tan^3 t = (\tan t)^3 = (\tan t)(\tan t)(\tan t)$$
$$\sec^4 t = (\sec t)^4 = (\sec t)(\sec t)(\sec t)(\sec t).$$

Let us first list all the fundamental identities and then discuss the proofs. The formulas to follow are true for all values of t in the domains of the indicated functions.

The Fundamental Identities

$$\csc t = \frac{1}{\sin t}, \qquad \sec t = \frac{1}{\cos t}, \qquad \cot t = \frac{1}{\tan t}$$

$$\tan t = \frac{\sin t}{\cos t}, \qquad \cot t = \frac{\cos t}{\sin t}$$

$$\sin^2 t + \cos^2 t = 1, \qquad 1 + \tan^2 t = \sec^2 t, \qquad 1 + \cot^2 t = \csc^2 t$$

The proofs follow directly from the definition of the trigonometric functions. Thus,

$$\csc t = \frac{1}{y} = \frac{1}{\sin t}, \qquad \sec t = \frac{1}{x} = \frac{1}{\cos t}, \qquad \cot t = \frac{x}{y} = \frac{1}{(y/x)} = \frac{1}{\tan t},$$

$$\tan t = \frac{y}{x} = \frac{\sin t}{\cos t}, \qquad \cot t = \frac{x}{y} = \frac{\cos t}{\sin t},$$

where we assume that no denominator is zero.

If (x, y) is a point on the unit circle U, then

$$y^2 + x^2 = 1$$

Since $y = \sin t$ and $x = \cos t$, this gives us

$$(\sin t)^2 + (\cos t)^2 = 1$$

or equivalently,

$$\sin^2 t + \cos^2 t = 1.$$

If $\cos t \neq 0$, then, dividing both sides of the last equation by $\cos^2 t$, we obtain

$$\frac{\sin^2 t}{\cos^2 t} + 1 = \frac{1}{\cos^2 t}$$

or

$$\left(\frac{\sin t}{\cos t}\right)^2 + 1 = \left(\frac{1}{\cos t}\right)^2.$$

Since $\tan t = \sin t / \cos t$ and $\sec t = 1/\cos t$ we see that

$$\tan^2 t + 1 = \sec^2 t.$$

The final fundamental identity is left as an exercise.

Example 3 Find the values of the trigonometric functions at the number t if
(a) $t = 0$; (b) $t = \pi/2$.

Solutions

(a) The rectangular coordinates of the point $P(0)$ on the unit circle are $(1, 0)$, and hence we take $x = 1$ and $y = 0$ in the definition of the trigonometric functions. This gives us

$$\sin 0 = 0 \quad \text{and} \quad \cos 0 = 1.$$

Next, using two fundamental identities,

$$\tan 0 = \frac{\sin 0}{\cos 0} = \frac{0}{1} = 0, \quad \sec 0 = \frac{1}{\cos 0} = \frac{1}{1} = 1.$$

Of course, we could also find $\tan 0$ and $\sec 0$ by direct substitution for x and y in the definition. The cosecant and cotangent functions are undefined at $t = 0$, since 0 appears in a denominator in the definitions of these functions.

(b) As shown in Figure 6.2, $P(\pi/2) = (0, 1)$, and hence we let $x = 0$ and $y = 1$ in the definition of the trigonometric functions, obtaining

$$\sin (\pi/2) = 1 \quad \text{and} \quad \cos (\pi/2) = 0.$$

Next, applying appropriate fundamental identities,

$$\csc(\pi/2) = \frac{1}{\sin(\pi/2)} = \frac{1}{1} = 1$$

$$\cot(\pi/2) = \frac{\cos(\pi/2)}{\sin(\pi/2)} = \frac{0}{1} = 0.$$

The tangent and secant functions are undefined. (Why?)   ∎

It is not difficult to determine the signs of the functional values of the trigonometric functions when $P(t)$ is in various quadrants. For example, if $P(t) = (x, y)$ is in quadrant II, then y is positive, x is negative, and we see from the definition that $\sin t$ and $\csc t$ are positive, whereas the other four functions are negative. The reader should check the remaining quadrants. The table below indicates the signs in all four quadrants.

Signs of the Trigonometric Functions

Quadrant containing $P(t) = (x, y)$	Positive functions	Negative functions
I	All	None
II	sin, csc	cos, sec, tan, cot
III	tan, cot	sin, csc, cos, sec
IV	cos, sec	sin, csc, tan, cot

Example 4   Find the quadrant containing $P(t)$ if both $\sin t < 0$ and $\cos t > 0$.

Solution   Referring to the preceding table, we see that $\sin t < 0$ if $P(t)$ is in quadrant III or IV and $\cos t > 0$ if $P(t)$ is in quadrant I or IV. Hence, for both conditions to be satisfied, $P(t)$ must be in quadrant IV.   ∎

Example 5   If $\sin t = \sqrt{3}/2$ and $\sec t = -2$, find the following:
(a) $\csc t$;   (b) $\cos t$;   (c) $\tan t$.

Solutions

(a)   Using a fundamental identity,

$$\csc t = \frac{1}{\sin t} = \frac{1}{(\sqrt{3}/2)} = \frac{2}{\sqrt{3}} = \frac{2\sqrt{3}}{3}.$$

(b)   Since $\sec t = 1/\cos t$, we also have $(\cos t)(\sec t) = 1$, and hence

$$\cos t = \frac{1}{\sec t} = \frac{1}{-2} = -\frac{1}{2}.$$

(c) Since $\sin t = \sqrt{3}/2$ and $\cos t = -1/2$,

$$\tan t = \frac{\sin t}{\cos t} = \frac{\sqrt{3}/2}{-1/2} = -\sqrt{3}.$$ ■

The identity $\sin^2 t + \cos^2 t = 1$ can be used to express $\sin t$ in terms of $\cot t$, or vice versa. For example, since

$$\sin^2 t = 1 - \cos^2 t$$

we have

$$\sin t = \pm\sqrt{1 - \cos^2 t}$$

where the "$+$" sign is used if $P(t)$ is in quadrant I or II, and the "$-$" sign is used if $P(t)$ is in quadrant III or IV. Similarly,

$$\cos t = \pm\sqrt{1 - \sin^2 t}$$

where "$+$" is used if $P(t)$ is in quadrant I or IV and "$-$" is used if $P(t)$ is in quadrant II or III. The formulas $1 + \tan^2 t = \sec^2 t$ and $1 + \cot^2 t = \csc^2 t$ can be used in like manner.

Example 6 If $\sin t = 3/5$ and $P(t)$ is in quadrant II, find the values of the other trigonometric functions.

Solution Since $\cos t$ is negative in quadrant II,

$$\cos t = -\sqrt{1 - \sin^2 t} = -\sqrt{1 - (3/5)^2} = -\sqrt{16/25} = -4/5.$$

Next we see that

$$\tan t = \frac{\sin t}{\cos t} = \frac{3/5}{-4/5} = -\frac{3}{4}.$$

Finally, applying the first three fundamental identities gives us

$$\csc t = \frac{5}{3}, \quad \sec t = -\frac{5}{4}, \quad \cot t = -\frac{4}{3}.$$ ■

EXERCISES 6.2

1 Prove that $1 + \cot^2 t = \csc^2 t$.

2 Prove that $\sin t = 1/\csc t$.

In Exercises 3–6 use fundamental identities to find the values of $\csc t$, $\sec t$, $\tan t$, and $\cot t$.

3 $\sin t = -3/5$, $\cos t = 4/5$

4 $\sin t = 8/17$, $\cos t = -15/17$

5 $\sin t = a/\sqrt{a^2 + b^2}$, $\cos t = b/\sqrt{a^2 + b^2}$, where $a^2 + b^2 \neq 0$

6 $\sin t = \sqrt{5}/5$, $\sec t = \sqrt{5}/2$

In Exercises 7–24 find the values of the trigonometric functions corresponding to the number t. (Refer to Figure 6.3 or 6.5 if necessary).

7 $t = -3\pi/2$ **8** $t = -\pi/2$

9 $t = 3\pi/4$ **10** $t = -3\pi/4$

11 $t = -\pi$ **12** $t = 5\pi$

13 $t = 7\pi/2$ **14** $t = 9\pi/2$

15 $t = 200\pi$ **16** $t = 375\pi$

17 $t = -5\pi/4$ **18** $t = 29\pi/4$

19 $t = \pi/3$ **20** $t = -\pi/6$

21 $t = 5\pi/6$ **22** $t = 4\pi/3$

23 $t = -2\pi/3$ **24** $t = -11\pi/6$

In each of Exercises 25–34 find the quadrant containing $P(t)$ if the given conditions are true.

25 $\cos t > 0$ and $\sin t < 0$

26 $\tan t < 0$ and $\cos t > 0$

27 $\sin t < 0$ and $\cot t > 0$

28 $\sec t > 0$ and $\tan t < 0$

29 $\csc t > 0$ and $\sec t < 0$

30 $\csc t > 0$ and $\cot t < 0$

31 $\sec t < 0$ and $\tan t > 0$

32 $\sin t < 0$ and $\sec t > 0$

33 $\cos t > 0$ and $\tan t > 0$

34 $\cos t < 0$ and $\csc t < 0$

In Exercises 35–43 use fundamental identities to find the values of all six trigonometric functions if the given conditions are true.

35 $\tan t = -\frac{3}{4}$ and $\sin t > 0$

36 $\cot t = \frac{3}{4}$ and $\cos t < 0$

37 $\sin t = -\frac{5}{13}$ and $\sec t > 0$

38 $\cos t = \frac{1}{2}$ and $\sin t < 0$

39 $\cos t = -\frac{1}{3}$ and $\sin t < 0$

40 $\csc t = 5$ and $\cot t < 0$

41 $\sec t = -4$ and $\csc t > 0$

42 $\sin t = \frac{2}{5}$ and $\cos t < 0$

43 Is there a real number t such that $7 \sin t = 9$? Explain.

44 Is there a real number t such that $3 \csc t = 1$? Explain.

45 If $f(t) = \cos t$ and $g(t) = t/4$, find the following.
(a) $f(g(\pi))$ (b) $g(f(\pi))$

46 If $f(t) = \tan t$ and $g(t) = t/4$, find the following.
(a) $f(g(\pi))$ (b) $g(f(\pi))$

47 Show that if $P(t)$ is not on the y-axis, then $(\cos t)(\sec t) = 1$.

48 Show that if $P(t)$ is not on a coordinate axis, then $(\tan t)(\cot t) = 1$.

CALCULATOR EXERCISES 6.2

In each of Exercises 1–6 use fundamental identities to approximate the values of all six trigonometric functions to four significant figures if the two given conditions are true.

1 $\sin t = 0.3145$, $\cos t = -0.9492$

2 $\cos t = 0.7314$, $\sin t = 0.6820$

3 $\cos t = 0.8910$, $\tan t = 0.5095$

4 $\sin t = 0.9744$, $\cot t = 0.2309$

5 $\cos t = -0.1219$, $\csc t > 0$

6 $\sin t = 0.9063$, $\tan t < 0$

6.3 THE VARIATION OF THE TRIGONOMETRIC FUNCTIONS

In this section we shall discuss the domains and ranges of each trigonometric function and make some preliminary observations about the variation of $\sin t$ and $\cos t$ as t varies through the set of real numbers.

The domain of each trigonometric function can be determined by referring to the definition in Section 6.2. If $P(t)$ has coordinates (x, y), then the sine and cosine functions always exist and, therefore, each has domain $\mathbb{R}$. In the definitions of the tangent and secant functions, x appears in the denominator and hence we must exclude values of t for which x is 0, that is, the values of t corresponding to the points $(0, 1)$ and $(0, -1)$ on the unit circle U. It follows that the domain of the tangent and secant functions consists of all numbers *except* those of the form $(\pi/2) + n\pi$, where n is an integer. In particular, we exclude $\pm \pi/2$, $\pm 3\pi/2$, $\pm 5\pi/2$, etc.

Since $\cot t = x/y$ and $\csc t = 1/y$, the domain of the cotangent and cosecant functions is the set of all real numbers except those numbers t for which the y-coordinate of $P(t)$ is 0. These include the numbers, 0, $\pm \pi$, $\pm 2\pi$, $\pm 3\pi$, and, in general, all numbers of the form $n\pi$, where n is an integer. In the future, when we work with $\tan t$, $\cot t$, $\sec t$, and $\csc t$, we will always assume that t is in the appropriate domain, even though this fact will not always be mentioned explicitly.

It was pointed out in Section 6.1 that, except for special values of t, the determination of the point $P(t)$ on U (and hence the values of the trigonometric functions) requires techniques studied in calculus. We can, however, make some general observations. For the time being, let us concentrate on the sine function. Referring to the unit circle in Figure 6.7, we see that if t varies from 0 to $\pi/2$, then the coordinates (x, y) of $P(t)$ vary from $(1, 0)$ to $(0, 1)$. In particular, the y-coordinate, that is, the value of the sine function, increases from 0 to 1. Moreover, this function takes on *all* values between 0 and 1. If we let t vary from $\pi/2$ to π, then the coordinates (x, y) of $P(t)$ vary from $(0, 1)$ to $(-1, 0)$, and hence the sine function, that is, the ordinate y of $P(t)$, decreases from 1 to 0. In similar fashion, we see that if t varies from π to $3\pi/2$, then $\sin t$ decreases from 0 to -1; and as t varies from $3\pi/2$ to 2π, $\sin t$ increases from -1 to 0. The following table partially indicates this behavior of $\sin t$ in the interval $[0, 2\pi]$. For a more complete description we would have to insert many more values of t and $\sin t$.

t	0	$\dfrac{\pi}{2}$	π	$\dfrac{3\pi}{2}$	2π
$\sin t$	0	1	0	-1	0

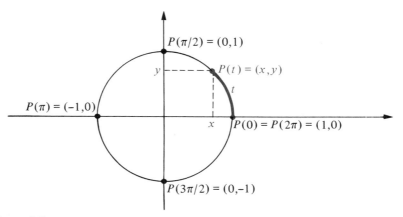

Figure 6.7

If we let t vary through the interval $[2\pi, 4\pi]$, then $P(t)$ traces the circle again and the identical pattern for $\sin t$ is repeated, that is,

$$\sin(t + 2\pi) = \sin t$$

for every t in $[0, 2\pi]$. The same is true for other intervals of length 2π. Indeed,

$$\sin(t + 2\pi n) = \sin t$$

since, as we observed in Section 6.1, $P(t) = P(t + 2\pi n)$ for every t and for every integer n. According to the next definition, this repetitive behavior is specified by saying that the sine function is *periodic*.

Definition

> A function f is **periodic** if there exists a positive real number k such that
>
> $$f(t + k) = f(t)$$
>
> for every t in the domain of f. The least such positive real number k, if it exists, is called the **period** of f.

The period of the sine function is 2π (see Exercise 3).

The variation of the cosine function in the interval $[0, 2\pi]$ can be determined by observing the behavior of the x-coordinate of the point $P(t)$ on U as t varies from 0 to 2π. Again referring to Figure 6.7, we see that the cosine function (that is, x) decreases from 1 to 0 in the interval $[0, \pi/2]$, decreases from 0 to -1 in $[\pi/2, \pi]$, increases from -1 to 0 in $[\pi, 3\pi/2]$, and increases from 0 to 1 in $[3\pi/2, 2\pi]$. The pattern is then repeated in successive intervals of length 2π. It follows that the cosine function is periodic with period 2π.

The preceding discussion suggests that the graphs of the sine and cosine functions may have the appearance of those in Figures 6.8 and 6.9, respectively. In the figures we have plotted several points of the form $(t, \sin t)$ and $(t, \cos t)$ using the functional values mentioned earlier. Since each function has period 2π, each graph repeats itself every 2π units along the horizontal axis. These sketches should be considered as very rough approximations, since we have not yet developed techniques for obtaining many intermediate points. Note, however, that we could use Example 1 of the preceding section to get the points $(\pi/4, \sin \pi/4)$ and $(\pi/4, \cos \pi/4)$, that is, $(\pi/4, \sqrt{2}/2)$.

Figure 6.8

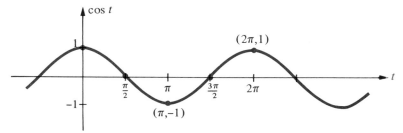

Figure 6.9

Similarly, Example 2 gives us the point $(\pi/6, 1/2)$ on the sine graph, and $(\pi/6, \sqrt{3}/2)$ on the cosine graph. We will study the graphs of the trigonometric functions more carefully in Section 6.7.

It can be shown that the secant and cosecant functions are also periodic with period 2π (see Exercise 6). The variations of these functions will be discussed in Section 6.7. At that time we shall see that the tangent and cotangent functions are periodic with period π.

Finally, let us briefly discuss the range of each trigonometric function. Since the pair (x, y) in the definition of the trigonometric functions gives the coordinates of a point on the unit circle U, we have

$$-1 \le x \le 1 \quad \text{and} \quad -1 \le y \le 1.$$

Since $\sin t = y$ and $\cos t = x$, it follows that the range of both the sine and cosine functions is the set of all numbers in the closed interval $[-1, 1]$.

If x is a nonzero real number such that $-1 \le x \le 1$, then $|x| \le 1$. Multiplying both sides of the last inequality by $1/|x|$, we obtain $1 \le 1/|x|$. Since $\sec t = 1/x$, we see that $1 \le |\sec t|$ for every real number t, and therefore the range of the secant function consists of all real numbers having absolute value greater than or equal to 1. The same is true for the range of the cosecant function.

It can be shown that $\tan t$ and $\cot t$ take on all real values (see Exercise 5). That will also become apparent in Section 6.7, when we study the graphs of the trigonometric functions.

EXERCISES 6.3

In each of Exercises 1 and 2 plot the points $(a, 0)$ on the x-axis if a satisfies both of the given conditions.

1 (a) The tangent and secant functions are undefined at a.

    (b) $-4\pi \le a \le 4\pi$

2 (a) The cotangent and cosecant functions are undefined at a.

    (b) $-4\pi \le a \le 4\pi$

3 Prove that the sine function has period 2π. (*Hint*: Assume that there is a positive real number k less than 2π such that $\sin(t + k) = \sin t$ for all t. Arrive at a contradiction by letting $t = 0$).

4 Prove that the cosine function has a period 2π.

5 Prove that the range of the tangent function is $\mathbb{R}$ by showing that if a is any real number, then there is a point $P(t)$ on U such that $\tan t = a$. (*Hint*: If $P(t) = (x, y)$ is on U, consider the equation $\tan t = y/x = a$, where $x^2 + y^2 = 1$.)

6 Prove that the cosecant and secant function each have period 2π. (*Hint*: Use Exercises 3 and 4 and two fundamental identities.)

6.4 ANGLES

The definitions of the trigonometric functions can also be based upon the notion of angles. This more traditional approach is quite common in applications of mathematics and hence should not be obscured by the definitions stated in Section 6.2. Indeed, for a thorough appreciation of the trigonometric functions, it is best to blend the two ideas. In this section we restrict our discussion to angles and their measurement. In Section 6.5 we shall demonstrate how angles can be used to find values of the trigonometric functions.

An angle is often regarded as the set of all points on two rays or half-lines, l_1 and l_2, having the same initial point O. If A and B are points on l_1 and l_2, respectively (see Figure 6.10), then we may refer to **angle AOB**. The same is true for finite line segments with a common endpoint. For trigonometric purposes it is convenient to regard angle AOB as generated by starting with the fixed ray l_1 with endpoint O and rotating it about O, in a plane, to a position specified by ray l_2. We call l_1 the **initial side**, l_2 the **terminal side**, and O the **vertex** of the angle. The amount or direction of rotation is not restricted in any way. Thus, we might let l_1 make several revolutions in either direction about O before coming to the position l_2.

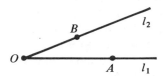

Figure 6.10

If a rectangular coordinate system is introduced, then the **standard position** of an angle is obtained by taking the vertex at the origin and letting l_1 coincide with the positive x-axis. If l_1 is rotated in a counterclockwise direction to position l_2, then the angle is considered **positive**, whereas if l_1 is rotated in a clockwise direction, the angle is **negative**. We often denote angles by lower-case Greek letters and specify the direction of rotation by means of a circular arc or spiral with an arrow attached. Figure 6.11 contains sketches of two positive angles α and β and a negative angle γ. If the terminal side of an angle is in a certain quadrant, we speak of the *angle* as being in that quadrant. In Figure 6.11 α is in quadrant II, β is in quadrant I, and γ is in quadrant III. If the terminal side coincides with a coordinate axis, then the angle is referred to as a

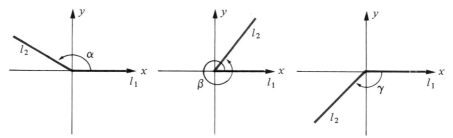

Figure 6.11

quadrantal angle. It is important to observe that there are many different angles in standard position which have the same terminal side. Any two such angles are called **coterminal**.

We shall next consider the problem of assigning a measure to a given angle. Let U be a unit circle with the center at the origin O of a rectangular coordinate system, and let θ be an angle in standard position. We regard θ as generated by rotating the positive x-axis about O. As the axis rotates to its terminal position, its point of intersection with U travels a certain distance t before arriving at its final position P, as illustrated in Figure 6.12. If t is considered positive for a counterclockwise rotation and negative for a clockwise rotation, then P is precisely the point on U that corresponds to the real number t. A natural way of assigning a measure to θ is to use the number t. When this is done, we say that θ **is an angle of t radians** and we write $\theta = t$ or $\theta = t$ *radians*. In particular, if $\theta = 1$, then θ is an angle that subtends an arc of unit length on the unit circle U. The notation $\theta = -7.5$ means that θ is the angle generated by a clockwise rotation in which the point of intersection of the terminal side of θ with the unit circle U travels 7.5 units. Several angles, measured in radians, are sketched in Figure 6.13.

Figure 6.12

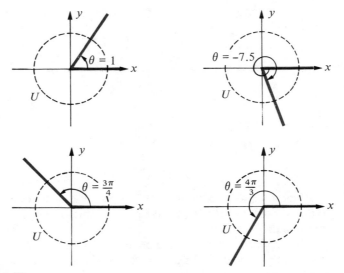

Figure 6.13

The radian measure of an angle can be found by using a circle of *any* radius. In the following discussion, the terminology **central angle** of a circle refers to an angle whose vertex is at the center of the circle. Thus, suppose that θ is a central angle of a circle of radius r and that θ subtends an arc of length s, where $0 \le s < 2\pi r$. To find the radian measure of θ let us place θ in standard position on a rectangular coordinate system and superimpose a unit circle U, as shown in Figure 6.14. If t is the length of arc subtended by θ on U, then by definition we may write $\theta = t$. From plane geometry,

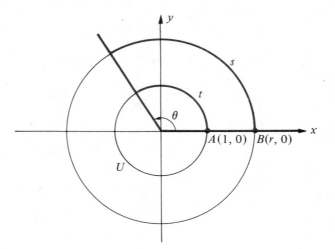

Figure 6.14

the ratio of the arcs in Figure 6.14 is the same as the ratio of the radii; that is,

$$\frac{t}{s} = \frac{1}{r} \quad \text{or} \quad t = \frac{s}{r}.$$

Substituting θ for t gives us the following result.

Theorem

> If a central angle θ of a circle of radius r subtends an arc of length s, then the radian measure of θ is given by
>
> $$\theta = \frac{s}{r}.$$

Example 1 A central angle θ subtends an arc 10 cm long on a circle of radius 4 cm. Find the radian measure of θ.

Solution   Substituting in the formula $\theta = s/r$ gives us the radian measure

$$\theta = \frac{10}{4} = 2.5. \qquad \blacksquare$$

The formula $\theta = s/r$ indicates that the radian measure of an angle is independent of the size of the circle. For example, if the radius of the circle is $r = 4$ cm and the arc subtended by a central angle θ is 8 cm, then the radian measure is

$$\theta = \frac{8 \text{ cm}}{4 \text{ cm}} = 2.$$

If the radius of the circle is 5 km and the subtended arc is 10 km then

$$\theta = \frac{10 \text{ km}}{5 \text{ km}} = 2.$$

These calculations indicate that the radian measure of an angle is dimensionless, and hence may be regarded as a real number. Indeed, it is for this reason that we usually employ the notation $\theta = t$ instead of $\theta = t$ radians.

Another unit of measurement for angles is the **degree**. If the angle is placed in standard position on a rectangular coordinate system, then an angle of 1 degree is, by definition, the measure of the angle formed by $\frac{1}{360}$ of a complete revolution in the counterclockwise direction. The symbol "$°$" is used to denote the degrees in the measure of an angle. In Figure 6.15 several angles measured in degrees are shown in standard position on a rectangular coordinate system. It is customary to refer to an angle of measure $90°$ as a **right angle**. An angle is **acute** if its degree measure is between $0°$ and $90°$. If its measure is between $90°$ and $180°$ an angle is **obtuse**.

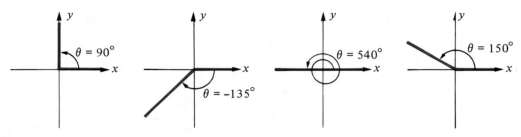

Figure 6.15

If smaller measurements than those afforded by the degree or radian are required, we can use tenths, hundredths, or thousandths of radians or degrees. If degrees are used, then another method is to divide each degree into 60 equal parts called **minutes** (denoted by ′) and each minute into 60 equal parts called **seconds** (denoted by ″). Thus, 1′ is $\frac{1}{60}$ of 1°, and 1″ is $\frac{1}{60}$ of 1′. A notation such as $\theta = 73°56'18''$ refers to an angle θ of measure 73 degrees, 56 minutes, and 18 seconds.

It is not difficult to transform angular measure from one system to another. If we consider the angle θ, in standard position, generated by one-half of a complete counterclockwise rotation, then $\theta = 180°$. Using the formula $\theta = s/r$, we see that the radian measure of θ is π. This gives us the following:

Relationships Between Degrees and Radians

$$180° = \pi \text{ radians}$$

$$1° = \frac{\pi}{180} \text{ radians}, \qquad 1 \text{ radian} = \left(\frac{180}{\pi}\right)°.$$

If we use the approximation $\pi \approx 3.14159$ and calculate $\pi/180$ and $180/\pi$, we obtain

$$1° \approx 0.01745 \text{ radians} \quad \text{and} \quad 1 \text{ radian} \approx 57.296°.$$

Example 2
(a) Find the radian measure of θ if $\theta = 150°$ and if $\theta = 225°$.
(b) Find the degree measure of θ if $\theta = 7\pi/4$ and if $\theta = \pi/3$.

Solutions
(a) Since there are $\pi/180$ radians in each degree, the number of radians in 150° can be found by multiplying 150 by $\pi/180$. Thus,

$$150° = 150\left(\frac{\pi}{180}\right) = \frac{5\pi}{6} \text{ radians.}$$

Similarly, $\qquad 225° = 225\left(\frac{\pi}{180}\right) = \frac{5\pi}{4} \text{ radians.}$

(b) The number of degrees in 1 radian is $180/\pi$. Consequently, to find the number of degrees in $7\pi/4$ radians, we multiply by $180/\pi$, obtaining

$$\frac{7\pi}{4} \text{ radians} = \frac{7\pi}{4}\left(\frac{180}{\pi}\right) = 315°.$$

In like manner, $\frac{\pi}{3} \text{ radians} = \frac{\pi}{3}\left(\frac{180}{\pi}\right) = 60°.$ ∎

Example 3 If the measure of an angle θ is 3 radians, find the approximate measure of θ in terms of degrees, minutes, and seconds.

Solution Since 1 radian $\approx 57.296°$, we have

$$3 \text{ radians} \approx 171.888° = 171° + 0.888°.$$

Since there are 60′ in each degree, the number of minutes in 0.888° is 60(0.888), or 53.28′. Hence,

$$3 \text{ radians} \approx 171°53.28'.$$

Finally, $0.28' = (0.28)60'' \approx 17''$. Therefore,

$$3 \text{ radians} \approx 171°53'17''.$$ ∎

Many hand-held calculators have a key which will automatically convert the degree measure of an angle to radians, and vice versa. The owner of a calculator should refer to the user's guide for information. Since entries are made in the decimal system, angles which are expressed in terms of degrees and minutes must be converted to decimal form before they are entered into the calculator. For example,

$$67°30' = 67.5° \qquad \text{since} \quad 30' = (30/60)° = 0.5°$$
$$123°45' = 123.75° \qquad \text{since} \quad 45' = (45/60)° = 0.75°$$
$$8°13' \approx 8.21667° \qquad \text{since} \quad 13' = (13/60)° \approx 0.21667°.$$

If seconds are involved, another appropriate decimal must be calculated. To illustrate, since there are 3,600 seconds in each degree, $18'' = (18/3600)° = 0.005'$ and hence,

$$67°30'18'' = 67.505°.$$

EXERCISES 6.4

In each of Exercises 1–12 place the angle with the indicated measure in standard position on a rectangular coordinate system, and find the measure of two positive angles and two negative angles which are coterminal with the given angle.

1	120°	2	240°
3	135°	4	315°
5	−30°	6	−150°
7	620°	8	570°

9    $5\pi/6$

10    $2\pi/3$

11    $-\pi/4$

12    $-5\pi/4$

In Exercises 13–24 find the radian measure that corresponds to the given degree measure.

13    $150°$

14    $120°$

15    $-60°$

16    $-135°$

17    $225°$

18    $210°$

19    $450°$

20    $630°$

21    $72°$

22    $54°$

23    $100°$

24    $95°$

In Exercises 25–36 find the degree measure that corresponds to the given radian measure.

25    $2\pi/3$

26    $5\pi/6$

27    $11\pi/6$

28    $4\pi/3$

29    $3\pi/4$

30    $11\pi/4$

31    $-7\pi/2$

32    $-5\pi/2$

33   7π

34   9π

35    $\pi/9$

36    $\pi/16$

In Exercises 37–40 find the approximate measure of θ in terms of degrees, minutes, and seconds.

37    $\theta = 2$

38    $\theta = 1.5$

39    $\theta = 5$

40    $\theta = 4$

41   A central angle θ subtends an arc 7 cm long on a circle of radius 4 cm. Approximate the measure of θ in   (a) radians; (b) degrees.

CALCULATOR EXERCISES 6.4

In Exercises 1–6 approximate, to four decimal places, the radian measure that corresponds to the given degree measure of θ, or the degree measure that corresponds to the given radian measure.

1    $\theta = 73°24'$

2    $\theta = 261°37'$

3    $\theta = 482°16'43''$

4    $\theta = \sqrt{317} + \pi$

5    $\theta = \ln(13.6521)$

6    $\theta = e^{\sqrt{\pi}}$

42   A central angle θ subtends an arc 3 feet long on a circle of radius 20 inches. Approximate the measure of θ in (a) radians; (b) degrees.

43   Approximate the length of an arc subtended by a central angle of measure $50°$ on a circle of radius 8 meters.

44   Approximate the length of an arc subtended by a central angle of 2.2 radians on a circle of radius 60 cm.

45   If a central angle θ of measure $20°$ subtends a circular arc of length 3 km, find the radius of the circle.

46   If a central angle θ of radian measure 4 subtends a circular arc 10 cm long, find the radius of the circle.

47   The distance between two points A and B on the earth is measured along a circle having center C at the center of the earth and radius equal to the distance from C to the surface. If the diameter of the earth is approximately 8,000 miles, approximate the distance between A and B if angle ACB has measure (a) $60°$; (b) $45°$; (c) $30°$; (d) $10°$; (e) $1°$.

48   Refer to Exercise 47. If angle ACB has measure $1'$, then the distance between A and B is called a **nautical mile**. Approximate the number of ordinary **(statute)** miles in a nautical mile.

49   Refer to Exercise 47. If two points A and B are 500 miles apart, find angle ACB in both degree measure and radian measure.

50   The **angular speed** of a wheel which is rotating at a constant rate is the angle generated by a line segment from the center of the wheel to the circumference, per unit of time. If a wheel of diameter 3 feet is rotating at a rate of 2,400 rpm (revolutions per minute), find the angular speed.

7–9   The diameter of the earth at the equator is estimated as 7,926.41 miles. Rework Exercises 47–49 using that measurement, assuming that points A and B are on the equator.

10   Refer to Exercise 50. If a wheel of diameter 1.683 meters is rotating at a rate of 2,875 rpm, find the angular speed.

6.5   TRIGONOMETRIC FUNCTIONS OF ANGLES

For certain applications it is convenient to change the domain of the trigonometric functions from a set of real numbers to a set of angles. This is very easy to do. If θ is an angle, we merely agree on the values $\sin\theta$, $\cos\theta$, and so on. The usual way of assigning such values is to use the radian measure of θ, as in the following definition.

Definition of Trigonometric Functions of Angles

If θ is an angle, and if the radian measure of θ is t, then the value of each trigonometric function at θ is its value at the real number t.

We see from the preceding definition that if t is the radian measure of θ, then

$$\sin\theta = \sin t, \quad \cos\theta = \cos t, \quad \tan\theta = \tan t,$$

and likewise for the other functions. For convenience we shall use the terminology *trigonometric functions* regardless of whether angles or real numbers are employed for the domain. To make the unit of angular measure clear we shall use the degree symbol and write $\sin 65°$, $\tan 150°$, etc., whenever the angle is measured in degrees. Numerals without any symbol attached, such as $\cos 3$ and $\csc(\pi/6)$, will indicate that radian measure is being used. This is not in conflict with our previous work where, for example, $\cos 3$ meant the value of the cosine function at the real number 3, since by definition, the cosine of an angle of measure 3 radians is identical with the cosine of the real number 3.

Example 1   Find $\sin 90°$, $\cos 45°$, and $\tan 720°$.

Solution   The angles are shown in standard position in Figure 6.16. The radian measures of the angles may be found as follows:

$$90° = 90\left(\frac{\pi}{180}\right) = \frac{\pi}{2}, \quad 45° = 45\left(\frac{\pi}{180}\right) = \frac{\pi}{4}, \quad 720° = 720\left(\frac{\pi}{180}\right) = 4\pi.$$

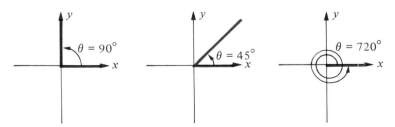

Figure 6.16

Using the definition of trigonometric functions of angles and Examples 1 and 2 of Section 6.2, we obtain

$$\sin 90° = \sin (\pi/2) = 1$$
$$\cos 45° = \cos (\pi/4) = \sqrt{2}/2$$
$$\tan 720° = \tan 4\pi = \tan 0 = 0.$$ ∎

In the next section we shall introduce techniques for approximating trigonometric functional values corresponding to *any* angle θ.

The values of the trigonometric functions at an angle θ may be determined by means of an arbitrary point on the terminal side of θ. To prove this, let θ be an angle in standard position and let $Q(a, b)$ be any point on the terminal side of θ, where $d(O, Q) = r > 0$. Figure 6.17 illustrates the case in which the terminal side lies in quadrant III; however, our discussion applies to any angle.

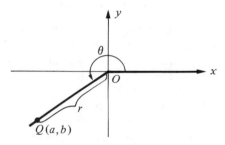

Figure 6.17

The point $Q(a, b)$ is not necessarily a point on the unit circle U, since r may be different from 1. Let $P(x, y)$ be the point on the terminal side of θ such that $d(O, P) = 1$. Hence, $P(x, y)$ is on the unit circle U. If t is the radian measure of θ, then by definition we have

$$\sin \theta = \sin t = y$$
$$\cos \theta = \cos t = x.$$

As in Figure 6.18, let us consider vertical lines through Q and P intersecting the x-axis at $Q'(a, 0)$ and $P'(x, 0)$, respectively. Since triangles OQQ' and OPP' are similar, we have

$$\frac{d(P', P)}{d(O, P)} = \frac{d(Q', Q)}{d(O, Q)} \quad \text{or} \quad \frac{|y|}{1} = \frac{|b|}{r}.$$

Since b and y always have the same sign, this gives us

$$y = \frac{b}{r} \quad \text{and hence} \quad \sin \theta = y = \frac{b}{r}.$$

In similar fashion we obtain

$$\cos \theta = \frac{a}{r}.$$

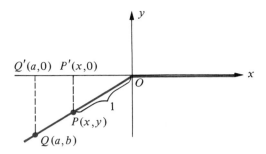

Figure 6.18

Using a fundamental identity,

$$\tan \theta = \frac{\sin \theta}{\cos \theta} = \frac{(b/r)}{(a/r)} = \frac{b}{a}.$$

The remaining three functions may be obtained by taking reciprocals. This gives us the following theorem.

Theorem on Trigonometric Functions as Ratios

> Let θ be an angle in standard position on a rectangular coordinate system and let $Q(a,b)$ be any point other than O on the terminal side of θ. If $d(O,Q) = r$, then
>
> $$\sin \theta = \frac{b}{r} \qquad\qquad \csc \theta = \frac{r}{b} \quad (\text{if } b \neq 0)$$
>
> $$\cos \theta = \frac{a}{r} \qquad\qquad \sec \theta = \frac{r}{a} \quad (\text{if } a \neq 0)$$
>
> $$\tan \theta = \frac{b}{a} \quad (\text{if } a \neq 0) \qquad \cot \theta = \frac{a}{b} \quad (\text{if } b \neq 0).$$

It can be shown, by using similar triangles, that the formulas given in this theorem are independent of the point $Q(a, b)$ that is chosen on the terminal side of θ. Note that if $r = 1$, then the formulas reduce to the definition of the trigonometric functions given in Section 6.2, with $a = x$ and $b = y$. This theorem has many applications in trigonometry. In Section 6.10 we shall use it to solve problems concerned with right triangles. Another reason for its importance is that it can often be used to obtain values of trigonometric functions. Indeed, note that it is sufficient to find *one* point (other than O) on the terminal side of an angle provided the angle is in standard position on a rectangular coordinate system.

Example 2   If θ is an angle in standard position on a rectangular coordinate system, and if the point $Q(-15, 8)$ is on the terminal side of θ, find the values of the trigonometric functions of θ.

Solution   By the Distance Formula, the distance r from the origin O to any point $Q(a, b)$ is $r = \sqrt{(a - 0)^2 + (b - 0)^2} = \sqrt{a^2 + b^2}$. Hence, for $Q(-15, 8)$, we have

$$r = \sqrt{(-15)^2 + 8^2} = \sqrt{225 + 64} = \sqrt{289} = 17.$$

Applying the last theorem with $a = -15$, $b = 8$, and $r = 17$, we obtain

$$\sin \theta = \frac{8}{17} \qquad \csc \theta = \frac{17}{8}$$

$$\cos \theta = -\frac{15}{17} \qquad \sec \theta = -\frac{17}{15}$$

$$\tan \theta = -\frac{8}{15} \qquad \cot \theta = -\frac{15}{8}. \qquad ■$$

Example 3   An angle θ is in standard position on a rectangular coordinate system. If its terminal side is in quadrant III and it lies on the line $y = 3x$, find the values of the trigonometric functions of θ.

Solution   The graph of $y = 3x$ is sketched in Figure 6.19, together with the initial and terminal sides of θ (in color). We begin by choosing a convenient point, say $Q(-1, -3)$, on the terminal side. The distance r from the origin to Q is

$$r = d(O, Q) = \sqrt{(-1)^2 + (-3)^2} = \sqrt{10}.$$

Applying the Theorem on Trigonometric Functions as Ratios with $a = -1$, $b = -3$, and $r = \sqrt{10}$ gives us

$$\sin \theta = \frac{-3}{\sqrt{10}} = \frac{-3\sqrt{10}}{10}, \qquad \csc \theta = \frac{\sqrt{10}}{-3} = -\frac{\sqrt{10}}{3}$$

$$\cos \theta = \frac{-1}{\sqrt{10}} = \frac{-\sqrt{10}}{10}, \qquad \sec \theta = \frac{\sqrt{10}}{-1} = -\sqrt{10}$$

$$\tan \theta = \frac{-3}{-1} = 3, \qquad \cot \theta = \frac{-1}{-3} = \frac{1}{3}.$$

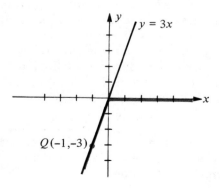

$Q(-1,-3)$

Figure 6.19   ■

In the proof of the Theorem on Trigonometric Functions as Ratios we considered θ as a nonquadrantal angle; however, the formulas we derived are also true if the terminal side of θ lies on either the x- or y-axis. This is illustrated by the next example.

Example 4   Find the values of the trigonometric functions of θ if $\theta = 270°$.

Solution   If we place θ in standard position, then the terminal side coincides with the negative y-axis. We next choose any point Q on the terminal side of θ. For simplicity we consider $Q(0, -1)$. In this case $r = 1$, $a = 0$, $b = -1$, and hence

$$\sin \theta = \frac{-1}{1} = -1 \qquad \csc \theta = \frac{1}{-1} = -1$$

$$\cos \theta = \frac{0}{1} = 0 \qquad \cot \theta = \frac{0}{-1} = 0.$$

The tangent and secant functions are undefined, since the meaningless expressions $\tan \theta = (-1)/0$ and $\sec \theta = 1/0$ arise when we substitute in the appropriate formulas.   ■

We shall conclude this section by showing that for acute angles, values of the trigonometric functions can be interpreted as ratios of the lengths of the sides of a right triangle. Recall that a triangle is called a **right triangle** if one of its angles is a right angle. If θ is an acute angle, then it can be regarded as an angle of a right triangle and we may refer to the lengths of the **hypotenuse**, the **opposite side**, and the **adjacent side** in the usual way. For convenience we shall use **hyp**, **opp**, and **adj**, respectively, to denote these numbers. Introducing a rectangular coordinate system as in Figure 6.20 we see that the lengths of the adjacent side and the opposite side for θ are the abscissa and ordinate, respectively, of a point Q on the terminal side of θ. By the Theorem on Trigonometric Functions as Ratios, we have

$$\sin \theta = \frac{\text{opp}}{\text{hyp}} \qquad \csc \theta = \frac{\text{hyp}}{\text{opp}}$$

$$\cos \theta = \frac{\text{adj}}{\text{hyp}} \qquad \sec \theta = \frac{\text{hyp}}{\text{adj}}$$

$$\tan \theta = \frac{\text{opp}}{\text{adj}} \qquad \cot \theta = \frac{\text{adj}}{\text{opp}}$$

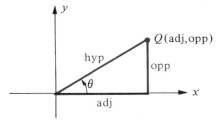

Figure 6.20

These formulas are very important in work with right triangles and should be memorized. The next example illustrates how they may be used.

Example 5 Find the values of $\sin \theta$, $\cos \theta$, and $\tan \theta$ for the following values of θ:
(a) $\theta = 60°$;   (b) $\theta = 30°$;   (c) $\theta = 45°$.

Solutions   Let us consider an equilateral triangle having sides of length 2. The median from one vertex to the opposite side bisects the angle at that vertex, as illustrated in (i) of Figure 6.21. By the Pythagorean Theorem, the length of this median is $\sqrt{3}$. Using the colored triangle, we obtain the following.

(a) $\sin 60° = \dfrac{\sqrt{3}}{2}$, $\cos 60° = \dfrac{1}{2}$, $\tan 60° = \dfrac{\sqrt{3}}{1} = \sqrt{3}$

(b) $\sin 30° = \dfrac{1}{2}$, $\cos 30° = \dfrac{\sqrt{3}}{2}$, $\tan 30° = \dfrac{1}{\sqrt{3}} = \dfrac{\sqrt{3}}{3}$

(c)   To find the functional values for $\theta = 45°$, let us consider an isosceles right triangle whose two equal sides have length 1, as illustrated in (ii) of Figure 6.21. Thus,

$$\sin 45° = \frac{1}{\sqrt{2}} = \frac{\sqrt{2}}{2} = \cos 45°, \quad \tan 45° = \frac{1}{1} = 1.$$

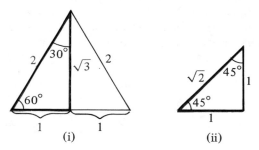

(i)   (ii)

Figure 6.21   ∎

The radian measures corresponding to 30°, 45°, and 60° are $\pi/6$, $\pi/4$, and $\pi/3$, respectively. Since the values of the trigonometric functions at a real number t are the same as those at an angle of t radians, Example 5 provides a convenient method for obtaining these special values. For convenience they are listed in tabular form in Table A on page 261, together with the values at 0 and $\pi/2$. The reader should check each entry (see Exercise 19). A dash indicates that the function is undefined. Two reasons for stressing these special values are (1) they are exact, and (2) they arise frequently in work involving trigonometry. It is a good idea either to memorize

Table A. *Special Values of the Trigonometric Functions*

t number, radians	t degrees	$\sin t$	$\cos t$	$\tan t$	$\cot t$	$\sec t$	$\csc t$
0	$0°$	0	1	0	—	1	—
$\dfrac{\pi}{6}$	$30°$	$\dfrac{1}{2}$	$\dfrac{\sqrt{3}}{2}$	$\dfrac{\sqrt{3}}{3}$	$\sqrt{3}$	$\dfrac{2\sqrt{3}}{3}$	2
$\dfrac{\pi}{4}$	$45°$	$\dfrac{\sqrt{2}}{2}$	$\dfrac{\sqrt{2}}{2}$	1	1	$\sqrt{2}$	$\sqrt{2}$
$\dfrac{\pi}{3}$	$60°$	$\dfrac{\sqrt{3}}{2}$	$\dfrac{1}{2}$	$\sqrt{3}$	$\dfrac{\sqrt{3}}{3}$	2	$\dfrac{2\sqrt{3}}{3}$
$\dfrac{\pi}{2}$	$90°$	1	0	—	0	—	1

Table A or to be able to find the values quickly, using triangles and points on the coordinate axes (compare with Examples 1 and 2 of Section 6.1).

We have now discussed two different approaches to the trigonometric functions. The definition introduced in Section 6.2 emphasizes the fact that the trigonometric functions have domains consisting of real numbers. Such functions are the building blocks for the subject of calculus. In addition, we shall see later that the unit circle approach is useful for discussing graphs and deriving various trigonometric identities. The development in terms of angles and ratios considered in this section has many applications in the sciences and engineering. Students should strive to become proficient in the use of both formulations of the trigonometric functions, since each will reinforce the other and make it easier to master more advanced aspects of trigonometry.

EXERCISES 6.5

In Exercises 1–18 use the Theorem on Trigonometric Functions as Ratios to find the values of the six trigonometric functions of θ if θ is in standard position and satisfies the given condition.

1 The point $P(4, -3)$ is on the terminal side of θ.

2 The point $P(-8, -15)$ is on the terminal side of θ.

3 The point $P(-2, -5)$ is on the terminal side of θ.

4 The point $P(-1, 2)$ is on the terminal side of θ.

5 The terminal side of θ is in quadrant II and lies on the line $y = -4x$.

6 The terminal side of θ is in quadrant IV and lies on the line $3y + 5x = 0$.

7 The terminal side of θ is in quadrant III and is parallel to the line $2y - 7x + 2 = 0$.

8 The terminal side of θ is in quadrant II and is parallel to the line through the points $A(1, 4)$ and $B(3, -2)$.

9 The terminal side of θ is in quadrant I and lies on a line having slope 4/3.

10 The terminal side of θ bisects the third quadrant.

11 $\theta = 450°$ **12** $\theta = 180°$

13 $\theta = 135°$ **14** $\theta = 225°$

15 $\theta = -45°$ **16** $\theta = 405°$

17 $\theta = 360°$ **18** $\theta = 630°$

19 Check each entry in Table A.

20 Prove geometrically that the formulas given in the Theorem on Trigonometric Functions as Ratios are independent of the point $Q(a, b)$ that is chosen on the terminal side of θ.

In Exercises 21–30 find the values of the trigonometric functions of θ if θ is the angle of the pictured right triangle.

21

22

23

24

25

26

27

28

29

30

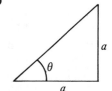

CALCULATOR EXERCISES 6.5

1 Use a calculator to find the values of the sine, cosine, and tangent functions at 30°, 45°, and 60°, and compare your answers with the entries in Table A. How do you account for any discrepancies?

2 Find the values of the six trigonometric functions of θ to two-decimal-place accuracy if θ is in standard position and the point $P(\ln 4.37, \sqrt{9.61}e^{1.35})$ is on the terminal side.

3 Find the values of the trigonometric functions of θ to three-decimal-place accuracy if θ is the angle of the right triangle shown below.

6.6 VALUES OF THE TRIGONOMETRIC FUNCTIONS

Several special values of the trigonometric functions were calculated in previous sections. Let us now turn to the problem of finding *all* values. Since the sine function has period 2π, it is sufficient to know the values of $\sin t$ for $0 \leq t \leq 2\pi$, because these values are repeated in every t-interval of length 2π. The same is true for the other trigonometric functions. As a matter of fact, the values of any trigonometric function can be determined if its values in the t-interval $[0, \pi/2]$ are known. In order to prove this, suppose that t is any real number and let $P(t)$ be the point on the unit circle U that corresponds to t. We shall make use of the following concept.

Definition

> The shortest arc length t' between $P(t)$ and the x-axis, on the unit circle U, is called the **reference number** associated with t.

Figure 6.22 illustrates arcs of length t' for positions of $P(t)$ in various quadrants. Note that $0 \leq t' < \pi/2$ for all values of t.

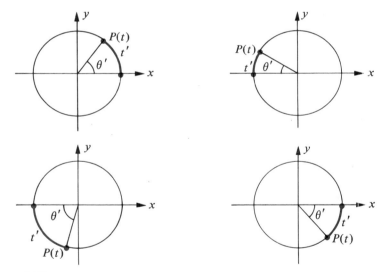

Figure 6.22

If we let θ' denote the angle with vertex O which subtends the arc of length t', as illustrated in Figure 6.22, then θ' is called the *reference angle associated with $P(t)$, or with any angle θ whose terminal side contains $P(t)$.* Thus, if θ is an angle in standard position and θ is not a quadrantal angle, we have the following definition.

Definition

> The **reference angle** associated with θ is the acute angle θ' that the terminal side of θ makes with the positive or negative x-axis.

Referring to Figure 6.22 and using the definition of trigonometric functions of angles stated in Section 6.5, we see that

$$\sin \theta' = \sin t', \quad \cos \theta' = \cos t', \quad \tan \theta = \tan t'.$$

In the remainder of our work we shall use the approximation $\pi \approx 3.1416$ for estimating reference numbers and angles.

Example 1   Approximate the reference number t' if

    (a) $t = 2$;   (b) $t = 4$;   (c) $t = 7\pi/4$;   (d) $t = -2\pi/3$.

Solutions   Each point $P(t)$ on the unit circle U, and the magnitude of its reference number t', are indicated in Figure 6.23. Referring to the figure we see that

    (a) $\qquad\qquad\qquad\qquad t' = \pi - 2 \approx 3.1416 - 2 = 1.1416$

    (b) $\qquad\qquad\qquad\qquad t' = 4 - \pi \approx 4 - 3.1416 = 0.8584$

    (c) $\qquad\qquad\qquad\qquad t' = 2\pi - (7\pi/4) = \pi/4$

    (d) $\qquad\qquad\qquad\qquad t' = \pi - (2\pi/3) = \pi/3$

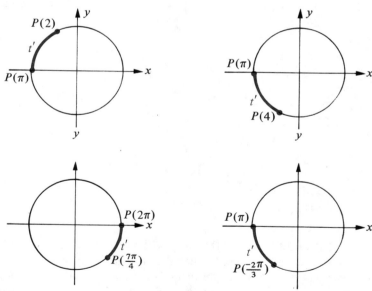

Figure 6.23 ∎

Example 2   Sketch the reference angle θ' for each of the following and express the measure of θ' in terms of both radians and degrees.

    (a) $\theta = 5\pi/6$ (b) $\theta = 315°$ (c) $\theta = -240°$

Solutions Each angle θ and its reference angle θ' are sketched in Figure 6.24 Evidently,

(a) $\qquad\qquad\qquad\qquad\qquad \theta' = \pi/6 = 30°$

(b) $\qquad\qquad\qquad\qquad\qquad \theta' = 45° = \pi/4$

(c) $\qquad\qquad\qquad\qquad\qquad \theta' = 60° = \pi/3$

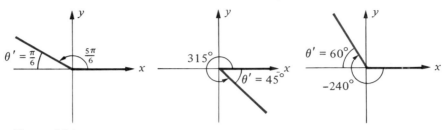

Figure 6.24 ∎

Let us now demonstrate how reference numbers or reference angles can be used to find values of the trigonometric functions. If the point $P(t)$ is not on a coordinate axis and t' is the reference number for t, then $0 < t' < \pi/2$. Suppose that $P(t)$ has rectangular coordinates (x, y), so that $P(t) = P(x, y)$. Consider the point $A(1, 0)$ and let $P'(x', y')$ be the point on U in quadrant I such that $\widehat{AP'} = t'$. Illustrations in which $P(x, y)$ lies in quadrants II, III, or IV are given in Figure 6.25.

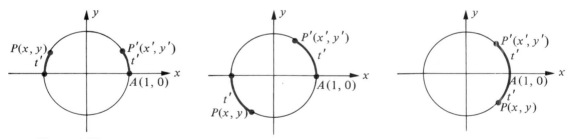

Figure 6.25

We see that in all cases

$$x' = |x| \quad \text{and} \quad y' = |y|$$

and hence we have
$$|\cos t| = |x| = x' = \cos t'$$
$$|\sin t| = |y| = y' = \sin t'.$$

It is easy to show that the absolute value of *every* trigonometric function at t is the same as its value at t'. For example,

$$|\tan t| = \left|\frac{y}{x}\right| = \frac{|y|}{|x|} = \frac{y'}{x'} = \tan t'.$$

In terms of angles, if θ' is the reference angle of θ, then

$$|\sin \theta| = \sin \theta', \quad |\cos \theta| = \cos \theta',$$

and similarly for the other trigonometric functions. We may, therefore, state the following rules.

Rules for Finding Values of the Trigonometric Functions

(i) To find the value of a trigonometric function at a number t, determine its value for the reference number t' associated with t and prefix the appropriate sign.

(ii) To find the value of a trigonometric function at an angle θ, determine its value for the reference angle θ' associated with θ and prefix the appropriate sign.

The "appropriate sign" can be determined from the table of signs in Section 6.2.

Example 3 Find $\sin(7\pi/4)$ and $\sec(-7\pi/6)$.

Solution   As indicated in Figure 6.26, the reference numbers of $7\pi/4$ and $-7\pi/6$ are $\pi/4$ and $\pi/6$, respectively. Hence, by the preceding rules and Table A of Section 6.5,

$$\sin \frac{7\pi}{4} = -\sin \frac{\pi}{4} = -\frac{\sqrt{2}}{2}$$

$$\sec \frac{-7\pi}{6} = -\sec \frac{\pi}{6} = -\frac{2\sqrt{3}}{3}.$$

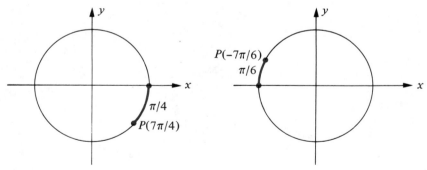

Figure 6.26   ∎

Example 4 Find each of the following:

(a) $\sin 150°$;   (b) $\tan 315°$;   (c) $\sec(-240°)$.

Solutions   The angles and their reference angles are shown in Figure 6.27. Using the rules for finding functional values, and Table A of Section 6.5 gives us

(a) $\sin 150° = \sin 30° = 1/2$

(b) $\qquad$ $\tan 315° = -\tan 45° = -1$

(c) $\qquad$ $\sec(-240°) = -\sec 60° = -2$

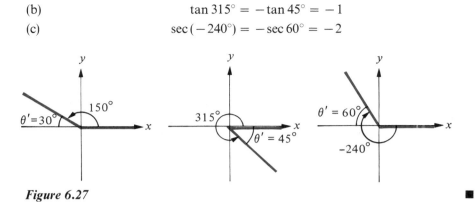

Figure 6.27   ■

By employing advanced techniques it is possible to compute, to any degree of accuracy, all the values of the trigonometric functions in the t-interval $[0, \pi/2]$ or, equivalently, in the degree interval $[0°, 90°]$. Table 2, part of which is reproduced here, gives approximations to such values.

Part of Table 2

t	t degrees	$\sin t$	$\cos t$	$\tan t$	$\cot t$	$\sec t$	$\csc t$		
.4887	**28°00′**	.4695	.8829	.5317	1.881	1.133	2.130	**62°00′**	1.0821
.4916	10	.4720	.8816	.5354	1.868	1.134	2.118	50	1.0792
.4945	20	.4746	.8802	.5392	1.855	1.136	2.107	40	1.0763
,4974	30	.4772	.8788	.5430	1.842	1.138	2.096	30	1.0734
.5003	40	.4797	.8774	.5467	1.829	1.140	2.085	20	1.0705
.5032	50	.4823	.8760	.5505	1.816	1.142	2.074	10	1.0676
.5061	**29°00′**	.4848	.8746	.5543	1.804	1.143	2.063	**61°00′**	1.0647
.5091	10	.4874	.8732	.5581	1.792	1.145	2.052	50	1.0617
.5120	20	.4899	.8718	.5619	1.780	1.147	2.041	40	1.0588
.5149	30	.4924	.8704	.5658	1.767	1.149	2.031	30	1.0559
.5178	40	.4950	.8689	.5696	1.756	1.151	2.020	20	1.0530
.5207	50	.4975	.8675	.5735	1.744	1.153	2.010	10	1.0501
.5236	**30°00′**	.5000	.8660	.5774	1.732	1.155	2.000	**60°00′**	1.0472
⋮	⋮	⋮	⋮	⋮	⋮	⋮	⋮	⋮	⋮
		$\cos t$	$\sin t$	$\cot t$	$\tan t$	$\csc t$	$\sec t$	t degrees	t

The domain of t in Table 2 is from 0 to 1.5708. The last number is a four-decimal-place approximation to $\pi/2$. Table 2 is arranged so that functional values which correspond to angles in degree measure may be found directly. Angular measures are given in 10′ intervals from 0° to 90°. The inclusion of the degree columns is the reason that t varies at intervals of approximately 0.0029, since 10′ ≈ 0.0029 radians.

To find values of trigonometric functions at a real number t, whenever $0 \leq t \leq 0.7854 \approx \pi/4$ or $0° \leq \theta \leq 45°$, the labels at the *top* of the columns in Table 2 should be used. For example,

$$\sin{(0.5003)} \approx 0.4797 \qquad \tan{28°30'} \approx 0.5430$$
$$\cos{(0.4945)} \approx 0.8802 \qquad \sec{29°00'} \approx 1.143$$
$$\cot{(0.5120)} \approx 1.780 \qquad \csc{29°40'} \approx 2.020.$$

However, if $0.7854 \leq t \leq 1.5708$, or if $45° \leq \theta \leq 90°$, then the labels at the *bottom* of the columns should be employed. For example,

$$\sin{(1.0705)} \approx 0.8774 \qquad \csc{62°00'} \approx 1.133$$
$$\cos{(1.0530)} \approx 0.4950 \qquad \cot{61°10'} \approx 0.5505$$
$$\tan{(1.0821)} \approx 1.881 \qquad \sec{60°30'} \approx 2.031.$$

The reason that the table can be so arranged follows from the fact, to be proved later, that

$$\sin t = \cos\left(\frac{\pi}{2} - t\right), \quad \cot t = \tan\left(\frac{\pi}{2} - t\right), \quad \csc t = \sec\left(\frac{\pi}{2} - t\right)$$

or equivalently,

$$\sin \theta = \cos(90° - \theta), \quad \cot \theta = \tan(90° - \theta), \quad \csc \theta = \sec(90° - \theta).$$

In particular,

$$\sin 29° = \cos(90° - 29°) = \cos 61°$$
$$\cot 28°20' = \tan(90° - 28°20') = \tan 61°40'$$

as shown in the table.

If it is necessary to find functional values when t lies *between* numbers given in the table, the method of linear interpolation used for logarithms may be employed. Similarly, given a value such as $\sin t = 0.6371$, we may refer to the body of Table 2 and use linear interpolation, if necessary, to obtain an approximation to t. If t is measured in degrees we round off to the nearest minute.

Example 5   Find approximations for the following:

(a) $\tan{(2.3824)}$;   (b) $\cos{(0.7)}$.

Solutions

(a) Since $P(2.3824)$ is in quadrant II, the reference number t' is $\pi - 2.3824$ or

$$t' \approx 3.1416 - 2.3824 = 0.7592.$$

Using the rules for finding functional values stated in this section, together with Table 2, we obtain

$$\tan{(2.3824)} \approx -\tan{(0.7592)} \approx -0.9490.$$

(b)   To find $\cos(0.7)$ we locate the number 0.7000 between successive values of t in Table 2 and interpolate as follows:

$$0.0029\left\{ 0.0019\left\{ \begin{array}{l} \cos(0.6981) \approx 0.7660 \\ \cos(0.7000) \approx \ ? \\ \cos(0.7010) \approx 0.7642 \end{array} \right\} d \right\} 0.0018$$

$$\frac{0.0019}{0.0029} = \frac{d}{0.0018}$$

or

$$d = \frac{19}{29}\,(0.0018) \approx 0.0012.$$

Hence, $\cos(0.7000) \approx 0.7660 - 0.0012 = 0.7648.$

Note that since the cosine function is decreasing in the given interval we must *subtract d* from 0.7660.   ■

Example 6   Approximate the smallest positive real number t such that $\sin t = 0.6635$.

Solution   We locate 0.6635 between successive entries in the sine column of Table 2 and interpolate as follows:

$$0.0029\left\{ d\left\{ \begin{array}{l} \sin(0.7243) \approx 0.6626 \\ \sin t \quad\ \ = 0.6635 \\ \sin(0.7272) = 0.6648 \end{array} \right\} 0.0009 \right\} 0.0022$$

$$\frac{d}{0.0029} = \frac{0.0009}{0.0022}$$

or

$$d = \frac{9}{22}\,(0.0029) \approx 0.0012.$$

Hence, $t \approx 0.7243 + 0.0012 = 0.7255.$ ■

Example 7   Approximate $\tan 155°44'$.

Solution   Since the angle is in quadrant :II, the reference angle is $180° - 155°44' = 24°16'$ and hence $\tan 155°44' = -\tan 24°16'$ (Why?). We now consult Table 2 and interpolate as follows:

$$10'\left\{ 6'\left\{ \begin{array}{l} \tan 24°10' \approx 0.4487 \\ \tan 24°16' = \ ? \\ \tan 24°20' \approx 0.4522 \end{array} \right\} d \right\} 0.0035$$

$$\frac{d}{0.0035} = \frac{6}{10} \quad \text{or} \quad d = \frac{6}{10}\,(0.0035) \approx 0.0021$$

$$\tan 24°16' \approx 0.4487 + 0.0021 = 0.4508$$

$$\tan 155°44' \approx -0.4508.$$ ■

Example 8   Approximate $\cos(-117°47')$.

Solution   The angle is in quadrant III. The reader should check that the reference angle is $62°13'$. Consequently, $\cos(-117°47') = -\cos 62°13'$. Interpolating, we have

$$10'\left\{3'\left\{\begin{array}{l}\cos 62°10' \approx 0.4669 \\ \cos 62°13' = ? \end{array}\right\}d \atop \cos 62°20' \approx 0.4643 \right\}0.0026$$

$$\frac{d}{0.0026} = \frac{3}{10} \quad \text{or} \quad d \approx 0.0008.$$

Since the cosine function is decreasing, we have

$$\cos 62°13' \approx 0.4669 - 0.0008 = 0.4661.$$

Hence, $\cos(-117°47') \approx -0.4661$.   ∎

Example 9   If $\sin \theta = -0.7963$, approximate the degree measure of all angles θ which are in the interval $[0°, 360°]$.

Solution   Let θ' be the reference angle, so that $\sin \theta' = 0.7963$. Interpolating in Table 2,

$$10'\left\{d\left\{\begin{array}{l}\sin 52°40' \approx 0.7951 \\ \sin \theta' \quad\;\; = 0.7963 \end{array}\right\}0.0012 \atop \sin 52°50' \approx 0.7969 \right\}0.0018$$

$$\frac{d}{10} = \frac{0.0012}{0.0018} \quad \text{or} \quad d \approx 7'$$

$$\theta' \approx 52°47'.$$

Since $\sin \theta$ is negative, θ lies in quadrant III or IV. Using the reference angle $52°47'$ we have

$$\theta \approx 180° + 52°47' = 232°47'$$
$$\theta \approx 360° - 52°47' = 307°13'.$$ ∎

As a final remark, Table 3 can be used to find functional values of t if t is a two-decimal-place approximation of a real number or the radian measure of an angle, and $0 \le t \le 1.57$. For example, referring to the table, we see that

$$\sin(0.95) \approx 0.8134$$
$$\cos(1.48) \approx 0.0907$$
$$\cot(0.50) \approx 1.830.$$

Interpolation can be used if t is given to three decimal places. We shall not include exercises which require the use of Table 3, but instead leave it to the discretion of the teacher and student as to what use should be made of that table.

EXERCISES 6.6

In Exercises 1–6 find the reference number t' if t has the given value.

1 (a) $3\pi/4$ (b) $4\pi/3$
   (c) $-\pi/6$

2 (a) $5\pi/6$ (b) $2\pi/3$
   (c) $-3\pi/4$

3 (a) $9\pi/4$ (b) $7\pi/6$
   (c) $-2\pi/3$

4 (a) $8\pi/3$ (b) $7\pi/4$
   (c) $-7\pi/6$

5 (a) 1.5 (b) 5

6 (a) 3.5 (b) -4

In Exercises 7–10 find the reference angle θ' if θ has the given measure.

7 (a) $240°$ (b) $340°$
   (c) $-110°$

8 (a) $165°$ (b) $275°$
   (c) $-202°$

9 (a) $130°40'$ (b) $-405°$
   (c) $-260°35'$

10 (a) $335°20'$ (b) $-620°$
   (c) $-185°40'$

In Exercises 11–16 find the exact values without the use of tables or calculators.

11 (a) $\sin 2\pi/3$ (b) $\sin 4\pi/3$

12 (a) $\cos 5\pi/6$ (b) $\cos 7\pi/6$

13 (a) $\tan(-5\pi/4)$ (b) $\cot 315°$

14 (a) $\sin 210°$ (b) $\csc(-150°)$

15 (a) $\csc 300°$ (b) $\sec(-120°)$

16 (a) $\tan(-135°)$ (b) $\sec 225°$

In Exercises 17–30 use Table 2 to approximate the given numbers.

17 $\sin(0.2676)$ **18** $\cos(0.3258)$

19 $\tan(0.9948)$ **20** $\sec(0.8988)$

21 $\csc(0.7738)$ **22** $\cot(0.7883)$

23 $\cos 38°30'$ **24** $\sin 73°20'$

25 $\cot 9°10'$ **26** $\csc 43°40'$

27 $\sec 168°50'$ **28** $\tan 207°10'$

29 $\sin 342°20'$ **30** $\cos 432°40'$

In Exercises 31–42 use interpolation in Table 2 to approximate the given numbers. If a suitable calculator is available, compare your approximations with those obtained by means of the calculator.

31 $\sin(0.46)$ **32** $\cos(0.82)$

33 $\tan 3$ **34** $\cot 6$

35 $\sec(1/4)$ **36** $\csc(1.54)$

37 $\cos 37°43'$ **38** $\sin 22°34'$

39 $\cot 62°27'$ **40** $\tan 57°16'$

41 $\csc 16°55'$ **42** $\sec 9°12'$

In Exercises 43–48 use interpolation to approximate the smallest positive number t for which the equality is true.

43 $\cos t = 0.8620$ **44** $\sin t = 0.6612$

45 $\tan t = 4.501$ **46** $\sec t = 3.641$

47 $\csc t = 1.436$ **48** $\cot t = 1.165$

In Exercises 49–56 use interpolation in Table 2 to approximate, to the nearest minute, the degree measure of all angles θ which lie in the interval $[0°, 360°]$.

49 $\sin\theta = 0.3672$ **50** $\cos\theta = 0.8426$

51 $\tan\theta = 0.5042$ **52** $\cot\theta = 1.348$

53 $\cos\theta = 0.3465$ **54** $\csc\theta = 1.219$

55 $\sec\theta = 1.385$ **56** $\sin\theta = 0.7534$

6.7   GRAPHS OF THE TRIGONOMETRIC FUNCTIONS

In our previous work with graphs of functions we used the symbols x and y as labels for the coordinate axes. In this chapter x has been used primarily for the abscissa of a point on the unit circle U, and hence in this section we shall use the symbol t for the horizontal axis. However, we shall continue to use y to denote the vertical axis. In the ty-coordinate system the graph of the sine function is the same as the graph of the equation $y = \sin t$.

It is not difficult to sketch the graphs of the trigonometric functions. For example, since $-1 \leq \sin t \leq 1$ for every real number t, the graph of the sine function lies between the horizontal lines $y = 1$ and $y = -1$. Moreover, the sine function is periodic with period 2π, and therefore it is sufficient to determine the graph for $0 \leq t \leq 2\pi$, since the same pattern is repeated in intervals of length 2π along the t-axis. In Section 6.3 we discussed the behavior of $\sin t$ in the interval $[0, 2\pi]$ by concentrating on the ordinate y of the point $P(t)$ as $P(t)$ traversed the unit circle U once in the counterclockwise direction (see Figure 6.7). The manner in which $\sin t$ varies in $[0, 2\pi]$ may be briefly summarized as follows.

Variation of t	Variation of $\sin t$
0 to $\pi/2$	0 to 1
$\pi/2$ to π	1 to 0
π to $3\pi/2$	0 to -1
$3\pi/2$ to 2π	-1 to 0

Using these facts we conjectured that the graph of the sine function might look like that in Figure 6.8. This guess may be further justified by calculating some intermediate values of $\sin t$, as indicated in the following table.

t	0	$\dfrac{\pi}{4}$	$\dfrac{\pi}{2}$	$\dfrac{3\pi}{4}$	π	$\dfrac{5\pi}{4}$	$\dfrac{3\pi}{2}$	$\dfrac{7\pi}{4}$	2π
$\sin t$	0	$\dfrac{\sqrt{2}}{2}$	1	$\dfrac{\sqrt{2}}{2}$	0	$\dfrac{-\sqrt{2}}{2}$	-1	$\dfrac{-\sqrt{2}}{2}$	0

To obtain a rough sketch we may plot the points $(t, \sin t)$ given by this table, draw a smooth curve through them, and extend the configuration to the right and left in

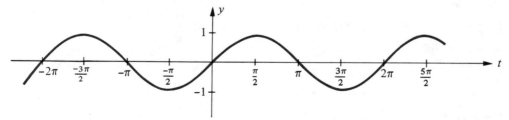

Figure 6.28 $y = \sin t$

periodic fashion. This gives us the portion of the graph shown in Figure 6.28. Of course, the graph does not terminate, but continues indefinitely to the right and left. If greater accuracy is desired, additional points could be plotted using, for example, $\sin \pi/6 = 1/2$, $\sin \pi/3 = \sqrt{3}/2 \approx 0.86$, etc. We could also use Table 2 or a calculator to obtain many additional points on the graph. We refer to the part of the graph corresponding to the interval $[0, 2\pi]$ as a **sine wave**.

The graph of the cosine function can be found in similar fashion, by studing the abscissa x of the point $P(t)$ in Figure 6.7 as $P(t)$ goes around U once in the counterclockwise direction. We may briefly summarize the behavior of $\cos t$ in the interval $[0, 2\pi]$ as follows.

Variation of t	Variation of $\cos t$
0 to $\pi/2$	1 to 0
$\pi/2$ to π	0 to -1
π to $3\pi/2$	-1 to 0
$3\pi/2$ to 2π	0 to 1

Using this information, together with the fact that the cosine function has period 2π, and plotting points as we did for the sine function, leads to the sketch in Figure 6.29. Note that the graph of $y = \cos t$ can be obtained by shifting the graph of $y = \sin t$ to the left a distance $\pi/2$.

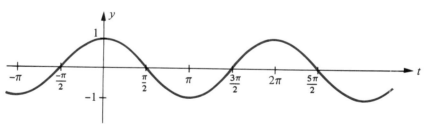

Figure 6.29 $y = \cos t$

Several values of the tangent function are given in the following table.

t	$-\dfrac{\pi}{3}$	$-\dfrac{\pi}{4}$	$-\dfrac{\pi}{6}$	0	$\dfrac{\pi}{6}$	$\dfrac{\pi}{4}$	$\dfrac{\pi}{3}$
$\tan t$	$-\sqrt{3}$	-1	$\dfrac{-\sqrt{3}}{3}$	0	$\dfrac{\sqrt{3}}{3}$	1	$\sqrt{3}$

The corresponding points are plotted in Figure 6.30. The values of $\tan t$ near $t = \pi/2$ demand special consideration. As t increases through positive values toward $\pi/2$, the point $P(t) = (x, y)$ on the unit circle U that corresponds to t approaches the point

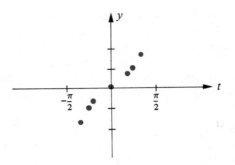

Figure 6.30

$(0, 1)$. Since, by definition, $\tan t = y/x$, it follows that if y approaches 1 and x approaches 0 (through positive values), then $\tan t$ takes on large positive values. This can also be seen by referring to Table 2. First recall that $\pi/2 \approx \frac{1}{2}(3.1416) = 1.5708$. Next, we see from Table 2 that

$$\tan(1.5650) \approx 171.9$$
$$\tan(1.5679) \approx 343.8.$$

The following additional values were obtained using a hand-held calculator.

$$\tan(1.5700) \approx \quad 1{,}255.8$$
$$\tan(1.5703) \approx \quad 2{,}014.8$$
$$\tan(1.5706) \approx \quad 5{,}093.5$$
$$\tan(1.5707) \approx \quad 10{,}381.3$$
$$\tan(1.57079) \approx 158{,}057.9$$

Notice how rapidly $\tan t$ increases as t gets close to $\pi/2$. Indeed, $\tan t$ can be made arbitrarily large by choosing t sufficiently close to $\pi/2$. Using the terminology introduced in Section 4.3, we say that $\tan t$ *increases without bound as t approaches $\pi/2$ through values less than $\pi/2$,* written

$$\tan t \to \infty \quad \text{as} \quad t \to \frac{\pi^-}{2}.$$

Similarly, if t approaches $-\pi/2$ through values greater than $-\pi/2$, then $\tan t$ *decreases without bound,*written

$$\tan t \to -\infty \quad \text{as} \quad t \to -\frac{\pi^+}{2}.$$

This behavior in the open interval $(-\pi/2, \pi/2)$ is illustrated in Figure 6.31. The lines $t = \pi/2$ and $t = -\pi/2$ are vertical asymptotes for the graph. It is not difficult to show that the same pattern is repeated in the open intervals $(\pi/2, 3\pi/2)$, $(3\pi/2, 5\pi/2)$, and similar intervals of length π, as shown in Figure 6.31. Thus, *the tangent function has period π.*

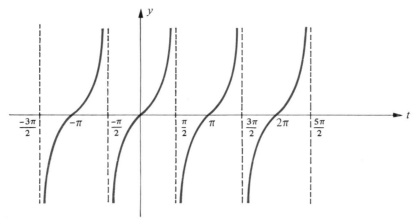

Figure 6.31 $y = \tan t$

The graphs of the sine and tangent functions are symmetric with respect to the origin, whereas the graph of the cosine function is symmetric with respect to the y-axis. These facts are consequences of the following formulas, each of which holds for every real number t in the domain of the indicated function.

Formulas for
Negatives

$$\sin(-t) = -\sin t$$
$$\cos(-t) = \cos t$$
$$\tan(-t) = -\tan t$$

To establish these formulas, let us recall from Section 6.1 that if t ranges from 0 to 2π, the point $P(t)$ traces the unit circle U once in the counterclockwise direction, whereas $P(-t)$ traces U once in the clockwise direction. Moreover, as illustrated in Figure 6.32, if $P(t)$ has coordinates (x, y) then $P(-t)$ has coordinates $(x, -y)$.

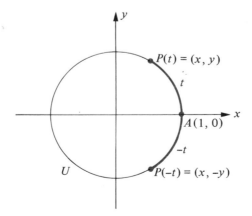

Figure 6.32

Applying the definition of the trigonometric functions (see Section 6.2) gives us

$$\sin(-t) = -y = -\sin t$$
$$\cos(-t) = x = \cos t$$

$$\tan(-t) = \frac{-y}{x} = -\frac{y}{x} = -\tan t$$

which is what we wished to prove. According to the tests for symmetry stated in Section 3.2, these identities imply that the graphs of the sine and tangent functions are symmetric with respect to the origin, whereas that of the cosine function is symmetric with respect to the y-axis.

The graphs of the remaining three trigonometric functions can now be easily obtained. For example, since $\csc t = 1/\sin t$, we may find an ordinate of a point on the graph of the cosecant function by taking the reciprocal of the corresponding ordinate of the sine graph. This is possible except for $t = n\pi$, where n is an integer, for in this case $\sin t = 0$ and hence $1/\sin t$ is undefined. As an aid to sketching the graph of the cosecant function it is convenient to sketch the graph of the sine function with dashes (see Figure 6.33) and then take reciprocals of ordinates to obtain points on the cosecant graph. Notice the manner in which the cosecant function increases or decreases without bound as t approaches $n\pi$, where n is an integer. The graph has vertical asymptotes as indicated in the figure.

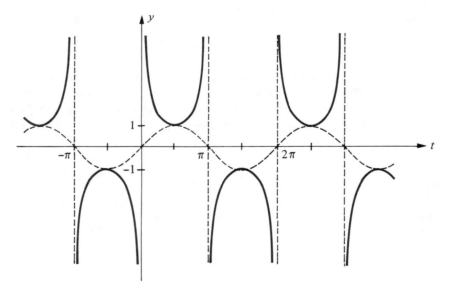

Figure 6.33 $y = \csc t$

Since $\sec t = 1/\cos t$ and $\cot t = 1/\tan t$, the graphs of the secant and cotangent functions may be obtained in similar fashion (see Figures 6.34 and 6.35). Their verifications are left as exercises.

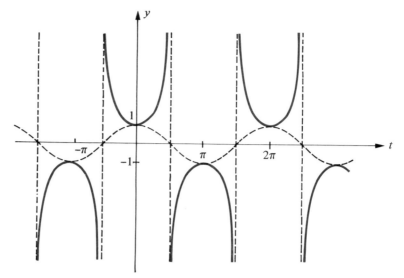

Figure 6.34 $y = \sec t$

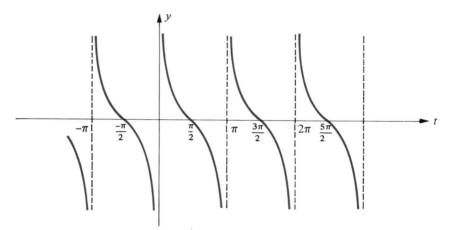

Figure 6.35 $y = \cot t$

Example 1 Sketch the graph of the function f if $f(t) = 2 \sin t$.

Solution Although the graph could be obtained by plotting points, note that for each abscissa t_1 the ordinate $f(t_1)$ is always twice that of the corresponding ordinate on the sine graph. A simple graphical technique is to sketch the graph of $y = \sin t$ with dashes and then double each ordinate to find points on the graph of $y = 2 \sin t$ (see Figure 6.36)

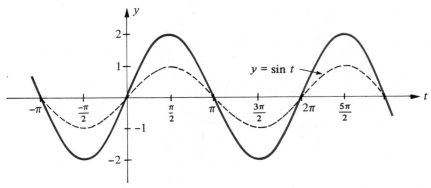

Figure 6.36 $y = 2 \sin t$ ∎

Graphs of the type given in the last example will be discussed more thoroughly in the next section.

EXERCISES 6.7

1 (a) Sketch the graph of the cosine function after plotting a sufficient number of points.
  (b) Sketch the graph of the secant function by taking reciprocals in part (a).
  (c) Describe the intervals between -2π and 2π in which the secant function is increasing.
  (d) Describe the intervals between -2π and 2π in which the secant function is decreasing.
  (e) With the aid of Table 2 or a calculator, discuss the behavior of $\sec t$ as t approaches $\pi/2$ through values less than $\pi/2$ and through values greater than $\pi/2$.
  (f) Discuss the symmetry of the graph of the secant function.

2 (a) Sketch the graph of the cotangent function after plotting a sufficient number of points.
  (b) In what intervals is the cotangent function increasing?
  (c) In what intervals is the cotangent function decreasing?
  (d) With the aid of Table 2 or a calculator, discuss the behavior of $\cot t$ as t approaches 0 through values greater than 0 and through values less than 0.
  (e) Discuss the symmetry of the graph of the cotangent function.

3 Establish the following identities.

  (a) $\csc(-t) = -\csc t$
  (b) $\sec(-t) = \sec t$
  (c) $\cot(-t) = -\cot t$

4 Prove that the graph of the cosecant function is symmetric with respect to the origin.

5 Practice sketching graphs of the sine, cosine, and tangent functions, taking different units of length on the horizontal and vertical axes. Continue this practice until you reach the stage at which, if you were awakened from a sound sleep in the middle of the night and asked to sketch one of these graphs, you could do so in less than thirty seconds.

6 Repeat Exercise 5 for the cosecant, secant, and cotangent functions.

In Exercises 7–16 use the method illustrated in the Example of this section to sketch the graph of f.

7 $f(t) = 4 \sin t$ **8** $f(t) = 3 \sin t$

9 $f(t) = \frac{1}{2} \sin t$ **10** $f(t) = \frac{1}{4} \sin t$

11 $f(t) = 2 \cos t$ **12** $f(t) = 4 \cos t$

13 $f(t) = \frac{1}{3} \cos t$ **14** $f(t) = \frac{1}{2} \cos t$

15 $f(t) = -\sin t$ **16** $f(t) = -\cos t$

CALCULATOR EXERCISES 6.7

If f is a function, the notation

$$f(t) \to L \quad \text{as} \quad t \to a^+$$

is sometimes used to express the fact that $f(t)$ gets very close to L as t gets close to a (through values greater than a). Use a hand-held calculator to support Exercises 1–6 by substituting the following values for t: $t = 0.1$, $t = 0.01$, $t = 0.001$, $t = 0.0001$.

1 $\dfrac{\sin t}{t} \to 1 \quad \text{as} \quad t \to 0^+$

2 $\dfrac{\tan t}{t} \to 1 \quad \text{as} \quad t \to 0^+$

3 $\dfrac{1 - \cos t}{t} \to 0 \quad \text{as} \quad t \to 0^+$

4 $\dfrac{\sin t}{1 + \cos t} \to 0 \quad \text{as} \quad t \to 0^+$

5 $t \cot t \to 1 \quad \text{as} \quad t \to 0^+$

6 $\dfrac{t + \tan t}{\sin t} \to 2 \quad \text{as} \quad t \to 0^+$

6.8 TRIGONOMETRIC GRAPHS

In this section we shall consider graphs of equations of the form

$$y = a \sin(bx + c)$$

where, a, b, c are real numbers. We shall also discuss graphs of similar equations that involve different trigonometric functions. Instead of using a ty-coordinate system as in Section 6.7, we shall now use the conventional xy-coordinate system. In this situation the letter x is used in place of t and hence is not to be regarded as the x used in the definition of the trigonometric functions. To sketch a graph we could begin by plotting many points; however, it is generally easier to use information about the graphs of the trigonometric functions discussed in the preceding section. Let us consider the special case in which $c = 0$, $b = 1$, and $a > 0$. Thus, we wish to sketch the graph of the equation

$$y = a \sin x.$$

As in our discussion of stretching of graphs (see Section 3.4), we may find the ordinate of a point on the graph by multiplying the corresponding ordinate on the graph of $y = \sin x$ by a. Thus, if $y = 2 \sin x$, we multiply by 2; if $y = \frac{1}{2} \sin x$, we multiply by $\frac{1}{2}$; and so on. The graph of $y = 2 \sin x$ is sketched in Figure 6.36. The graph of $y = \frac{1}{2} \sin x$ is sketched in Figure 6.37, where for comparison we have indicated the graph of $y = \sin x$ with dashes.

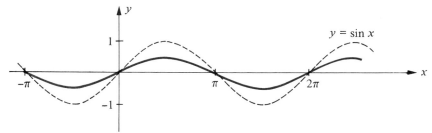

Figure 6.37 $y = \frac{1}{2} \sin x$

Example 1   Sketch the graph of the equation $y = 3\cos x$.

Solution   The graph of the given equation may be obtained from the graph of $y = \cos x$ by multiplying ordinates of points by 3. We first sketch $y = \cos x$ with dashes and then we triple the ordinates. This gives us the sketch in Figure 6.38.

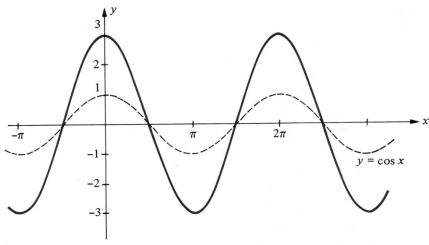

Figure 6.38 $y = 3\cos x$ ■

If $a < 0$, then the ordinates of points on the graph of $y = a\sin x$ are negatives of the corresponding ordinates of points on the graph of $y = |a|\sin x$, as illustrated in the next example.

Example 2   Sketch the graph of $y = -2\sin x$.

Solution   As a guide we first sketch the graph of $y = \sin x$ with dashes and then we multiply each ordinate by -2. This gives us the sketch shown in Figure 6.39. The sketch may also be obtained by multiplying ordinates of the graph of

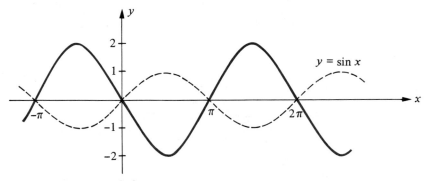

Figure 6.39 $y = -2\sin x$

$y = 2 \sin x$ (see Figure 6.36) by -1. Thus, the graph of $y = -2 \sin x$ is a *reflection through the x-axis* of the graph of $y = 2 \sin x$.   ■

If $f(x) = a \sin(bx + c)$, then the **amplitude** of f (or of its graph) is defined as the largest ordinate of points on the graph. If $a > 0$, the largest ordinate occurs if $\sin(bx + c) = 1$, whereas if $a < 0$, we must take $\sin(bx + c) = -1$. In either case the amplitude is $|a|$. For example, if $f(x) = \frac{1}{2} \sin x$, then from Figure 6.36 we see that the amplitude is $\frac{1}{2}$. If $f(x) = -2 \sin x$, then from Figure 6.38 we see that the amplitude is 2. The same is true if f is given by $f(x) = a \cos(bx + c)$.

The next theorem provides some information about periods.

Theorem

> If $f(x) = \sin bx$, where $b \neq 0$, then the period of f is $2\pi/|b|$.

Proof   If $b > 0$, then as bx varies from 0 to 2π, we obtain exactly one sine wave for the graph of f. The conclusion of the theorem follows from the fact that bx ranges from 0 to 2π if and only if x ranges from 0 to $2\pi/b$. A similar proof may be given if $b < 0$.   ■

Example 3   Find the period and sketch the graph of f if:

(a) $f(x) = \sin 2x$;   (b) $f(x) = \sin \frac{1}{2} x$.

Solutions   Both functions have the form given in the preceding theorem. Hence, in part (a) the period is $2\pi/2 = \pi$, which means that there is exactly one sine wave of amplitude 1 corresponding to the interval $[0, \pi]$. The graph is sketched in Figure 6.40, where for convenience we have used different scales on the x- and y-axes. For part (b) the period is $2\pi/(1/2) = 4\pi$ and hence there is one sine wave of amplitude 1 corresponding to the interval $[0, 4\pi]$, as illustrated in Figure 6.41.

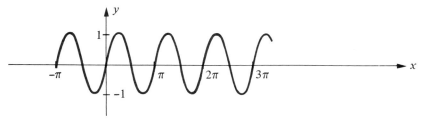

Figure 6.40 $y = \sin 2x$

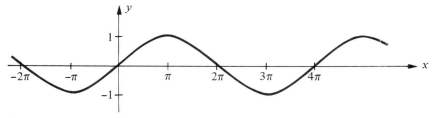

Figure 6.41 $y = \sin \frac{1}{2} x$ ■

Given $f(x) = \sin bx$, where $b > 0$, then, roughly speaking, if b is large, $2\pi/b$ is small and the sine waves are close together. As a matter of fact, there are b sine waves in an interval of 2π units. However, if b is small, then $2\pi/b$ is large and the waves are shallow. For example, if $y = \sin(1/10)x$, then $1/10$ of a sine wave occurs in the interval $[0, 2\pi]$, and an interval 20π units long is required for one complete wave.

If $b < 0$, we can use the fact that $\sin(-t) = -\sin t$ to obtain the graph. To illustrate, the graph of $y = \sin(-2x)$ is the same as the graph of $y = -\sin 2x$.

By combining the above remarks we can arrive at a technique for sketching the graph of a function f defined by $f(x) = a \sin bx$. The graph has the basic sine wave pattern. However, the amplitude is $|a|$ and the period is $2\pi/|b|$. If $a < 0$ or $b < 0$, we make adjustments on the signs of ordinates, as discussed earlier.

Example 4   Sketch the graph of f if $f(x) = 3 \sin 2x$.

Solution   From the preceding discussion we see that the amplitude of f is 3 and the period is $2\pi/2 = \pi$. The graph is readily obtained by first sketching the graph of $y = \sin 2x$ with dashes and then multiplying the ordinates of each point by 3. This leads to the sketch in Figure 6.42.

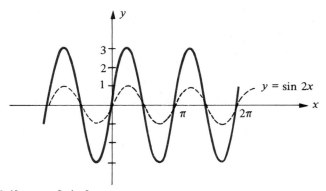

Figure 6.42 $y = 3 \sin 2x$ ∎

Example 5   Sketch the graph of f if $f(x) = 2 \sin(-3x)$.

Solution   Since $f(x) = -2 \sin 3x$, we see that the amplitude is 2 and the period is $2\pi/3$. Thus, there is one sine wave in every interval of length $2\pi/3$. The minus sign indicates a reflection through the x-axis. If we consider the interval $[0, 2\pi/3]$ and sketch a sine wave of amplitude 2 (reflected through the x-axis), the shape of the graph is apparent. The configuration given in the interval $[0, 2\pi/3]$ is carried along periodically, as illustrated in Figure 6.43.   ∎

Similar discussions can be given if f is defined by $f(x) = a \cos bx$ or by $f(x) = a \tan bx$. In the latter case the period is $\pi/|b|$ since the tangent function has period π. There is no largest ordinate for points on the graph of the tangent function and hence we do not refer to its amplitude; however, we may still use the process of multiplying tangent ordinates by a in order to obtain points on the graph of $y = a \tan bx$.

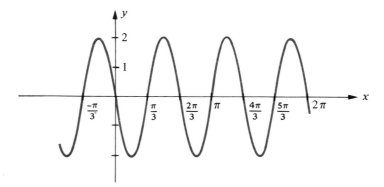

Figure 6.43 $y = -2 \sin 3x$

Let us conclude this section by considering the general situation

$$f(x) = a \sin (bx + c).$$

We have already observed that the amplitude is $|a|$. We also know that one sine wave for the graph of $y = \sin (bx + c)$ is obtained if $bx + c$ varies from 0 to 2π, that is, if bx ranges from $-c$ to $2\pi - c$. In turn, the latter variation is obtained by letting x range from $-c/b$ to $(2\pi - c)/b$. If $-c/b > 0$, this amounts graphically to *shifting* the graph of $y = a \sin bx$ to the right $-c/b$ units. If $-c/b < 0$, the shift is to the left. The number $-c/b$ is sometimes called the **phase shift** associated with the function. Similar remarks can be made for the other functions. It is unnecessary to remember a general formula for finding the phase shift. In any specific problem the interval which contains one complete sine wave can be found by solving the two equations

$$bx + c = 0 \quad \text{and} \quad bx + c = 2\pi$$

for x, as illustrated in the next example.

Example 6   Sketch the graph of $f(x) = 3 \sin \left(2x - \dfrac{\pi}{2} \right)$.

Solution   The equation is of the form discussed in this section, with $a = 3$, $b = 2$, and $c = -\pi/2$. It follows from our discussion that the graph has the sine wave pattern with amplitude 3 and period $2\pi/2 = \pi$. In order to obtain an interval containing exactly one sine wave, we let $2x - (\pi/2)$ range from 0 to 2π. The endpoints of the interval can be found by solving the two equations

$$2x - \frac{\pi}{2} = 0 \quad \text{and} \quad 2x - \frac{\pi}{2} = 2\pi.$$

This gives us

$$x = \frac{\pi}{4} \quad \text{and} \quad x = \frac{5\pi}{4}.$$

Thus, one sine wave of amplitude 3 will occur in the interval $[\pi/4, 5\pi/4]$. Sketching that wave and then repeating it to the right and left gives us the graph in Figure 6.44.

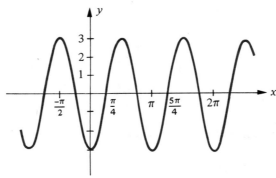

Figure 6.44 $y = 3\sin\left(2x - \dfrac{\pi}{2}\right)$ ■

EXERCISES 6.8

3-21

1 Without plotting many points, sketch the graph and determine the amplitude and period of each function f defined as follows:
   (a) $f(x) = 4\sin x$ (b) $f(x) = \sin 4x$
   (c) $f(x) = \frac{1}{4}\sin x$ (d) $f(x) = \sin(x/4)$
   (e) $f(x) = 2\sin(x/4)$ (f) $f(x) = \frac{1}{2}\sin 4x$
   (g) $f(x) = -4\sin x$ (h) $f(x) = \sin(-4x)$

2 Sketch the graphs of the functions involving the cosine which are analogous to those defined in Exercise 1.

3 Without plotting many points, sketch the graph of each function f defined as follows, determining the amplitude and period in each case:
   (a) $f(x) = 3\cos x$ (b) $f(x) = \cos 3x$
   (c) $f(x) = \frac{1}{3}\cos x$ (d) $f(x) = \cos(x/3)$
   (e) $f(x) = 2\cos(x/3)$ (f) $f(x) = \frac{1}{3}\cos(2x)$
   (g) $f(x) = -3\cos x$ (h) $f(x) = \cos(-3x)$

4 Sketch the graphs of the functions involving the sine which are analogous to those defined in Exercise 3.

Sketch the graphs of the equations in Exercises 5–44.

5 $y = \sin\left(x - \dfrac{\pi}{2}\right)$ 6 $y = \cos\left(x + \dfrac{\pi}{4}\right)$

7 $y = 3\sin\left(x - \dfrac{\pi}{2}\right)$ 8 $y = 4\cos\left(x + \dfrac{\pi}{4}\right)$

9 $y = \cos\left(x + \dfrac{\pi}{3}\right)$ 10 $y = \sin\left(x - \dfrac{\pi}{3}\right)$

11 $y = 4\cos\left(x + \dfrac{\pi}{3}\right)$ 12 $y = 5\sin\left(x - \dfrac{\pi}{3}\right)$

13 $y = \sin(3x + \pi)$ 14 $y = \cos(3x - \pi)$

15 $y = -2\sin(3x + \pi)$ 16 $y = 6\cos(3x - \pi)$

17 $y = 5\sin\left(3x - \dfrac{\pi}{2}\right)$ 18 $y = -4\cos\left(2x + \dfrac{\pi}{3}\right)$

19 $y = 6\sin \pi x$ 20 $y = 3\cos\frac{1}{2}\pi x$

21 $y = 2\cos\dfrac{\pi}{2}x$ 22 $y = 4\sin 3\pi x$

23 $y = \frac{1}{2}\sin 2\pi x$ 24 $y = \frac{1}{2}\cos\dfrac{\pi}{2}x$

25 $y = \frac{1}{2}\sec x$ 26 $y = \frac{1}{4}\csc x$

27 $y = -2\csc x$ 28 $y = -3\sec x$

29 $y = \sec 2x$ 30 $y = \csc 3x$

31 $y = \csc\left(x - \dfrac{\pi}{4}\right)$ **32** $y = \sec\left(x + \dfrac{\pi}{3}\right)$

33 $y = \tan(x/2)$ **34** $y = \cot 2x$

35 $y = \tan\left(x + \dfrac{\pi}{2}\right)$ **36** $y = \cot\left(x - \dfrac{\pi}{4}\right)$

37 $y = \frac{1}{2}\cot x$ **38** $y = -2\tan x$

39 $y = \tan(-x)$ **40** $y = -\cot x$

41 $y = 2\sin\left(x - \dfrac{\pi}{2}\right)$ **42** $y = \cos\left(x - \dfrac{\pi}{2}\right)$

43 $y = \sin\left(x + \dfrac{\pi}{2}\right)$ **44** $y = \cos\left(x + \dfrac{\pi}{2}\right)$

CALCULATOR EXERCISES 6.8

1–6 The graphs in Examples 1–6 of this section were obtained *qualitatively*, in the sense that we determined the general shape by referring to other graphs and not by plotting points. Verify them *quantitatively*, by calculating the coordinates of many points on the graphs.

6.9 ADDITIONAL GRAPHICAL TECHNIQUES

In mathematical applications it is common to encounter functions that are defined in terms of sums and products of expressions such as

$$f(x) = \sin 2x + \cos x \quad \text{or} \quad f(x) = 2^{-x}\sin x.$$

When working with sums, the graphical technique called **addition of ordinates** is useful. The method applies not only to trigonometric expressions but to arbitrary expressions as well. If f is a sum of two functions g and h having the same domain X, then

$$f(x) = g(x) + h(x)$$

for every x in X. A sketch of the graph of f may be obtained from the graphs of g and h as follows. We begin by sketching the graphs of the equations $y = g(x)$ and $y = h(x)$ on the same coordinate axes, as illustrated by the dashes in Figure 6.45.

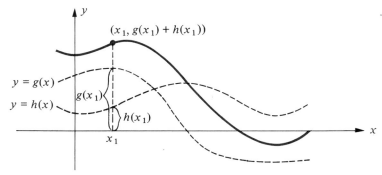

Figure 6.45 $y = g(x) + h(x)$

Since $f(x_1) = g(x_1) + h(x_1)$ for every x_1 in X, the ordinate of the point on the graph of $y = g(x) + h(x)$ with abscissa x_1 is the *sum* of the corresponding ordinates of points on the graphs of g and h. If we draw a vertical line at the point with coordinates $(x_1, 0)$, then the ordinates $g(x_1)$ and $h(x_1)$ may be added geometrically by means of a compass or ruler, as is illustrated in Figure 6.45. If either $g(x_1)$ or $h(x_1)$ is negative, then a *subtraction* of ordinates may be employed. By using this technique a sufficient number of times, we obtain a sketch of the graph of f.

Example 1   If $f(x) = \cos x + \sin x$, use the method of addition of ordinates to sketch the graph of f.

Solution   We begin by sketching (with dashes) the graphs of the equations $y = \cos x$ and $y = \sin x$. Next, for various numbers x_1, we add ordinates geometrically. After a sufficient number of ordinates are added and a pattern emerges, we draw a smooth curve through the points, as indicated by the sketch in Figure 6.46. As a check, it would be worthwhile to plot some points on the graph by substituting numbers for x. We shall leave such verifications to the reader. It can be seen from the graph that the function f is periodic with period 2π.

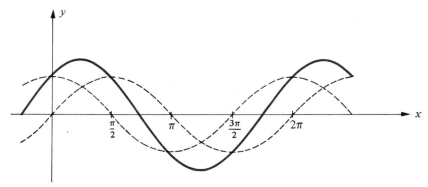

Figure 6.46 $y = \cos x + \sin x$ ■

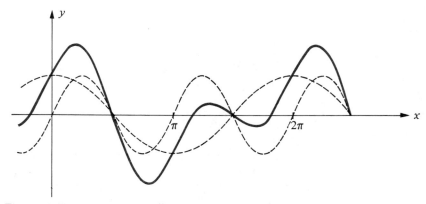

Figure 6.47 $y = \cos x + \sin 2x$

Example 2   Sketch the graph of the equation $y = \cos x + \sin 2x$.

Solution   We sketch (with dashes) the graphs of the equations $y = \cos x$ and $y = \sin 2x$ on the same coordinate axes and use the method of addition of ordinates. The graph is illustrated in Figure 6.47. Evidently f is periodic with period 2π.   ∎

The next example illustrates the fact that if f is the product of two functions g and h, then the graphs of g and h can often be used to supply information about the graph of f.

Example 3   Sketch the graph of f if $f(x) = 2^{-x} \sin x$.

Solution   The function f is the product of two functions g and h, where $g(x) = 2^{-x}$ and $h(x) = \sin x$. Using properties of absolute values, we see that $|f(x)| = |2^{-x}| \, |\sin x|$, and since $|\sin x| \le 1$ and $2^{-x} > 0$, it follows that $|f(x)| \le 2^{-x}$ for all x. Consequently,

$$-2^{-x} \le f(x) \le 2^{-x}$$

for all x, which implies that the graph of f lies between the graphs of the equations $y = -2^{-x}$ and $y = 2^{-x}$. The graph of f will coincide with one of these graphs when $|\sin x| = 1$, that is, when $x = (\pi/2) + n\pi$, where n is an integer. Since $2^{-x} > 0$, the x-intercepts on the graph of f occur at $\sin x = 0$, that is, at $x = n\pi$. With this information we obtain the sketch shown in Figure 6.48.

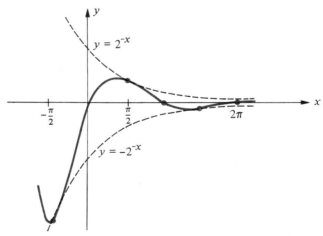

Figure 6.48 $y = 2^{-x} \sin x$ ∎

The graph of $y = 2^{-x} \sin x$ sketched in Figure 6.48 is called a **damped sine wave** and 2^{-x} is called the **damping factor**. By using different damping factors, we may obtain other variations of such compressed or expanded sine waves. The analysis of such graphs is important in physics and engineering.

EXERCISES 6.9

Use addition of ordinates to sketch the graphs of the equations in Exercises 1–20.

1 $y = \cos x + 3 \sin x$ **2** $y = \sin x + 3 \cos x$

3 $y = 2 \cos x + 3 \sin x$ **4** $y = 2 \sin x + 3 \cos x$

5 $y = \sin x + \cos 2x$ **6** $y = 2 \cos x + \sin 2x$

7 $y = \cos x - \sin x$ **8** $y = 2 \sin x - \cos x$

9 $y = 2 \cos x - \frac{1}{2} \sin 2x$ **10** $y = 2 \cos x + \cos \frac{1}{2} x$

11 $y = 1 + \sin x$ **12** $y = 2 + \tan x$

13 $y = \frac{1}{2} x + \sin x$ **14** $y = 1 + \cos x$

15 $y = 2^x + \sin x$ **16** $y = \csc x - 1$

17 $y = 2 + \sec x$ **18** $y = x - \sin x$

19 $y = \cos x - 1$ **20** $y = |x + 1| + \cos 2x$

In each of Exercises 21–30 sketch the graph of f.

21 $f(x) = 2^{-x} \cos x$ **22** $f(x) = 2^x \sin x$

23 $f(x) = |x| \sin x$ **24** $f(x) = (1 + x^2) \cos x$

25 $f(x) = \frac{1}{2} x^2 \sin x$ **26** $f(x) = \cot x \tan x$

27 $f(x) = |\sin x|$ **28** $f(x) = |\cos x| + 1$

29 $f(x) = \sin |x|$ **30** $f(x) = \cos |x + 1|$

CALCULATOR EXERCISES 6.9

Verify some of the graphs in this section quantitatively, by calculating the coordinates of many points.

6.10 APPLICATIONS INVOLVING RIGHT TRIANGLES

The early development of trigonometry was concerned with measurements of angles and finding lengths of sides of triangles. Solving problems of that type are no longer the most important applications; however, trigonometric techniques are still used to answer questions about triangles which arise in physical situations. We shall restrict our discussion in this section to right triangles. Triangles which do not contain a right angle will be considered in the next chapter.

In this section and Chapter Seven we shall often use the following notation. The vertices of a triangle will often be denoted by A, B, and C. The angles of the triangle at A, B, and C will be denoted by α, β, and γ, respectively, and the lengths of the sides opposite these angles by a, b, and c, respectively. The triangle itself will often be referred to as *triangle ABC*. To *solve* a triangle means to find all of its parts, that is, the lengths of the three sides and the measures of the three angles. If the triangle is a right triangle and if one of the acute angles and a side are known, or if two sides are given, then the formulas in Section 6.5 which express the trigonometric functions as ratios of sides of a triangle may be used to find the remaining parts.

Example 1   If, in triangle ABC, $\gamma = 90°$, $\alpha = 34°$, and $b = 10.5$, approximate the remaining parts of the triangle.

Solution Since the sum of the angles is $180°$ it follows that $\beta = 56°$. Referring to Figure 6.49 and using the fact that the tangent equals opp/adj, we have

$$\tan 34° = \frac{a}{10.5} \quad \text{or} \quad a = (10.5)\tan 34°.$$

Substituting from Table 2 we obtain

$$a \approx (10.5)(0.6745) \approx 7.1.$$

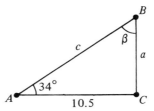

Figure 6.49

Side c can be found using either the cosine or secant functions. Since the cosine equals adj/hyp, we may write

$$\cos 34° = \frac{10.5}{c} \quad \text{or} \quad c = \frac{10.5}{\cos 34°}.$$

Using Table 2 and dividing we obtain

$$c \approx \frac{10.5}{0.8290} \approx 12.7.$$

The division involved in finding c could have been avoided through the use of the secant, which equals hyp/adj. Thus,

$$\sec 34° = \frac{c}{10.5} \quad \text{or} \quad c = (10.5)\sec 34°.$$

Referring to Table 2 and multiplying gives us

$$c \approx (10.5)(1.2062) = 12.6651 \approx 12.7. \qquad \blacksquare$$

As illustrated in Example 1, when working with triangles we shall round off answers. One reason for doing this is that in applications, the lengths of sides of triangles and measures of angles are usually found by mechanical devices, and hence are only approximations to the exact values. Consequently, a number such as 10.5 in Example 1 is assumed to have been rounded off to the nearest tenth. One cannot expect more accuracy in the calculated values for the remaining sides and, therefore, they should also be rounded off to the nearest tenth.

In some problems a large number of digits, such as 13,647.29, may be given for a number. If Table 2 is used, a number of this type should be rounded off to four significant figures and written as 13,650 before beginning any calculations. The situation here is similar to that discussed in Chapter Five with regard to the use of the table of logarithms. Since the values of the trigonometric functions given in Table 2 have been rounded off to four significant figures, we cannot expert more than four-figure accuracy in our computations.

As a final remark on approximations, answers should also be rounded off when Table 2 is used to find angles. In general, we shall use the following rules: if the sides of a triangle are known to four significant figures, then measures of angles calculated from Table 2 should be rounded off to the nearest minute; if the sides are known to three significant figures, then calculated measures of angles should be rounded off to the nearest multiple of ten minutes; and if the sides are known to only two significant figures, then calculations should be rounded off to the nearest degree. In order to justify these rules, we would have to make a much deeper analysis of problems involving approximate data.

Example 2 If, in triangle ABC, $\gamma = 90°$, $a = 12.3$, and $b = 31.6$, find the remaining parts.

Solution Referring to the triangle illustrated in Figure 6.50, we have

$$\tan \alpha = \frac{12.3}{31.6} \approx 0.3892.$$

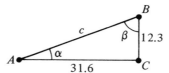

Figure 6.50

By the rule stated in the preceding paragraph, α should be rounded off to the nearest multiple of 10′. From Table 2, we see that $\alpha \approx 21°20'$. Consequently,

$$\beta \approx 90° - 21°20' = 68°40'.$$

Again referring to Figure 6.50,

$$\sec \alpha = \frac{c}{31.6} \quad \text{or} \quad c = (31.6)\sec \alpha$$

and hence, $c \approx (31.6)\sec 21°20' \approx (31.6)(1.0736) \approx 33.9.$

Side c can also be found using the cosine. Referring to Figure 6.50, we see that

$$\cos \alpha = \frac{31.6}{c}$$

and hence,
$$c = \frac{31.6}{\cos 21°20'} \approx \frac{31.6}{0.9315} \approx 33.9.$$ ■

Right angles are useful in solving various types of applied problems. The following examples give two illustrations; others will be found in the exercises.

Example 3   From a point on level ground 135 feet from the foot of a tower, the angle of elevation of the top of the tower is 57°20′. Find the height of the tower.

Solution   The **angle of elevation** is the angle which the line of sight makes with the horizontal. If we let d denote the height of the tower, then the given facts are represented by the triangle in Figure 6.51. Referring to the figure we see that

$$\tan 57°20' = \frac{d}{135} \quad \text{or} \quad d = (135)\tan 57°20'.$$

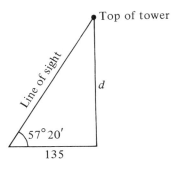

Figure 6.51

Substituting from Table 2 gives us

$$d = (135)(1.560) = 210.6 \approx 211.$$ ■

Example 4   From the top of a building which overlooks the ocean, a man sees a boat sailing directly toward him. If the man is 100 feet above sea level and if the angle of depression of the boat changes from 25° to 40° during the period of observation, find the approximate distance the boat travels during that time.

Solution   The **angle of depression** is the angle between the line of sight and the horizontal. Let A and B be the positions of the boat which correspond to the 25° and 40° angles, respectively. Suppose that the man is at point D and let C be the point 100 feet directly below him. Let d denote the distance the boat travels and let g denote the distance from B to C. This gives us the drawing in Figure 6.52 where α and β denote angles DAC and DBC, respectively. It follows that $\alpha = 25°$ and $\beta = 40°$. (Why?)

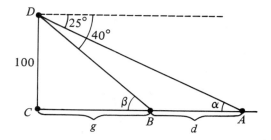

Figure 6.52

From triangle BCD we have

$$\cot \beta = \frac{g}{100} \quad \text{or} \quad g = 100 \cot \beta.$$

From triangle DAC we have

$$\cot \alpha = \frac{d+g}{100} \quad \text{or} \quad d+g = 100 \cot \alpha.$$

Consequently,

$$\begin{aligned}
d &= 100 \cot \alpha - g \\
&= 100 \cot \alpha - 100 \cot \beta \\
&= 100 \left(\cot \alpha - \cot \beta \right) \\
&= 100 \left(\cot 25° - \cot 40° \right) \\
&\approx 100(2.145 - 1.192) \\
&= 100(0.953) = 95.3
\end{aligned}$$

Hence, $d \approx 95$ feet.   ∎

In certain navigation and surveying problems the **direction**, or **bearing**, from a point P to a point Q is often specified by stating the acute angle which the half-line from P through Q varies to the east or west of the north-south line. Figure 6.53 illustrates four such lines. The north-south and east-west lines are labeled NS and WE, respectively. The bearing from P to Q_1 is 25° east of north and is denoted by N25°E. We also refer to the **direction** N25°E, meaning the direction from P to Q_1. The bearings from P to Q_2, Q_3, and Q_4 are represented in a similar manner in the figure.

In air navigation, directions and bearings are specified by measuring from the north in a clockwise direction. In this particular situation, a positive measure is assigned to the angle instead of the negative measure to which we are accustomed for clockwise rotations. Thus, referring to (ii) of Figure 6.53, we see that the direction of PQ is 40°, whereas the direction of PR is 300°.

We shall use these notations in Exercises 35–37.

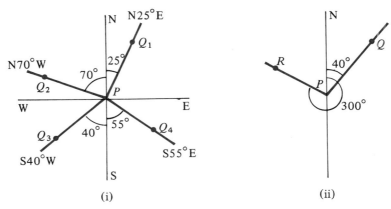

Figure 6.53

EXERCISES 6.10

Given the indicated parts of triangle ABC with $\gamma = 90°$, approximate the remaining parts of each triangle in Exercises 1–16.

1 $\alpha = 30°$, $b = 20$ **2** $\beta = 45°$, $b = 35$

3 $\beta = 52°$, $a = 15$ **4** $\alpha = 37°$, $b = 24$

5 $\alpha = 17°40'$, $a = 4.50$

6 $\beta = 64°20'$, $a = 20.1$

7 $\beta = 71°51'$, $b = 240.0$

8 $\alpha = 31°10'$, $a = 510$

9 $a = 25$, $b = 45$ **10** $a = 31$, $b = 9.0$

11 $c = 5.8$, $b = 2.1$ **12** $a = 0.42$, $c = 0.68$

13 $\alpha = 37°46'$, $b = 512.0$

14 $\beta = 10°17'$, $b = 68.40$

15 $a = 614$, $c = 806$ **16** $c = 37.4$, $b = 21.6$

17 Approximate the angle of elevation of the sun if a boy 5.0 feet tall casts a shadow 4.0 feet long on level ground.

18 From a point 15 meters above level ground an observer measures the angle of depression of an object on the ground as 68°. Approximately how far is the object from the point on the ground directly beneath the observer?

19 The string on a kite is taut and makes an angle of 54°20′ with the horizontal. Find the approximate height of the kite above level ground if 85.0 meters of string are out and the end of the string is held 1.50 meters above the ground.

20 The side of a regular pentagon is 24.0 cm long. Approximate the radius of the circumscribed circle.

21 From a point P on level ground the angle of elevation of the top of a tower is 26°50′. From a point 25.0 meters closer to the tower and on the same line with P and the base of the tower, the angle of elevation of the top is 53°30′. Approximate the height of the tower.

22 A ladder 20 feet long leans against the side of a building. If the angle between the ladder and the building is 22°, approximately how far is the bottom of the ladder from the building? If the distance from the bottom of the ladder to the building is increased by 3.0 feet, approximately how far does the top of the ladder move down the building?

23 As a weather balloon rises vertically its angle of elevation from a point P on the level ground 110 km from the point Q directly underneath the balloon changes from 19°20′ to 31°50′. Approximately how far does the balloon rise during this period?

24 From a point A, 8.20 meters above level ground, the angle of elevation of the top of a building is 31°20′ and the angle of depression of the base of the building is 12°50′. Approximate the height of the building.

25 In order to find the distance d between two points P and Q on opposite shores of a lake, a surveyor locates a point R which is 50.0 meters from P and such that RP is perpendicular to PQ. Next, using a transit, angle PRQ is found to measure 72°40′. What is d?

26 A guy wire is attached to the top of a radio antenna and to a point on horizontal ground which is 40.0 meters from the base of the antenna. If the wire makes an angle of 58°20′ with the ground, approximately how long is the wire?

27 An octagon is inscribed in a circle of radius 12.0 cm. Approximate the perimeter of the octagon.

28 A builder wishes to construct a ramp 24 feet long which rises to a height of 5.0 feet above level ground. Approximate the angle that the ramp should make with the horizontal.

29 A rocket is fired at sea level and climbs at a constant angle of 75° through a distance of 10,000 ft. Approximate its altitude to the nearest foot.

30 A CB antenna is located on the top of a garage 16 feet tall. From a point on level ground which is 100 feet from a point directly below the antenna, the antenna subtends an angle of 12°. Approximate the length of the antenna.

31 An airplane flying at an altitude of 10,000 feet passes directly over a fixed object on the ground. One minute later the angle of depression of the object is 42°. Approximate the speed of the airplane to the nearest mph.

32 A motorist is travelling along a level highway at a speed of 60 km per hour directly toward a distant mountain. She observes that between 1:00 P.M. and 1:10 P.M. the angle of elevation of

the top of the mountain changes from 10° to 70°. Approximate the height of the mountain.

33 An airplane pilot wishes to make his approach to an airstrip at an angle of 10° with the horizontal. If he is flying at an altitude of 5,000 feet, approximately how far from the airstrip should he begin his descent? (Give answer to the nearest 100 feet.)

34 In order to measure the height h of a cloud cover, a spotlight is directed vertically upward from the ground. From a point on level ground which is d meters from the spotlight, the angle of elevation θ of the light image on the clouds is then measured. Find a formula which expresses h in terms of d and θ. As a special case, approximate h if $d = 1000$ meters and $\theta = 59°0′$.

35 A ship leaves port at 1:00 P.M. and sails in the direction N34°W at a rate of 24 miles per hour. Another ship leaves port at 1:30 P.M. and sails in the direction N56°E at a rate of 18 miles per hour. Approximately how far apart are the ships at 3:00 P.M.?

36 From an observation point A a forest ranger sights a fire in the direction S48°20′W. From a point B, 5 miles due west of A, another ranger sights the same fire in the direction S54°10′E. Approximate, to the nearest mile, the distance of the fire from A.

37 An airplane flying at a speed of 360 miles per hour flies from a point A in the direction 137° for 30 minutes and then flies in the direction 227° for 45 minutes. Approximate, to the nearest mile, the distance from the airplane to A.

38 Generalize Exercise 19 to the case where the angle is α, the number of meters of string out is d, and the end of the string is held c meters above the ground. Express the height h of the kite in terms of α, d, and c.

39 Generalize Exercise 23 to the case where the distance from P to Q is d km and the angle of elevation changes from α to β.

40 Generalize Exercise 24 to the case where point A is d meters above ground and the angles of elevation and depression are α and β, respectively. Express the height h of the building in terms of d, α, and β.

CALCULATOR EXERCISES 6.10

Given the indicated parts of triangle ABC with $\gamma = 90°$, approximate the remaining parts of each triangle in Exercises 1–10. Round off answers to four significant figures.

1 $\alpha = 41.27°$, $a = 314.6$

2 $\beta = 24.96°$, $b = 209.3$

3 $\beta = 37.06°$, $a = 0.4613$

4 $\alpha = 17.69°$, $b = 1.307$

5 $\beta = 2.71°$, $b = 7149$

6 $\alpha = 84.07°$, $a = 0.1024$

7 $a = 46.87$, $b = 13.12$

8 $a = 6.948$, $b = 8.371$

9 $b = 2,462$, $c = 5,074$

10 $a = 88.12$, $c = 94.06$

6.11 HARMONIC MOTION

Trigonometric functions are important in the investigation of vibratory or oscillatory motion, such as the motion of a particle in a vibrating guitar string or in a spring that has been compressed or elongated, and then released to oscillate back and forth. The fundamental type of particle displacement inherent in these illustrations is termed *harmonic motion*. As an aid to introducing this concept, let us consider a point P moving at a constant rate around a circle of radius a with center at the origin O of a rectangular coordinate system. Suppose the initial position of P is $A(a, 0)$ and θ is the angle generated by the ray OP after t units of time (see Figure 6.54).

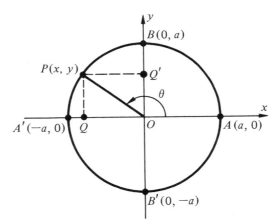

Figure 6.54

The **angular speed** ω of OP is, by definition, the rate at which the measure of θ changes per unit time. To state that P moves around the circle at a constant rate is equivalent to stating that the angular speed ω is constant. If ω is constant, then $\theta = \omega t$. To illustrate, if $\omega = \pi/6$ radians per second, then $\theta = (\pi/6)t$. In this case, at $t = 1$

second, $\theta = (\pi/6)(1) = \pi/6$, and P is one-third of the way from $A(a, 0)$ to $B(0, a)$. At $t = 2$ seconds, $\theta = (\pi/6)(2) = \pi/3$, and P is two-thirds of the way from A to B. At $t = 6$ seconds, $\theta = (\pi/6)(6) = \pi$, and P is at $A'(-a, 0)$, etc.

If the coordinates of P are (x, y), then $\sin \theta = y/a$ and $\cos \theta = x/a$. Multiplying both sides of these equations by a and using the fact that $\theta = \omega t$, gives us

$$x = a \cos \omega t, \quad y = a \sin \omega t.$$

The last two equations specify the position (x, y) of P at any time t. Let us next consider the point $Q(x, 0)$, which is called the **projection** of P on the x-axis. The position of Q is given by $x = a \cos \omega t$. As P moves around the circle several times, the point Q oscillates back and forth between $A(a, 0)$ and $A'(-a, 0)$. Similarly, the point $Q'(0, y)$ is called the **projection** of P on the y-axis, and its position is given by $y = a \sin \omega t$. As P moves around the circle, Q' oscillates between $B'(0, -a)$ and $B(0, a)$. The motions of Q and Q' are of the type described in the next definition.

Definition

> A point which moves on a coordinate line such that its distance d from the origin at time t is given by either $d = a \cos \omega t$ or $d = a \sin \omega t$, where a and ω are real numbers, is said to be in **simple harmonic motion**.

In the preceding definition, the **amplitude** of the motion is the maximum displacement $|a|$ of the point from the origin. The **period** is the time $2\pi/\omega$ required for one complete oscillation. The **frequency** $\omega/2\pi$ is the number of oscillations per unit of time.

A physical interpretation of simple harmonic motion can be obtained by considering a spring with an attached weight, which is oscillating vertically relative to a coordinate line, as illustrated in Figure 6.55. The number d represents the coordinate

Figure 6.55

of a fixed point Q in the weight, and we assume that the amplitude a of the motion is constant. In this case there is no frictional force retarding the motion. If friction is present, then the amplitude decreases with time, and the motion is said to be damped.

Example 1 Suppose the weight shown in Figure 6.55 is oscillating according to the law

$$d = 10 \cos\left(\frac{\pi}{6} t\right)$$

where t is measured in seconds and d in centimeters. Discuss the motion of the weight.

Solution By definition, the motion is simple harmonic. The amplitude is $a = 10$ cm. Since $\omega = \pi/6$, the period is $2\pi/\omega = 2\pi/(\pi/6) = 12$. Thus, it requires 12 seconds for one complete oscillation. The frequency is $\omega/2\pi = (\pi/6)/2\pi = 1/12$, that is, $1/12$ of an oscillation takes place each second. The following table indicates the position of Q at various times.

t	0	1	2	3	4	5	6
$(\pi/6)t$	0	$\pi/6$	$\pi/3$	$\pi/2$	$2\pi/3$	$5\pi/6$	π
d	10	$5\sqrt{3} \approx 8.7$	5	0	-5	$-5\sqrt{3} \approx -8.7$	-10

Note that the initial position of Q is 10 cm above the origin O. It then moves downward, gaining speed until it reaches O. In particular, Q travels approximately 1.3 cm during the first second, 3.7 cm during the next second, and 5 cm during the third second. It then slows down until it reaches a point 10 cm below O at the end of 6 seconds. It is left to the reader to verify that the direction of motion is then reversed, and the weight moves upward, gaining speed until it reaches O, after which it slows down until it returns to its original position at the end of 12 seconds. The direction of motion is then reversed again, and the same pattern is repeated indefinitely. ∎

The preceding example is typical of simple harmonic motion. If the initial angle AOP in Figure 6.54 is ϕ at $t = 0$, then the position (x, y) of P on the circle is given by

$$x = a \cos(\omega t + \phi), \qquad y = a \sin(\omega t + \phi).$$

Points which vary on a coordinate line according to either of these formulas are also said to be in simple harmonic motion.

Simple harmonic motion takes place in many different types of wave motion, such as water waves, sound waves, radio waves, light waves, and distortional waves which are present in vibrating bodies. As a specific example, consider the waves made by holding one end of a long rope, as in Figure 6.56, and then causing it to vibrate by raising and lowering the hand in simple harmonic motion. Waves appear to move along the rope, traveling away from the hand. By placing a fixed, vertical, measuring device a distance c from the end, as shown in the figure, one may study the upward

Figure 6.56

and downward motion of a specific particle P in the rope. If a suitable horizontal axis is chosen, it can be shown that P is in simple harmonic motion, as if it were attached to a spring. A similar motion takes place in water ripples on a large pond.

The definition of simple harmonic motion is usually extended to include situations where d is *any* mathematical or physical quantity (not necessarily a distance). In the next illustration d is itself an angle.

Historically, the first type of harmonic motion to be scientifically investigated involved pendulums. Figure 6.57 illustrates a *simple pendulum* consisting of a bob of mass m attached to one end of a string, with the other end of the string attached to a fixed point P.

Figure 6.57

In the ideal case it is assumed that the string is weightless, that there is no air resistance, and the only force acting on the bob is gravity. If the bob is displaced sideways and released, the pendulum oscillates back and forth in a vertical plane. Let α denote the *angular displacement* at time t (see Figure 6.57). If the bob moves through a small arc (say $|\alpha| < 5°$), then it can be shown, by using physical laws, that

$$\alpha = \beta \cos(\omega t + \alpha_0)$$

where α_0 is the initial displacement, ω is the frequency of oscillation of the angle α, and β is the (angular) amplitude of oscillation. This means that the *angle* α is in simple harmonic motion and hence we refer to the motion as *angular* simple harmonic motion.

As a final illustration, in electric circuits an alternating electromotive force (emf) and the current may vary harmonically. For example, the emf e (measured in volts) at time t may be given by

$$e = E \sin \omega t$$

where E is the maximum value of e. If an emf of this type is impressed on a circuit containing only a resistance R, then by Ohm's Law, the current i at time t is

$$i = \frac{e}{R} = \frac{E}{R} \sin \omega t = I \sin \omega t$$

where $I = E/R$. A schematic drawing of an electric circuit of this type is illustrated in Figure 6.58.

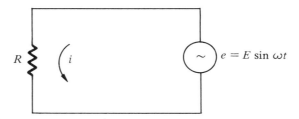

Figure 6.58

In this case the maximum value I of i occurs at the same time as the maximum value E of e. In other situations these maximum values may occur at different times, in which case we say there is a **phase difference** between e and i. For example, if $e = E \sin \omega t$, we could have either

(a) $i = I \sin (\omega t - \phi)$,   or   (b) $i = I \sin (\omega t + \phi)$

where $\phi > 0$. In case (a), the current is said to **lag** the emf by an amount ϕ/ω, and the graph of i can be obtained by shifting the graph of $i = I \sin \omega t$ and amount ϕ/ω to the *right*. In case (b) we shift the graph to the *left*, and the current is said to **lead** the emf by an amount ϕ/ω.

There are numerous other physical illustrations involving simple harmonic motion. Interested students may find further information in books on physics and engineering.

EXERCISES 6.11

1 A wheel of diameter 40 centimeters is rotating about an axle at a rate of 100 revolutions per minute. If a coordinate system is introduced as in Figure 6.55, where P is a point on the rim of the wheel, find:

(a) the angular speed of OP.
(b) the position (x, y) of P after t minutes.

2 A wheel of radius 2 feet is rotating about an axle, and the angular speed of a ray from the center of the wheel to a point P on the rim is $5\pi/6$ radians per second.

(a) How many revolutions does the wheel make in 10 minutes?

(b) If a coordinate system is introduced as in Figure 6.54, find the position (x, y) of P after t seconds.

In each of Exercises 3–6, the given formula specifies the position of a point P moving harmonically on a vertical axis, where t is in seconds and d is in centimeters. Determine the amplitude, period, and frequency, and describe the motion of the point during one complete oscillation (starting at $t = 0$).

3 $d = 10 \sin 6\pi t$ **4** $d = \dfrac{1}{3} \cos \dfrac{\pi}{4} t$

5 $d = 4 \cos \dfrac{3\pi}{2} t$ **6** $d = 6 \sin \dfrac{2\pi}{3} t$

7 A point P in simple harmonic motion has a period of 3 seconds and an amplitude of 5

centimeters. Express the motion of P by means of an equation of the form $d = a \cos \omega t$.

8 A point P in simple harmonic motion has a frequency of 1/2 oscillation per minute and amplitude of 4 feet. Express the motion of P by means of an equation of the form $d = a \sin \omega t$.

9 The electromotive force e and current i in a certain alternating current circuit are given by

$$e = 220 \sin 360\pi t,$$

$$i = 20 \sin\left(360\pi t - \frac{\pi}{4}\right).$$

Sketch the graphs of e and i on the same coordinate axes and determine the lag or lead.

10 Rework Exercise 9 if

$$e = 110 \sin 120\pi t,$$

$$i = 15 \sin\left(120\pi t + \frac{\pi}{3}\right).$$

6.12 REVIEW

Concepts

Define or discuss each of the following.

1 The point $P(t)$ on the unit circle U

2 The trigonometric functions

3 The Fundamental Identities

4 Domains and ranges of the trigonometric functions

5 Periodic function; period; periods of the trigonometric functions

6 Graphs of the trigonometric functions

7 Angles

8 Standard position of an angle

9 Initial and terminal sides

10 Coterminal angles

11 Positive and negative angles

12 Radian measure

13 Degree measure

14 Acute and obtuse angles

15 The relationship between radian measure and degree measure

16 Trigonometric functions of angles

17 Trigonometric functions as ratios

18 Reference numbers and angles

19 Finding values of trigonometric functions using reference numbers and angles

20 The graphs of $f(x) = a \sin(bx + c)$ and $f(x) = a \cos(bx + c)$

21 Addition of ordinates

22 Trigonometric solutions of right triangles

Exercises

In Exercises 1–3, $P(t)$ denotes the point on the unit circle U that corresponds to the real number t.

1 Find the rectangular coordinates of $P(7\pi)$, $P(-5\pi/2)$, $P(9\pi/2)$, $P(-3\pi/4)$, $P(18\pi)$, and $P(\pi/6)$.

2 If $P(t)$ has coordinates $(-3/5, -4/5)$, find the coordinates of $P(t + 3\pi)$, $P(t - \pi)$, $P(-t)$, and $P(2\pi - t)$.

3 Find the quadrant containing $P(t)$ if:
(a) $\sec t < 0$ and $\sin t > 0$.
(b) $\cot t > 0$ and $\csc t < 0$.
(c) $\cos t > 0$ and $\tan t < 0$.

4 Find the values of the remaining trigonometric functions if:
(a) $\sin t = -4/5$ and $\cos t = 3/5$.
(b) $\csc t = \sqrt{13}/2$ and $\cot t = -3/2$.

5 Without the use of tables, find the values of the trigonometric functions corresponding to each of the following real numbers.
(a) $9\pi/2$ (b) $-5\pi/4$
(c) 0 (d) $11\pi/6$

6 Find the radian measures that correspond to the following degree measures: $330°$, $405°$, $-150°$, $240°$, $36°$.

7 Find the degree measures that correspond to the following radian measures: $9\pi/2$, $-2\pi/3$, $7\pi/4$, 5π, $\pi/5$.

8 Find the reference number t' if t equals: $5\pi/4$, $-5\pi/6$, $9\pi/8$.

9 Find the reference angle θ' if θ has measure: $245°$, $137°10'$, $892°$.

10 A central angle θ subtends an arc 20 cm long on a circle of radius 2 meters. What is the radian measure of θ?

11 Find the values of the six trigonometric functions of θ if θ is in standard position and satisfies the stated condition.
(a) The point $(30, -40)$ is on the terminal side of θ.

(b) The terminal side of θ is in quadrant II and is parallel to the line $2x + 3y + 6 = 0$.
(c) $\theta = -90°$.

12 Find each of the following without the use of tables or calculators.
(a) $\cos 225°$ (b) $\tan 150°$
(c) $\sin(-\pi/6)$ (d) $\sec(4\pi/3)$
(e) $\cot(7\pi/4)$ (f) $\csc(300°)$

13 Use interpolation in Table 2 to approximate the following.
(a) $\cos 31°47'$ (b) $\tan 71°24'$
(c) $\csc 64°3'$

14 Use interpolation to approximate, to the nearest minute, the degree measure of all angles θ which are in the interval $[0°, 360°]$ if:
(a) $\sin \theta = 0.8237$.
(b) $\cos \theta = -0.7281$.
(c) $\cot \theta = 1.238$.

In each of Exercises 15–22, find the amplitude and period, and sketch the graph of f.

15 $f(x) = 5\cos x$ **16** $f(x) = \tfrac{2}{3}\sin x$

17 $f(x) = -\tan x$ **18** $f(x) = 4\sin 3x$

19 $f(x) = 3\cos 4x$ **20** $f(x) = -4\sin(x/3)$

21 $f(x) = 2\sec(x/2)$ **22** $f(x) = \tfrac{1}{2}\csc(2x)$

Sketch the graph of the equation in each of Exercises 23–29.

23 $y = 2\sin\left(x - \dfrac{2\pi}{3}\right)$ **24** $y = -4\cos\left(x + \dfrac{\pi}{6}\right)$

25 $y = 5\cos\left(2x + \dfrac{\pi}{2}\right)$ **26** $y = \tan\left(x - \dfrac{\pi}{4}\right)$

27 $y = 2\sin x + \sin 2x$ **28** $y = 1 + x + \cos x$

29 $y = 2^{-x}\sin 2x$

30 Given the following parts of triangle ABC with $\gamma = 90°$, approximate the remaining parts.
(a) $\beta = 60°$, $b = 40$
(b) $\alpha = 54°40'$, $b = 220$
(c) $a = 62$, $b = 25$

ANALYTIC TRIGONOMETRY

In this chapter we shall examine various algebraic and geometric aspects of trigonometry. In Sections 7.1 through 7.5 the emphasis is on identities and equations. In Section 7.6 we consider inverse functions for the trigonometric functions. Methods for solving oblique triangles are explained in Sections 7.7 and 7.8. The chapter concludes with a discussion of vectors. In the course of our work we shall derive many important trigonometric identities and formulas. For convenient reference they are listed on the inside covers of the text.

7.1 THE FUNDAMENTAL IDENTITIES

The Fundamental Identities were introduced in Section 6.2. At that time we were primarily interested in their numerical significance. In this chapter they will be employed in a more general fashion. In particular, we shall see how the Fundamental Identities may be used to help simplify complicated expressions, derive other important formulas, and solve equations which involve trigonometric functions. Let us begin by restating them and working two numerical examples of the type considered in the previous chapter.

The Fundamental Identities

$$\csc t = \frac{1}{\sin t}, \qquad \sec t = \frac{1}{\cos t}, \qquad \cot t = \frac{1}{\tan t}$$

$$\tan t = \frac{\sin t}{\cos t}, \qquad \cot t = \frac{\cos t}{\sin t}$$

$$\sin^2 t + \cos^2 t = 1, \qquad 1 + \tan^2 t = \sec^2 t, \qquad 1 + \cot^2 t = \csc^2 t$$

Example 1 If $\sec \theta = 3$ and θ is a fourth quadrant angle, find $\tan \theta$.

Solution   Using the identity $1 + \tan^2 \theta = \sec^2 \theta$ we obtain

$$\tan \theta = \pm \sqrt{\sec^2 \theta - 1}.$$

Since θ is in quadrant IV, $\tan \theta$ is negative, and hence

$$\tan \theta = -\sqrt{\sec^2 \theta - 1} = -\sqrt{(3)^2 - 1} = -\sqrt{8} = -2\sqrt{2}. \qquad \blacksquare$$

Example 2   If $\csc t = 3/2$ and $\tan t < 0$ use fundamental identities to find the values of the remaining trigonometric functions of t.

Solution   Since $\csc t = 1/\sin t$, it follows that

$$\sin t = \frac{1}{\csc t} = \frac{1}{(3/2)} = 2/3.$$

Since $\csc t > 0$ and $\tan t < 0$, the angle (or point on the unit circle) corresponding to t is in quadrant II. Thus, $\cos t$ is negative and we may write

$$\begin{aligned}
\cos t &= -\sqrt{1 - \sin^2 t} \\
&= -\sqrt{1 - (2/3)^2} \\
&= -\sqrt{1 - (4/9)} = -\sqrt{5/9} = -\sqrt{5}/3.
\end{aligned}$$

Consequently, $\sec t = \dfrac{1}{\cos t} = -\dfrac{3}{\sqrt{5}} = \dfrac{-3\sqrt{5}}{5}.$

Next we see that

$$\tan t = \frac{\sin t}{\cos t} = \frac{(2/3)}{(-\sqrt{5}/3)} = -\frac{2}{\sqrt{5}} = \frac{-2\sqrt{5}}{5}$$

and $\cot t = \dfrac{1}{\tan t} = -\dfrac{\sqrt{5}}{2}.$

This gives us all of the functional values.

There are other ways of arriving at the values. For example, we could have started with

$$\cot^2 t = \csc^2 t - 1$$

and, since t is in quadrant II,

$$\begin{aligned}
\cot t &= -\sqrt{\csc^2 t - 1} \\
&= -\sqrt{(3/2)^2 - 1} \\
&= -\sqrt{(9/4) - 1} \\
&= -\sqrt{5/4} = -\sqrt{5}/2.
\end{aligned}$$

Other identities could also have been employed. $\blacksquare$

The next example illustrates how fundamental identities may be used to express one of the trigonometric functions in terms of another. You are asked to provide similar illustrations in Exercises 33–38.

Example 3   Find a formula which expresses $\tan t$ in terms of $\cos t$.

Solution   We may write

$$\tan t = \frac{\sin t}{\cos t} = \frac{\pm\sqrt{1 - \cos^2 t}}{\cos t}$$

where the sign depends on the particular value of t.   ∎

The following example illustrates how fundamental identities may be used to simplify trigonometric expressions. In the next section much more will be done along these lines.

Example 4   Use fundamental identities to simplify the expression

$$\sec t - \sin t \tan t.$$

Solution   Give reasons for each of the following steps:

$$\sec t - \sin t \tan t = \frac{1}{\cos t} - \sin t \left(\frac{\sin t}{\cos t}\right)$$

$$= \frac{1}{\cos t} - \frac{\sin^2 t}{\cos t}$$

$$= \frac{1 - \sin^2 t}{\cos t}$$

$$= \frac{\cos^2 t}{\cos t}$$

$$= \cos t.$$ ∎

Example 5   Use fundamental identities to transform

$$\frac{\sin^2 t}{1 + \cos t} \quad \text{into} \quad 1 - \cos t.$$

Solution   We may proceed as follows:

$$\frac{\sin^2 t}{1 + \cos t} = \frac{1 - \cos^2 t}{1 + \cos t}$$

$$= \frac{(1 + \cos t)(1 - \cos t)}{1 + \cos t}$$

$$= 1 - \cos t.$$ ∎

EXERCISES 7.1

In each of Exercises 1–8 use fundamental identities to find the values of the remaining trigonometric functions of t.

1 $\cos t = 1/2, \quad \sin t = -\sqrt{3}/2$

2 $\cos t = -\sqrt{3}/2, \quad \csc t = 2$

3 $\sin t = \sqrt{2}/2, \quad \sec t = -\sqrt{2}$

4 $\cos t = 0, \quad \sin t = 1$

5 $\cot t = -\sqrt{3}, \quad \csc t = -2$

6 $\sec t = \sqrt{2}, \quad \csc t = -\sqrt{2}$

7 $\cos t = -1, \quad \sin t = 0$

8 $\cot t = -1, \quad \sec t = -\sqrt{2}$

In Exercises 9–16 use fundamental identities to find the values of all six trigonometric functions if the given conditions are true.

9 $\tan t = 3/4$ and $\sin t < 0$

10 $\cot t = -3/4$ and $\cos t > 0$

11 $\sec t = -13/5$ and $\tan t < 0$

12 $\cos t = -1/2$ and $\sin t > 0$

13 $\sin t = -2/3$ and $\sec t > 0$

14 $\sec t = -5$ and $\csc t < 0$

15 $\csc t = 8$ and $\cos t < 0$

16 $\cos t = 0$ and $\sin t = -1$

In Exercises 17–32 use fundamental identities to transform the first expression into the second.

17 $\cos t \csc t, \quad \cot t$

18 $\tan t \csc t, \quad \sec t$

19 $\tan x \cos x, \quad \sin x$

20 $\cot x \sin x, \quad \cos x$

21 $\dfrac{\sec \theta}{\tan \theta}, \quad \csc \theta$

22 $\dfrac{\cot \theta}{\csc \theta}, \quad \cos \theta$

23 $\dfrac{\cos \alpha}{\sec \alpha}, \quad \cos^2 \alpha$

24 $\dfrac{\sec t}{\csc t}, \quad \tan t$

25 $(1 + \cot^2 t) \tan^2 t, \quad \sec^2 t$

26 $(1 - \sin^2 \theta) \sec^2 \theta, \quad 1$

27 $\sin x (\csc x - \sin x), \quad \cos^2 x$

28 $(\sec t - \cos t) \cos t, \quad \sin^2 t$

29 $\dfrac{\sin^2 \theta}{1 - \sin^2 \theta}, \quad \tan^2 \theta$

30 $\dfrac{1 - \sin^2 \alpha}{1 - \cos^2 \alpha}, \quad \cot^2 \alpha$

31 $\dfrac{1 + \tan^2 x}{\tan^2 x}, \quad \csc^2 x$

32 $\dfrac{\cos^2 t}{1 - \sin t}, \quad 1 + \sin t$

33 Express each trigonometric function in terms of $\sin t$.

34 Express each trigonometric function in terms of $\cos t$.

35 Express $\sin t$ in terms of $\tan t$.

36 Express $\cos t$ in terms of $\cot t$.

37 Express $\sec t$ in terms of $\csc t$.

38 Express $\csc t$ in terms of $\tan t$.

39 Prove that if $0 < t < \pi/2$, then

$$\log \csc t = -\log \sin t.$$

40 Prove that if θ is an acute angle, then

$$\log \tan \theta = \log \sin \theta - \log \cos \theta.$$

CALCULATOR EXERCISES 7.1

Use direct calculations to demonstrate that

$$\tan t = \frac{\sin t}{\cos t} \quad \text{and} \quad \sin^2 t + \cos^2 t = 1$$

for the values of t given in Exercises 1–8.

1 $1.62°$

2 $37°15'$

3 $\pi/20$

4 137.48

5 $\sin 2$

6 $\sqrt{\pi} + \log 9$

7 $e^{1.6}$

8 $\ln 13.42$

9 If a calculator has keys for sin, cos, and tan, but not for csc, sec, and cot, how are values for the latter functions calculated?

10 Calculate $(\cos 5)^a$, where $a = \log (\tan (\sin 2))$.

7.2   TRIGONOMETRIC IDENTITIES

Any mathematical expression that contains symbols such as $\sin x$, $\cos \beta$, $\tan v$, etc., where the letters x, β, and v are variables, is referred to as a **trigonometric expression**. Trigonometric expressions may be very simple or very complicated. Some examples are

$$x + \sin x, \quad \frac{\cos(3y + 1)}{x^2 + \tan^2(z - y^2)}, \quad \frac{\sqrt{\theta} + 2^{\sin \theta}}{\sec(\cot \theta)}.$$

As usual, we assume that the domain of each variable is the set of real numbers (or angles) for which the expressions are meaningful. As indicated in Section 7.1 many trigonometric expressions can be simplified or changed in form by employing fundamental identities. The next example is another illustration of how this may be accomplished.

Example 1   Simplify the expression $(\sec \theta + \tan \theta)(1 - \sin \theta)$.

Solution   Supply reasons for the following steps:

$$(\sec \theta + \tan \theta)(1 - \sin \theta) = \left(\frac{1}{\cos \theta} + \frac{\sin \theta}{\cos \theta}\right)(1 - \sin \theta)$$

$$= \left(\frac{1 + \sin \theta}{\cos \theta}\right)(1 - \sin \theta)$$

$$= \frac{1 - \sin^2 \theta}{\cos \theta}$$

$$= \frac{\cos^2 \theta}{\cos \theta}$$

$$= \cos \theta \qquad\qquad ∎$$

There are other ways to treat the expression in Example 1. We could begin by multiplying the two factors and then proceed to simplify and combine terms. The method we employed—of changing all expressions to expressions which involve only sines and cosines—is often worthwhile. However, that technique does not always lead to the shortest possible simplification.

A **trigonometric equation** is an equation which contains trigonometric expressions. When working with trigonometric equations, we may express solutions either in terms of real numbers or in terms of angles.

We encountered several trigonometric identities in our previous work. Some of them, such as the Fundamental Identities, are basic and should be memorized. Other identities are introduced merely to supply practice in manipulating trigonometric expressions and therefore should not be memorized. We shall take the latter point of view in this section. Thus, the identities to be investigated are unimportant in their own right. The important thing is the manipulative practice that is gained. The ability to carry out trigonometric manipulations is essential for problems which are encountered in advanced courses in mathematics and science.

We often use the phrase "verify an identity" instead of "prove that an equation is an identity." When verifying an identity, we shall not use the definition of the trigonometric functions to return to the coordinates (x, y) of a point on the unit circle U. Although many of the identities could be proved in that way, it is not the type of practice we desire. Rather, we shall use fundamental identities and algebraic manipulations to change the form of trigonometric expressions in a manner similar to the solution of Example 1. The preferred method of showing that an equation is an identity is to transform one side into the other, as illustrated in the next three examples. The reader should supply reasons for all steps in the solutions.

Example 2   Verify the identity $\dfrac{\tan t + \cos t}{\sin t} = \sec t + \cot t$.

Solution   We shall transform the left side into the right side. Thus,

$$\frac{\tan t + \cos t}{\sin t} = \frac{\tan t}{\sin t} + \frac{\cos t}{\sin t}$$

$$= \frac{\left(\dfrac{\sin t}{\cos t}\right)}{\sin t} + \cot t$$

$$= \frac{1}{\cos t} + \cot t$$

$$= \sec t + \cot t.$$

Example 3   Verify the identity $\sec \alpha - \cos \alpha = \sin \alpha \tan \alpha$.

Solution

$$\sec \alpha - \cos \alpha = \frac{1}{\cos \alpha} - \cos \alpha$$

$$= \frac{1 - \cos^2 \alpha}{\cos \alpha}$$

$$= \frac{\sin^2 \alpha}{\cos \alpha} = \sin \alpha \left(\frac{\sin \alpha}{\cos \alpha}\right)$$

$$= \sin \alpha \tan \alpha.$$

Example 4   Verify the identity $\dfrac{\cos x}{1 - \sin x} = \dfrac{1 + \sin x}{\cos x}$.

Solution   We begin by multiplying the numerator and denominator of the fraction on the left by $1 + \sin x$. Thus,

$$\frac{\cos x}{1 - \sin x} = \frac{\cos x}{1 - \sin x} \cdot \frac{1 + \sin x}{1 + \sin x}$$

$$= \frac{\cos x (1 + \sin x)}{1 - \sin^2 x}$$

$$= \frac{\cos x (1 + \sin x)}{\cos^2 x}$$

$$= \frac{1 + \sin x}{\cos x}. \qquad \blacksquare$$

In another technique for showing that an equation $p = q$ is an identity, we begin by transforming the left side p into another expression s, making sure that each step is *reversible* in the sense that it is possible to transform s back into p by reversing the procedure which has been used. In this case the equation $p = s$ is an identity. Next, as a *separate* exercise, it is shown that the right side q can also be transformed to the expression s by means of reversible steps and hence that $q = s$ is an identity. It then follows that $p = q$ is an identity. This method is illustrated in the next example.

Example 5   Verify the identity $(\tan \theta - \sec \theta)^2 = \dfrac{1 - \sin \theta}{1 + \sin \theta}$.

Solution   We shall verify the identity by showing that each side of the equation can be transformed into the same expression. Starting with the left side we may write

$$(\tan \theta - \sec \theta)^2 = \tan^2 \theta - 2 \tan \theta \sec \theta + \sec^2 \theta$$

$$= \frac{\sin^2 \theta}{\cos^2 \theta} - \frac{2 \sin \theta}{\cos^2 \theta} + \frac{1}{\cos^2 \theta}$$

$$= \frac{\sin^2 \theta - 2 \sin \theta + 1}{\cos^2 \theta}.$$

The right side of the given equation may be changed by multiplying numerator and denominator by $1 - \sin \theta$. Thus,

$$\frac{1 - \sin \theta}{1 + \sin \theta} = \frac{1 - \sin \theta}{1 + \sin \theta} \cdot \frac{1 - \sin \theta}{1 - \sin \theta}$$

$$= \frac{1 - 2 \sin \theta + \sin^2 \theta}{1 - \sin^2 \theta}$$

$$= \frac{1 - 2 \sin \theta + \sin^2 \theta}{\cos^2 \theta}$$

which is the same expression as that obtained for $(\tan \theta - \sec \theta)^2$. Since all steps are reversible, it follows that the given equation is an identity. ∎

EXERCISES 7.2

Verify the following identities.

1 $\cos \theta \sec \theta = 1$

2 $\tan \alpha \cot \alpha = 1$

3 $\sin \theta \sec \theta = \tan \theta$

4 $\sin \alpha \cot \alpha = \cos \alpha$

5 $\dfrac{\csc x}{\sec x} = \cot x$

6 $\cot \beta \sec \beta = \csc \beta$

7 $(1 + \cos \alpha)(1 - \cos \alpha) = \sin^2 \alpha$

8 $\cos^2 x (\sec^2 x - 1) = \sin^2 x$

9 $\cos^2 t - \sin^2 t = 2 \cos^2 t - 1$

10 $(\tan \theta + \cot \theta) \tan \theta = \sec^2 \theta$

11 $\dfrac{\sin t}{\csc t} + \dfrac{\cos t}{\sec t} = 1$

12 $1 - 2 \sin^2 x = 2 \cos^2 x - 1$

13 $(1 + \sin \alpha)(1 - \sin \alpha) = \dfrac{1}{\sec^2 \alpha}$

14 $(1 - \sin^2 t)(1 + \tan^2 t) = 1$

15 $\sec \beta - \cos \beta = \tan \beta \sin \beta$

16 $\dfrac{\sin w + \cos w}{\cos w} = 1 + \tan w$

17 $\dfrac{\csc^2 \theta}{1 + \tan^2 \theta} = \cot^2 \theta$

18 $\sin x + \cos x \cot x = \csc x$

19 $\sin t (\csc t - \sin t) = \cos^2 t$

20 $\cot t + \tan t = \csc t \sec t$

21 $\csc \theta - \sin \theta = \cot \theta \cos \theta$

22 $\cos \theta (\tan \theta + \cot \theta) = \csc \theta$

23 $\dfrac{\sec^2 u - 1}{\sec^2 u} = \sin^2 u$

24 $(\tan u + \cot u)(\cos u + \sin u) = \sec u + \csc u$

25 $(\cos^2 x - 1)(\tan^2 x + 1) = 1 - \sec^2 x$

26 $(\cot \alpha + \csc \alpha)(\tan \alpha - \sin \alpha) = \sec \alpha - \cos \alpha$

27 $\sec t \csc t + \cot t = \tan t + 2 \cos t \csc t$

28 $\dfrac{1 + \cos^2 y}{\sin^2 y} = 2 \csc^2 y - 1$

29 $\sec^2 \theta \csc^2 \theta = \sec^2 \theta + \csc^2 \theta$

30 $\dfrac{\sec x - \cos x}{\tan x} = \dfrac{\tan x}{\sec x}$

31 $\dfrac{1 + \cos t}{\sin t} + \dfrac{\sin t}{1 + \cos t} = 2 \csc t$

32 $\tan^2 \alpha - \sin^2 \alpha = \tan^2 \alpha \sin^2 \alpha$

33 $\dfrac{1 + \tan^2 v}{\tan^2 v} = \csc^2 v$

34 $\dfrac{\sec \theta + \csc \theta}{\sec \theta - \csc \theta} = \dfrac{\sin \theta + \cos \theta}{\sin \theta - \cos \theta}$

35 $\dfrac{1 + \sin x}{1 - \sin x} - \dfrac{1 - \sin x}{1 + \sin x} = 4 \tan x \sec x$

36 $\dfrac{1}{1 - \cos \gamma} + \dfrac{1}{1 + \cos \gamma} = 2 \csc^2 \gamma$

37 $\dfrac{1 + \csc \beta}{\sec \beta} - \cot \beta = \cos \beta$

38 $\dfrac{\cos x \cot x}{\cot x - \cos x} = \dfrac{\cot x + \cos x}{\cos x \cot x}$

39 $(\sec u - \tan u)(\csc u + 1) = \cot u$

40 $\dfrac{\cot \theta - \tan \theta}{\sin \theta + \cos \theta} = \csc \theta - \sec \theta$

41 $\dfrac{\cot \alpha - 1}{1 - \tan \alpha} = \cot \alpha$

42 $\dfrac{1 + \sec \beta}{\tan \beta + \sin \beta} = \csc \beta$

43 $\csc^4 t - \cot^4 t = \cot^2 t + \csc^2 t$

44 $\cos^4 \theta + \sin^2 \theta = \sin^4 \theta + \cos^2 \theta$

45 $\dfrac{\cos \beta}{1 - \sin \beta} = \sec \beta + \tan \beta$

46 $\dfrac{1}{\csc y - \cot y} = \csc y + \cot y$

47 $\dfrac{\tan^2 x}{\sec x + 1} = \dfrac{1 - \cos x}{\cos x}$

48 $\dfrac{\cot x}{\csc x + 1} = \dfrac{\csc x - 1}{\cot x}$

49 $\dfrac{\cot u - 1}{\cot u + 1} = \dfrac{1 - \tan u}{1 + \tan u}$

50 $\dfrac{1 + \sec x}{\sin x + \tan x} = \csc x$

51 $\sin^4 r - \cos^4 r = \sin^2 r - \cos^2 r$

52 $\sin^4 \theta + 2 \sin^2 \theta \cos^2 \theta + \cos^4 \theta = 1$

53 $\tan^4 k - \sec^4 k = 1 - 2 \sec^2 k$

54 $\sec^4 u - \sec^2 u = \tan^4 u + \tan^2 u$

55 $(\sec t + \tan t)^2 = \dfrac{1 + \sin t}{1 - \sin t}$

56 $\sec^2 \gamma + \tan^2 \gamma = (1 - \sin^4 \gamma)(\sec^4 \gamma)$

57 $(\sin^2 \theta + \cos^2 \theta)^3 = 1$

58 $\dfrac{\sin t}{1 - \cos t} = \csc t + \cot t$

59 $\dfrac{1 + \csc \beta}{\cot \beta + \cos \beta} = \sec \beta$

60 $\dfrac{\sin z \tan z}{\tan z - \sin z} = \dfrac{\tan z + \sin z}{\sin z \tan z}$

61 $\left(\dfrac{\sin^2 x}{\tan^4 x}\right)^3 \left(\dfrac{\csc^3 x}{\cot^6 x}\right)^2 = 1$

62 $\dfrac{\cos^3 x - \sin^3 x}{\cos x - \sin x} = 1 + \sin x \cos x$

63 $\dfrac{\sin \theta + \cos \theta}{\tan^2 \theta - 1} = \dfrac{\cos^2 \theta}{\sin \theta - \cos \theta}$

64 $(\csc t - \cot t)^4 (\csc t + \cot t)^4 = 1$

65 $(a \cos t - b \sin t)^2 + (a \sin t + b \cos t)^2$
      $= a^2 + b^2$

66 $\sin^6 v + \cos^6 v = 1 - 3 \sin^2 v \cos^2 v$

67 $\dfrac{\sin \alpha \cos \beta + \cos \alpha \sin \beta}{\cos \alpha \cos \beta - \sin \alpha \sin \beta} = \dfrac{\tan \alpha + \tan \beta}{1 - \tan \alpha \tan \beta}$

68 $\dfrac{\tan u - \tan v}{1 + \tan u \tan v} = \dfrac{\cot v - \cot u}{1 + \cot u \cot v}$

69 $\sqrt{\dfrac{1 - \cos t}{1 + \cos t}} = \dfrac{1 - \cos t}{|\sin t|}$

70 $\sqrt{\dfrac{1 - \sin \theta}{1 + \sin \theta}} = \dfrac{|\cos \theta|}{1 + \sin \theta}$

71 $\dfrac{\sin \alpha}{1 + \cos \alpha} + \dfrac{1 + \cos \alpha}{\sin \alpha} = 2 \csc \alpha$

72 $\dfrac{\csc x}{1 + \csc x} - \dfrac{\csc x}{1 - \csc x} = 2 \sec^2 x$

73 $\dfrac{1}{\tan \beta + \cot \beta} = \sin \beta \cos \beta$

74 $\dfrac{\cot y - \tan y}{\sin y \cos y} = \csc^2 y - \sec^2 y$

75 $\sec \theta + \csc \theta - \cos \theta - \sin \theta$
      $= \sin \theta \tan \theta + \cos \theta \cot \theta$

76 $\sin^3 t + \cos^3 t = (1 - \sin t \cos t)(\sin t + \cos t)$

77 $(1 - \tan^2 \phi)^2 = \sec^4 \phi - 4 \tan^2 \phi$

78 $\cos^4 w + 1 - \sin^4 w = 2 \cos^2 w$

79 $\dfrac{\tan x}{1 - \cot x} + \dfrac{\cot x}{1 - \tan x} = 1 + \sec x \csc x$

80 $\dfrac{\cos \gamma}{1 - \tan \gamma} + \dfrac{\sin \gamma}{1 - \cot \gamma} = \cos \gamma + \sin \gamma$

81 $\sin(-t) \sec(-t) = -\tan t$

82 $\dfrac{\cot(-v)}{\csc(-v)} = \cos v$

83 $\log 10^{\tan t} = \tan t$

84 $10^{\log |\sin t|} = |\sin t|$

85 $\log \cot x = -\log \tan x$

86 $\log \sec \theta = -\log \cos \theta$

87 $-\log |\sec \theta - \tan \theta| = \log |\sec \theta + \tan \theta|$

88 $\log |\csc x - \cot x| = -\log |\csc x + \cot x|$

Show that the equations in Exercises 89–100 are not identities. (*Hint*: Find one number in the domain of t or θ for which the equation is false.)

89 $\cos t = \sqrt{1 - \sin^2 t}$

90 $\sqrt{\sin^2 t + \cos^2 t} = \sin t + \cos t$

91 $\sqrt{\sin^2 t} = \sin t$

92 $\sec t = \sqrt{\tan^2 t + 1}$

93 $(\sin \theta + \cos \theta)^2 = \sin^2 \theta + \cos^2 \theta$

94 $\log(1/\sin t) = 1/(\log \sin t)$

95 $\cos(-t) = -\cos t$

96 $\sin(t + \pi) = \sin t$

97 $\cos(\sec t) = 1$

98 $\cot(\tan \theta) = 1$

99 $\sin^2 t - 4 \sin t - 5 = 0$

100 $3 \cos^2 \theta + \cos \theta - 2 = 0$

7.3 TRIGONOMETRIC EQUATIONS

If a trigonometric equation is not an identity, then techniques similar to those used for algebraic equations may be employed to find the solutions. The main difference here is that we usually solve for $\sin x$, $\cos \theta$, and so on, and then find x and θ as illustrated below.

Example 1 Solve the equation $\sin \theta \tan \theta = \sin \theta$.

Solution Each of the following is equivalent to the given equation:

$$\sin \theta \tan \theta - \sin \theta = 0$$
$$(\sin \theta)(\tan \theta - 1) = 0.$$

To find the solutions we set each factor on the left equal to zero, obtaining

$$\sin \theta = 0 \quad \text{and} \quad \tan \theta = 1.$$

The solutions of the equation $\sin \theta = 0$ consist of all multiples of π, that is, $\theta = n\pi$, where n is any integer.

Since the tangent function has period π, it is sufficient to find the solutions of the equation $\tan \theta = 1$ which are in the interval $[0, \pi)$, for once they are known, we may obtain all the others by adding multiples of π. The only solution of $\tan \theta = 1$ in $[0, \pi)$ is $\pi/4$, and hence every solution has the form

$$\theta = \frac{\pi}{4} + n\pi$$

where n is an integer.

It follows that the solutions of the given equation consist of all numbers of the form

$$n\pi \quad \text{and} \quad \frac{\pi}{4} + n\pi$$

where n is any integer. Some particular solutions are 0, $\pm \pi$, $\pm 2\pi$, $\pm 3\pi$, $\pi/4$, $5\pi/4$, $-3\pi/4$, and $-7\pi/4$. ■

Note that in Example 1 it would have been incorrect to begin by dividing both sides by $\sin \theta$, for this manipulation would lose the solutions of $\sin \theta = 0$.

Example 2 Solve the equation $2 \sin^2 t - \cos t - 1 = 0$.

Solution   We first change the equation to an equation which involves only $\cos t$, and then we factor as follows:

$$2(1 - \cos^2 t) - \cos t - 1 = 0$$
$$-2 \cos^2 t - \cos t + 1 = 0$$
$$2 \cos^2 t + \cos t - 1 = 0$$
$$(2 \cos t - 1)(\cos t + 1) = 0.$$

As in Example 1, the solutions of the last equation are the solutions of

$$2 \cos t - 1 = 0 \quad \text{and} \quad \cos t + 1 = 0$$

or equivalently, $\cos t = \frac{1}{2} \quad \text{and} \quad \cos t = -1$.

It is sufficient to find the solutions which are in the interval $[0, 2\pi)$, for once they are known, all solutions may be found by adding multiples of 2π.

If $\cos t = \frac{1}{2}$, then the reference number (or reference angle) is $\pi/3$ (or $60°$). Since $\cos t$ is positive, the point $P(t)$ on U (or the angle t) lies in quadrant I or quadrant IV. Hence, in the interval $[0, 2\pi)$, we have

$$t = \frac{\pi}{3} \quad \text{or} \quad t = 2\pi - \frac{\pi}{3} = \frac{5\pi}{3}.$$

If $\cos t = -1$, then $t = \pi$.

It follows that the solutions of the given equation are

$$\frac{\pi}{3} + 2\pi n, \quad \frac{5\pi}{3} + 2\pi n, \quad \pi + 2\pi n$$

where n is any integer. If we wish to express the solutions in terms of degrees, we may write

$$60° + 360°n, \quad 300° + 360°n, \quad 180° + 360°n. \quad ■$$

Example 3 Find the solutions of $4 \sin^2 x \tan x - \tan x = 0$ which are in the interval $[0, 2\pi)$.

Solution   Factoring the left side we obtain

$$(\tan x)(4\sin^2 x - 1) = 0.$$

Setting each factor equal to zero gives us

$$\tan x = 0 \quad \text{and} \quad \sin^2 x = \tfrac{1}{4}$$

or equivalently,

$$\tan x = 0, \quad \sin x = \tfrac{1}{2}, \quad \text{and} \quad \sin x = -\tfrac{1}{2}.$$

The equation $\tan x = 0$ has solutions 0 and π in the interval $[0, 2\pi)$. The equation $\sin x = \tfrac{1}{2}$ has solutions $\pi/6$ and $5\pi/6$. The equation $\sin x = -\tfrac{1}{2}$ leads to the numbers between π and 2π which have reference number $\pi/6$. These are

$$\pi + \frac{\pi}{6} = \frac{7\pi}{6} \quad \text{and} \quad 2\pi - \frac{\pi}{6} = \frac{11\pi}{6}.$$

Hence, the solutions of the given equation in the interval $[0, 2\pi)$ are

$$0, \pi, \frac{\pi}{6}, \frac{5\pi}{6}, \frac{7\pi}{6}, \frac{11\pi}{6}. \qquad \blacksquare$$

Example 4   Find the solutions of the equation $\csc^4 2u - 4 = 0$.

Solution   Factoring the left side we obtain

$$(\csc^2 2u - 2)(\csc^2 2u + 2) = 0.$$

Setting each factor equal to 0 leads to

$$\csc^2 2u = 2 \quad \text{and} \quad \csc^2 2u = -2.$$

The equation $\csc^2 2u = -2$ has no solutions. (Why?) The solutions of $\csc^2 2u = 2$ consist of the solutions of the two equations

$$\csc 2u = \sqrt{2} \quad \text{and} \quad \csc 2u = -\sqrt{2}.$$

If $\csc 2u = \sqrt{2}$, then

$$2u = \frac{\pi}{4} + 2\pi n \quad \text{and} \quad 2u = \frac{3\pi}{4} + 2\pi n$$

where n is any integer. Dividing both sides of the last two equations by 2 gives us the solutions

$$u = \frac{\pi}{8} + \pi n \quad \text{and} \quad u = \frac{3\pi}{8} + \pi n.$$

Similarly, from $\csc 2u = -\sqrt{2}$ we obtain

$$2u = \frac{5\pi}{4} + 2\pi n \quad \text{and} \quad 2u = \frac{7\pi}{4} + 2\pi n$$

where n is any integer. Dividing both sides of the last two equations by 2 gives us

$$u = \frac{5\pi}{8} + \pi n \quad \text{and} \quad u = \frac{7\pi}{8} + \pi n.$$

Collecting the above information, we see that all solutions are given by the formula

$$u = \frac{\pi}{8} + \frac{\pi}{4}n, \quad \text{where } n \text{ is any integer.} \qquad \blacksquare$$

Example 5   Approximate the solutions of the equation

$$5 \sin \theta \tan \theta - 10 \tan \theta + 3 \sin \theta - 6 = 0$$

in the degree interval $[0°, 360°)$.

Solution   The equation may be factored by grouping terms as follows:

$$5 \tan \theta (\sin \theta - 2) + 3(\sin \theta - 2) = 0$$
$$(5 \tan \theta + 3)(\sin \theta - 2) = 0.$$

Since the equation $\sin \theta = 2$ has no solutions (Why?), the solutions of the given equation are the same as those of

$$\tan \theta = -\tfrac{3}{5} = -0.6000.$$

Approximations to θ may be found by interpolating in Table 2. We first approximate the reference angle θ' as follows:

$$10' \left\{ d \begin{cases} \tan 30°50' \approx 0.5969 \\ \tan \theta' \quad\;\; = 0.6000 \end{cases} 0.0031 \\ \quad\; \tan 31°00' \approx 0.6009 \end{cases} \right\} 0.0040$$

$$\frac{d}{10} = \frac{31}{40} \quad \text{or} \quad d = \frac{31}{4} \approx 8'$$

$$\theta' \approx 30°58'.$$

Since θ lies in quadrant II or quadrant IV, this implies that

$$\theta \approx 180° - 30°58' = 149°2' \quad \text{or} \quad \theta \approx 360° - 30°58' = 329°2'. \qquad \blacksquare$$

A technique for solving trigonometric equations with the aid of a calculator will be discussed later in the chapter, after inverse trigonometric functions have been introduced.

EXERCISES 7.3

In each of Exercises 1–16 find all solutions of the given equation.

1 $2\cos t + 1 = 0$

2 $\cot\theta + 1 = 0$

3 $\tan^2 x = 1$

4 $4\cos\theta - 2 = 0$

5 $(\cos\theta - 1)(\sin\theta + 1) = 0$

6 $2\cos x = \sqrt{3}$

7 $\sec^2\alpha - 4 = 0$

8 $3 - \tan^2\beta = 0$

9 $\sqrt{3} + 2\sin\beta = 0$

10 $4\sin^2 x - 3 = 0$

11 $\cot^2 x - 3 = 0$

12 $(\sin t - 1)\cos t = 0$

13 $(2\sin\theta + 1)(2\cos\theta + 3) = 0$

14 $(2\sin u - 1)(\cos u - \sqrt{2}) = 0$

15 $\sin 2x(\csc 2x - 2) = 0$

16 $\tan\alpha + \tan^2\alpha = 0$

In each of Exercises 17–36 find the solutions of the equation in the interval $[0, 2\pi)$ and also find the degree measure of each solution.

17 $2 - 8\cos^2 t = 0$

18 $\cot^2\theta - \cot\theta = 0$

19 $2\sin^2 u = 1 - \sin u$

20 $2\cos^2 t + 3\cos t + 1 = 0$

21 $\tan^2 x \sin x = \sin x$

22 $\sec\beta\csc\beta = 2\csc\beta$

23 $2\cos^2\gamma + \cos\gamma = 0$

24 $\sin x - \cos x = 0$

25 $\sin^2\theta + \sin\theta - 6 = 0$

26 $2\sin^2 u + \sin u - 6 = 0$

27 $1 - \sin t = \sqrt{3}\cos t$

28 $\cos\theta - \sin\theta = 1$

29 $\cos\alpha + \sin\alpha = 1$

30 $2\tan t - \sec^2 t = 0$

31 $\tan\theta + \sec\theta = 1$

32 $\cot\alpha + \tan\alpha = \csc\alpha\sec\alpha$

33 $2\sin^3 x + \sin^2 x - 2\sin x - 1 = 0$

34 $\sec^5\theta = 4\sec\theta$

35 $2\tan t\csc t + 2\csc t + \tan t + 1 = 0$

36 $2\sin v\csc v - \csc v = 4\sin v - 2$

In each of Exercises 37–40 use Table 2 to approximate, to the nearest multiple of ten minutes, the solutions of the given equation in the interval $[0°, 360°)$.

37 $\sin^2 t - 4\sin t + 1 = 0$

38 $\tan^2\theta + 3\tan\theta + 2 = 0$

39 $12\sin^2 u - 5\sin u - 2 = 0$

40 $5\cos^2\alpha + 3\cos\alpha - 2 = 0$

7.4 THE ADDITION AND SUBTRACTION FORMULAS

In this section we shall derive formulas involving trigonometric functions of $u + v$ or $u - v$, where u and v represent any real numbers or angles. These formulas are known as **addition** or **substraction** formulas, respectively.

Let t_1 and t_2 be any real numbers, and let $P(t_1)$ and $P(t_2)$ be the corresponding points on the unit circle U. We shall denote these points in terms of rectangular coordinates by $P_1(x_1, y_1)$ and $P_2(x_2, y_2)$, respectively. Let us next consider the real number $t_1 - t_2$ and denote the rectangular coordinates of $P(t_1 - t_2)$ by $P_3(x_3, y_3)$. Applying the definition of the trigonometric functions we have

$$(*) \qquad \begin{aligned} \cos t_1 &= x_1, & \cos t_2 &= x_2, & \cos(t_1 - t_2) &= x_3, \\ \sin t_1 &= y_1, & \sin t_2 &= y_2, & \sin(t_1 - t_2) &= y_3, \end{aligned}$$

where the symbol (∗) has been used for later reference to these formulas. Our goal is to obtain a formula for $\cos(t_1 - t_2)$ in terms of functional values of t_1 and t_2. For convenience let us assume that t_1 and t_2 are in the interval $[0, 2\pi]$ and $0 \le t_1 - t_2 \le t_2$. In this case, $t_2 \le t_1$ and hence if A is the point $(1, 0)$ on U, then the length t_1 of $\overset{\frown}{AP_1}$ is greater than or equal to the length t_2 of $\overset{\frown}{AP_2}$. Also the length $t_1 - t_2$ of $\overset{\frown}{AP_3}$ is less than the length t_2 of $\overset{\frown}{AP_2}$. Figure 7.1 illustrates one arrangement of points P_1 and P_2 under these conditions.

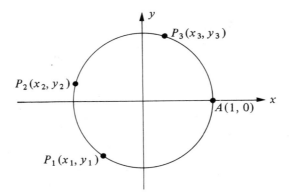

Figure 7.1

It is possible to extend our discussion to cover all values of t. Since the arc lengths of both $\overset{\frown}{P_2P_1}$ and $\overset{\frown}{AP_3}$ equal $t_1 - t_2$, the line segments P_2P_1 and AP_3 also have the same length; that is,

$$d(A, P_3) = d(P_1, P_2).$$

Using the Distance Formula, the last equation may be written

$$\sqrt{(x_3 - 1)^2 + (y_3 - 0)^2} = \sqrt{(x_2 - x_1)^2 + (y_2 - y_1)^2}.$$

Squaring both sides and expanding the terms underneath the radical gives us

$$x_3^2 - 2x_3 + 1 + y_3^2 = x_2^2 - 2x_1x_2 + x_1^2 + y_2^2 - 2y_1y_2 + y_1^2.$$

Since P_1, P_2, and P_3 lie on U, and since an equation for U is $x^2 + y^2 = 1$, we may substitute 1 for each of $x_1^2 + y_1^2$, $x_2^2 + y_2^2$, and $x_3^2 + y_3^2$. Doing this and simplifying, we obtain

$$2 - 2x_3 = 2 - 2x_1x_2 - 2y_1y_2$$

which reduces to

$$x_3 = x_1x_2 + y_1y_2.$$

Substituting from the formulas stated in (∗) at the beginning of this section gives us the following:

$$\cos(t_1 - t_2) = \cos t_1 \cos t_2 + \sin t_1 \sin t_2.$$

In order to make the formula less cumbersome, we shall eliminate subscripts by changing the symbols for the variables from t_1 and t_2 to u and v, respectively. Our identity then takes on the following form:

**Subtraction
Formula for
Cosine**

$$\cos(u - v) = \cos u \cos v + \sin u \sin v.$$

Example 1   Find the exact value of $\cos 15°$.

Solution   If we write $15° = 60° - 45°$, we may use the last formula with $u = 60°$ and $v = 45°$ as follows:

$$\cos 15° = \cos(60° - 45°)$$
$$= \cos 60° \cos 45° + \sin 60° \sin 45°$$
$$= \left(\frac{1}{2}\right)\left(\frac{\sqrt{2}}{2}\right) + \left(\frac{\sqrt{3}}{2}\right)\left(\frac{\sqrt{2}}{2}\right)$$
$$= \frac{\sqrt{2} + \sqrt{6}}{4}.$$ ∎

It is easy to obtain a formula for $\cos(u + v)$. We simply write $u + v = u - (-v)$ and employ the formula derived previously. Thus,

$$\cos(u + v) = \cos[u - (-v)]$$
$$= \cos u \cos(-v) + \sin u \sin(-v).$$

It was shown in Section 6.7 that $\cos(-v) = \cos v$ and $\sin(-v) = -\sin v$ for all v, and hence

**Addition
Formula for
Cosine**

$$\cos(u + v) = \cos u \cos v - \sin u \sin v.$$

Example 2   Find $\cos(7\pi/12)$ by using $\pi/3$ and $\pi/4$.

Solution   Writing $7\pi/12 = (\pi/3) + (\pi/4)$ and using the formula for $\cos(u + v)$ gives
us

$$\cos \frac{7\pi}{12} = \cos\left(\frac{\pi}{3} + \frac{\pi}{4}\right)$$
$$= \cos \frac{\pi}{3} \cos \frac{\pi}{4} - \sin \frac{\pi}{3} \sin \frac{\pi}{4}$$
$$= \frac{1}{2} \frac{\sqrt{2}}{2} - \frac{\sqrt{3}}{2} \frac{\sqrt{2}}{2}$$
$$= \frac{\sqrt{2} - \sqrt{6}}{4}.$$ ∎

Similar identities are true for the sine function. Let us first establish the following identities, which are of some interest in themselves.

$$\cos\left(\frac{\pi}{2} - u\right) = \sin u$$

$$\sin\left(\frac{\pi}{2} - u\right) = \cos u$$

$$\tan\left(\frac{\pi}{2} - u\right) = \cot u$$

The first of these identities may be proved as follows:

$$\cos\left(\frac{\pi}{2} - u\right) = \cos\frac{\pi}{2} \cdot \cos u + \sin\frac{\pi}{2} \cdot \sin u$$

$$= 0 \cdot \cos u + 1 \cdot \sin u$$

$$= \sin u.$$

To obtain the second identity we substitute $(\pi/2) - v$ for u in the first identity, obtaining

$$\cos\left[\frac{\pi}{2} - \left(\frac{\pi}{2} - v\right)\right] = \sin\left(\frac{\pi}{2} - v\right)$$

or equivalently,

$$\cos v = \sin\left(\frac{\pi}{2} - v\right).$$

The third identity may be established as follows:

$$\tan\left(\frac{\pi}{2} - u\right) = \frac{\sin\left(\frac{\pi}{2} - u\right)}{\cos\left(\frac{\pi}{2} - u\right)} = \frac{\cos u}{\sin u} = \cot u.$$

If θ denotes the degree measure of an angle, we may write

$$\sin(90° - \theta) = \cos\theta$$
$$\cos(90° - \theta) = \sin\theta$$
$$\tan(90° - \theta) = \cot\theta$$

If θ is acute, then θ and $90° - \theta$ are **complementary**, since their sum is $90°$. It is customary to refer to the sine and cosine functions as **cofunctions** of one another. Similarly, the tangent and cotangent functions are cofunctions, as are the secant and cosecant. Consequently, the last three formulas constitute a partial description of the fact that *any functional value of θ equals the cofunction of the complementary angle 90° − θ.*

The following identities may now be established.

Addition and Subtraction Formulas for Sine and Tangent

$$\sin(u + v) = \sin u \cos v + \cos u \sin v$$
$$\sin(u - v) = \sin u \cos v - \cos u \sin v$$
$$\tan(u + v) = \frac{\tan u + \tan v}{1 - \tan u \tan v}$$
$$\tan(u - v) = \frac{\tan u - \tan v}{1 + \tan u \tan v}$$

We shall prove the first and third and leave the proofs of the remaining two as exercises. The reader should supply reasons for each of the following steps.

$$\sin(u + v) = \cos\left[\frac{\pi}{2} - (u + v)\right]$$

$$= \cos\left[\left(\frac{\pi}{2} - u\right) - v\right]$$

$$= \cos\left(\frac{\pi}{2} - u\right)\cos v + \sin\left(\frac{\pi}{2} - u\right)\sin v$$

$$= \sin u \cos v + \cos u \sin v.$$

To verify the formula for $\tan(u + v)$ we begin as follows:

$$\tan(u + v) = \frac{\sin(u + v)}{\cos(u + v)}$$

$$= \frac{\sin u \cos v + \cos u \sin v}{\cos u \cos v - \sin u \sin v}.$$

Next, dividing numerator and denominator by $\cos u \cos v$ (assuming, of course, that $\cos u \cos v \neq 0$), we obtain

$$\tan(u + v) = \frac{\left(\dfrac{\sin u}{\cos u}\right)\left(\dfrac{\cos v}{\cos v}\right) + \left(\dfrac{\cos u}{\cos u}\right)\left(\dfrac{\sin v}{\cos v}\right)}{\left(\dfrac{\cos u}{\cos u}\right)\left(\dfrac{\cos v}{\cos v}\right) - \left(\dfrac{\sin u}{\cos u}\right)\left(\dfrac{\sin v}{\cos v}\right)}$$

$$= \frac{\tan u + \tan v}{1 - \tan u \tan v}.$$

Example 3   Given $\sin \alpha = 4/5$, where α is an angle in quadrant I, and $\cos \beta = -12/13$, where β is in quadrant II, find $\sin(\alpha + \beta)$, $\tan(\alpha + \beta)$, and the quadrant containing $\alpha + \beta$.

Solution   It is convenient to represent α and β geometrically, as illustrated in Figure 7.2. There is no loss of generality in picturing α and β as positive angles between 0 and 2π as we have done in the figure. Since $\sin \alpha = 4/5$, the point $(3, 4)$ is on the terminal side of α. (Why?) Similarly, since $\cos \beta = -12/13$, we may choose the point $(-12, 5)$ on the terminal side of β. Referring to Figure 7.2 and using the Theorem on Trigonometric Functions as Ratios (see Section 6.5) we obtain

$$\cos \alpha = \tfrac{3}{5}, \quad \tan \alpha = \tfrac{4}{3}, \quad \sin \beta = \tfrac{5}{13}, \quad \text{and} \quad \tan \beta = -\tfrac{5}{12}.$$

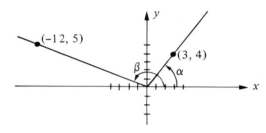

Figure 7.2

Using addition formulas gives us

$$\sin(\alpha + \beta) = \sin \alpha \cos \beta + \cos \alpha \sin \beta$$

$$= \left(\frac{4}{5}\right)\left(-\frac{12}{13}\right) + \left(\frac{3}{5}\right)\left(\frac{5}{13}\right)$$

$$= -\frac{33}{65}.$$

$$\tan(\alpha + \beta) = \frac{\tan \alpha + \tan \beta}{1 - \tan \alpha \tan \beta}$$

$$= \frac{\dfrac{4}{3} + \left(-\dfrac{5}{12}\right)}{1 - \left(\dfrac{4}{3}\right)\left(-\dfrac{5}{12}\right)}$$

$$= \frac{33}{56}.$$

Since $\sin(\alpha + \beta)$ is negative and $\tan(\alpha + \beta)$ is positive, it follows that $\alpha + \beta$ lies in quadrant III.   ∎

The addition formulas are very important in the study of calculus, where the type of simplification illustrated in the next example is required.

Example 4   If $f(x) = \sin x$ and $h \neq 0$, prove that

$$\frac{f(x + h) - f(x)}{h} = \sin x \left(\frac{\cos h - 1}{h} \right) + \cos x \left(\frac{\sin h}{h} \right).$$

Solution   Using the definition of f and the addition formula for the sine function we obtain

$$\frac{f(x + h) - f(x)}{h} = \frac{\sin (x + h) - \sin x}{h}$$

$$= \frac{\sin x \cos h + \cos x \sin h - \sin x}{h}$$

$$= \frac{\sin x (\cos h - 1) + \cos x \sin h}{h}$$

$$= \sin x \left(\frac{\cos h - 1}{h} \right) + \cos x \left(\frac{\sin h}{h} \right). \qquad \blacksquare$$

Example 5   Prove that for every x,

$$a \cos Bx + b \sin Bx = A \cos (Bx - C)$$

where $A = \sqrt{a^2 + b^2}$ and $\tan C = b/a$.

Solution   Using the formula for $\cos (u - v)$ with $u = Bx$ and $v = C$ gives us

$$A \cos (Bx - C) = A(\cos Bx \cos C + \sin Bx \sin C).$$

We shall complete the proof by determining conditions on a and b such that

$$a \cos Bx + b \sin Bx = A \cos Bx \cos C + A \sin Bx \sin C.$$

The last formula is true for every x if and only if

$$a = A \cos C \quad \text{and} \quad b = A \sin C$$

(to verify this let $x = 0$). Consequently,

$$a^2 + b^2 = A^2 \cos^2 C + A^2 \sin^2 C$$
$$= A^2(\cos^2 C + \sin^2 C) = A^2$$

and we may choose $\qquad A = \sqrt{a^2 + b^2}.$

Finally, we note that

$$\frac{b}{a} = \frac{A \sin C}{A \cos C} = \frac{\sin C}{\cos C} = \tan C.$$

This establishes the identity. $\qquad \blacksquare$

Example 6 The graph of $f(x) = \cos x + \sin x$ was sketched in Section 6.9 by adding ordinates (see Figure 6.44). Use the formula derived in Example 5 to find the amplitude, period, and phase shift of f.

Solution Letting $a = 1$, $b = 1$, and $B = 1$ in Example 5 we obtain

$$A = \sqrt{a^2 + b^2} = \sqrt{1 + 1} = \sqrt{2} \quad \text{and} \quad \tan C = b/a = 1/1 = 1.$$

Since $\tan C = 1$ we may choose $C = \pi/4$. Substitution in the formula gives us

$$\cos x + \sin x = \sqrt{2} \cos\left(x - \frac{\pi}{4}\right).$$

It follows from our work in Section 6.8 that the amplitude is $\sqrt{2}$, the period is 2π, and the phase shift is $\pi/4$. The reader may check these facts by referring to Figure 6.44. ∎

EXERCISES 7.4

3-42

Write the expressions in Exercises 1–4 in terms of cofunctions of complementary angles.

1 (a) $\sin 46°37'$ (b) $\cos 73°12'$
    (c) $\tan \pi/6$

2 (a) $\tan 24°12'$ (b) $\sin 89°41'$
    (c) $\cos \pi/3$

3 (a) $\cos 7\pi/20$ (b) $\sin(1/4)$
    (c) $\tan 1$

4 (a) $\sin \pi/12$ (b) $\cos(0.64)$
    (c) $\tan \sqrt{2}$

Find the exact functional values in Exercises 5–10.

5 (a) $\cos \pi/4 + \cos \pi/3$ (b) $\cos 7\pi/12$

6 (a) $\sin 2\pi/3 + \sin \pi/4$ (b) $\sin 11\pi/12$

7 (a) $\tan 60° + \tan 225°$ (b) $\tan 285°$

8 (a) $\cos 135° - \cos 60°$ (b) $\cos 75°$

9 (a) $\sin 3\pi/4 - \sin \pi/6$ (b) $\sin 7\pi/12$

10 (a) $\tan 3\pi/4 - \tan \pi/6$ (b) $\tan 7\pi/12$

Express each of Exercises 11–16 in terms of one function of one angle.

11 $\cos 48° \cos 23° + \sin 48° \sin 23°$

12 $\cos 13° \cos 50° - \sin 13° \sin 50°$

13 $\cos 10° \sin 5° - \sin 10° \cos 5°$

14 $\sin 57° \cos 4° + \cos 57° \sin 4°$

15 $\cos 3 \sin(-2) - \cos 2 \sin 3$

16 $\sin(-5)\cos 2 + \cos 5 \sin(-2)$

17 If α and β are acute angles such that $\cos \alpha = 4/5$ and $\tan \beta = 8/15$, find $\cos(\alpha + \beta)$, $\sin(\alpha + \beta)$, and the quadrant containing $\alpha + \beta$.

18 If $\sin \alpha = -4/5$ and $\sec \beta = 5/3$, where α is a third-quadrant angle and β is a first-quadrant angle, find $\sin(\alpha + \beta)$, $\tan(\alpha + \beta)$, and the quadrant containing $\alpha + \beta$.

19 If $\tan \alpha = -7/24$ and $\cot \beta = 3/4$, where α is in the second quadrant and β is in the third quadrant, find $\sin(\alpha + \beta)$, $\cos(\alpha + \beta)$, $\tan(\alpha + \beta)$, $\sin(\alpha - \beta)$, $\cos(\alpha - \beta)$, and $\tan(\alpha - \beta)$.

20 If $P(t_1)$ and $P(t_2)$ are in quadrant III, and if $\cos t_1 = -2/5$ and $\cos t_2 = -3/5$, find $\sin(t_1 - t_2)$, $\cos(t_1 - t_2)$, and the quadrant containing $P(t_1 - t_2)$.

Verify the identities in Exercises 21–40.

21 $\sin\left(x + \dfrac{\pi}{2}\right) = \cos x$

22 $\cos\left(x + \dfrac{\pi}{2}\right) = -\sin x$

23 $\cos\left(\theta + \dfrac{3\pi}{2}\right) = \sin\theta$

24 $\sin\left(\alpha - \dfrac{3\pi}{2}\right) = \cos\alpha$

25 $\sin\left(\theta + \dfrac{\pi}{4}\right) = \left(\dfrac{\sqrt{2}}{2}\right)(\sin\theta + \cos\theta)$

26 $\cos\left(\theta + \dfrac{\pi}{4}\right) = \left(\dfrac{\sqrt{2}}{2}\right)(\cos\theta - \sin\theta)$

27 $\tan\left(u + \dfrac{\pi}{4}\right) = \dfrac{1 + \tan u}{1 - \tan u}$

28 $\tan\left(x - \dfrac{\pi}{4}\right) = \dfrac{\tan x - 1}{\tan x + 1}$

29 $\tan\left(u + \dfrac{\pi}{2}\right) = -\cot u$

30 $\cot\left(t - \dfrac{\pi}{3}\right) = \dfrac{\sqrt{3}\tan t + 1}{\tan t - \sqrt{3}}$

31 $\sin(u + v) \cdot \sin(u - v) = \sin^2 u - \sin^2 v$

32 $\cos(u + v) \cdot \cos(u - v) = \cos^2 u - \sin^2 v$

33 $\cos(u + v) + \cos(u - v) = 2\cos u\cos v$

34 $\sin(u + v) + \sin(u - v) = 2\sin u\cos v$

35 $\sin 2u = 2\sin u\cos u$ (*Hint*: $2u = u + u$)

36 $\cos 2u = \cos^2 u - \sin^2 u$

37 $\dfrac{\sin(u + v)}{\cos(u - v)} = \dfrac{\tan u + \tan v}{1 + \tan u\tan v}$

38 $\dfrac{\cos(u + v)}{\cos(u - v)} = \dfrac{1 - \tan u\tan v}{1 + \tan u\tan v}$

39 $\dfrac{\sin(u + v)}{\sin(u - v)} = \dfrac{\tan u + \tan v}{\tan u - \tan v}$

40 $\tan u + \tan v = \dfrac{\sin(u + v)}{\cos u\cos v}$

41 Express $\sin(u + v + w)$ in terms of functions of u, v, and w. (*Hint*: Write $\sin(u + v + w) = \sin[(u + v) + w]$ and use an addition formula.)

42 Express $\tan(u + v + w)$ in terms of functions of u, v, and w.

43 Derive the formula
$$\cot(u + v) = \dfrac{\cot u\cot v - 1}{\cot u + \cot v}.$$

44 If α and β are complementary angles, prove that
$$\sin^2\alpha + \sin^2\beta = 1.$$

45 Derive the subtraction formulas for the sine and tangent functions

46 Prove each of the following:
(a) $\sec(\pi/2 - u) = \csc u$
(b) $\csc(\pi/2 - u) = \sec u$
(c) $\cot(\pi/2 - u) = \tan u$

47 If $f(x) = \cos x$, prove that
$$\dfrac{f(x + h) - f(x)}{h}$$
$$= \cos x\left(\dfrac{\cos h - 1}{h}\right) - \sin x\left(\dfrac{\sin h}{h}\right).$$

48 If $f(x) = \tan x$, prove that
$$\dfrac{f(x + h) - f(x)}{h}$$
$$= \sec^2 x\left(\dfrac{\sin h}{h}\right)\dfrac{1}{\cos h - \sin h\tan x}.$$

In Exercises 49 and 50 find the solutions of the given equation which are in the interval $[0, 2\pi)$ and also find the degree measure of each solution.

49 $\sin 4t\cos t = \sin t\cos 4t$

50 $\cos 5t\cos 3t = 2^{-1} + \sin(-5t)\sin 3t$

In each of Exercises 51–54 use the formula given in Example 5 to express f in terms of the cosine function and determine the amplitude, period, and phase shift.

51 $f(x) = \sqrt{3}\cos 2x + \sin 2x$

52 $f(x) = \cos 4x + \sqrt{3}\sin 4x$

53 $f(x) = 2\cos 3x - 2\sin 3x$

54 $f(x) = -5\cos 10x + 5\sin 10x$

CALCULATOR EXERCISES 7.4

1 Demonstrate, by direct calculations, that the addition formulas for the sine, cosine, and tangent functions are true for the following values of u and v: $u = 1.46$, $v = 8.27$.

2 If u and v are in the interval $[0, \pi]$ such that $\cos u = -0.4163$ and $\sin v = 0.7216$, find $\cos (u + v)$ and $\sin (u + v)$.

7.5 MULTIPLE-ANGLE FORMULAS

In the preceding section we were interested primarily in trigonometric identities which involve values of $u \pm v$. In this section we shall obtain formulas for values of nu, where n represents a certain integer or rational number. The formulas are referred to as **multiple-angle formulas**. In particular, the following identities are called the **double-angle formulas**, because of the expression $2u$ which appears.

Double-Angle Formulas

$$\sin 2u = 2 \sin u \cos u$$
$$\cos 2u = \cos^2 u - \sin^2 u = 1 - 2 \sin^2 u = 2 \cos^2 u - 1$$
$$\tan 2u = \frac{2 \tan u}{1 - \tan^2 u}$$

These identities may be proved by letting $u = v$ in the appropriate addition formulas. If we use the formula for $\sin (u + v)$, then

$$\sin 2u = \sin (u + u)$$
$$= \sin u \cos u + \cos u \sin u$$
$$= 2 \sin u \cos u.$$

Similarly, using the formula for $\cos (u + v)$ gives us

$$\cos 2u = \cos (u + u)$$
$$= \cos u \cos u - \sin u \sin u$$
$$= \cos^2 u - \sin^2 u.$$

The other two forms for $\cos 2u$ may be obtained by using the fundamental identity $\sin^2 u + \cos^2 u = 1$. Thus,

$$\cos 2u = \cos^2 u - \sin^2 u$$
$$= (1 - \sin^2 u) - \sin^2 u$$
$$= 1 - 2 \sin^2 u.$$

Similarly, if we substitute for $\sin^2 u$ instead of $\cos^2 u$, we obtain

$$\cos 2u = \cos^2 u - (1 - \cos^2 u)$$
$$= 2 \cos^2 u - 1.$$

The formula for $\tan 2u$ may be obtained by taking $u = v$ in the formula for $\tan (u + v)$.

Example 1   Find $\sin 2\alpha$ and $\cos 2\alpha$ if $\sin \alpha = \frac{4}{5}$ and α is in quadrant I.

Solution   As in Example 3 of the preceding section, $\cos \alpha = 3/5$. Substitution in double-angle formulas gives us

$$\sin 2\alpha = 2 \sin \alpha \cos \alpha = 2\left(\frac{4}{5}\right)\left(\frac{3}{5}\right) = \frac{24}{25}$$

$$\cos 2\alpha = \cos^2 \alpha - \sin^2 \alpha = \frac{9}{25} - \frac{16}{25} = -\frac{7}{25}.$$ ∎

Example 2   Express $\cos 3\theta$ in terms of $\cos \theta$.

Solution

$$\begin{aligned}
\cos 3\theta &= \cos (2\theta + \theta) \\
&= \cos 2\theta \cos \theta - \sin 2\theta \sin \theta \\
&= (2\cos^2 \theta - 1)\cos \theta - (2 \sin \theta \cos \theta) \sin \theta \\
&= 2\cos^3 \theta - \cos \theta - 2\cos \theta \sin^2 \theta \\
&= 2\cos^3 \theta - \cos \theta - 2\cos \theta (1 - \cos^2 \theta) \\
&= 2\cos^3 \theta - \cos \theta - 2\cos \theta + 2\cos^3 \theta \\
&= 4\cos^3 \theta - 3\cos \theta.
\end{aligned}$$ ∎

The next three identities are useful for simplifying certain expressions involving powers of trigonometric functions.

$$\sin^2 u = \frac{1 - \cos 2u}{2}; \qquad \cos^2 u = \frac{1 + \cos 2u}{2}; \qquad \tan^2 u = \frac{1 - \cos 2u}{1 + \cos 2u}$$

The first and second of these identities may be verified by solving the equations

$$\cos 2u = 1 - 2 \sin^2 u \quad \text{and} \quad \cos 2u = 2\cos^2 u - 1$$

for $\sin^2 u$ and $\cos^2 u$, respectively. The third identity may be obtained from the first two by using the fact that $\tan^2 u = \sin^2 u / \cos^2 u$.

Example 3   Verify the identity $\sin^2 x \cos^2 x = \frac{1}{8}(1 - \cos 4x)$.

Solution   Using the preceding identities for $\sin^2 u$ and $\cos^2 u$ with $u = x$, we obtain

$$\sin^2 x \cos^2 x = \left(\frac{1 - \cos 2x}{2}\right)\left(\frac{1 + \cos 2x}{2}\right)$$

$$= \frac{1}{4}(1 - \cos^2 2x)$$

$$= \frac{1}{4}\sin^2 2x.$$

Finally, using the formula $\sin^2 u = (1 - \cos 2u)/2$ with $u = 2x$, gives us

$$\sin^2 x \cos^2 x = \frac{1}{4}\left(\frac{1 - \cos 4x}{2}\right)$$

$$= \frac{1}{8}(1 - \cos 4x).$$

Another method of proof is to use the fact that $\sin 2x = 2 \sin x \cos x$ and hence that

$$\sin x \cos x = \frac{1}{2}\sin 2x.$$

Squaring both sides, we have

$$\sin^2 x \cos^2 x = \frac{1}{4}\sin^2 2x.$$

The remainder of the solution is the same as that given above.   ■

Example 4   Express $\cos^4 t$ in terms of values of cos with exponent 1.

Solution   Supply reasons for the following steps:

$$\cos^4 t = (\cos^2 t)^2$$

$$= \left(\frac{1 + \cos 2t}{2}\right)^2$$

$$= \frac{1}{4}(1 + 2\cos 2t + \cos^2 2t)$$

$$= \frac{1}{4}\left(1 + 2\cos 2t + \frac{1 + \cos 4t}{2}\right)$$

$$= \frac{3}{8} + \frac{1}{2}\cos 2t + \frac{1}{8}\cos 4t.$$ ■

Substituting $v/2$ for u in the three formulas for $\sin^2 u$, $\cos^2 u$, and $\tan^2 u$ gives us

$$\sin^2 \frac{v}{2} = \frac{1 - \cos v}{2}$$

$$\cos^2 \frac{v}{2} = \frac{1 + \cos v}{2}$$

$$\tan^2 \frac{v}{2} = \frac{1 - \cos v}{1 + \cos v}.$$

If we take the square root of both sides of the latter equations and use the fact that $\sqrt{a^2} = |a|$ for every real number a, the following identities result. They are called *half-angle formulas* because of the expression $v/2$ which appears.

Half-Angle Formulas

$$\left| \sin \frac{v}{2} \right| = \sqrt{\frac{1 - \cos v}{2}}$$

$$\left| \cos \frac{v}{2} \right| = \sqrt{\frac{1 + \cos v}{2}}$$

$$\left| \tan \frac{v}{2} \right| = \sqrt{\frac{1 - \cos v}{1 + \cos v}}$$

The absolute value signs may be eliminated if more information is known about $v/2$. For example, if the point $P(v/2)$ on the unit circle U is in either quadrant I or II (or, equivalently, if the angle determined by $v/2$ lies in one of these quadrants), then $\sin v/2$ is positive and we may write

$$\sin \frac{v}{2} = \sqrt{\frac{1 - \cos v}{2}}.$$

However, if $v/2$ leads to a point (or angle) in either quadrant III or IV, then

$$\sin \frac{v}{2} = -\sqrt{\frac{1 - \cos v}{2}}.$$

Similar remarks are true for the other formulas.

An alternate form for $\tan v/2$ can be obtained. Multiplying numerator and denominator of the radicand in the third half-angle formula by $1 - \cos v$ gives us

$$\left| \tan \frac{v}{2} \right| = \sqrt{\frac{1 - \cos v}{1 + \cos v} \cdot \frac{1 - \cos v}{1 - \cos v}}$$

$$= \sqrt{\frac{(1 - \cos v)^2}{\sin^2 v}}$$

$$= \frac{1 - \cos v}{|\sin v|}.$$

The absolute value sign is unnecessary in the numerator since $1 - \cos v$ is never negative. It can be shown that $\tan v/2$ and $\sin v$ always have the same sign. For example, if $0 < v < \pi$, then $0 < v/2 < \pi/2$, and hence both $\sin v$ and $\tan v/2$ are positive. If $\pi < v < 2\pi$, then $\pi/2 < v/2 < \pi$, and hence both $\sin v$ and $\tan v/2$ are negative. It is possible to generalize these remarks to all values of v for which the expressions $\tan v/2$ and $(1 - \cos v)/|\sin v|$ have meaning. This gives us the first of the next two identities. The second identity for $\tan(v/2)$ may be obtained by multiplying numerator and denominator of the radicand in the third half-angle formula by $1 + \cos v$.

Half-Angle Formulas for Tangent

$$\tan\frac{v}{2} = \frac{1 - \cos v}{\sin v}, \qquad \tan\frac{v}{2} = \frac{\sin v}{1 + \cos v}$$

Example 5   Find the exact values of $\sin 22.5°$ and $\cos 22.5°$.

Solution   Using the formula for $\sin(v/2)$ and the fact that $22.5°$ is in quadrant I, we have

$$\sin 22.5° = \sin\frac{45°}{2} = \sqrt{\frac{1 - \cos 45°}{2}}$$

$$= \sqrt{\frac{1 - \sqrt{2}/2}{2}}$$

$$= \frac{\sqrt{2 - \sqrt{2}}}{2}$$

$$\cos 22.5° = \sqrt{\frac{1 + \cos 45°}{2}}$$

$$= \sqrt{\frac{1 + \sqrt{2}/2}{2}}$$

$$= \frac{\sqrt{2 + \sqrt{2}}}{2}.$$ ∎

Example 6   If $\tan\alpha = -4/3$, where α is in quadrant IV, find $\tan\alpha/2$.

Solution   If we choose the point $(3, -4)$ on the terminal side of α as illustrated in Figure 7.3, then $\sin\alpha = -4/5$ and $\cos\alpha = 3/5$. (Why?) Applying a half-angle formula, we have

$$\tan\frac{\alpha}{2} = \frac{1 - \cos\alpha}{\sin\alpha} = \frac{1 - \frac{3}{5}}{-\frac{4}{5}} = -\frac{1}{2}.$$

Figure 7.3   ∎

Example 7 Find the solutions of the equation $\cos 2x + \cos x = 0$ which are in the interval $[0, 2\pi)$. Express the solutions both in radian measure and degree measure.

Solution We first use a double-angle formula to write the equation in terms of $\cos x$ and then solve by factoring as follows:

$$\cos 2x + \cos x = 0$$
$$(2\cos^2 x - 1) + \cos x = 0$$
$$2\cos^2 x + \cos x - 1 = 0$$
$$(2\cos x - 1)(\cos x + 1) = 0$$

Setting each factor equal to zero we obtain

$$\cos x = \frac{1}{2} \quad \text{and} \quad \cos x = -1.$$

The solutions of the last two equations (and hence of the given equation) in $[0, 2\pi)$ are

$$\pi/3, \quad 5\pi/3, \quad \text{and} \quad \pi$$

The corresponding degree measures are

$$60°, \quad 300°, \quad \text{and} \quad 180° \qquad \blacksquare$$

EXERCISES 7.5

In Exercises 1–4 find the exact values of $\sin 2\theta$, $\cos 2\theta$, and $\tan 2\theta$ subject to the given conditions.

1 $\cos\theta = \frac{3}{5}$ and θ acute

2 $\cot\theta = \frac{4}{3}$ and $180° < \theta < 270°$

3 $\sec\theta = -3$ and $90° < \theta < 180°$

4 $\sin\theta = -\frac{4}{5}$ and $270° < \theta < 360°$

In Exercises 5–8 find the exact values of $\sin\theta/2$, $\cos\theta/2$, and $\tan\theta/2$ subject to the given conditions.

5 $\sec\theta = \frac{5}{4}$ and θ acute

6 $\csc\theta = -\frac{5}{3}$ and $-90° < \theta < 0°$

7 $\tan\theta = 1$ and $-180° < \theta < -90°$

8 $\sec\theta = -4$ and $180° < \theta < 270°$

In Exercises 9 and 10, use half-angle formulas to find exact values for the given numbers.

9 (a) $\cos 67°30'$ (b) $\sin 15°$
    (c) $\tan 3\pi/8$

10 (a) $\cos 165°$ (b) $\sin 157°30'$
    (c) $\tan \pi/8$

Verify the identities stated in Exercises 11–28.

11 $\sin 10\theta = 2\sin 5\theta \cos 5\theta$

12 $\cos^2 3x - \sin^2 3x = \cos 6x$

13 $4\sin\dfrac{x}{2}\cos\dfrac{x}{2} = 2\sin x$

14 $\dfrac{\sin^2 2\alpha}{\sin^2 \alpha} = 4 - 4\sin^2 \alpha$

15 $(\sin t + \cos t)^2 = 1 + \sin 2t$

16 $\csc 2u = \frac{1}{2}\sec u \csc u$

17 $\sin 3u = \sin u(3 - 4\sin^2 u)$

18 $\sin 4t = 4\cos t \sin t(1 - 2\sin^2 t)$

19 $\cos 4\theta = 8\cos^4\theta - 8\cos^2\theta + 1$

20 $\cos 6t = 32\cos^6 t - 48\cos^4 t + 18\cos^2 t - 1$

21 $\sin^4 t = \frac{3}{8} - \frac{1}{2}\cos 2t + \frac{1}{8}\cos 4t$

22 $\cos^4 x - \sin^4 x = \cos 2x$

23 $\sec 2\theta = \dfrac{\sec^2\theta}{2 - \sec^2\theta}$

24 $\cot 2u = \dfrac{\cot^2 u - 1}{2\cot u}$

25 $2\sin^2 2t + \cos 4t = 1$

26 $\tan\theta + \cot\theta = 2\csc 2\theta$

27 $\tan 3u = \dfrac{(3 - \tan^2 u)\tan u}{1 - 3\tan^2 u}$

28 $\dfrac{1 + \sin 2v + \cos 2v}{1 + \sin 2v - \cos 2v} = \cot v$

Write the expressions in Exercises 29 and 30 in terms of values of cos with exponent 1.

29 $\cos^4(\theta/2)$ **30** $\sin^4 2x$

In Exercises 31–38 find all solutions of the given equations in the interval $[0, 2\pi)$. Express the solutions both in radian measure and degree measure.

31 $\sin 2t + \sin t = 0$ **32** $\cos t - \sin 2t = 0$

33 $\cos u + \cos 2u = 0$ **34** $\cos 2\theta - \tan\theta = 1$

35 $\tan 2x = \tan x$ **36** $\tan 2t - 2\cos t = 0$

37 $\sin\frac{1}{2}u + \cos u = 1$

38 $2 - \cos^2 x = 4\sin^2\frac{1}{2}x$

39 If $a > 0$, $b > 0$, and $0 < u < \pi/2$, prove that

$$a\sin u + b\cos u = \sqrt{a^2 + b^2}\,\sin(u + v)$$

where $0 < v < \pi/2$, $\sin v = b/\sqrt{a^2 + b^2}$, and $\cos v = a/\sqrt{a^2 + b^2}$.

40 Use Exercise 39 to express $8\sin u + 15\cos u$ in the form $c\sin(u + v)$.

CALCULATOR EXERCISES 7.5

1 Approximate $\sin 2\theta$ if θ is a fourth-quadrant angle and $\sin\theta = -0.4637$.

2 Approximate $\cos(\theta/2)$ if θ is a third-quadrant angle and $\tan\theta = 3.8105$.

In each of Exercises 3–6 demonstrate, by direct calculations, that the indicated formula is true for the stated value of x.

3 $\sin 2x = 2\sin x\cos x$; $x = 3.624$

4 $\cos 2x = \cos^2 x - \sin^2 x$; $x = \sqrt{2\pi} + e^{\sqrt{7}}$

5 $\tan 2x = 2\tan x/(1 - \tan^2 x)$; $x = \log(13.6)^5$

6 $|\sin(x/2)| = \sqrt{(1 - \cos x)/2}$; $x = \tan 0.1634$

7.6   PRODUCT AND FACTORING FORMULAS

The following identities may be used to change the form of certain trigonometric expressions.

Product
Formulas

$$\sin(u + v) + \sin(u - v) = 2\sin u \cos v$$
$$\sin(u + v) - \sin(u - v) = 2\cos u \sin v$$
$$\cos(u + v) + \cos(u - v) = 2\cos u \cos v$$
$$\cos(u - v) - \cos(u + v) = 2\sin u \sin v$$

Each of these identities is a consequence of our work in Section 7.4. For example, to verify the first product formula, we merely add the left- and right-hand sides of the identities we obtained for $\sin(u + v)$ and $\sin(u - v)$. The remaining product formulas are obtained in similar fashion.

By starting with the expression on the *right* side of each product formula, we can express certain products as sums, as illustrated in the next example.

Example 1 Express each of the following as a sum.

(a) $\sin 4\theta \cos 3\theta$ (b) $\sin 3x \sin x$

Solution

(a) Using the first product formula with $u = 4\theta$ and $v = 3\theta$ gives us

$$2\sin 4\theta \cos 3\theta = \sin(4\theta + 3\theta) + \sin(4\theta - 3\theta)$$

or

$$\sin 4\theta \cos 3\theta = \tfrac{1}{2}\sin 7\theta + \tfrac{1}{2}\sin\theta.$$

This relationship could also have been obtained by using the second product formula.

(b) Using the fourth product formula with $u = 3x$ and $v = x$ gives us

$$2\sin 3x \sin x = \cos(3x - x) - \cos(3x + x)$$

or

$$\sin 3x \sin x = \tfrac{1}{2}\cos 2x - \tfrac{1}{2}\cos 4x. \qquad\blacksquare$$

The Product Formulas may also be employed to express a sum as a product. In order to obtain a form which can be applied more easily, we shall change the notation as follows. If we let

$$u + v = a \quad \text{and} \quad u - v = b$$

then $(u + v) + (u - v) = a + b$, which simplifies to

$$u = \frac{a + b}{2}.$$

Similarly, from $(u + v) - (u - v) = a - b$ we obtain

$$v = \frac{a - b}{2}.$$

If we now substitute for $u + v$ and $u - v$ on the left sides of the Product Formulas, and for u and v on the right sides, we obtain the following.

**Factoring
Formulas**

$$\sin a + \sin b = 2 \sin \frac{a+b}{2} \cos \frac{a-b}{2}$$

$$\sin a - \sin b = 2 \cos \frac{a+b}{2} \sin \frac{a-b}{2}$$

$$\cos a + \cos b = 2 \cos \frac{a+b}{2} \cos \frac{a-b}{2}$$

$$\cos b - \cos a = 2 \sin \frac{a+b}{2} \sin \frac{a-b}{2}$$

Example 2 Express $\sin 5x - \sin 3x$ as a product.

Solution Using the second factoring formula with $a = 5x$ and $b = 3x$,

$$\sin 5x - \sin 3x = 2 \cos \frac{5x+3x}{2} \sin \frac{5x-3x}{2}$$

$$= 2 \cos 4x \sin x. \qquad \blacksquare$$

Example 3 Verify the identity $\dfrac{\sin 3t + \sin 5t}{\cos 3t - \cos 5t} = \cot t$.

Solution Using the first and fourth factoring formulas,

$$\frac{\sin 3t + \sin 5t}{\cos 3t - \cos 5t} = \frac{2 \sin \dfrac{3t+5t}{2} \cos \dfrac{3t-5t}{2}}{2 \sin \dfrac{5t+3t}{2} \sin \dfrac{5t-3t}{2}}$$

$$= \frac{2 \sin 4t \cos(-t)}{2 \sin 4t \sin t}$$

$$= \frac{\cos(-t)}{\sin t}$$

$$= \frac{\cos t}{\sin t}$$

$$= \cot t \qquad \blacksquare$$

Example 4 Find the solutions of the equation $\cos t - \sin 2t - \cos 3t = 0$.

Solution Using the fourth factoring formula,

$$\cos t - \cos 3t = 2 \sin \frac{3t+t}{2} \sin \frac{3t-t}{2}$$

$$= 2 \sin 2t \sin t.$$

Hence, the given equation is equivalent to

$$2 \sin 2t \sin t - \sin 2t = 0$$

or $$\sin 2t (2 \sin t - 1) = 0.$$

Thus, the solutions of the given equation are the solutions of

$$\sin 2t = 0 \quad \text{and} \quad \sin t = \frac{1}{2}.$$

The first of these equations has the solution

$$2t = n\pi, \quad \text{or} \quad t = \frac{\pi}{2} n$$

where n is any integer. The second equation has solutions

$$t = \frac{\pi}{6} + 2n\pi \quad \text{and} \quad t = \frac{5\pi}{6} + 2n\pi$$

where n is any integer. ∎

The addition formulas may also be employed to derive the so-called **reduction formulas**. The latter formulas can be used to write expressions such as

$$\sin\left(\theta + n \cdot \frac{\pi}{2}\right) \quad \text{and} \quad \cos\left(\theta + n \cdot \frac{\pi}{2}\right)$$

where n is any integer, in terms of only $\sin \theta$ or $\cos \theta$. Similar formulas are true for the other trigonometric functions. Instead of deriving general reduction formulas we shall illustrate several special cases in the next example.

Example 5 Express $\sin (\theta - 3\pi/2)$ and $\cos (\theta + \pi)$ in terms of a function of θ.

Solution We may proceed as follows.

$$\sin\left(\theta - \frac{3\pi}{2}\right) = \sin \theta \cos \frac{3\pi}{2} - \cos \theta \sin \frac{3\pi}{2}$$

$$= \sin \theta \cdot (0) - \cos \theta \cdot (-1)$$

$$= \cos \theta$$

$$\cos (\theta + \pi) = \cos \theta \cos \pi - \sin \theta \sin \pi$$

$$= \cos \theta \cdot (-1) - \sin \theta \cdot (0)$$

$$= -\cos \theta. \quad ∎$$

We have derived many important identities in this chapter. In order to make then readily available for reference, they are listed on the inside of the covers of the text.

EXERCISES 7.6

In each of Exercises 1–8 express the product as a sum or difference.

1 $2 \sin 9\theta \cos 3\theta$

2 $2 \cos 5\theta \sin 5\theta$

3 $\sin 7t \sin 3t$

4 $\sin(-4x)\cos 8x$

5 $\cos 6u \cos(-4u)$

6 $\sin 4t \sin 6t$

7 $3 \cos x \sin 2x$

8 $5 \cos u \sin 5u$

In each of Exercises 9–16 write the expression as a product.

9 $\sin 6\theta + \sin 2\theta$

10 $\sin 4\theta - \sin 8\theta$

11 $\cos 5x - \cos 3x$

12 $\cos 5t + \cos 6t$

13 $\sin 3t - \sin 7t$

14 $\cos \theta - \cos 5\theta$

15 $\cos x + \cos 2x$

16 $\sin 8t + \sin 2t$

Verify the identities in Exercises 17–24.

17 $\dfrac{\sin 4t + \sin 6t}{\cos 4t - \cos 6t} = \cot t$

18 $\dfrac{\sin \theta + \sin 3\theta}{\cos \theta + \cos 3\theta} = \tan 2\theta$

19 $\dfrac{\sin u + \sin v}{\cos u + \cos v} = \tan \dfrac{u+v}{2}$

20 $\dfrac{\sin u - \sin v}{\cos u - \cos v} = -\cot \dfrac{u+v}{2}$

21 $\dfrac{\sin u - \sin v}{\sin u + \sin v} = \dfrac{\tan \frac{1}{2}(u-v)}{\tan \frac{1}{2}(u+v)}$

22 $\dfrac{\cos u - \cos v}{\cos u + \cos v} = \tan \frac{1}{2}(u+v)\tan \frac{1}{2}(u-v)$

23 $\sin 2x + \sin 4x + \sin 6x = 4\cos x \cos 2x \sin 3x$

24 $\dfrac{\cos t + \cos 4t + \cos 7t}{\sin t + \sin 4t + \sin 7t} = \cot 4t$

25 Express $(\sin ax)(\cos bx)$ as a sum.

26 Express $(\cos mu)(\cos nu)$ as a sum.

In Exercises 27–30 find the solutions of the given equations.

27 $\sin 5t + \sin 3t = 0$

28 $\sin t + \sin 3t = \sin 2t$

29 $\cos x = \cos 3x$

30 $\cos 4x - \cos 3x = 0$

In Exercises 31–36 verify the given reduction formulas by using addition formulas.

31 $\sin(\theta + \pi) = -\sin \theta$

32 $\sin\left(\theta + \dfrac{3\pi}{2}\right) = -\cos \theta$

33 $\cos\left(\theta - \dfrac{5\pi}{2}\right) = \sin \theta$

34 $\cos(\theta - 3\pi) = -\cos \theta$

35 $\tan(\pi - \theta) = -\tan \theta$

$\left(Hint: \tan(\pi - \theta) = \dfrac{\sin(\pi - \theta)}{\cos(\pi - \theta)}\right)$

36 $\tan(\theta + \pi/2) = -\cot \theta$

CALCULATOR EXERCISES 7.6

1 Demonstrate, by direct calculations, that the first product formula is true if $u = 4.631$ and $v = -8.059$.

2 Demonstrate that the first factoring formula is true if $a = 10^{\sqrt{148}}$ and $b = \log 735$.

7.7   THE INVERSE TRIGONOMETRIC FUNCTIONS

The concept of inverse function was defined in Section 3.6. Recall that if f is a one-to-one function with domain A and range B, then, as illustrated in (i) of Figure 7.4, for each number u in B there is exactly one number v in A such that $f(v) = u$. In this case we define the function f^{-1} from B to A by letting $f^{-1}(u) = v$ (see (ii) of Figure 7.4).

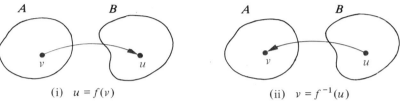

(i) $u = f(v)$ (ii) $v = f^{-1}(u)$

Figure 7.4

We call f^{-1} the **inverse function** of f. The function f^{-1} *reverses* the correspondence from A to B given by f in the sense that

$$v = f^{-1}(u) \quad \text{if and only if} \quad f(v) = u$$

for every u in B and v in A. Since the notation for the variables is immaterial, we shall use the traditional symbols x and y, and write

(I)
$$\boxed{y = f^{-1}(x) \quad \text{if and only if} \quad f(y) = x}$$

for every x in the domain of f^{-1} and every y in the domain of f.

In the remainder of this section we shall use the relationship in (I) to obtain the *inverse trigonometric functions*. As pointed out in Section 3.6, the inverse function f^{-1} is characterized by the identities

(II)
$$\boxed{\begin{aligned} &f^{-1}(f(y)) = y \text{ for every } y \text{ in the domain of } f \\ &f(f^{-1}(x)) = x \text{ for every } x \text{ in the domain of } f^{-1}. \end{aligned}}$$

Since the trigonometric functions are not one-to-one, they do not have inverse functions. However, by restricting the domains, it is possible to obtain functions which behave in the same way as the trigonometric functions (over the smaller domains) and which do possess inverse functions. Given a trigonometric function we shall choose a subset S of the domain throughout which the function is either increasing or decreasing, since this will guarantee the existance of an inverse. Of course, we must also choose S in such a way that the function takes on all of its values.

Let us first consider the sine function, whose domain is $\mathbb{R}$ and whose range is the set of real numbers in the interval $[-1, 1]$. The sine function is not one-to-one since, for example, numbers such as $\pi/6$, $5\pi/6$, and $-7\pi/6$ give rise to the same functional value, $\frac{1}{2}$. If we restrict y to the interval $[-\pi/2, \pi/2]$, then as y varies from $-\pi/2$ to $\pi/2$, $\sin y$ takes on each value between -1 and 1 once and only once. Hence the new function obtained by restricting the domain to $[-\pi/2, \pi/2]$ has an inverse function. Applying the relationship in (I) leads to the following definition.

Definition

> The **inverse sine function**, denoted by $\sin^{-1}$, is defined by
>
> $$y = \sin^{-1} x \quad \text{if and only if} \quad \sin y = x$$
>
> where $-1 \le x \le 1$ and $-\pi/2 \le y \le \pi/2$.

It is also customary to refer to $\sin^{-1}$ as the **arcsine function** and to use the notation $\arcsin x$ in place of $\sin^{-1} x$. The expression $\arcsin x$ is used because if $y = \arcsin x$, then $\sin y = x$ that is, y is a number (or an *arc*length) whose sine is x. Since both notations $\sin^{-1}$ and $\arcsin$ are commonly used in mathematics and its applications, we shall employ both of them in our work. Note that by definition we have

$$-\frac{\pi}{2} \le \sin^{-1} x \le \frac{\pi}{2}$$

or equivalently,

$$-\frac{\pi}{2} \le \arcsin x \le \frac{\pi}{2}.$$

Example 1   Sketch the graph of $y = \sin^{-1} x$.

Solution   By definition, the graph of $y = \sin^{-1} x$ is the same as the graph of $\sin y = x$, except that the variables are restricted as follows:

$$-\frac{\pi}{2} \le y \le \frac{\pi}{2} \quad \text{and} \quad -1 \le x \le 1.$$

Several coordinates of points on the graph are given in the following table.

y	$-\dfrac{\pi}{2}$	$-\dfrac{\pi}{3}$	$-\dfrac{\pi}{6}$	0	$\dfrac{\pi}{6}$	$\dfrac{\pi}{3}$	$\dfrac{\pi}{2}$
x	-1	$-\dfrac{\sqrt{3}}{2}$	$-\dfrac{1}{2}$	0	$\dfrac{1}{2}$	$\dfrac{\sqrt{3}}{2}$	1

The graph is sketched in Figure 7.5.

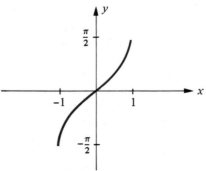

Figure 7.5 $y = \arcsin x = \sin^{-1} x$

Using the relationships stated in (II) gives us the following important identities:

$$\sin^{-1}(\sin y) = y \quad \text{if} \quad -\frac{\pi}{2} \le y \le \frac{\pi}{2}$$
$$\sin(\sin^{-1} x) = x \quad \text{if} \quad -1 \le x \le 1.$$

These may also be written in the form

$$\arcsin(\sin y) = y \quad \text{and} \quad \sin(\arcsin x) = x$$

provided that y and x are suitably restricted. ∎

Example 2   Find $\sin^{-1}(\sqrt{2}/2)$ and $\arcsin(-1/2)$.

Solution   If $y = \sin^{-1}(\sqrt{2}/2)$, then $\sin y = \sqrt{2}/2$, and consequently, $y = \pi/4$. Note that it is essential to choose y in the interval $[-\pi/2, \pi/2]$. A number such as $3\pi/4$ is incorrect, even though $\sin(3\pi/4) = \sqrt{2}/2$.

In like manner, if $y = \arcsin(-1/2)$, then $\sin y = -1/2$. Since, by definition, we must choose y in the interval $[-\pi/2, \pi/2]$, it follows that $y = -\pi/6$. ∎

Example 3   Find $\sin^{-1}(\tan 3\pi/4)$.

Solution   If we let

$$y = \sin^{-1}\left(\tan \frac{3\pi}{4}\right) = \sin^{-1}(-1)$$

then, by definition, $\qquad\qquad \sin y = -1.$

Consequently, $y = -\pi/2$. ∎

The other trigonometric functions may also be used to introduce inverse functions. The procedure is first to determine a convenient subset of the domain so that a one-to-one function is obtained and then to apply the formulas in (I).

If the domain of the cosine function is restricted to the interval $[0, \pi]$, and if y increases from 0 to π, then $\cos y$ decreases and takes on each value between -1 and 1 once and only once. This leads to the following.

Definition

The **inverse cosine function**, denoted by $\cos^{-1}$, is defined by

$$y = \cos^{-1} x \quad \text{if and only if} \quad \cos y = x$$

where $-1 \le x \le 1$ and $0 \le y \le \pi$.

The inverse cosine function is also referred to as the **arccosine function** and the notation arccos x is used interchangeably with $\cos^{-1} x$.

Applying the formulas in (II) we see that if $0 \le y \le \pi$ and $-1 \le x \le 1$, then

$$\cos(\cos^{-1} x) = \cos(\arccos x) = x$$
$$\cos^{-1}(\cos y) = \arccos(\cos y) = y$$

By definition, the graph of $y = \cos^{-1} x$ is the same as the graph of $\cos y = x$, where $0 \le y \le \pi$. It is left as an exercise to verify that the graph has the appearance indicated in Figure 7.6.

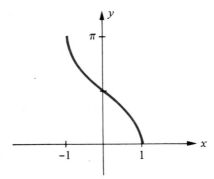

Figure 7.6 $y = \arccos x = \cos^{-1} x$

Example 4

(a) Find $\cos^{-1}(-\sqrt{3}/2)$.

(b) Approximate $\cos^{-1}(-0.7951)$.

Solution

(a) If we let $y = \cos^{-1}(-\sqrt{3}/2)$, then $\cos y = -\sqrt{3}/2$. The reference number for y is $\pi/6$. (Why?) By definition, y must be chosen in the interval $[0, \pi]$, and hence $y = \pi - (\pi/6) = 5\pi/6$. Thus,

$$\cos^{-1}(-\sqrt{3}/2) = 5\pi/6.$$

(b) If $y = \cos^{-1}(-0.7951)$, then $\cos y = -0.7951$. If y' is the reference number for y then $\cos y' = 0.7951$. From Table 2 we see that $y' \approx 0.6516$. According to the definition of the inverse cosine function, y must be chosen between 0 and π. Hence,

$$y = \pi - y' \approx 3.1416 - 0.6516.$$

That is,
$$\cos^{-1}(-0.7951) \approx 2.4900. \qquad \blacksquare$$

If we restrict the domain of the tangent function to the open interval $(-\pi/2, \pi/2)$, then a one-to-one correspondence is obtained. We may, therefore, adopt the following definition.

Definition

The **inverse tangent function** or **arctangent function**, denoted by $\tan^{-1}$ or arctan, is defined by

$$y = \tan^{-1} x = \arctan x \quad \text{if and only if} \quad \tan y = x$$

where x is any real number and $-\pi/2 < y < \pi/2$.

Note that the domain of the arctangent function is all of $\mathbb{R}$ and the range is the open interval $(-\pi/2, \pi/2)$. In Exercise 42 you are asked to verify the graph of the inverse tangent function sketched in Figure 7.7.

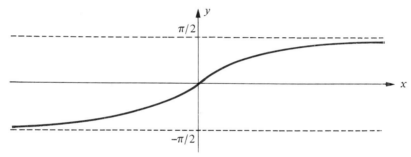

Figure 7.7 $y = \tan^{-1} x = \arctan x$

An analogous procedure may be used for the remaining trigonometric functions. The functions we have defined are generally referred to as the **inverse trigonometric functions**.

Example 5   Without using tables or calculators find $\sec(\arctan \frac{2}{3})$.

Solution   If $y = \arctan(2/3)$, then $\tan y = 2/3$. Since $\sec^2 y = 1 + \tan^2 y$ and $0 < y < \pi/2$, we may write

$$\sec y = \sqrt{1 + \tan^2 y} = \sqrt{1 + \left(\frac{2}{3}\right)^2}$$

$$= \sqrt{1 + \frac{4}{9}} = \sqrt{\frac{13}{9}} = \frac{\sqrt{13}}{3}.$$

Hence, $\sec(\arctan \frac{2}{3}) = \sec y = \sqrt{13}/3$. ∎

The next two examples illustrate some of the manipulations that can be carried out with the inverse trigonometric functions.

Example 6   Evaluate $\sin(\arctan\frac{1}{2} - \arccos\frac{4}{5})$.

Solution   If we let $u = \arctan(1/2)$ and $v = \arccos(4/5)$, then $\tan u = 1/2$ and $\cos v = 4/5$. We wish to find $\sin(u - v)$. Since u and v are in the interval $(0, \pi/2)$, they may be considered as the radian measure of positive acute ang es, and other functional values of u and v may be found by referring to the right triangles in Figure 7.8. This gives us $\sin u = 1/\sqrt{5}$, $\cos u = 2/\sqrt{5}$, and $\sin v = 3/5$. Consequently,

$$\sin(u - v) = \sin u \cos v - \cos u \sin v$$

$$= \frac{1}{\sqrt{5}}\frac{4}{5} - \frac{2}{\sqrt{5}}\frac{3}{5}$$

$$= \frac{-2}{5\sqrt{5}} = \frac{-2\sqrt{5}}{25}.$$

 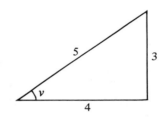

Figure 7.8   ∎

Example 7   Write $\cos(\sin^{-1} x)$ as an algebraic expression in x.

Solution   Let $y = \sin^{-1} x$, so that $\sin y = x$. We wish to find $\cos y$. Since $-\pi/2 \le y \le \pi/2$, it follows that $\cos y \ge 0$, and hence

$$\cos y = \sqrt{1 - \sin^2 y} = \sqrt{1 - x^2}.$$

Consequently, $\cos(\sin^{-1} x) = \sqrt{1 - x^2}$.   ∎

Example 8   Verify the identity

$$\frac{1}{2}\cos^{-1} x = \tan^{-1}\sqrt{\frac{1 - x}{1 + x}}, \quad \text{where } |x| < 1.$$

Solution   Let $y = \cos^{-1} x$. We wish to show that

$$\frac{y}{2} = \tan^{-1}\sqrt{\frac{1 - x}{1 + x}}.$$

By a half-angle formula (see Section 7.5),

$$\left|\tan \frac{y}{2}\right| = \sqrt{\frac{1 - \cos y}{1 + \cos y}}.$$

Since $y = \cos^{-1} x$ and $|x| < 1$ it follows that $0 < y < \pi$ and hence that $0 < (y/2) < \pi/2$. Consequently, $\tan(y/2) > 0$ and we may drop the absolute value sign, obtaining

$$\tan \frac{y}{2} = \sqrt{\frac{1 - \cos y}{1 + \cos y}}.$$

Using the fact that $\cos y = x$ gives us

$$\tan \frac{y}{2} = \sqrt{\frac{1 - x}{1 + x}}$$

or equivalently,

$$\frac{y}{2} = \tan^{-1} \sqrt{\frac{1 - x}{1 + x}}$$

which is what we wished to show.   ∎

Most of the trigonometric equations considered in Section 7.3 had solutions which were rational multiples of π, such as $\pi/3$, $3\pi/4$, π, and so on. If solutions are not of that type we can sometimes use inverse functions to express them in exact form, as illustrated in the next example.

Example 9   Find the solutions of the equation

$$5 \sin^2 t + 3 \sin t - 1 = 0$$

which are in the interval $\left[-\pi/2, \pi/2\right]$.

Solution   The given equation may be regarded as a quadratic equation in $\sin t$. Applying the Quadratic Formula gives us

$$\sin t = \frac{-3 \pm \sqrt{9 + 20}}{10} = \frac{-3 \pm \sqrt{29}}{10}.$$

Hence, by the definition of the inverse sine function, with $v = t$ and $u = (-3 \pm \sqrt{29})/10$, we obtain the solutions

$$t = \sin^{-1} \frac{1}{10}(-3 + \sqrt{29})$$

and

$$t = \sin^{-1} \frac{1}{10}(-3 - \sqrt{29}).$$ ∎

If approximations to the solutions in Example 9 are desired, one may refer to Table 2. If a hand-held calculator is available, the task is much simpler. To illustrate, using a typical calculator we first calculate

$$\frac{1}{10}(-3 + \sqrt{29}) \approx 0.2385.$$

If we next press the inverse sine key the result is

$$t \approx \sin^{-1}(0.2385) \approx 0.2408.$$

In like manner,

$$t = \sin^{-1}\frac{1}{10}(-3 - \sqrt{29}) \approx \sin^{-1}(-0.8385) \approx -0.9946.$$

EXERCISES 7.7

Find the numbers in Exercises 1–10 without the use of tables or calculators.

1 (a) $\sin^{-1}(1/2)$ (b) $\cos^{-1}(1/2)$

2 (a) $\sin^{-1}\sqrt{3}/2$ (b) $\sin^{-1}(-\sqrt{3}/2)$

3 (a) $\arcsin 0$ (b) $\arccos 0$

4 (a) $\cos^{-1}\sqrt{2}/2$ (b) $\cos^{-1}(-\sqrt{2}/2)$

5 (a) $\arcsin 1$ (b) $\arccos 1$

6 (a) $\sin^{-1}(-1)$ (b) $\cos^{-1}(-1)$

7 (a) $\tan^{-1}\sqrt{3}$ (b) $\arctan(-\sqrt{3})$

8 (a) $\tan^{-1}(-1)$ (b) $\arccos(-1/2)$

9 (a) $\arcsin(-\sqrt{2}/2)$ (b) $\tan^{-1}(-\sqrt{3}/3)$

10 (a) $\tan^{-1}0$ (b) $\arctan 1$

In Exercises 11–14 use Table 2 to approximate the given numbers.

11 $\sin^{-1}(-0.6494)$ 12 $\cos^{-1}(-0.9112)$

13 $\arctan(2.1775)$ 14 $\arcsin(0.8004)$

Determine the numbers in Exercises 15–26 without the use of tables or calculators.

15 $\sin\left[\cos^{-1}\frac{1}{2}\right]$ 16 $\cos\left[\sin^{-1}0\right]$

17 $\sin\left[\arccos\frac{4}{5}\right]$ 18 $\tan\left[\tan^{-1}5\right]$

19 $\arcsin(\sin 5\pi/4)$ 20 $\cos^{-1}(\cos 3\pi/4)$

21 $\cos\left[\sin^{-1}\frac{4}{5} + \tan^{-1}\frac{3}{4}\right]$

22 $\sin\left[\arcsin\frac{1}{2} + \arccos 0\right]$

23 $\tan\left[\arctan\frac{4}{3} + \arccos\frac{8}{17}\right]$

24 $\cos\left[2\sin^{-1}\frac{15}{17}\right]$ 25 $\sin\left[2\arccos\left(-\frac{3}{5}\right)\right]$

26 $\cos\left[\frac{1}{2}\tan^{-1}\frac{8}{15}\right]$

In each of Exercises 27–30 rewrite the given expression as an algebraic expression in u.

27 $\sin(\tan^{-1}x)$ 28 $\tan(\arccos x)$

29 $\cos(\frac{1}{2}\arccos x)$ 30 $\cos(2\tan^{-1}x)$

Verify the identities in Exercises 31–38.

31 $\sin^{-1}x + \cos^{-1}x = \pi/2$
    (*Hint*: let $\alpha = \sin^{-1}x$, $\beta = \cos^{-1}x$, and consider $\sin(\alpha + \beta)$.)

32 $\arctan x + \arctan(1/x) = \pi/2.\ x > 0$

33 $\arcsin\dfrac{2x}{1 + x^2} = 2\arctan x,\ |x| < 1$

34 $2\cos^{-1}x = \cos^{-1}(2x^2 - 1)$, where $0 \le x \le 1$

35 $\arcsin(-x) = -\arcsin x$

36 $\arccos(-x) = \pi - \arccos x$

37 $\sin^{-1}x = \tan^{-1}\dfrac{x}{\sqrt{1 - x^2}}$

38 $\tan^{-1} u + \tan^{-1} v = \tan^{-1} \dfrac{u + v}{1 - uv}$, where $|u| < 1$ and $|v| < 1$.

39 Define $\cot^{-1}$ by restricting the domain of cot to the interval $(0, \pi)$.

40 Define $\sec^{-1}$ by restricting the domain of sec to $[0, \pi/2) \cup [\pi, 3\pi/2)$.

41 Verify the sketch in Figure 7.6.

42 Verify the sketch in Figure 7.7.

Sketch the graphs of the equations in Exercises 43–52.

43 $y = \sin^{-1} 2x$

44 $y = \cos^{-1}(x/2)$

45 $y = \frac{1}{2} \sin^{-1} x$

46 $y = 2 \cos^{-1} x$

47 $y = 2 \tan^{-1} x$

48 $y = \tan^{-1} 2x$

49 $y = 2 + \tan^{-1} x$

50 $y = \sin^{-1}(x + 1)$

51 $y = \sin(\arccos x)$

52 $y = \sin(\sin^{-1} x)$

Prove that the equations in Exercises 53 and 54 are not identities.

53 $\tan^{-1} x = \dfrac{1}{\tan x}$

54 $(\arcsin x)^2 + (\arccos x)^2 = 1$

In each of Exercises 55–60, use inverse trigonometric functions to state the solutions of the given equation in the given interval.

55 $2 \tan^2 t + 9 \tan t + 3 = 0; \ (-\pi/2, \pi/2)$

56 $3 \sin^2 t + 7 \sin t + 3 = 0; \ [-\pi/2, \pi/2]$

57 $15 \cos^4 x - 14 \cos^2 x + 3 = 0; \ [0, \pi]$

58 $3 \tan^4 \theta - 19 \tan^2 \theta + 2 = 0; \ (-\pi/2, \pi/2)$

59 $6 \sin^3 \theta + 18 \sin^2 \theta - 5 \sin \theta - 15 = 0;$ $(-\pi/2, \pi/2)$

60 $6 \sin 2x - 8 \cos x + 9 \sin x - 6 = 0;$ $(-\pi/2, \pi/2)$

CALCULATOR EXERCISES 7.7

1–6 Approximate the solutions of the equations in Exercises 55–60 to four decimal places.

7 Show, by actual calculations, that $\sin(\arcsin x) = x$ for the following values of x:
(a) 0.4631 (b) $\log 3.64$
(c) $1/\sqrt{\pi}$

8 Show that $\tan^{-1}(\tan x) = x$ for the following values of x:
(a) 74.85 (b) $\log(\pi^2 + 94.7)$
(c) $10^{6.39}$

Exercises 9 and 10 employ the notation introduced in the Calculator Exercises of Section 6.7. Use a calculator to support the given statements by substituting the following values for t: 0.1, 0.01, 0.001, 0.0001.

9 $\dfrac{\arcsin 2x}{\arcsin x} \to 2$ as $x \to 0^+$

10 $\csc x \tan^{-1} x \to 1$ as $x \to 0^+$

7.8 THE LAW OF SINES

A triangle that does not contain a right angle is referred to as an **oblique triangle**. Since it is always possible to divide such triangles into two right triangles, methods developed in Chapter Six may be used for solving them; however, sometimes it is cumbersome to proceed in that manner. In this, and the next section, formulas are obtained which aid in simplifying solutions of oblique triangles.

If two angles and a side of a triangle are known, or if two sides and an angle opposite one of them are known, then the remaining parts of the triangle may be found by means of the formula given in this section. We shall use the letters A, B, C, a, b, c, α, β, and γ for parts of triangles as they were used in Section 6.10. Given triangle ABC, let us place angle α in standard position on a rectangular coordinate system so that B is on the positive x-axis. The case where α is obtuse is illustrated in Figure 7.9. The type of argument we shall give may also be used if α is acute.

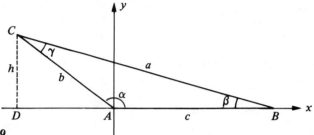

Figure 7.9

Consider the line through C parallel to the y-axis and intersecting the x-axis at point D. Suppose that $d(C, D) = h$, so that the ordinate of C is h. It follows that

$$\sin \alpha = \frac{h}{b} \quad \text{or} \quad h = b \sin \alpha.$$

Referring to triangle BDC we see that

$$\sin \beta = \frac{h}{a} \quad \text{or} \quad h = a \sin \beta.$$

Consequently, $a \sin \beta = b \sin \alpha$

which may be written $\dfrac{a}{\sin \alpha} = \dfrac{b}{\sin \beta}.$

If α is taken in standard position, so that C is on the positive x-axis, then by the same reasoning,

$$\frac{a}{\sin \alpha} = \frac{c}{\sin y}.$$

The last two equalities give us the following result.

The Law of Sines

> If ABC is any oblique triangle labeled in the usual manner, then
>
> $$\frac{a}{\sin \alpha} = \frac{b}{\sin \beta} = \frac{c}{\sin \gamma}.$$

The next example illustrates a method of applying the Law of Sines to the case in which two angles and a side of a triangle are known. We shall use the rules for rounding off answers discussed in Section 6.10.

Example 1 Given triangle ABC with $\alpha = 48°20'$, $\gamma = 57°30'$, and $b = 47.3$, approximate the remaining parts.

Solution The triangle is represented in Figure 7.10. Since the sum of the angles is $180°$,

$$\beta = 180° - (57°30' + 48°20') = 74°10'.$$

Applying the Law of Sines,

$$a = \frac{b \sin \alpha}{\sin \beta} = \frac{(47.3) \sin 48°20'}{\sin 74°10'}.$$

Consulting Table 2,

$$a \approx \frac{(47.3)(0.7470)}{(0.9621)} \approx 36.7.$$

Similarly,

$$c \approx \frac{b \sin \gamma}{\sin \beta} = \frac{(47.3) \sin 57°30'}{\sin 74°10'}$$

$$\approx \frac{(47.3)(0.8434)}{(0.9621)} \approx 41.5.$$

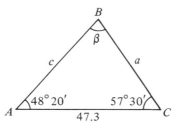

Figure 7.10

Data such as that given in Example 1 always gives us a unique triangle ABC. However, if two sides and an angle opposite one of them are given, a unique triangle is not always determined. To illustrate, suppose that two numbers a and b are to be lengths of sides of a triangle ABC. In addition, suppose that there is given an angle α which is to be opposite the side of length a. Let us consider the case in which α is acute. Place α in standard position on a rectangular coordinate system and consider the line segment AC of length b on the terminal side of α, as shown in Figure 7.11.

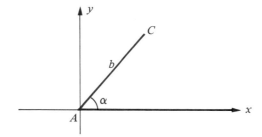

Figure 7.11

The third vertex, B, should be somewhere on the x-axis. Since the length a of the side opposite α is given, B may be found by striking off a circular arc of length a with center at C. There are four possible outcomes for the construction, as illustrated in Figure 7.12, where the coordinate axes have been deleted.

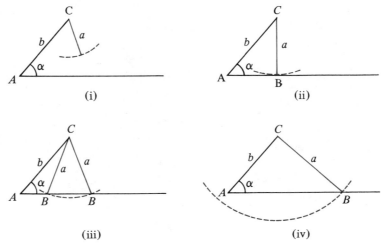

Figure 7.12

The four possibilities may be listed as follows:

(i)   The arc does not intersect the x-axis and no triangle is formed.

(ii) The arc is tangent to the x-axis and a right triangle is formed.

(iii) The arc intersects the positive x-axis in two distinct points and two triangles are formed.

(iv) The arc intersects both the positive and nonpositive parts of the x-axis and one triangle is formed.

Since the distance from C to the x-axis is $b \sin \alpha$ (Why?), we see that (i) occurs if $a < b \sin \alpha$, (ii) occurs if $a = b \sin \alpha$, (iii) occurs if $b \sin \alpha < a < b$, and (iv) occurs if $a \geq b$. It is unnecessary to memorize these facts, since in any specific problem the case that occurs will become evident when the solution is attempted. For example, in solving the equation

$$\frac{a}{\sin \alpha} = \frac{b}{\sin \beta}$$

suppose that we obtain $|\sin \beta| > 1$. This will indicate that no triangle exists. If the equation $\sin \beta = 1$ is obtained, then $\beta = 90°$ and hence case (ii) occurs. On the other hand, if $|\sin \beta| < 1$, then there are two possible choices for the angle β. By checking both possibilities, it will become apparent if (iii) or (iv) occurs.

If the measure of α is greater than 90°, then, as in Figure 7.13, a triangle exists if and only if $a > b$. Needless to say, our discussion is independent of the symbols we have used; that is, we might be given b, c, β, or a, c, γ, and so on.

Figure 7.13

Since different possibilities may arise, the case in which two sides and an angle opposite one of them are given is sometimes called the **ambiguous case**.

Example 2   Solve triangle ABC if $\alpha = 67°$, $c = 125$, and $a = 100$.

Solution   Using $\sin \gamma = \dfrac{c \sin \alpha}{a}$, we obtain

$$\sin \gamma = \frac{(125) \sin 67°}{100}$$

$$\approx \frac{(125)(0.9205)}{100}$$

$$\approx 1.1506.$$

Since $\sin \gamma > 1$, there is no triangle with the given parts.   ■

Example 3   Approximate the remaining parts of triangle ABC if $a = 12.4$, $b = 8.7$, and $\beta = 36°40'$.

Solution   Using $\sin \alpha = \dfrac{a \sin \beta}{b}$ we have

$$\sin \alpha = \frac{(12.4) \sin 36°40'}{8.7}$$

$$\approx \frac{(12.4)(0.5972)}{8.7} \approx 0.8512.$$

There are two possible angles α between 0° and 180° such that $\sin \alpha \approx 0.8512$. If we let α' denote the reference angle for α, then from Table 2 we obtain

$$\alpha' \approx 58°20'.$$

Consequently, the two possibilities for α are

$$\alpha_1 \approx 58°20' \quad \text{and} \quad \alpha_2 \approx 121°40'.$$

If we let γ_1 and γ_2 denote the third angle of the triangle corresponding to the angles α_1 and α_2, respectively, then

$$\gamma_1 \approx 180° - (36°40' + 58°20') = 85°$$

and

$$\gamma_2 \approx 180° - (36°40' + 121°40') = 21°40'.$$

Thus, there are two possible triangles which have the given parts. They are the triangles A_1BC and A_2BC shown in Figure 7.14.

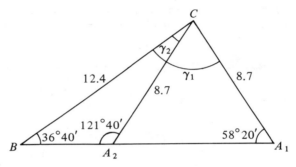

Figure 7.14

If c_1 is the side opposite γ_1 in triangle A_1BC, then

$$c_1 = \frac{a \sin \gamma_1}{\sin \alpha_1}$$

$$c_1 \approx \frac{(12.4) \sin 85°}{\sin 58°20'}$$

$$\approx \frac{(12.4)(0.9962)}{0.8511} \approx 14.5.$$

Hence, the solution for triangle A_1BC is $\alpha_1 \approx 58°20'$, $\gamma_1 \approx 85°$, and $c_1 \approx 14.5$.

Similarly, if c_2 is the side opposite angle γ_2 in triangle A_2BC, we have

$$c_2 = \frac{a \sin \gamma_2}{\sin \alpha_2}$$

or

$$c_2 \approx \frac{(12.4) \sin 21°40'}{\sin 121°40'} \approx \frac{(12.4)(0.3692)}{0.8511} \approx 5.4.$$

Consequently, the solution for triangle A_2BC is $\alpha_2 \approx 121°40'$, $\gamma_2 \approx 21°40'$, and $c_2 \approx 5.4$.   ∎

EXERCISES 7.8

3, 6, 9, 12

In each of Exercises 1–12 approximate the remaining parts of triangle ABC.

1 $\alpha = 41°$, $\gamma = 77°$, $a = 10.5$

2 $\beta = 20°$, $\gamma = 31°$, $b = 210$

3 $\alpha = 27°40'$, $\beta = 52°10'$, $a = 32.4$

4 $\alpha = 42°10'$, $\gamma = 61°20'$, $b = 19.7$

5 $\beta = 50°50'$, $\gamma = 70°30'$, $c = 537$

6 $\alpha = 7°10'$, $\beta = 11°40'$, $a = 2.19$

7 $\alpha = 65°10'$, $a = 21.3$, $b = 18.9$

8 $\beta = 30°$, $b = 17.9$, $a = 35.8$

9 $\gamma = 53°20'$, $a = 140$, $c = 115$

10 $\alpha = 27°30'$, $c = 52.8$, $a = 28.1$

11 $\beta = 113°10'$, $b = 248$, $c = 195$

12 $\gamma = 81°$, $c = 11$, $b = 12$

13 It is desired to find the distance between two points A and B lying on opposite banks of a river. A line segment AC of length 240 yards is laid off and the measures of angles BAC and ACB are found to be $63°20'$ and $54°10'$, respectively. Approximate the distance from A to B.

14 In order to determine the distance between two points A and B, a surveyor chooses a point C which is 375 yards from A and 530 yards from B. If angle BAC has measure $49°30'$, approximate the required distance.

15 When the angle of elevation of the sun is $64°$, a telegraph pole that is tilted at an angle of $12°$ directly away from the sun casts a shadow 34 feet long on level ground. Approximate the length of the pole.

16 A straight road makes an angle of $15°$ with the horizontal. When the angle of elevation of the sun is $57°$, a vertical pole at the side of the road casts a shadow 75 feet long directly down the road. Approximate the length of the pole.

17 The angles of elevation of a balloon from two points A and B on level ground are $24°10'$ and $47°40'$, respectively. If A and B are 8.4 miles apart and the balloon is between A and B in the same vertical plane, approximate, to the nearest tenth of a mile, the height of the balloon above the ground.

18 An airport A is 480 miles due east of airport B. A pilot flew in the direction $235°$ from A to C and then in the direction $320°$ from C to B. Approximate, to the nearest mile, the total distance the pilot flew.

19 A forest ranger at an observation point A sights a fire in the direction N27°10′E. Another ranger at an observation point B, 6.0 miles due east of A, sights the same fire at N52°40′W. Approximate, to the nearest tenth of a mile, the distance from each of the observation points to the fire.

20 A surveyor notes that the direction from point A to point B is S63°W and the direction from A to C is S38°W. If the distance from A to B is 239 yards and the distance from B to C is 374 yards, approximate the distance from A to C.

21 A straight road makes an angle of $22°$ with the horizontal. From a certain point P on the road the angle of elevation of an airplane is $57°$. At the same instant, from another point 100 meters farther up the road, the angle of elevation is $63°$. Approximate, to the nearest meter, the distance from P to the airplane.

22 A point P on level ground is 3.0 km due north of a point Q. If a person jogged in the direction N25°E from Q to a point R, and then from R to P in the direction S40°W, approximate, to the nearest km, the total distance jogged.

CALCULATOR EXERCISES 7.8

In the following, approximate the remaining parts of triangle ABC.

1 $\beta = 25.6°$, $\gamma = 34.7°$, $b = 184.8$

2 $\alpha = 6.24°$, $\beta = 14.08°$, $a = 4.56$

3 $\alpha = 103.45°$, $\gamma = 27.19°$, $b = 38.84$

4 $\gamma = 47.74°$, $a = 131.08$, $c = 97.84$

5 $\beta = 121.624°$, $b = 0.283$, $c = 0.178$

6 $\alpha = 32.32°$, $c = 574.3$, $a = 263.6$

7.9   THE LAW OF COSINES

If two sides and the included angle or the three sides of a triangle are given, then the Law of Sines cannot be applied directly. We may, however, use the following result.

The Law of Cosines

> If ABC is any triangle labeled in the usual manner, then
>
> $$a^2 = b^2 + c^2 - 2bc \cos \alpha$$
> $$b^2 = a^2 + c^2 - 2ac \cos \beta$$
> $$c^2 = a^2 + b^2 - 2ab \cos \gamma.$$

Instead of memorizing each of the formulas given in this law, it is more convenient to remember the following statement, which takes all of them into account.

> The square of the length of any side of a triangle equals the sum of the squares of the lengths of the other two sides minus twice the product of the lengths of the other two sides and the cosine of the angle between them.

We shall use the Distance Formula to establish the Law of Cosines. We again place α in standard position on a rectangular coordinate system, as illustrated in Figure 7.15.

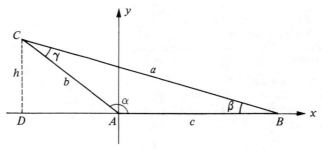

Figure 7.15

Although α is pictured as an obtuse angle, our development is also true if α is acute. Since the segment AB has length c, the coordinates of B are $(c, 0)$. It follows from the Theorem on Trigonometric Functions as Ratios (see Section 6.5), that the coordinates of C are $(b \cos \alpha, b \sin \alpha)$. We also have $a = d(B, C)$ and hence $a^2 = [d(B, C)]^2$. Using the Distance Formula leads to the following equations:

$$
\begin{aligned}
a^2 &= (b \cos \alpha - c)^2 + (b \sin \alpha - 0)^2 \\
&= b^2 \cos^2 \alpha - 2bc \cos \alpha + c^2 + b^2 \sin^2 \alpha \\
&= b^2(\cos^2 \alpha + \sin^2 \alpha) + c^2 - 2bc \cos \alpha \\
&= b^2 + c^2 - 2bc \cos \alpha.
\end{aligned}
$$

This gives us the first formula stated in the Law of Cosines. The second and third formulas may be obtained by placing β and γ, respectively, in standard position on a rectangular coordinate system and using a similar procedure.

Example 1   Approximate the remaining parts of triangle ABC if $a = 5.0$, $c = 8.0$, and $\beta = 77°10'$.

Solution   Using the Law of Cosines,

$$b^2 = (5.0)^2 + (8.0)^2 - 2(5.0)(8.0)\cos 77°10'$$
$$\approx 25 + 64 - (80)(0.2221)$$
$$\approx 71.2.$$

Consequently, $b \approx \sqrt{71.2} \approx 8.44 \approx 8.4$.

We next use the Law of Sines to find α. Thus,

$$\sin\alpha = \frac{a\sin\beta}{b} \approx \frac{(5.0)\sin 77°10'}{8.44} \approx \frac{(5.0)(0.9750)}{8.44} \approx 0.5776.$$

Consulting Table 2 we see that $\alpha \approx 35°20'$. Hence,

$$\gamma \approx 180° - (77°10' + 35°20') = 67°30'. \qquad \blacksquare$$

After the length of the third side was found in Example 1, the Law of Sines was used to find a second angle of the triangle. Whenever this procedure is followed, it is best to find the angle opposite the shortest side (as we did), since that angle is always acute. Of course, the Law of Cosines can also be used to find the remaining angles.

Example 2   Given sides $a = 90$, $b = 70$, and $c = 40$ of triangle ABC, approximate angles α, β, and γ.

Solution   According to the first formula stated in the Law of Cosines,

$$\cos\alpha = \frac{b^2 + c^2 - a^2}{2bc}$$
$$= \frac{4900 + 1600 - 8100}{5600}$$
$$\approx -0.2857.$$

From Table 2 we see that the reference angle for α is approximately $73°20'$. Hence, $\alpha \approx 180° - 73°20' = 106°40'$.

Similarly, from the second formula stated in the Law of Cosines,

$$\cos \beta = \frac{a^2 + c^2 - b^2}{2ac}$$

$$= \frac{8100 + 1600 - 4900}{7200}$$

$$\approx 0.6667$$

and hence $\beta \approx 48°10'$. Finally,

$$\gamma \approx 180° - (106°40' + 48°10') = 25°10'. \qquad \blacksquare$$

After finding the first angle in the preceding solution, we could have used the Law of Sines to find a second angle. Whenever this procedure is followed, it is best to use the Law of Cosines to find the angle opposite the longer side, that is, the largest angle of the triangle. This will guarantee that the remaining two angles are acute.

EXERCISES 7.9

In Exercises 1–10 approximate the remaining parts of triangle ABC.

1 $\alpha = 60°$, $b = 20$, $c = 30$

2 $\gamma = 45°$, $b = 10.0$, $a = 15.0$

3 $\beta = 150°$, $a = 150$, $c = 30.0$

4 $\beta = 73°50'$, $c = 14.0$, $a = 87.0$

5 $\gamma = 115°10'$, $a = 1.10$, $b = 2.10$

6 $\alpha = 23°40'$, $c = 4.30$, $b = 70.0$

7 $a = 2.0$, $b = 3.0$, $c = 4.0$

8 $a = 10$, $b = 15$, $c = 12$

9 $a = 25.0$, $b = 80.0$, $c = 60.0$

10 $a = 20.0$, $b = 20.0$, $c = 10.0$

11 A parallelogram has sides of length 30 inches and 70 inches. If one of the angles has measure 65°, approximate the length of each diagonal.

12 The angle at one corner of a triangular plot of ground has measure 73°40'. If the sides that meet at this corner are 175 feet and 150 feet long, approximate the length of the third side.

13 A vertical pole 40 feet tall stands on a hillside that makes an angle of 17° with the horizontal. Approximate the minimal length of rope that will reach from the top of the pole to a point directly down the hill 72 feet from the base of the pole?

14 To find the distance between two points A and B, a surveyor chooses a point C which is 420 yards from A and 540 yards from B. If angle ACB has measure 63°10', approximate the distance.

15 Two automobiles leave from the same point and travel along straight highways which differ in direction by 84°. If their speeds are 60 miles per hour and 45 miles per hour, respectively, approximately how far apart will they be at the end of 20 minutes?

16 A triangular plot of land has sides of length 420 feet, 350 feet, and 180 feet. Find the smallest angle between the sides.

17 A ship leaves P at 1:00 P.M. and travels S35°E at the rate of 24 miles per hour. Another ship leaves P at 1:30 P.M. and travels S20°W at 18 miles per hour. Approximately how far apart are the ships at 3:00 P.M.?

18   An airplane flies 165 miles from point A in the direction 130° and then travels in the direction 245° for 80 miles. Approximately how far is the airplane from A?

19   A jogger, running at a constant speed of one mile every eight minutes, runs in the direction S40°E for 20 minutes and then in the direction N20°E for the next 16 minutes. Approximate, to the nearest tenth of a mile, the distance from the jogger to the starting point.

20   Two points P and Q on level ground are on opposite sides of a building. In order to find the distance d between the points, a surveyor chooses a point R which is 300 feet from P and 438 feet from Q, and then determines that angle PRQ has measure 37°40′. Approximate d.

21   A motor boat traveled along a triangular course of sides 2 km, 4 km, and 3 km, respectively. If the first side was traversed in the direction N20°W and the second in the southwest direction, approximate, to the nearest minute, the direction that the third side was traversed.

22   If a rhombus has sides of length 100 cm and if the angle at one of the vertices is 70°, approximate the lengths of the diagonals to the nearest tenth of a centimeter.

23   Show that for every triangle ABC,
(a) $a^2 + b^2 + c^2 = 2(bc\cos\alpha + ac\cos\beta + ab\cos\gamma)$.

(b) $\dfrac{\cos\alpha}{a} + \dfrac{\cos\beta}{b} + \dfrac{\cos\gamma}{c} = \dfrac{a^2 + b^2 + c^2}{2abc}$.

24   Show that if ABC is a right triangle, then one part of the Law of Cosines reduces to the Pythagorean Theorem.

CALCULATOR EXERCISES 7.9

Approximate the remaining parts of triangle ABC in Exercises 1–6.

1 $\alpha = 48.3°$, $b = 24.7$, $c = 52.8$

2 $\beta = 137.8°$, $a = 178.2$, $c = 431.4$

3 $\gamma = 6.85°$, $a = 0.846$, $b = 0.364$

4 $a = 435$, $b = 482$, $c = 78$

5 $a = 5,150$, $b = 1814$, $c = 3,429$

6 $a = 0.64$, $b = 0.27$, $c = 0.49$

7.10   VECTORS

In previous work we assigned directions to certain lines, such as the x- and y-axes. In similar fashion, a **directed line segment** is a line segment to which a direction has been assigned. Another name for a directed line segment is a **vector**. If a vector extends from a point A (called the **initial point**) to a point B (called the **terminal point**), it is customary to place an arrowhead at B and use $\overrightarrow{AB}$ to represent the vector (see Figure 7.16). The length of the directed line segment is called the **magnitude** of the vector $\overrightarrow{AB}$ and is denoted by $|\overrightarrow{AB}|$. The vectors $\overrightarrow{AB}$ and $\overrightarrow{CD}$ are considered **equal**, and we write $\overrightarrow{AB} = \overrightarrow{CD}$ if and only if they have the same magnitude and direction, as illustrated in Figure 7.16. Consequently, vectors may be translated from one position to another, provided neither the magnitude nor direction is changed. Vectors of this type are often referred to as **free vectors**.

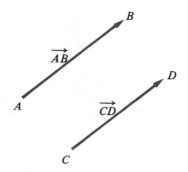

Figure 7.16

Many physical concepts may be represented by vectors. To illustrate, suppose an airplane is descending at a constant rate of 100 mph and the line of flight makes an angle of 20° with the horizontal. Both of these facts are represented by the vector in (i) of Figure 7.17, where it is assumed that units have been chosen so that the magnitude is 100. As shown in the figure, we will use a boldface letter such as **v** to denote a vector whose endpoints are not specified. The vector **v** in this illustration is called a **velocity vector**. The magnitude |**v**| of **v** is the speed of the airplane.

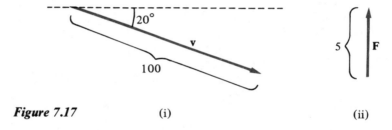

Figure 7.17   (i)   (ii)

As a second illustration, suppose a person pulls directly upward on an object with a force of 5 kg, as would be the case in lifting a 5 kg weight. We may indicate this fact by the vector **F** in (ii) of Figure 7.17. A vector which represents a pull or push of some type is called a **force vector**.

Another use for vectors is to let $\overrightarrow{AB}$ represent the path of a point (or some physical particle) as it moves along the line segment from A to B. We then refer to $\overrightarrow{AB}$ as a

Figure 7.18

displacement of the point (or particle). As illustrated in Figure 7.18, a displacement $\vec{AB}$ followed by a displacement $\vec{BC}$ will lead to the same point as the single displacement $\vec{AC}$. The latter vector is called the **sum** of the first two, and we write

$$\vec{AC} = \vec{AB} + \vec{BC}.$$

Since we are working with free vectors, any two vectors can be added by placing the initial point of one on the terminal point of the other and then proceeding as in Figure 7.18.

Another way to find the sum of two vectors is to consider vectors which are equal to the given ones and have the same initial point, as illustrated by $\vec{PQ}$ and $\vec{PR}$ in Figure 7.19. If we construct the parallelogram $RPQS$ with adjacent sides $\vec{PR}$ and $\vec{PQ}$, then since $\vec{PR} = \vec{QS}$, it follows that

$$\vec{PS} = \vec{PQ} + \vec{PR}.$$

If $\vec{PQ}$ and $\vec{PR}$ are two forces acting at P, then it can be shown experimentally that $\vec{PS}$ is the **resultant** force, that is, the single force that produces the same effect as the two combined forces.

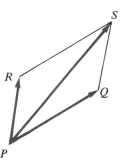

Figure 7.19

If c is a real number and $\vec{AB}$ is a vector, then the product $c\vec{AB}$ is defined as a vector whose magnitude is $|c|$ times the magnitude of $\vec{AB}$ and whose direction is the same as $\vec{AB}$ if $c > 0$, and opposite that of $\vec{AB}$ if $c < 0$. Geometric illustrations are given in Figure 7.20. We refer to c as a **scalar** and $c\vec{AB}$ as a **scalar multiple** of $\vec{AB}$.

Figure 7.20

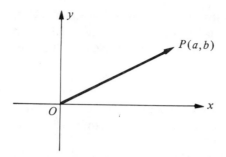

Figure 7.21

Let us next introduce a coordinate plane and assume that all vectors under discussion are in that plane. Since the position of a vector may be changed, provided the magnitude and direction are not altered, we may place the initial point of each vector at the origin. The terminal point of a typical vector $\overrightarrow{OP}$ may then be assigned coordinates (a, b), as shown in Figure 7.21. Conversely, every ordered pair (a, b) determines the vector $\overrightarrow{OP}$, where P has coordinates (a, b). We thus obtain a one-to-one correspondence between vectors and ordered pairs. This allows us to regard a vector in a plane as an ordered pair of real numbers instead of a directed line segment. To avoid confusion with the notation for open intervals or points, we shall use the symbol $\langle a, b \rangle$ for an ordered pair which represents a vector. Moreover, we shall refer to $\langle a, b \rangle$ as a vector and denote it by a boldface letter. The numbers a and b are called the **components** of the vector $\langle a, b \rangle$. The magnitude of $\langle a, b \rangle$ is, by definition, the distance from the origin to the point $P(a, b)$. This may also be stated as follows.

Definition

> The **magnitude** $|\mathbf{v}|$ of the vector $\mathbf{v} = \langle a, b \rangle$ is given by $|\mathbf{v}| = \sqrt{a^2 + b^2}$.

Example 1   Sketch the vectors corresponding to each of the following, and find the magnitude and the smallest positive angle θ from the positive x-axis to each vector.

(a) $\mathbf{a} = \langle -3, -3 \rangle$ (b) $\mathbf{b} = \langle 0, -2 \rangle$ (c) $\mathbf{c} = \langle 4/5, 3/5 \rangle$

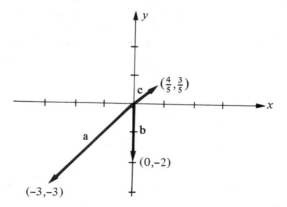

Figure 7.22

Solutions   The vectors are sketched in Figure 7.22. Applying the definition of magnitude, and our knowledge of trigonometry, we obtain:

(a) $|\mathbf{a}| = \sqrt{9+9} = 3\sqrt{2}; \quad \theta = 5\pi/4$

(b) $|\mathbf{b}| = \sqrt{0+4} = 2; \quad \theta = 3\pi/2$

(c) $|\mathbf{c}| = \sqrt{(16/25)+(9/25)} = 1; \quad \theta = \arctan 3/4.$ ■

If, as in Figure 7.23, we consider vectors $\overrightarrow{OQ}$ and $\overrightarrow{OR}$ corresponding to $\langle a, b\rangle$ and $\langle c, d\rangle$, respectively, and if we let $\overrightarrow{OS}$ be the vector corresponding to $\langle a + c, b + d\rangle$, then it can be shown that O, Q, S, and R are vertices of a parallelogram. It follows that

$$\overrightarrow{OQ} + \overrightarrow{OR} = \overrightarrow{OS}.$$

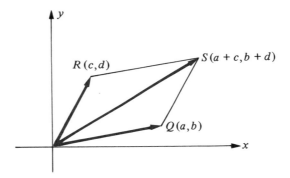

Figure 7.23

Expressing this fact in terms of ordered pairs gives us the following rule for addition of vectors.

Addition of Vectors

$$\langle a, b\rangle + \langle c, d\rangle = \langle a + c, b + d\rangle$$

Although we shall not prove it, the rule corresponding to a scalar multiple $k\overrightarrow{OP}$ of a vector is as follows.

Scalar Multiples of Vectors

$$k\langle a, b\rangle = \langle ka, kb\rangle$$

Example 2   If $\mathbf{a} = \langle 2, 1\rangle$, find $3\mathbf{a}$ and $-2\mathbf{a}$, and represent all three vectors geometrically.

Solution   If $\mathbf{a} = \langle 2, 1\rangle$, then, by the preceding rule for scalar multiples, $3\mathbf{a} = \langle 6, 3\rangle$ and $-2\mathbf{a} = \langle -4, -2\rangle$. The geometric representations are shown in Figure 7.24.

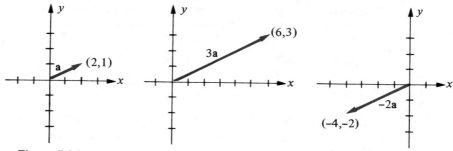

Figure 7.24

Example 3   If $\mathbf{a} = \langle 3, -2 \rangle$ and $\mathbf{b} = \langle -6, 7 \rangle$, find $\mathbf{a} + \mathbf{b}$, $4\mathbf{a}$, and $2\mathbf{a} + 3\mathbf{b}$.

Solution   Using the rules for addition and scalar multiples of vectors,

$$\mathbf{a} + \mathbf{b} = \langle 3, -2 \rangle + \langle -6, 7 \rangle = \langle -3, 5 \rangle$$
$$4\mathbf{a} = 4\langle 3, -2 \rangle = \langle 12, -8 \rangle$$
$$2\mathbf{a} + 3\mathbf{b} = \langle 6, -4 \rangle + \langle -18, 21 \rangle = \langle -12, 17 \rangle.$$

By definition, the **zero vector 0** corresponds to $\langle 0, 0 \rangle$. Also if $\mathbf{a} = \langle a, b \rangle$, then we define $-\mathbf{a} = \langle -a, -b \rangle$. Using these definitions we may establish the following properties, where $\mathbf{a}$, $\mathbf{b}$, and $\mathbf{c}$ denote arbitrary vectors.

Properties of Addition of Vectors

$$\mathbf{a} + \mathbf{b} = \mathbf{b} + \mathbf{a}$$
$$\mathbf{a} + (\mathbf{b} + \mathbf{c}) = (\mathbf{a} + \mathbf{b}) + \mathbf{c}$$
$$\mathbf{a} + \mathbf{0} = \mathbf{a}$$
$$\mathbf{a} + (-\mathbf{a}) = \mathbf{0}$$

The proof of each property follows readily from the rule for addition of vectors and properties of real numbers. For example, if $\mathbf{a} = \langle a_1, a_2 \rangle$ and $\mathbf{b} = \langle b_1, b_2 \rangle$, then, since $a_1 + b_1 = b_1 + a_1$ and $a_2 + b_2 = b_2 + a_2$,

$$\mathbf{a} + \mathbf{b} = \langle a_1 + b_1, a_2 + b_2 \rangle$$
$$= \langle b_1 + a_1, b_2 + a_2 \rangle$$
$$= \mathbf{b} + \mathbf{a}.$$

The remainder of the proof is left as an exercise (see Exercise 15). The reader should also give geometric interpretations for each property.

The operation of **subtraction** of vectors, denoted by "$-$", is defined as follows.

Subtraction of Vectors

$$\mathbf{a} - \mathbf{b} = \mathbf{a} + (-\mathbf{b})$$

If we use the ordered pair notation for **a** and **b**, then since $-\mathbf{b} = \langle -b_1, -b_2 \rangle$,

$$\mathbf{a} - \mathbf{b} = \langle a_1, a_2 \rangle + \langle -b_1, -b_2 \rangle$$

and hence,

$$\boxed{\mathbf{a} - \mathbf{b} = \langle a_1 - b_1, a_2 - b_2 \rangle.}$$

Thus, to find $\mathbf{a} - \mathbf{b}$, we merely subtract the components of **b** from the corresponding components of **a**.

Example 4 If $\mathbf{a} = \langle 2, -4 \rangle$ and $\mathbf{b} = \langle 6, 7 \rangle$, find $3\mathbf{a} - 5\mathbf{b}$.

Solution We proceed as follows:

$$3\mathbf{a} - 5\mathbf{b} = \langle 6, -12 \rangle - \langle 30, 35 \rangle = \langle -24, -47 \rangle. \qquad \blacksquare$$

The following properties can be proved for any vectors **a**, **b** and real numbers c, d.

Properties of Scalar Multiples of Vectors

$$\begin{aligned} c(\mathbf{a} + \mathbf{b}) &= c\mathbf{a} + c\mathbf{b} \\ (c + d)\mathbf{a} &= c\mathbf{a} + d\mathbf{a} \\ (cd)\mathbf{a} &= c(d\mathbf{a}) = d(c\mathbf{a}) \\ 1\mathbf{a} &= \mathbf{a} \\ 0\mathbf{a} &= \mathbf{0} = c\mathbf{0} \end{aligned}$$

We shall prove the first property and leave the remaining proofs as an exercise (see Exercise 16). Letting $\mathbf{a} = \langle a_1, a_2 \rangle$ and $\mathbf{b} = \langle b_1, b_2 \rangle$, we have

$$\begin{aligned} c(\mathbf{a} + \mathbf{b}) &= c\langle a_1 + b_1, a_2 + b_2 \rangle \\ &= \langle ca_1 + cb_1, ca_2 + cb_2 \rangle \\ &= \langle ca_1, ca_2 \rangle + \langle cb_1, cb_2 \rangle \\ &= c\mathbf{a} + c\mathbf{b}. \end{aligned}$$

The special vectors **i** and **j** are defined as follows:

Definition of i and j

$$\boxed{\mathbf{i} = \langle 1, 0 \rangle, \qquad \mathbf{j} = \langle 0, 1 \rangle.}$$

The vectors **i** and **j** can be used to obtain an alternate way of denoting vectors. Specifically, if $\mathbf{a} = \langle a_1, a_2 \rangle$, then we may write

$$\begin{aligned} \mathbf{a} &= \langle a_1, 0 \rangle + \langle 0, a_2 \rangle \\ &= a_1 \langle 1, 0 \rangle + a_2 \langle 0, 1 \rangle \end{aligned}$$

that is,

$$\boxed{\mathbf{a} = \langle a_1, a_2 \rangle = a_1 \mathbf{i} + a_2 \mathbf{j}.}$$

The vector sum on the right in the last formula is called a **linear combination** of **i** and **j**. If this notation is employed, then previous rules for addition, subtraction, and multiplication by a scalar may be written as follows, where $\mathbf{b} = \langle b_1, b_2 \rangle = b_1\mathbf{i} + b_2\mathbf{j}$:

$$(a_1\mathbf{i} + a_2\mathbf{j}) + (b_1\mathbf{i} + b_2\mathbf{j}) = (a_1 + b_1)\mathbf{i} + (a_2 + b_2)\mathbf{j}$$
$$(a_1\mathbf{i} + a_2\mathbf{j}) - (b_1\mathbf{i} + b_2\mathbf{j}) = (a_1 - b_1)\mathbf{i} + (a_2 - b_2)\mathbf{j}$$
$$c(a_1\mathbf{i} + a_2\mathbf{j}) = (ca_1)\mathbf{i} + (ca_2)\mathbf{j}.$$

These formulas show that linear combinations of **i** and **j** may be regarded as ordinary algebraic sums.

Example 5 If $\mathbf{a} = 5\mathbf{i} + \mathbf{j}$ and $\mathbf{b} = 4\mathbf{i} - 7\mathbf{j}$, express $3\mathbf{a} - 2\mathbf{b}$ as a linear combination of **i** and **j**.

Solution

$$3\mathbf{a} - 2\mathbf{b} = 3(5\mathbf{i} + \mathbf{j}) - 2(4\mathbf{i} - 7\mathbf{j})$$
$$= (15\mathbf{i} + 3\mathbf{j}) - (8\mathbf{i} - 14\mathbf{j})$$
$$= 7\mathbf{i} + 17\mathbf{j}$$ ∎

A **unit vector** is a vector of magnitude 1. The vectors **i** and **j** are unit vectors, as is the vector **c** in Example 1 of this section.

The formula $\mathbf{a} = a_1\mathbf{i} + a_2\mathbf{j}$ for the vector $\mathbf{a} = \langle a_1, a_2 \rangle$ has a useful geometric interpretation. Vectors corresponding to **i**, **j**, and **a** are illustrated in (i) of Figure 7.25. Since **i** and **j** are unit vectors, $a_1\mathbf{i}$ and $a_2\mathbf{j}$ may be represented by horizontal and vertical vectors of magnitudes $|a_1|$ and $|a_2|$, respectively, as illustrated in (ii) of Figure 7.25. The vector **a** may be regarded as the sum of these vectors. For this reason a_1 is called the **horizontal component** and a_2 the **vertical component** of the vector **a**.

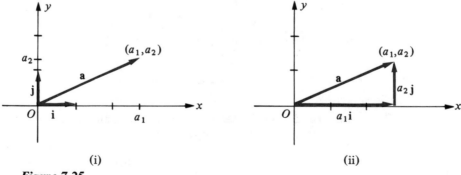

(i)   (ii)

Figure 7.25

Let θ be an angle in standard position, measured from the positive x-axis to the terminal side of the vector corresponding to $\mathbf{a} = \langle a_1, a_2 \rangle$, as illustrated in Figure 7.26.

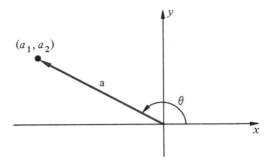

Figure 7.26

Since $\cos \theta = a_1/|\mathbf{a}|$ and $\sin \theta = a_2/|\mathbf{a}|$ we obtain the following useful formulas.

Horizontal and Vertical Components of $\mathbf{a} = \langle a_1, a_2 \rangle$

$$a_1 = |\mathbf{a}| \cos \theta, \qquad a_2 = |\mathbf{a}| \sin \theta$$

Note, in addition, that $\tan \theta = a_2/a_1$. An application of these formulas is given in the next example.

Example 6 Two forces $\overrightarrow{PQ}$ and $\overrightarrow{PR}$ of magnitudes 5.0 kg and 8.0 kg, respectively, act at a point P. If the direction of $\overrightarrow{PQ}$ is N20°E, and the direction of $\overrightarrow{PR}$ is N65°E, approximate the direction of the resultant vector $\overrightarrow{PS}$ to the nearest degree, and find $|\overrightarrow{PS}|$ to the nearest tenth.

Solution In order to use components, let us introduce a coordinate system with origin at the point P, as in Figure 7.27.

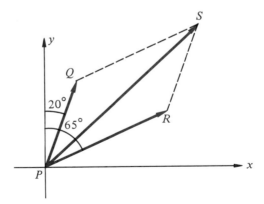

Figure 7.27

Note that the angles from the x-axis to $\overrightarrow{PQ}$ and $\overrightarrow{PR}$ have measures 70° and 25°, respectively. Using the formulas for horizontal and vertical components gives us

$$\overrightarrow{PQ} = \langle 5 \cos 70°, 5 \sin 70° \rangle, \qquad \overrightarrow{PR} = \langle 8 \cos 25°, 8 \sin 25° \rangle.$$

Since $\overrightarrow{PS} = \overrightarrow{PQ} + \overrightarrow{PR}$, it follows that

$$\overrightarrow{PS} = \langle 5\cos 70° + 8\cos 25°, \, 5\sin 70° + 8\sin 25° \rangle.$$

Referring to Table 2 gives us the approximation

$$\overrightarrow{PS} \approx \langle 8.9604, 8.0793 \rangle.$$

Consequently

$$|\overrightarrow{PS}| \approx \sqrt{(8.9604)^2 + (8.0793)^2} \approx 12.1 \text{ kg.}$$

If θ is the angle from the positive x-axis to the resultant $\overrightarrow{PS} = \langle c_1, c_2 \rangle$, then

$$\tan \theta = \frac{c_2}{c_1} \approx \frac{8.0793}{8.9604} \approx 0.9017.$$

Referring to Table 2, $\theta \approx 42°$ and hence the direction of $\overrightarrow{PS}$ is approximately N48°E. ∎

The technique used in Example 6 has many practical applications (see, for example, Exercises 43–48).

EXERCISES 7.10

In each of Exercises 1–10 find $\mathbf{a} + \mathbf{b}, \mathbf{a} - \mathbf{b}, 4\mathbf{a} + 5\mathbf{b}$, and $4\mathbf{a} - 5\mathbf{b}$.

1 $\mathbf{a} = \langle 2, -3 \rangle, \quad \mathbf{b} = \langle 1, 4 \rangle$

2 $\mathbf{a} = \langle -2, 6 \rangle, \quad \mathbf{b} = \langle 2, 3 \rangle$

3 $\mathbf{a} = -\langle 7, -2 \rangle, \quad \mathbf{b} = 4\langle -2, 1 \rangle$

4 $\mathbf{a} = 2\langle 5, -4 \rangle, \quad \mathbf{b} = -\langle 6, 0 \rangle$

5 $\mathbf{a} = \mathbf{i} + 2\mathbf{j}, \quad \mathbf{b} = 3\mathbf{i} - 5\mathbf{j}$

6 $\mathbf{a} = -3\mathbf{i} + \mathbf{j}, \quad \mathbf{b} = -3\mathbf{i} + \mathbf{j}$

7 $\mathbf{a} = -(4\mathbf{i} - \mathbf{j}), \quad \mathbf{b} = 2(\mathbf{i} - 3\mathbf{j})$

8 $\mathbf{a} = 8\mathbf{j}, \quad \mathbf{b} = (-3)(-2\mathbf{i} + \mathbf{j})$

9 $\mathbf{a} = 2\mathbf{j}, \quad \mathbf{b} = -3\mathbf{i}$

10 $\mathbf{a} = \mathbf{0}, \quad \mathbf{b} = \mathbf{i} + \mathbf{j}$

In Exercises 11–14 sketch vectors corresponding to $\mathbf{a}, \mathbf{b}, \mathbf{a} + \mathbf{b}, 2\mathbf{a}$, and $-3\mathbf{b}$.

11 $\mathbf{a} = 3\mathbf{i} + 2\mathbf{j}, \quad \mathbf{b} = -\mathbf{i} + 5\mathbf{j}$

12 $\mathbf{a} = -5\mathbf{i} + 2\mathbf{j}, \quad \mathbf{b} = \mathbf{i} - 3\mathbf{j}$

13 $\mathbf{a} = \langle -4, 6 \rangle, \quad \mathbf{b} = \langle -2, 3 \rangle$

14 $\mathbf{a} = \langle 2, 0 \rangle, \quad \mathbf{b} = \langle -2, 0 \rangle$

15 Prove the second, third, and fourth properties of addition of vectors.

16 Prove the second through the fifth properties of scalar multiples of vectors.

Prove each of the properties in Exercises 17–24, where $\mathbf{a} = \langle a_1, a_2 \rangle$, $\mathbf{b} = \langle b_1, b_2 \rangle$, and c is any real number.

17 $(-1)\mathbf{a} = -\mathbf{a}$

18 $(-c)\mathbf{a} = -c\mathbf{a}$

19 $-(\mathbf{a} + \mathbf{b}) = -\mathbf{a} - \mathbf{b}$

20 $c(\mathbf{a} - \mathbf{b}) = c\mathbf{a} - c\mathbf{b}$

21 If $\mathbf{a} + \mathbf{b} = \mathbf{0}$, then $\mathbf{b} = -\mathbf{a}$.

22 If $\mathbf{a} + \mathbf{b} = \mathbf{a}$, then $\mathbf{b} = \mathbf{0}$.

23 If $c\mathbf{a} = \mathbf{0}$ and $c \neq 0$, then $\mathbf{a} = \mathbf{0}$.

24 If $c\mathbf{a} = \mathbf{0}$ and $\mathbf{a} \neq \mathbf{0}$, then $c = 0$.

25 If $\mathbf{v} = \langle a, b \rangle$, prove each of the following.
 (a) The magnitude of $2\mathbf{v}$ is twice the magnitude of $\mathbf{v}$.
 (b) The magnitude of $\frac{1}{2}\mathbf{v}$ is one-half the magnitude of $\mathbf{v}$.
 (c) The magnitude of $-2\mathbf{v}$ is twice the magnitude of $\mathbf{v}$.
 (d) If k is any real number, then the magnitude of $k\mathbf{v}$ is $|k|$ times the magnitude of $\mathbf{v}$.

26 If $\mathbf{v} = \langle a, b \rangle$ and $\mathbf{w} = \langle c, d \rangle$, give a geometric interpretation for $\mathbf{v} - \mathbf{w}$.

If $\mathbf{v} = a_1\mathbf{i} + a_2\mathbf{j}$ and $\mathbf{u} = b_1\mathbf{i} + b_2\mathbf{j}$, the **inner product** $\mathbf{v} \cdot \mathbf{u}$ is defined by $\mathbf{v} \cdot \mathbf{u} = a_1 b_1 + a_2 b_2$. Prove the properties in Exercises 27–30.

27 $\mathbf{v} \cdot \mathbf{u} = \mathbf{u} \cdot \mathbf{v}$

28 $\mathbf{v} \cdot \mathbf{v} = |\mathbf{v}|^2$

29 $(c\mathbf{v}) \cdot \mathbf{u} = c(\mathbf{v} \cdot \mathbf{u})$, for every real number c

30 $\mathbf{v} \cdot (\mathbf{u} + \mathbf{w}) = \mathbf{v} \cdot \mathbf{u} + \mathbf{v} \cdot \mathbf{w}$, where $\mathbf{w} = c_1\mathbf{i} + c_2\mathbf{j}$

In Exercises 31–38, find the magnitude of $\mathbf{a}$ and the smallest positive angle θ from the positive x-axis to the vector $\overrightarrow{OP}$ corresponding to $\mathbf{a}$.

31 $\mathbf{a} = \langle 3, -3 \rangle$ **32** $\mathbf{a} = \langle -2, -2\sqrt{3} \rangle$

33 $\mathbf{a} = \langle -5, 0 \rangle$ **34** $\mathbf{a} = \langle 0, 10 \rangle$

35 $\mathbf{a} = -4\mathbf{i} + 5\mathbf{j}$ **36** $\mathbf{a} = 10\mathbf{i} - 10\mathbf{j}$

37 $\mathbf{a} = -18\mathbf{j}$ **38** $\mathbf{a} = 2\mathbf{i} - 3\mathbf{j}$

In each of Exercises 39–42, (a) and (b) represent the magnitudes and directions of two forces acting at a point P. Approximate the magnitude of the resultant (to two significant figures) and its direction (to the the nearest degree).

39 (a) 90 kg, N75°W   (b) 60 kg, S5°E

40 (a) 20 kg, S17°W   (b) 50 kg, N82°W

41 (a) 6.0 lb, 110°   (b) 2.0 lb, 215°

42 (a) 70 lb, 320°   (b) 40 lb, 30°

43 An airplane with an airspeed of 200 mph is flying in the direction 50°, and a 40 mph wind is blowing directly from the west. As in Figure 7.28, these may be represented by vectors $\mathbf{p}$ and $\mathbf{w}$ of magnitudes 200 and 40, respectively. The direction of the resultant $\mathbf{p} + \mathbf{w}$ gives the **true course**

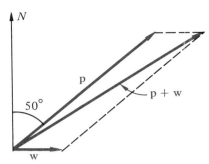

Figure 7.28

of the airplane relative to the ground, and the magnitude $|\mathbf{p} + \mathbf{w}|$ is called the **ground speed** of the airplane. Approximate the true course to the nearest degree and the ground speed to the nearest mph.

44 An airplane is flying in the direction 140° with an airspeed of 500 mph, and a 30 mph wind is blowing from the southwest, in the direction 65°. Approximate the true course (to the nearest degree) and the ground speed (to the nearest mph).

45 An airplane pilot wishes to maintain a true course in the direction 250°, with a ground speed of 400 mph. If the wind is blowing directly north at 50 mph, approximate the required airspeed (to the nearest mph) and compass heading (to the nearest degree).

46 An airplane is flying in the direction 20° with an airspeed of 300 mph. Its ground speed and true course are 350 mph and 30°, respectively. Approximate the direction and speed of the wind.

47 The current in a river flows directly from the west at a rate of 1.5 ft/sec. A person who rows a boat at a rate of 4 ft/sec in still water wishes to row directly across the river. Approximate, to the nearest 10′, the direction in which the person should row.

48 In order for a motor boat moving at a speed of 30 mph to travel directly north across a river, it must aim at a point which has a bearing of N15°E. If the current is flowing directly west, approximate the rate at which it flows, to the nearest mph.

7.11   REVIEW

Concepts

Define or discuss each of the following:

1   The Fundamental Identities

2   Verifying identities

3   Trigonometric equations

4   The addition formulas

5   The double-angle formulas

6   The half-angle formulas

7   The product formulas

8   The factoring formulas

9   Reduction formulas

10   The inverse trigonometric functions

11   The Law of Sines

12   The Law of Cosines

13   Vector

14   Magnitude of a vector

15   Addition of vectors

16   Scalar multiple of a vector

17   Vectors as ordered pairs

18   Components of a vector

19   Subtraction of vectors

20   Unit vector

Exercises

Verify the identities in Exercise 1–16.

1   $(\cot^2 x + 1)(1 - \cos^2 x) = 1$

2   $\cos \theta + \sin \theta \tan \theta = \sec \theta$

3   $\dfrac{(\sec^2 \theta - 1)\cot \theta}{\tan \theta \sin \theta + \cos \theta} = \sin \theta$

4   $(\tan x + \cot x)^2 = \sec^2 x \csc^2 x$

5   $\dfrac{1}{1 + \sin t} = (\sec t - \tan t)\sec t$

6   $\dfrac{\sin(\alpha - \beta)}{\cos(\alpha + \beta)} = \dfrac{\tan \alpha - \tan \beta}{1 - \tan \alpha \tan \beta}$

7   $\dfrac{2 \cot u}{\csc^2 u - 2} = \tan 2u$

8   $\cos^2 \dfrac{v}{2} = \dfrac{1 + \sec v}{2 \sec v}$

9   $\dfrac{\tan^3 \phi - \cot^3 \phi}{\tan^2 \phi + \csc^2 \phi} = \tan \phi - \cot \phi$

10   $\dfrac{\sin u + \sin v}{\csc u + \csc v} = \dfrac{1 - \sin u \sin v}{-1 + \csc u \csc v}$

11   $\cos\left(x - \dfrac{5\pi}{2}\right) = \sin x$

12   $\tan\left(x + \dfrac{3\pi}{4}\right) = \dfrac{\tan x - 1}{\tan x + 1}$

13   $\frac{1}{4}\sin 4\beta = \sin \beta \cos^3 \beta - \cos \beta \sin^3 \beta$

14   $\tan \frac{1}{2}\theta = \csc \theta - \cot \theta$

15   $\sin 8\theta =$
$8 \sin \theta \cos \theta(1 - 2 \sin^2 \theta)(1 - 8 \sin^2 \theta \cos^2 \theta)$

16   $\arctan x = \frac{1}{2}\arctan \dfrac{2x}{1 - x^2}$, where $|x| \le 1$

In each of Exercises 17–28 find the solutions of the given equation which are in the interval $[0, 2\pi)$, and also find the degree measure of each solution.

17   $2 \cos^3 \theta - \cos \theta = 0$

18   $2 \cos \alpha + \tan \alpha = \sec \alpha$

19   $\sin \theta = \tan \theta$

20   $\csc^5 \theta - 4\csc \theta = 0$

21   $2 \cos^3 t + \cos^2 t - 2 \cos t - 1 = 0$

22   $\cos x \cot^2 x = \cos x$

23   $\sin \beta + 2\cos^2 \beta = 1$

24   $\cos 2x + 3 \cos x + 2 = 0$

25 $2 \sec u \sin u + 2 = 4 \sin u + \sec u$

26 $\sin 2u = \sin u$

27 $2 \cos^2 \frac{1}{2}\theta - 3 \cos \theta = 0$

28 $\sec 2x \csc 2x = 2 \csc 2x$

In Exercises 29–32 find the exact values without the use of tables or calculators.

29 $\cos 75°$ **30** $\tan 285°$

31 $\sin 195°$ **32** $\csc \pi/8$

If θ and ϕ are acute angles such that $\csc \theta = 5/3$ and $\cos \phi = 8/17$, find the numbers in Exercises 33–41 without the use of tables or calculators.

33 $\sin(\theta + \phi)$ **34** $\cos(\theta + \phi)$

35 $\tan(\theta - \phi)$ **36** $\sin(\phi - \theta)$

37 $\sin 2\phi$ **38** $\cos 2\phi$

39 $\tan 2\theta$ **40** $\sin \theta/2$

41 $\tan \theta/2$

42 Express $\cos(\alpha + \beta + \gamma)$ in terms of functions of α, β, and γ.

43 Express each of the following products as a sum or difference.

    (a) $\sin 7t \sin 4t$
    (b) $\cos(u/4) \cos(-u/6)$
    (c) $6 \cos 5x \sin 3x$

44 Express each of the following as a product.
    (a) $\sin 8u + \sin 2u$
    (b) $\cos 3\theta - \cos 8\theta$
    (c) $\sin(t/4) - \sin(t/5)$

Find the numbers in Exercises 45–53 without the use of tables or calculators.

45 $\cos^{-1}\left(\dfrac{-\sqrt{3}}{2}\right)$ **46** $\sin^{-1}\left(\dfrac{\sqrt{2}}{2}\right)$

47 $\arccos(-1)$ **48** $\arctan\left(\dfrac{-\sqrt{3}}{3}\right)$

49 $\sin \arccos\left(\dfrac{-\sqrt{3}}{2}\right)$

50 $\cos\left[\sin^{-1}\dfrac{15}{17} - \sin^{-1}\dfrac{8}{17}\right]$

51 $\cos\left[2 \sin^{-1} \frac{4}{5}\right]$ **52** $\sin\left[\sin^{-1} \frac{2}{3}\right]$

53 $\cos^{-1}(\sin 0)$

Sketch the graphs of the equations in Exercises 54–56.

54 $y = \cos^{-1} 3x$ **55** $y = 4 \sin^{-1} x$

56 $y = 1 - \sin^{-1} x$

Without using tables or calculators, find the remaining parts of triangle ABC in Exercises 57–60.

57 $\alpha = 60°$, $\beta = 45°$, $b = 100$

58 $\gamma = 30°$, $a = 2\sqrt{3}$, $c = 2$

59 $\alpha = 60°$, $b = 6$, $c = 7$

60 $a = 2$, $b = 3$, $c = 4$

61 If $\mathbf{a} = \langle -4, 5 \rangle$ and $\mathbf{b} = \langle 2, -8 \rangle$, sketch vectors corresponding to $\mathbf{a} + \mathbf{b}$, $\mathbf{a} - \mathbf{b}$, $2\mathbf{a}$, and $-\frac{1}{2}\mathbf{b}$.

62 If $\mathbf{a} = 2\mathbf{i} + 5\mathbf{j}$ and $\mathbf{b} = 4\mathbf{i} - \mathbf{j}$, find the vectors or numbers corresponding to:
    (a) $4\mathbf{a} + \mathbf{b}$ (b) $2\mathbf{a} - 3\mathbf{b}$
    (c) $|\mathbf{a} - \mathbf{b}|$ (d) $|\mathbf{a}| - |\mathbf{b}|$.

SYSTEMS OF EQUATIONS AND INEQUALITIES

In certain types of mathematical problems it is necessary to work simultaneously with more than one equation in several variables. We then refer to the given equations as a system of equations. It is usually desirable to find the solutions which are common to all equations in the system. In this chapter we shall investigate special types of systems of equations and develop methods for finding their common solutions. Of particular importance are the matrix techniques introduced for systems of linear equations. We shall also touch briefly on systems of inequalities and on linear programming.

8.1 SYSTEMS OF EQUATIONS

An ordered pair (a, b) is a **solution** of an equation in two variables x and y if a true statement is obtained when a and b are substituted for x and y, respectively. Two equations in x and y are **equivalent** if they have exactly the same solutions. For example, the equation

$$x^2 - 4y - 2 = 3x + 2y + 6$$

is equivalent to $\qquad x^2 - 6y - 3x - 8 = 0.$

As we know, the **graph** of an equation in x and y consists of all points in a coordinate plane which correspond to solutions.

By a **system** of two equations in x and y we mean any two equations in those variables. An ordered pair (a, b) is called a **solution of the system** if (a, b) is a solution of both equations. It follows that the points which correspond to the solutions are precisely the points at which the graphs of the two equations intersect.

As a concrete example, consider the system

$$x^2 - y = 0, \quad y - 2x - 3 = 0.$$

366

The following table exhibits some solutions of the equation $x^2 - y = 0$.

x	-2	-1	0	1	2	3	4
y	4	1	0	1	4	9	16

The graph is the parabola sketched in Figure 8.1. A similar table for $y - 2x - 3 = 0$ is

x	-2	-1	0	1	2	3	4
y	-1	1	3	5	7	9	11

The graph is the line in Figure 8.1. Note that the pairs $(3, 9)$ and $(-1, 1)$ are solutions of both equations and hence are solutions of the system. We shall see later that they are the *only* solutions. As pointed out before, the two pairs which are solutions of the system represent points of intersection of the two graphs.

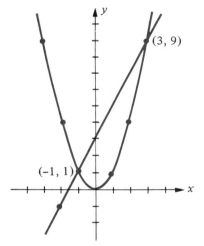

Figure 8.1
$$\begin{cases} x^2 - y = 0 \\ y - 2x - 3 = 0 \end{cases}$$

The graphical analysis we have used is obviously of little value if the solutions of the system involve rational numbers such as 37/849, or irrational numbers, or if the equations are more complicated than those we have considered. Let us illustrate how the solutions of the preceding system can be found without reference to graphs. We shall begin by solving the first equation for y in terms of x, obtaining $y = x^2$. It follows that if an ordered pair (x, y) is a solution of the system, then it must be of the form (x, x^2). In particular, (x, x^2) must be a solution of the equation $y - 2x - 3 = 0$, that is,

$$x^2 - 2x - 3 = 0.$$

Factoring gives us
$$(x - 3)(x + 1) = 0$$

and hence either $x = 3$ or $x = -1$. The corresponding values for y (obtained from $y = x^2$) are 9 and 1, respectively. Thus, the only possible solutions of the system are the ordered pairs $(3, 9)$ and $(-1, 1)$. That these are, indeed, solutions can be seen by checking each equation of the system.

To recapitulate, in order to find the solutions algebraically, we began by solving one equation of the system for y in terms of x, and then we substituted for y in the other equation, obtaining an equation in one variable, x. The solutions of the latter equation were the only possible x-values for the solutions of the system. The corresponding y-values were found by means of the equation which expressed y in terms of x. This algebraic technique is called the *method of substitution*. Sometimes it is more convenient to begin by solving one equation for x in terms of y, and then substituting for x in the other equation. In general, the steps used in this process may be listed as follows.

The Method of Substitution

(i) Solve one of the equations for one variable in terms of the other variable.

(ii) Substitute the expression obtained in step (i) in the other equation, obtaining an equation in one variable.

(iii) Find the solutions of the equation obtained in step (ii).

(iv) Use the solutions from step (iii) together with the expression obtained in step (i) to find the solutions of the system.

In the next example we find the solutions of the system considered at the beginning of this section by first solving one equation for x in terms of y.

Example 1   Find the solutions of the system

$$x^2 - y = 0, \quad y - 2x - 3 = 0.$$

Solution   We may solve the second equation for x in terms of y as follows:

$$2x = y - 3, \quad \text{or} \quad x = \frac{y - 3}{2}.$$

Substituting for x in the first equation gives us

$$\left(\frac{y - 3}{2}\right)^2 - y = 0$$

$$\frac{y^2 - 6y + 9}{4} - y = 0$$

$$y^2 - 6y + 9 - 4y = 0$$

$$y^2 - 10y + 9 = 0.$$

Factoring we obtain $(y - 9)(y - 1) = 0$

which has solutions $y = 9$ and $y = 1$. If we now substitute for y in the equation $x = (y - 3)/2$ we get the corresponding values $x = 3$ and $x = -1$. Hence, as before, the solutions of the system are $(3, 9)$ and $(-1, 1)$. ∎

Example 2   Find the solutions of the system $x^2 + y^2 = 25$, $x^2 + y = 19$. Sketch the graph of each equation, showing the points of intersection.

Solution   Solving the second equation for y we obtain $y = 19 - x^2$. Substituting for y in the first equation leads to the following chain of equivalent equations:

$$x^2 + (19 - x^2)^2 = 25$$
$$x^4 - 37x^2 + 336 = 0$$
$$(x^2 - 16)(x^2 - 21) = 0.$$

The solutions of the last equation are 4, -4, $\sqrt{21}$, and $-\sqrt{21}$. The corresponding y values are found by substituting for x in the equation $y = 19 - x^2$. Substitution of 4 or -4 for x gives us $y = 3$, whereas substitution of $\sqrt{21}$ or $-\sqrt{21}$ gives us $y = -2$. Hence, the only possible solutions of the system are

$$(4, 3), \quad (-4, 3), \quad (\sqrt{21}, -2), \quad \text{and} \quad (-\sqrt{21}, -2).$$

It can be seen by direct substitution in each of the given equations that all four pairs are solutions. The graph of $x^2 + y^2 = 25$ is a circle of radius 5 with center at the origin, and the graph of $y = 19 - x^2$ is a parabola with a vertical axis. The graphs and their points of intersection are illustrated in Figure 8.2.

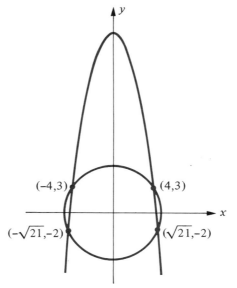

Figure 8.2 $\begin{cases} x^2 + y^2 = 25 \\ x^2 + y = 19 \end{cases}$

∎

Example 3   Find the solutions of the system $y - \log(x + 3) = 1$, $2 - y = \log x$.

Solution   From the second equation we obtain $y = 2 - \log x$. Substituting in the first equation and simplifying, we obtain

$$\log x + \log(x + 3) = 1.$$

This leads to the following chain of equivalent equations (supply the reasons):

$$\log x(x + 3) = 1$$
$$x(x + 3) = 10$$
$$x^2 + 3x - 10 = 0$$
$$(x + 5)(x - 2) = 0.$$

Hence, the only possible solutions are $x = -5$ and $x = 2$. If we substitute -5 for x in the equation $y = 2 - \log x$, we arrive at the logarithm of a negative number. Since $\log x$ is defined only if $x > 0$, this value of x is extraneous. If we substitute 2 for x we get $y = 2 - \log 2$. Checking, we see that the system has one solution $(2, 2 - \log 2)$.   ∎

We can also consider equations in three variables x, y, and z, such as

$$x^2 y + x \log z + 3^y = 4z^3.$$

We say that the equation has a **solution** (a, b, c) if substitution of a, b, and c for x, y and z, respectively, produces a true statement. We refer to (a, b, c) as an **ordered triple** of real numbers. Equivalent equations are defined as before, A system of equations in three variables and the corresponding solutions are defined as in the two-variable case. In like manner, we can consider systems of *any* number of equations in *any* number of variables.

The method of substitution can be extended to these more complicated systems. For example, given three equations in three variables, suppose that it is possible to solve one of the equations for one variable in terms of the remaining two variables. By substituting that expression in each of the other equations, we obtain a system of two equations in two variables. The solutions of the latter system can then be used to find the solutions of the original system, as illustrated in the following example.

Example 4   Find the solutions of the system $x - y + z = 2$, $xyz = 0$, $2y + z = 1$.

Solution   Solving the third equation for z gives us

$$z = 1 - 2y.$$

Substituting for z in the first two equations of the system we obtain

$$\begin{cases} x - y + (1 - 2y) = 2 \\ xy(1 - 2y) = 0 \end{cases}$$

or equivalently,

$$\begin{cases} x - 3y - 1 = 0 \\ xy(1 - 2y) = 0. \end{cases}$$

We now find the solutions of the latter system. Solving the first equation for x in terms of y gives us

$$x = 3y + 1.$$

Substituting $3y + 1$ for x in the second equation $xy(1 - 2y) = 0$, we obtain

$$(3y + 1)y(1 - 2y) = 0$$

which has, for its solutions, the numbers $-\frac{1}{3}, 0$, and $\frac{1}{2}$. These are the only possible y-values for the solutions of the system. To obtain the corresponding x-values, we use the equation $x = 3y + 1$, obtaining $x = 0, 1$, and $\frac{5}{2}$, respectively. Finally, returning to $z = 1 - 2y$, we obtain the z-values $\frac{5}{3}, 1$, and 0. It follows that the solutions of the original system consist of the ordered triples $(0, -\frac{1}{3}, \frac{5}{3}), (1, 0, 1)$, and $(\frac{5}{2}, \frac{1}{2}, 0)$. ∎

$\jmath-3\upsilon$

EXERCISES 8.1

In Exercises 1–34 use the method of substitution to find the solutions of each system of equations.

1 $\begin{cases} y = x^2 - 4 \\ y = 2x - 1 \end{cases}$

2 $\begin{cases} y = x^2 + 1 \\ x + y = 3 \end{cases}$

3 $\begin{cases} y^2 = 1 - x \\ x + 2y = 1 \end{cases}$

4 $\begin{cases} y^2 = x \\ x + 2y + 3 = 0 \end{cases}$

5 $\begin{cases} 2y = x^2 \\ y = 4x^3 \end{cases}$

6 $\begin{cases} x - y^3 = 1 \\ 2x = 9y^2 + 2 \end{cases}$

7 $\begin{cases} x + 2y = -1 \\ 2x - 3y = 12 \end{cases}$

8 $\begin{cases} 3x - 4y + 20 = 0 \\ 3x + 2y + 8 = 0 \end{cases}$

9 $\begin{cases} 2x - 3y = 1 \\ -6x + 9y = 4 \end{cases}$

10 $\begin{cases} 4x - 5y = 2 \\ 8x - 10y = -5 \end{cases}$

11 $\begin{cases} x + 3y = 5 \\ x^2 + y^2 = 25 \end{cases}$

12 $\begin{cases} 3x - 4y = 25 \\ x^2 + y^2 = 25 \end{cases}$

13 $\begin{cases} x^2 + y^2 = 8 \\ y - x = 4 \end{cases}$

14 $\begin{cases} x^2 + y^2 = 25 \\ 3x + 4y = -25 \end{cases}$

15 $\begin{cases} x^2 + y^2 = 9 \\ y - 3x = 2 \end{cases}$

16 $\begin{cases} x^2 + y^2 = 16 \\ y + 2x = -1 \end{cases}$

17 $\begin{cases} x^2 + y^2 = 16 \\ 2y - x = 4 \end{cases}$

18 $\begin{cases} x^2 + y^2 = 1 \\ y + 2x = -3 \end{cases}$

19 $\begin{cases} (x - 1)^2 + (y + 2)^2 = 10 \\ x + y = 1 \end{cases}$

20 $\begin{cases} xy = 2 \\ 3x - y + 5 = 0 \end{cases}$

21 $\begin{cases} y = 20/x^2 \\ y = 9 - x^2 \end{cases}$

22 $\begin{cases} x = y^2 - 4y + 5 \\ x - y = 1 \end{cases}$

23 $\begin{cases} y^2 - 4x^2 = 4 \\ 9y^2 + 16x^2 = 140 \end{cases}$

24 $\begin{cases} 25y^2 - 16x^2 = 400 \\ 9y^2 - 4x^2 = 36 \end{cases}$

25 $\begin{cases} x^2 - y^2 = 4 \\ x^2 + y^2 = 12 \end{cases}$ **26** $\begin{cases} 6x^3 - y^3 = 1 \\ 3x^3 + 4y^3 = 5 \end{cases}$

27 $\begin{cases} x + 2y - z = -1 \\ 2x - y + z = 9 \\ x + 3y + 3z = 6 \end{cases}$

28 $\begin{cases} 2x - 3y - z^2 = 0 \\ x - y - z^2 = -1 \\ x^2 - xy = 0 \end{cases}$

29 $\begin{cases} y = 5^x + 4 \\ y = 5^{2x} - 2 \end{cases}$ **30** $\begin{cases} x + 3(2^y) = 2^{2y} \\ x - 2^{y+1} = 6 \end{cases}$

31 $\begin{cases} \log(3r + 1) = s + 5 \\ \log(r - 2) = s + 4 \end{cases}$

32 $\begin{cases} \log u = v - 3 \\ 5 = v + \log(u - 2) \end{cases}$

33 $\begin{cases} \log(3x + 5) - \log y = 1 \\ 5x - 2y = 6 \end{cases}$

34 $\begin{cases} y + 2 = \log(2x + 5) \\ y - 1 = \log x \end{cases}$

Solve Exercises 35–40 by introducing several variables and using a suitable system of equations.

35 Find two positive integers whose difference is 4 and whose squares differ by 88. (*Hint*: Let x and y denote the two integers.)

36 Find two real numbers whose difference and product both equal 4.

37 The perimeter of a rectangle is 40 inches and its area is 96 square inches. Find the length and width.

38 Generalize Exercise 37 to the case where the perimeter is P and area is A. Express the length and width in terms of P and A. What restrictions are necessary?

39 Find three numbers whose sum and product are 20 and 60, respectively, such that one of the numbers equals the sum of the other two.

40 Find the values of b for which the system

$$\begin{cases} x^2 + y^2 = 4 \\ y = x + b \end{cases}$$

has solutions consisting of (a) one real number, (b) two real numbers, (c) no real numbers. Interpret the three cases geometrically.

8.2 SYSTEMS OF LINEAR EQUATIONS IN TWO VARIABLES

An equation of the form $ax + by + c = 0$ (or, equivalently, $ax + by = -c$) is called a *linear equation* in x and y. Similarly, a linear equation in three variables x, y, and z is an equation of the form $ax + by + cz = d$, where the coefficients are real numbers. Linear equations in any number of variables may be defined in like manner. If a large number of variables occurs, a subscript notation is usually employed. If we let $x_1, x_2, \ldots, x_n$ denote variables, where n is any positive integer, then an expression of the form

$$a_1 x_1 + a_2 x_2 + \cdots + a_n x_n = a$$

where $a_1, a_2, \ldots, a_n$ and a are real numbers, is called a **linear equation in n variables** with real coefficients.

In modern applications of mathematics, perhaps the most common systems of equations are those in which all the equations are linear. In this section we shall only consider systems of two linear equations in two variables. Systems involving more than two variables are discussed in Section 8.3.

The method used to solve a system of equations in several variables is similar to that used for one equation in one variable, in the sense that the given system is replaced by a chain of equivalent systems until a stage is reached from which the solutions are easily obtained. Some general rules which allow us to transform a system of equations into an equivalent system are given in the next theorem. Since we prefer not to specify the number or types of variables, we shall use the notation $p = 0$ and $q = 0$ for typical equations in a system. The reader should bear in mind that the symbols p and q represent expressions in the variables under consideration. For example, if $p = 3x - 2y + 5z - 1$ and $q = 6x + y - 7z - 4$, then the equations $p = 0$ and $q = 0$ have the form

$$3x - 2y + 5z - 1 = 0, \quad 6x + y - 7z - 4 = 0.$$

Transformations that Lead to Equivalent Systems of Equations

The following transformations do not change the solutions of a system of equations.

(i) Interchanging the position of any two equations.

(ii) Multiplying both sides of an equation in the system by a nonzero real number.

(iii) Replacing an equation $q = 0$ of the system by $kp + q = 0$, where $p = 0$ is any other equation in the system and k is any real number.

Proof

It is easy to show that (i) and (ii) do not change the solutions of the system and therefore we shall omit the proofs.

To prove (iii), we first note that a solution of the original system is a solution of both equations $p = 0$ and $q = 0$. Accordingly, each of the expressions p and q equals zero if the variables are replaced by appropriate real numbers, and hence the expression $kp + q$ will also equal zero. Since none of the other equations has been changed, this shows that any solution of the original system is also a solution of the transformed system obtained by replacing the equation $q = 0$ by $kp + q = 0$.

Conversely, given a solution of the transformed system, both of the expressions $kp + q$ and p equal zero if the variables are replaced by appropriate numbers. This implies, however, that $(kp + q) - kp$ equals 0. Since $(kp + q) - kp = q$, we see that q must also equal zero when the substitution is made. Thus, a solution of the transformed system is also a solution of the original system. This completes the proof. ∎

For convenience, we shall describe the process of using (iii) of the preceding theorem by the phrase "add to one equation of the system k times any other equation of the system." Of course, to *add* two equations means to add corresponding sides.

To *multiply* an equation by k means to multiply both sides of the equation by k. The process may also be applied if 0 is not on one side of each equation, as illustrated in the next example.

Example 1 Find the solutions of the system

$$\begin{cases} x + 3y = -1 \\ 2x - y = 5. \end{cases}$$

Solution By (ii) of the last theorem we may multiply the second equation by 3. This gives us the equivalent system

$$\begin{cases} x + 3y = -1 \\ 6x - 3y = 15. \end{cases}$$

Next, by (iii) of the theorem, we may add to the second equation 1 times the first. This gives us

$$\begin{cases} x + 3y = -1 \\ 7x \quad\quad = 14. \end{cases}$$

We see from the last equation that the only possible value for x is 2. The corresponding value for y may be found by substituting for x in the first equation. This gives us the following:

$$2 + 3y = -1, \quad 3y = -3, \quad y = -1.$$

Consequently, the given system has one solution, $(2, -1)$.

There are other methods of solution. For example, we could begin by multiplying the first equation by -2, obtaining

$$\begin{cases} -2x - 6y = 2 \\ 2x - y = 5. \end{cases}$$

If we next add the first equation to the second, we get

$$\begin{cases} -2x - 6y = 2 \\ \quad\quad -7y = 7. \end{cases}$$

The last equation implies that $y = -1$. Substitution for y in the first equation gives us

$$-2x - 6(-1) = 2, \quad -2x = -4, \quad x = 2.$$

Again we see that the solution is $(2, -1)$.

The graphs of the two equations, showing the point of intersection $(2, -1)$, are sketched in Figure 8.3.

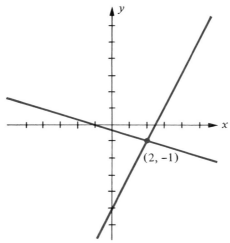

Figure 8.3 $\begin{cases} x + 3y = -1 \\ 2x - y = 5. \end{cases}$ ∎

The technique used in Example 1 is called the **method of elimination**, since it involves eliminating a variable from one of the equations. The method of elimination usually leads to solutions in fewer steps than the method of substitution discussed in Section 8.1.

Example 2 Find the solutions of the system

$$\begin{cases} 3x + y = 6 \\ 6x + 2y = 12. \end{cases}$$

Solution Multiplying the second equation by $\frac{1}{2}$ gives us

$$\begin{cases} 3x + y = 6 \\ 3x + y = 6. \end{cases}$$

Consequently, (a, b) is a solution if and only if $3a + b = 6$, that is, $b = 6 - 3a$. It follows that the solutions consist of all ordered pairs of the form $(a, 6 - 3a)$, where a is a real number. If we wish to find particular solutions we may substitute various values for a. For example, a few solutions are $(0, 6)$, $(1, 3)$, $(3, -3)$, $(-2, 12)$, and $(\sqrt{2}, 6 - 3\sqrt{2})$. ∎

Example 3 Find the solutions of the system

$$\begin{cases} 3x + y = 6 \\ 6x + 2y = 20. \end{cases}$$

Solution   If we add to the second equation -2 times the first equation we obtain the equivalent system

$$\begin{cases} 3x + y = 6 \\ 0 = 8. \end{cases}$$

Since the last equation is never true, the system has no solutions. Of course, this means that the graphs of the given equations do not intersect (see Figure 8.4).

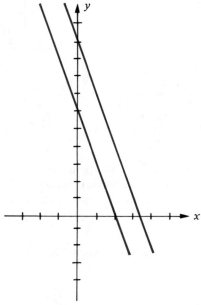

Figure 8.4 $\begin{cases} 3x + y = 6 \\ 6x + 2y = 20 \end{cases}$ ■

Every system of two linear equations in two variables x and y with real coefficients has the form

$$\begin{cases} a_1 x + a_2 y = a \\ b_1 x + b_2 y = b \end{cases}$$

where the coefficients are real numbers. We know from our work in Chapter Three that the graph of each equation in the system is a line. It follows that precisely *one* of the following three possibilities must occur:

    (i)   The lines are identical.

    (ii)   The lines are parallel.

    (iii)   The lines intersect in one and only one point.

Since the pairs which are solutions of the system exhibit the coordinates of the points of intersection of the graphs of the two equations, we may interpret statements (i)–(iii) in the following way.

Theorem

> For each system of two linear equations in two variables, one and only one of the following statements is true:
>
> (i)   The system has an infinite number of solutions.
> (ii) The system has no solutions.
> (iii) The system has exactly one solution.

If (i) occurs, then every solution of one equation of the system is also a solution of the other, and we say that the equations are **dependent**.

If (ii) occurs, there is no solution and the system is said to be **inconsistent**.

If (iii) occurs, there is a unique solution and the system is called **consistent**.

In practice, there should be little difficulty in determining which of the three cases occurs. In case (i) we have a solution similar to that for Example 2, where one of the equations can be transformed into the other. In case (ii) the lack of a solution is indicated by an absurdity such as the statement $0 = 8$, which appeared in Example 3. The case of the unique solution (iii) will be apparent when suitable transformations are applied to the system, as illustrated in Example 1.

Some applied problems can be solved by introducing systems of two linear equations. The Guidelines stated on pages 56 and 57 may still be used provided that, in place of Guideline 2, we introduce *two* letters to denote unknown quantities and, in place of Guideline 5, we formulate a system of two linear equations which involve the two unknowns. The technique is illustrated in the next example.

Example 4   A motor boat, operating at full throttle, made a trip 4 miles upstream (against a constant current) in 15 minutes. The return trip (with the same current and full throttle) took 12 minutes. Find the speed of the current and the equivalent speed of the boat in still water.

Solution   We shall begin by introducing letters to denote the unknown quantities. Thus, let

$$x = \text{speed of boat (in mph)}$$
$$y = \text{speed of current (in mph)}.$$

We plan to use the formula $d = rt$, where d denotes the distance traveled, r the rate, and t the time. Since the current slows the boat as it travels upstream, but adds to its speed as it travels downstream, we obtain

$$\text{upstream rate} = x - y \text{ (in mph)}$$
$$\text{downstream rate} = x + y \text{ (in mph)}.$$

The time (in hours) traveled in each direction is

$$\text{upstream time} = 15/60 = 1/4 \text{ hr}$$
$$\text{downstream time} = 12/60 = 1/5 \text{ hr.}$$

The distance is 4 miles for each trip. Substituting in $d = rt$ gives us the system

$$\begin{cases} 4 = (x - y)(1/4) \\ 4 = (x + y)(1/5) \end{cases}$$

or equivalently,

$$\begin{cases} x - y = 16 \\ x + y = 20. \end{cases}$$

Adding the last two equations we see that $2x = 36$, or $x = 18$. Consequently, $y = 20 - x = 20 - 18 = 2$. Hence, the speed of the boat in still water is 18 mph and the speed of the current is 2 mph. ∎

EXERCISES 8.2

Find the solutions of the systems in Exercises 1–20.

1 $\begin{cases} 2x + 3y = 2 \\ x - 2y = 8 \end{cases}$

2 $\begin{cases} 4x + 5y = 13 \\ 3x + y = -4 \end{cases}$

3 $\begin{cases} 2x + 5y = 16 \\ 3x - 7y = 24 \end{cases}$

4 $\begin{cases} 7x - 8y = 9 \\ 4x + 3y = -10 \end{cases}$

5 $\begin{cases} 3r + 4s = 3 \\ r - 2s = -4 \end{cases}$

6 $\begin{cases} 9u + 2v = 0 \\ 3u - 5v = 17 \end{cases}$

7 $\begin{cases} 5x - 6y = 4 \\ 3x + 7y = 8 \end{cases}$

8 $\begin{cases} 2x + 8y = 7 \\ 3x - 5y = 4 \end{cases}$

9 $\begin{cases} \frac{1}{3}c + \frac{1}{2}d = 5 \\ c - \frac{2}{3}d = -1 \end{cases}$

10 $\begin{cases} \frac{1}{2}t - \frac{1}{5}v = \frac{3}{2} \\ \frac{2}{3}t + \frac{1}{4}v = \frac{5}{12} \end{cases}$

11 $\begin{cases} \sqrt{3}x - \sqrt{2}y = 2\sqrt{3} \\ 2\sqrt{2}x + \sqrt{3}y = \sqrt{2} \end{cases}$

12 $\begin{cases} 0.11x - 0.03y = 0.25 \\ 0.12x + 0.05y = 0.70 \end{cases}$

13 $\begin{cases} 2x - 3y = 5 \\ -6x + 9y = 12 \end{cases}$

14 $\begin{cases} 3p - q = 7 \\ -12p + 4q = 3 \end{cases}$

15 $\begin{cases} 3m - 4n = 2 \\ -6m + 8n = -4 \end{cases}$

16 $\begin{cases} x - 5y = 2 \\ 3x - 15y = 6 \end{cases}$

17 $\begin{cases} 2y - 5x = 0 \\ 3y + 4x = 0 \end{cases}$

18 $\begin{cases} 3x + 7y = 9 \\ y = 5 \end{cases}$

19 $\begin{cases} \dfrac{2}{x} + \dfrac{3}{y} = -2 \\ \dfrac{4}{x} - \dfrac{5}{y} = 1 \end{cases}$ (*Hint*: Let $x' = 1/x$ and $y' = 1/y$.)

20 $\begin{cases} \dfrac{3}{x-1} + \dfrac{4}{y+2} = 2 \\ \dfrac{6}{x-1} - \dfrac{7}{y+2} = -3 \end{cases}$

21 Determine a and b so that the straight line with equation $ax + by = 10$ passes through the points $P_1(-3, 4)$ and $P_2(3, 1)$.

22 Generalize Exercise 21 to the case where the graph of $ax + by = 10$ passes through the points $P_1(x_1, y_1)$ and $P_2(x_2, y_2)$. (Express a and b in terms of x_1, y_1, x_2, and y_2.) Is it necessary to restrict P_1 or P_2 in any way?

Solve each of Exercises 23–30 by employing a system of linear equations.

23 The price of admission for a certain event was $2.25 for adults and $1.50 for children. If 450 tickets were sold for a total of $777.75, how many of each kind were purchased?

24 A chemist wishes to obtain 80 cc of a 30% acid solution by mixing a 50% solution and a 25% solution. How many cc of each solution should be used?

25 The sum of the digits of a certain two-digit number is 14. If the digits are reversed, the number is increased by 18. Find the number.

26 A man rows a boat 500 feet upstream against a constant current in 10 minutes. He then rows downstream (with the same current), covering 300 feet in 5 minutes. Find the speed of the current and the equivalent rate at which he can row in still water.

27 A collection of dimes and quarters amounts to $7.00. If there are seven more dimes than quarters, find the number of quarters.

28 A woman has $10,000 invested in two funds which pay simple interest rates of 8% and 7%, respectively. If she receives yearly interest of $772, how much is invested in each fund?

29 A man receives income from two investments at simple interest rates of $6\frac{1}{2}\%$ and 8%, respectively. He has twice as much invested at $6\frac{1}{2}\%$ as at 8%. If his annual income from the two investments is $698.25, find how much is invested at each rate.

30 Two hoses running at the same time can fill a swimming pool in 4 hours. If both hoses run for 2 hours and the first is then shut off, it takes 3 more hours for the second to fill the pool. How long does it take each hose to fill the pool by itself?

31 A silversmith has two alloys, the first containing 35% silver, and the second 60% silver. How much of each should be melted and combined to obtain 100 gm of an alloy containing 50% silver?

32 A merchant wishes to mix peanuts costing $2.00 per pound with cashews costing $3.50 per pound to obtain 60 pounds of a mixture costing $2.65 per pound. How many pounds of each variety should be mixed?

33 An airplane, flying with a tail wind, travels 1200 miles in 2 hours. The return trip, against the wind, takes $2\frac{1}{2}$ hours. Find the speed of the plane and the speed of the wind (assume that both rates are constant).

34 A stationery company sells two types of notebooks to college bookstores, the first wholesaling for 50¢ and the second for 70¢. The company receives an order for 500 notebooks, together with a check for $286. If the order fails to specify the number of each type, how should the company fill the order?

35 As a ball rolls down an inclined plane, its velocity $v(t)$ (in cm/sec) at time t (in sec) is given by $v(t) = v_0 + at$, where v_0 is the initial velocity and a is the acceleration (in cm/sec^2). If $v(2) = 16$ and $v(5) = 25$, find v_0 and a.

36 If an object is projected vertically upward from an altitude of s_0 feet with an initial velocity of v_0 ft/sec, then its distance above the ground after t seconds is given by

$$s(t) = -16t^2 + v_0 t + s_0.$$

If $s(1) = 84$ and $s(2) = 116$, find v_0 and s_0.

8.3 SYSTEMS OF LINEAR EQUATIONS IN MORE THAN TWO VARIABLES

For systems of equations containing more than two variables we can use either the method of substitution explained in Section 8.1 or the techniques developed in Section 8.2. Let us illustrate this by solving a system in two different ways, first by the method of substitution and second by the method of elimination. You will see that the method of elimination is the shorter and more straightforward technique for finding solutions.

Example 1   Find the solutions of the system

$$\begin{cases} x - 2y + 3z = 4 \\ 2x + y - 4z = 3 \\ -3x + 4y - z = -2. \end{cases}$$

Solution 1   Using the method of substitution we solve the first equation for x, obtaining

$$x = 4 + 2y - 3z.$$

Substituting for x in the second and third equations gives us

$$\begin{cases} 2(4 + 2y - 3z) + y - 4z = 2 \\ -3(4 + 2y - 3z) + 4y - z = -2 \end{cases}$$

which simplifies to

$$\begin{cases} y - 2z = -1 \\ y - 4z = -5. \end{cases}$$

If we solve the first equation for y, obtaining $y = 2z - 1$, and then substitute for y in the second equation, we find that

$$(2z - 1) - 4z = -5 \quad \text{or} \quad -2z = -4.$$

The last equation has the solution $z = 2$. To find y we return to $y = 2z - 1$, obtaining $y = 2(2) - 1 = 3$. The corresponding value of x is found by substituting for y and z in the equation $x = 4 + 2y - 3z$. This gives us $x = 4 + 2(3) - 3(2) = 4$. Hence, the given system has only one solution, the ordered triple $(4, 3, 2)$. We shall leave it to the reader to show that this triple is a solution of *each* of the given equations. Of course, other approaches to the problem could be used, such as solving the third equation for z and substituting in the first and second equations, obtaining two equations in x and y.

Solution 2   We begin by eliminating x from the second and third equations. Adding to the second equation -2 times the first equation gives us the equivalent system

$$\begin{cases} x - 2y + 3z = 4 \\ 5y - 10z = -5 \\ -3x + 4y - z = -2. \end{cases}$$

Next, adding to the third equation 3 times the first equation, we obtain the equivalent system

$$\begin{cases} x - 2y + 3z = 4 \\ 5y - 10z = -5 \\ -2y + 8z = 10. \end{cases}$$

To simplify computations we multiply the second equation by $\frac{1}{5}$, obtaining

$$\begin{cases} x - 2y + 3z = 4 \\ y - 2z = -1 \\ -2y + 8z = 10. \end{cases}$$

We now eliminate y from the third equation by adding to it 2 times the second equation. This gives us the system

$$\begin{cases} x - 2y + 3z = 4 \\ y - 2z = -1 \\ 4z = 8. \end{cases}$$

The solutions of the last system are easy to obtain. From the third equation it is clear that the only possible z-value is 2. Substituting 2 for z in the second equation, $y - 2z = -1$, we get $y - 2(2) = -1$, or $y = 3$. Finally, the x-value is found by substituting for y and z in the first equation. This gives us $x - 2(3) + 3(2) = 4$ and hence $x = 4$. Thus, there is one solution, $(4, 3, 2)$. ∎

The general form for a system of three linear equations in three variables with real coefficients is

$$\begin{cases} a_1 x + a_2 y + a_3 z = a \\ b_1 x + b_2 y + b_3 z = b \\ c_1 x + c_2 y + c_3 z = c \end{cases}$$

where the coefficients are real numbers. It can be shown that the preceding system has either a *unique solution*, an *infinite number of solutions*, or *no solutions*. As with the case of two equations in two variables, the terminology used to describe these cases is *consistent*, *dependent*, or *inconsistent*, respectively.

Similar remarks and methods of solution hold for systems of four linear equations in four variables or, for that matter, systems of n linear equations in n variables, where n is any positive integer. Using the method of substitution we can always solve one of the equations for one variable in terms of the remaining variables. By substituting in each of the remaining equations we obtain a system containing one less equation in one less variable. We keep repeating this process until we obtain a form from which the solution may be easily found. Of course, the method of elimination can also be used.

Sometimes it is necessary to work with systems in which the number of equations is not the same as the number of variables. Such systems can be attacked by using methods similar to those we have discussed.

Example 2 Find the solutions of the system

$$\begin{cases} 2x + 3y + 4z = 1 \\ 3x + 4y + 5z = 3. \end{cases}$$

Solution We can eliminate x from the second equation by adding to it $-3/2$ times the first equation, obtaining the equivalent system

$$\begin{cases} 2x + 3y + 4z = 1 \\ -\tfrac{1}{2}y - z = \tfrac{3}{2}. \end{cases}$$

To clear of fractions we multiply the second equation by -2, obtaining

$$\begin{cases} 2x + 3y + 4z = 1 \\ y + 2z = -3. \end{cases}$$

There are an infinite number of solutions (b, c) for the second equation. Indeed, if we write $y = -3 - 2z$, then by substituting any number c for z, the corresponding value b of y is given by $b = -3 - 2c$. For each such solution (b, c) of $y + 2z = -3$ there corresponds a number a such that the triple (a, b, c) is a solution of the first equation, that is,

$$2a + 3b + 4c = 1.$$

Solving the latter equation for a we obtain

$$a = \tfrac{1}{2}(1 - 3b - 4c) = \tfrac{1}{2} - \tfrac{3}{2}b - 2c.$$

Since $b = -3 - 2c$, we have

$$a = \tfrac{1}{2} - \tfrac{3}{2}(-3 - 2c) - 2c$$

which reduces to $\qquad a = 5 + c.$

Consequently, the solutions of the system consist of all ordered triples of the form

$$(5 + c, -3 - 2c, c)$$

where c is any real number. These solutions may be checked by substituting $5 + c$ for x, $-3 - 2c$ for y, and c for z in the two given equations. We can obtain any number of solutions for the system by substituting specific real numbers for c. For example, if $c = 0$, we obtain $(5, -3, 0)$; if $c = 2$, we have $(7, -7, 2)$; if $c = -1/2$, we get $(9/2, -2, -1/2)$; etc. ∎

Another method of solution for Example 2 is to rewrite the original system as

$$\begin{cases} 2x + 3y = 1 - 4z \\ 3x + 4y = 3 - 5z. \end{cases}$$

If a number c is substituted for z, we obtain a system of two equations in two variables x and y, and the solutions can be found as in Section 8.2. For a different approach we could solve for two *other* variables in terms of the third. This method can be extended. For example, given two equations in four variables, say x, y, z, and w, it may be possible to solve for two of the variables, say x and y, in terms of the remaining variables z and w. If we then substitute numbers c and d for z and w, respectively, and calculate the corresponding values for x and y, we obtain the solutions.

The general system of three linear equations in x, y, z discussed prior to Example 2 is called **homogeneous** if $a = b = c = 0$. The same terminology is used for more than three equations. A system of homogeneous equations always has the **trivial solution** obtained by substituting zero for all the variables. Nontrivial solutions sometimes exist. The procedure for finding solutions is the same as that used for nonhomogeneous systems.

Example 3 Find the solutions of the system

$$\begin{cases} x - y + 4z = 0 \\ 2x + y - z = 0 \\ -x - y + 2z = 0. \end{cases}$$

Solution   Adding to the second equation -2 times the first equation and then adding the first equation to the third gives us the equivalent system

$$\begin{cases} x - y + 4z = 0 \\ \quad 3y - 9z = 0 \cdot \\ \quad -2y + 6z = 0. \end{cases}$$

Multiplying the second equation by $\frac{1}{3}$ and the third equation by $-\frac{1}{2}$ leads to the same equation, $y - 3z = 0$. Hence, the given system is equivalent to

$$\begin{cases} x - y + 4z = 0 \\ \quad y - 3z = 0. \end{cases}$$

The solutions may now be found as in the previous example. If we substitute any value c for z, then the second equation gives us $y = 3c$. Substituting c for z and $3c$ for y in the first equation, we obtain $x - 3c + 4c = 0$, or $x = -c$. Thus, the solutions consist of all ordered triples of the form $(-c, 3c, c)$, where c is any real number. ∎

Example 4 Find the solutions of the system

$$\begin{cases} x + y + z = 0 \\ x - y + z = 0 \\ x - y - z = 0. \end{cases}$$

Solution   By the usual process we obtain the equivalent system

$$\begin{cases} x + y + z = 0 \\ \quad -2y \qquad = 0 \\ \quad -2y - 2z = 0. \end{cases}$$

Multiplying the second and third equations by $-\frac{1}{2}$ gives us

$$\begin{cases} x + y + z = 0 \\ \quad\quad y \quad\quad = 0 \\ \quad\quad y + z = 0. \end{cases}$$

From the second equation of this system we see that the only possible y-value is 0. Substituting 0 for y in the last equation we obtain $0 + z = 0$, or $z = 0$. Finally, substituting 0 for both y and z in the first equation we see that x also must equal 0. Hence, the only solution for the system is the trivial one, $(0, 0, 0)$. ∎

The next example is an illustration of an applied problem which can be solved by means of a system of three linear equations.

Example 5   A merchant wishes to mix two grades of peanuts costing $1.50 and $2.50 per pound, respectively, with cashews costing $4.00 per pound, to obtain 130 pounds of a mixture costing $3.00 per pound. If the merchant also wants the amount of cheaper grade peanuts to be twice that of the better grade, how many pounds of each variety should be mixed?

Solution   Let us introduce three variables as follows:

$$x = \text{pounds of peanuts at \$1.50 per pound}$$
$$y = \text{pounds of peanuts at \$2.50 per pound}$$
$$z = \text{pounds of cashews at \$4.00 per pound.}$$

We refer to the statement of the problem and write the following system of equations

$$\begin{cases} x + y + z = 130 \\ 1.50x + 2.50y + 4.00z = 3.00(130) \\ \quad\quad\quad\quad x = 2y. \end{cases}$$

We shall use a combination of the methods of substitution and elimination to find the solution. Let us begin by substituting $2y$ for x in the first two equations, obtaining

$$\begin{cases} 2y + y + z = 130 \\ 1.5(2y) + 2.5y + 4z = 3(130) \end{cases}$$

or equivalently,

$$\begin{cases} 3y + \quad z = 130 \\ 5.5y + 4z = 390. \end{cases}$$

We next eliminate z from the second equation by adding to it -4 times the first equation. This gives us

$$\begin{cases} 3y + z = \quad 130 \\ -6.5y = -130. \end{cases}$$

It follows that

$$y = \frac{-130}{-6.5} = 20$$

$$x = 2y = 2(20) = 40$$

$$z = 130 - x - y = 130 - 40 - 20 = 70.$$

Thus, the merchant should use 40 pounds of the $1.50 peanuts, 20 pounds of the $2.50 peanuts, and 70 pounds of cashews. The check is left to the reader. ∎

EXERCISES 8.3

3-27 3RD

Solve the systems in Exercises 1–26.

1 $\begin{cases} x - 2y - 3z = -1 \\ 2x + y + z = 6 \\ x + 3y - 2z = 13 \end{cases}$

2 $\begin{cases} x + 3y - z = -3 \\ 3x - y + 2z = 1 \\ 2x - y + z = -1 \end{cases}$

3 $\begin{cases} 5x + 2y - z = -7 \\ x - 2y + 2z = 0 \\ 3y + z = 17 \end{cases}$

4 $\begin{cases} 4x - y + 3z = 6 \\ -8x + 3y - 5z = -6 \\ 5x - 4y = -9 \end{cases}$

5 $\begin{cases} 2x + 6y - 4z = 1 \\ x + 3y - 2z = 4 \\ 2x + y - 3z = -7 \end{cases}$

6 $\begin{cases} 2x - y = 5 \\ 5y + 3z = -2 \\ x - 7z = 3 \end{cases}$

7 $\begin{cases} 2x - 3y + 2z = -3 \\ -3x + 2y + z = 1 \\ 4x + y - 3z = 4 \end{cases}$

8 $\begin{cases} 2x - 3y + z = 2 \\ 3x + 2y - z = -5 \\ 5x - 2y + z = 0 \end{cases}$

9 $\begin{cases} x + 3y + z = 0 \\ x + y - z = 0 \\ x - 2y - 4z = 0 \end{cases}$

10 $\begin{cases} 2x - y + z = 0 \\ x - y - 2z = 0 \\ 2x - 3y + z = 0 \end{cases}$

11 $\begin{cases} 2x + y + z = 0 \\ x - 2y - 2z = 0 \\ x + y + z = 0 \end{cases}$

12 $\begin{cases} x + y - 2z = 0 \\ x - y - 4z = 0 \\ z + y = 0 \end{cases}$

13 $\begin{cases} 3x - 2y + 5z = 7 \\ x + 4y - z = -2 \end{cases}$

14 $\begin{cases} 2x - y + 4z = 8 \\ -3x + y - 2z = 5 \end{cases}$

15 $\begin{cases} 4x - 2y + z = 5 \\ 3x + y - 4z = 0 \end{cases}$

16 $\begin{cases} 5x + 2y - z = 10 \\ y + z = -3 \end{cases}$

17 $\begin{cases} x + 2y - z - 3w = 2 \\ 3x + y - 2z - w = 6 \\ x + y + 3z - 2w = -3 \\ 4x - 3y - z - 2w = -8 \end{cases}$

18 $\begin{cases} x - 2y - 5z + w = -1 \\ 2x - y + z + w = 1 \\ 3x - 2y - 4z - 2w = 1 \\ x + y + 3z - 2w = 2 \end{cases}$

19 $\begin{cases} 5x + 2z = 1 \\ y - 3z = 2 \\ 2x + y = 3 \end{cases}$

20 $\begin{cases} 2x - 5y = 4 \\ 3y + 2z = -3 \\ 7x - 3z = 1 \end{cases}$

21 $\begin{cases} 4x - 3y = 1 \\ 2x + y = -7 \\ -x + y = -1 \end{cases}$

22 $\begin{cases} 2x + 3y = -2 \\ x + y = 1 \\ x - 2y = 13 \end{cases}$

23 $\begin{cases} 2x + 3y = 5 \\ x - 3y = 4 \\ x + y = -2 \end{cases}$ **24** $\begin{cases} 4x - y = 2 \\ 2x + 2y = 1 \\ 4x - 5y = 3 \end{cases}$

25 $\begin{cases} \dfrac{3}{x} - \dfrac{3}{y} + \dfrac{1}{z} = -1 \\ \dfrac{1}{x} + \dfrac{4}{y} + \dfrac{1}{z} = 5 \\ \dfrac{2}{x} + \dfrac{1}{y} - \dfrac{4}{z} = 0 \end{cases}$ **26** $\begin{cases} \dfrac{1}{x} + \dfrac{1}{y} - \dfrac{1}{z} = 4 \\ \dfrac{2}{x} - \dfrac{1}{y} + \dfrac{1}{z} = 2 \\ \dfrac{4}{x} + \dfrac{2}{y} + \dfrac{3}{z} = -3 \end{cases}$

27 A collection of nickels, dimes, and quarters amounts to $8.00. If there are 72 coins in all and there are twice as many nickels as there are quarters, find the number of dimes.

28 A three-digit number is equal to 31 times the sum of its digits. If the three digits are reversed, the resulting number exceeds the given number by 99. The tens digit is three less than the sum of the hundreds and units digits. Find the number.

29 A chemist has three solutions containing a certain acid. The first contains 10% acid, the second 30%, and the third 50%. He wishes to use all three solutions to obtain a mixture of 50 liters containing 32% acid, using twice as much of the 50% solution as the 30% solution. How many liters of each solution should be used?

30 A swimming pool can be filled by three pipes A, B, and C. Pipe A can fill the pool by itself in 8 hours. If pipes A and C are used together, the pool can be filled in 6 hours. If B and C are used together, it takes 10 hours. How long does it take to fill the pool if all three pipes are used?

31 A company has three machines A, B, and C that are each capable of producing a certain item. However, because of a lack of skilled operators, only two of the machines can be used simultaneously. The following table indicates production over a three-day period using various combinations of the machines.

Machines used	Hours used	Items produced
A and B	6	4500
A and C	8	3600
B and C	7	4900

How long would it take each machine, if used alone, to produce 1000 items?

32 In electrical circuits, the formula $1/R = (1/R_1) + (1/R_2)$ is used to find the total resistance R if two resistances R_1 and R_2 are connected in parallel. Given three resistances A, B, and C, suppose that the total resistance is 48 ohms if A and B are connected in parallel; 80 ohms if B and C are connected in parallel, and 60 ohms if A and C are connected in parallel. Find A, B, and C.

33 A supplier of lawn products has three types G_1, G_2, and G_3 of grass fertilizer having nitrogen contents of 30%, 20%, and 15%, respectively. It is planned to mix them, obtaining 600 pounds of fertilizer with a 25% nitrogen content. In addition, the mixture is to contain 100 pounds more of type G_3 than of type G_2. How much of each type should be used?

34 If a particle moves along a coordinate line with a constant acceleration a (cm/sec^2), then at time t (sec) its distance $s(t)$ (cm) from the origin is given by

$$s(t) = \frac{1}{2}at^2 + v_0 t + s_0$$

where v_0 and s_0 are the velocity and distance from the origin, respectively, at $t = 0$. If the distances of the particle from the origin at $t = 1/2$, $t = 1$, and $t = 3/2$ are 7, 11, and 17, respectively, find a, v_0, and s_0.

35 Find a quadratic function f such that $f(1) = 3$, $f(2) = 5/2$, and $f(-1) = 1$.

36 Given $f(x) = ax^3 + bx + c$, determine a, b, and c such that the graph of f passes through the points $(-3, -12)$, $(-1, 22)$, and $(2, 13)$.

37 Find an equation of the circle which passes through the three points $P_1(2, 1)$, $P_2(-1, -4)$, and $P_3(3, 0)$. (*Hint*: An equation of the circle has the form $x^2 + y^2 + ax + by + c = 0$.)

38 Determine a, b, and c such that the graph of the equation

$$y = ax^2 + bx + c$$

passes through the points $P_1(3, -1)$, $P_2(1, -7)$, and $P_3(-2, 14)$.

8.4 MATRIX SOLUTIONS OF SYSTEMS OF LINEAR EQUATIONS

In Example 1 of Section 8.3 we used two different methods to solve the system

$$\begin{cases} x - 2y + 3z = 4 \\ 2x + y - 4z = 3 \\ -3x + 4y - z = -2. \end{cases}$$

If we analyze the second method (the method of elimination), we see that the symbols used for the variables really have little effect on the solution of the problem. The *coefficients* of the variables are the important things to consider. Thus, if different symbols such as r, s, and t are used for the variables in that example, we obtain the system

$$\begin{cases} r - 2s + 3t = 4 \\ 2r + s - 4t = 3 \\ -3r + 4s - t = -2. \end{cases}$$

The method of elimination could then proceed exactly as before. Since this is true, it is possible to use a process which reduces the amount of labor involved. Specifically, we introduce a notation for keeping track of the coefficients in such a way that the variables do not have to be written down. Referring to the preceding system and checking that corresponding variables appear underneath one another, and terms not involving variables are to the right of the equals sign, we list the numbers which are involved in the equations in the following manner:

$$\begin{bmatrix} 1 & -2 & 3 & 4 \\ 2 & 1 & -4 & 3 \\ -3 & 4 & -1 & -2 \end{bmatrix}$$

If some variable had not appeared in one of the equations, we would have used a zero in the appropriate position. An array of numbers of this type is called a **matrix**. The **rows** of the matrix are the numbers which appear next to one another *horizontally*. Thus, the first row is $1 - 2\ 3\ 4$, the second row is $2\ 1\ -4\ 3$, and the third row is $-3\ 4 -1\ -2$. The **columns** of the matrix are the sets of numbers which form a *vertical* pattern. For example, the second column consists of the numbers -2, 1, 4 (in that order); the fourth column consists of 4, 3, -2; and so on.

Before discussing a method of solving a system of equations by means of a matrix, let us introduce a general definition of matrix. It will be convenient to use a **double subscript notation** for the symbols which represent the numbers in the matrix. Specifically, the symbol a_{ij} will denote the number which appears in row i and column j. We call i the **row subscript** and j the **column subscript** of a_{ij}.

Definition

> If m and n are positive integers, then an **$m \times n$ matrix** over the real number system is an array of the form
>
> $$\begin{bmatrix} a_{11} & a_{12} & a_{13} & \cdots & a_{1n} \\ a_{21} & a_{22} & a_{23} & \cdots & a_{2n} \\ a_{31} & a_{32} & a_{33} & \cdots & a_{3n} \\ \vdots & \vdots & \vdots & & \vdots \\ a_{m1} & a_{m2} & a_{m3} & \cdots & a_{mn} \end{bmatrix}$$
>
> where each a_{ij} is a real number.

The symbol $m \times n$ in the definition is read "m by n." It is also possible to consider matrices where the elements a_{ij} belong to other mathematical systems; however, we shall not do so in this book. The rows and columns of a matrix are defined as before. Thus, the matrix in the definition has m rows and n columns. The reader should carefully examine the notation for a matrix, observing, for example, that the symbol a_{23} appears in row 2 and column 3, whereas a_{32} appears in row 3 and column 2. Each a_{ij} is called an **element of the matrix**. The elements $a_{11}, a_{22}, a_{33}, \ldots$ are called the **main diagonal elements**. If $m \neq n$, the matrix is sometimes referred to as a **rectangular matrix**. If $m = n$, we refer to the matrix as a **square matrix of order n**.

It is possible to develop a comprehensive theory for matrices. In Section 8.5 we shall define *equality* of matrices, the *sum* of two matrices, the *product* of two matrices, and the *inverse* of a matrix. These ideas are very important and useful in mathematics and applications. In particular, matrix techniques are well suited for the types of problems which are solved by computers. Our main purpose in introducing matrices at this time is for use as a tool in solving systems of linear equations.

Let us return to the system of equations considered at the beginning of this section. The matrix we formed at that time is called the **matrix of the system**. In certain cases we may wish to consider only the *coefficients* of the variables of a system of linear equations. The corresponding matrix is called the **coefficient matrix**. The coefficient matrix for the system considered earlier is

$$\begin{bmatrix} 1 & -2 & 3 \\ 2 & 1 & -4 \\ -3 & 4 & -1 \end{bmatrix}.$$

In order to distinguish the matrix of the system from the coefficient matrix, the former is sometimes called the **augmented matrix**. After forming the matrix of the system, we work with the rows of the matrix *just as though they were equations*. The only items missing are the symbols for the variables, the addition signs used between terms, and the equals signs. We simply keep in mind that the numbers in the first column are the coefficients of the first variable, the numbers in the second column are the coefficients of the second variable, and so on. Our rules for transforming a matrix are formulated in such a way that they always produce a matrix of an equivalent system of equations. As a matter of fact, we have the following theorem for matrices, which is a perfect analogue of the theorem on transformations of systems stated on page 373.

Matrix Row Transformation Theorem

> Given a matrix of a system of linear equations, each of the following transformations results in a matrix of an equivalent system of linear equations:
>
> (i)   Interchanging any two rows.
>
> (ii) Multiplying all of the elements in a row by the same nonzero real number k.
>
> (iii) Adding to the elements in a row k times the corresponding elements of any other row, where k is any real number.

For convenience we shall refer to (ii) as the process of "multiplying a row by the number k" and (iii) will be described by saying "add k times any other row to a row." Rules (i)–(iii) of the preceding theorem are called the **elementary row transformations** of a matrix.

Let us find the solution of the system

$$\begin{cases} r - 2s + 3t = 4 \\ 2r + s - 4t = 3 \\ -3r + 4s - t = -2 \end{cases}$$

by means of the preceding theorem. The reader should compare our work with the second solution of Example 1 of Section 8.3. We shall use arrows to indicate that the form of the matrices has been changed by means of elementary row transformations, and we shall justify each transformation by stating the reason for each step. Thus,

$$\begin{bmatrix} 1 & -2 & 3 & 4 \\ 2 & 1 & -4 & 3 \\ -3 & 4 & -1 & -2 \end{bmatrix} \rightarrow \begin{bmatrix} 1 & -2 & 3 & 4 \\ 0 & 5 & -10 & -5 \\ -3 & 4 & -1 & -2 \end{bmatrix}$$

Add to the second row -2 times the first row.

$$\rightarrow \begin{bmatrix} 1 & -2 & 3 & 4 \\ 0 & 5 & -10 & -5 \\ 0 & -2 & 8 & 10 \end{bmatrix}$$

Add to the third row 3 times the first row.

$$\rightarrow \begin{bmatrix} 1 & -2 & 3 & 4 \\ 0 & 1 & -2 & -1 \\ 0 & -2 & 8 & 10 \end{bmatrix}$$

Multiply row 2 by $\frac{1}{5}$.

$$\rightarrow \begin{bmatrix} 1 & -2 & 3 & 4 \\ 0 & 1 & -2 & -1 \\ 0 & 0 & 4 & 8 \end{bmatrix}$$

Add to the third row 2 times the second row.

The last matrix has zeros everywhere below the main diagonal and is said to be in **echelon form**. We now use that matrix to return to the system of equations

$$\begin{cases} r - 2s + 3t = 4 \\ s - 2t = -1 \\ 4t = 8 \end{cases}$$

which is equivalent to the given system. The solutions may now be found as was done in the previous section.

The method used above is completely general. Beginning with *any* system of linear equations, we can use elementary row transformations of type (iii) many times, and also interchange rows if necessary, to introduce zeros everywhere in the first column after the first row; that is, we can obtain a matrix of an equivalent system which has the form

$$\begin{bmatrix} a_{11} & a_{12} & a_{13} & \cdots \\ 0 & b_{22} & b_{23} & \cdots \\ 0 & c_{32} & c_{33} & \cdots \\ 0 & \cdot & \cdot & \cdots \\ \cdot & \cdot & \cdot & \cdots \\ \cdot & \cdot & \cdot & \cdots \\ \cdot & \cdot & \cdot & \cdots \\ 0 & \cdot & \cdot & \cdots \end{bmatrix}$$

where the dots indicate that possibly more numbers appear. If the latter matrix has a nonzero element in the second column and after the first row, then again using elementary row transformations of type (iii), we may transform the matrix into the form

$$\begin{bmatrix} a_{11} & a_{12} & a_{13} & \cdots \\ 0 & k_{22} & k_{23} & \cdots \\ 0 & 0 & l_{33} & \cdots \\ \cdot & \cdot & \cdot & \cdots \\ \cdot & \cdot & \cdot & \cdots \\ \cdot & \cdot & \cdot & \cdots \\ 0 & 0 & \cdot & \cdots \end{bmatrix}$$

where only zeros appear below k_{22} in the second column. This process is continued until we obtain the matrix of a system which is in echelon form. That will be true when all elements below the main diagonal are zero. The final matrix is then used to obtain a system of equations which is equivalent to the original system. The solutions are then found as in previous sections. Let us illustrate the technique by solving a system of four linear equations.

Example 1   Find the solutions of the system

$$\begin{cases} x & -2z + 2w = & 1 \\ -2x + 3y + 4z & = -1 \\ y + z - w = & 0 \\ 3x + y - 2z - w = & 3. \end{cases}$$

Solution   Notice that we have arranged the equations so that the same variables appear in vertical columns. We shall begin with the matrix of the system and proceed to the solution.

$$\begin{bmatrix} 1 & 0 & -2 & 2 & 1 \\ -2 & 3 & 4 & 0 & -1 \\ 0 & 1 & 1 & -1 & 0 \\ 3 & 1 & -2 & -1 & 3 \end{bmatrix} \rightarrow \begin{bmatrix} 1 & 0 & -2 & 2 & 1 \\ 0 & 3 & 0 & 4 & 1 \\ 0 & 1 & 1 & -1 & 0 \\ 0 & 1 & 4 & -7 & 0 \end{bmatrix}$$

Add to the second row 2 times the first row and then add to the fourth row -3 times the first row.

$$\rightarrow \begin{bmatrix} 1 & 0 & -2 & 2 & 1 \\ 0 & 1 & 1 & -1 & 0 \\ 0 & 3 & 0 & 4 & 1 \\ 0 & 1 & 4 & -7 & 0 \end{bmatrix}$$

Interchange rows 2 and 3.

$$\rightarrow \begin{bmatrix} 1 & 0 & -2 & 2 & 1 \\ 0 & 1 & 1 & -1 & 0 \\ 0 & 0 & -3 & 7 & 1 \\ 0 & 0 & 3 & -6 & 0 \end{bmatrix}$$

Add to the third row -3 times the second row and then add to the fourth row -1 times the second row.

$$\rightarrow \begin{bmatrix} 1 & 0 & -2 & 2 & 1 \\ 0 & 1 & 1 & -1 & 0 \\ 0 & 0 & -3 & 7 & 1 \\ 0 & 0 & 0 & 1 & 1 \end{bmatrix}$$

Add row 3 to row 4.

The final matrix corresponds to the system of equations

$$\begin{cases} x & -2z + 2w = 1 \\ y + z - w = 0 \\ -3z + 7w = 1 \\ w = 1. \end{cases}$$

From the last equation we see that $w = 1$. Substituting in the third equation, we get $-3z + 7(1) = 1$, or $z = 2$. Using the second equation, we get $y + 2 - 1 = 0$, or $y = -1$. Finally, from the first equation, we have $x - 2(2) + 2(1) = 1$, or $x = 3$. Hence, the system has only one solution, $x = 3$, $y = -1$, $z = 2$, and $w = 1$. ∎

EXERCISES 8.4

1–20 Use matrices to solve Exercises 1–20 of Section 8.3.

21–28 Use matrices to solve Exercises 1–8 of Section 8.2.

Solve Exercises 29 and 30 by means of matrices.

29
$$\begin{cases} 2x - y - 2z + 2s - 5t = 2 \\ x + 3y - 2z + s - 2t = -5 \\ -x + 4y + 2z - 3s + 8t = -4 \\ 3x - 2y - 4z + s - 3t = -3 \\ 4x - 6y + z - 2s + t = 10 \end{cases}$$

30
$$\begin{cases} 3x + 2y + z + 3u + v + w = 1 \\ 2x + y - 2z + 3u - v + 4w = 6 \\ 6x + 3y + 4z - u + 2v + w = -6 \\ x + y + z + u - v - w = 8 \\ -2x - 2y + z - 3u + 2v - 3w = -10 \\ x - 3y + 2z + u + 3v + w = -1 \end{cases}$$

8.5 THE ALGEBRA OF MATRICES

It is possible to develop a comprehensive theory for matrices which has many mathematical and scientific applications. In this section we shall discuss several basic algebraic properties of matrices which serve as the starting point for such a theory. Our work will be restricted to matrices whose elements are real numbers.

In order to conserve space it is sometimes convenient to denote an $m \times n$ matrix A of the type displayed in the definition given in Section 8.4 by the symbol (a_{ij}). If (b_{ij}) denotes another $m \times n$ matrix B, then we say that A and B are **equal** and we write

$$A = B \quad \text{if and only if} \quad a_{ij} = b_{ij}$$

for every i and j. For example,

$$\begin{bmatrix} 1 & 0 & 5 \\ \sqrt[3]{8} & 3^2 & -2 \end{bmatrix} = \begin{bmatrix} (-1)^2 & 0 & \sqrt{25} \\ 2 & 9 & -2 \end{bmatrix}.$$

If $A = (a_{ij})$ and $B = (b_{ij})$ are $m \times n$ matrices, then their **sum** $A + B$ is defined as the $m \times n$ matrix $C = (c_{ij})$, where $c_{ij} = a_{ij} + b_{ij}$ for all i and j. Thus, to add two matrices we add the elements which appear in corresponding positions in each matrix. Note that two matrices can be added only if they have the same number of rows and the same number of columns. Using the parentheses notation, we have

$$(a_{ij}) + (b_{ij}) = (a_{ij} + b_{ij}).$$

Although we have used the symbol $+$ in two different ways, there is little chance for confusion, since whenever $+$ appears between symbols for matrices it refers to matrix addition, and when $+$ is used between real numbers it denotes their sum. An example of the sum of two 3×2 matrices is

$$\begin{bmatrix} 4 & -5 \\ 0 & 4 \\ -6 & 1 \end{bmatrix} + \begin{bmatrix} 3 & 2 \\ 7 & -4 \\ -2 & 1 \end{bmatrix} = \begin{bmatrix} 7 & -3 \\ 7 & 0 \\ -8 & 2 \end{bmatrix}.$$

It is not difficult to prove that addition of matrices is both commutative and associative, that is,

$$A + B = B + A, \quad A + (B + C) = (A + B) + C$$

for $m \times n$ matrices A, B, and C.

The **$m \times n$ zero matrix**, denoted by O, is the matrix with m rows and n columns in which every element is 0. It is an identity element relative to addition, since

$$A + O = A$$

for every $m \times n$ matrix A. For example,

$$\begin{bmatrix} a_{11} & a_{12} \\ a_{21} & a_{22} \\ a_{31} & a_{32} \end{bmatrix} + \begin{bmatrix} 0 & 0 \\ 0 & 0 \\ 0 & 0 \end{bmatrix} = \begin{bmatrix} a_{11} & a_{12} \\ a_{21} & a_{22} \\ a_{31} & a_{32} \end{bmatrix}.$$

The **additive inverse** $-A$ of the matrix $A = (a_{ij})$ is, by definition, the matrix $(-a_{ij})$ obtained by changing the sign of each element of A. For example,

$$-\begin{bmatrix} 2 & -3 & 4 \\ -1 & 0 & 5 \end{bmatrix} = \begin{bmatrix} -2 & 3 & -4 \\ 1 & 0 & -5 \end{bmatrix}.$$

It follows that for every $m \times n$ matrix A,

$$A + (-A) = O.$$

Subtraction of two $m \times n$ matrices is defined by

$$A - B = A + (-B).$$

Using the parentheses notation for matrices, this implies that

$$(a_{ij}) - (b_{ij}) = (a_{ij}) + (-b_{ij}) = (a_{ij} - b_{ij}).$$

Thus, to subtract two matrices, we merely subtract the elements in the same positions.

The **product** of a real number c and an $m \times n$ matrix $A = (a_{ij})$ is defined by

$$cA = (ca_{ij}).$$

Thus, to find cA, we multiply each element of A by c. For example,

$$3\begin{bmatrix} 4 & -1 \\ 2 & 3 \end{bmatrix} = \begin{bmatrix} 12 & -3 \\ 6 & 9 \end{bmatrix}.$$

The following results may be established, where A and B are $m \times n$ matrices, and c and d are any real numbers.

$$c(A + B) = cA + cB$$
$$(c + d)A = cA + dA$$
$$(cd)A = c(dA).$$

The following definition of the product of two matrices may appear unusual to the beginning student; however, there are many applications which justify the form of the definition. In order to define the product AB of two matrices A and B, *the number of columns of A must be the same as the number of rows of B.* Suppose that $A = (a_{ij})$ is $m \times n$ and $B = (b_{ij})$ is $n \times p$. To determine the element c_{ij} of the product, we single out row i of A and column j of B as follows:

$$\begin{bmatrix} a_{11} & a_{12} & \cdots & a_{1n} \\ & & & \\ & & & \\ & & & \\ a_{i1} & a_{i2} & \cdots & a_{in} \\ & & & \\ & & & \\ & & & \\ a_{m1} & a_{m2} & \cdots & a_{mn} \end{bmatrix} \begin{bmatrix} b_{11} & \cdots & b_{1j} & \cdots & b_{1p} \\ b_{21} & \cdots & b_{2j} & \cdots & b_{2p} \\ & & & & \\ & & & & \\ & & & & \\ b_{n1} & \cdots & b_{nj} & \cdots & b_{np} \end{bmatrix}$$

Next we multiply pairs of elements and then add them, according to the following formula:

$$c_{ij} = a_{i1}b_{1j} + a_{i2}b_{2j} + \cdots + a_{in}b_{nj}.$$

For example, the element c_{11} in the first row and the first column of AB is given by

$$c_{11} = a_{11}b_{11} + a_{12}b_{21} + \cdots + a_{1n}b_{n1}.$$

The element c_{12} in the first row and second column of AB is

$$c_{12} = a_{11}b_{12} + a_{12}b_{22} + \cdots + a_{1n}b_{n2}.$$

By definition, the product AB has the same number of rows as A and the same number of columns as B. In particular, if A is $m \times n$ and B is $n \times p$, then AB is $m \times p$. This is illustrated by the following product of a 2×3 matrix and a 3×4 matrix.

$$\begin{bmatrix} 1 & 2 & -3 \\ 4 & 0 & -2 \end{bmatrix} \begin{bmatrix} 5 & -4 & 2 & 0 \\ -1 & 6 & 3 & 1 \\ 7 & 0 & 4 & 8 \end{bmatrix} = \begin{bmatrix} -18 & 8 & -4 & -22 \\ 6 & -16 & 0 & -16 \end{bmatrix}.$$

Here are some typical computations of the elements c_{ij} in the product:

$$c_{11} = (1)(5) + (2)(-1) + (-3)(7) = 5 - 2 - 21 = -18$$
$$c_{13} = (1)(2) + (2)(3) + (-3)(4) = 2 + 6 - 12 = -4$$
$$c_{23} = (4)(2) + (0)(3) + (-2)(4) = 8 + 0 - 8 = 0$$
$$c_{24} = (4)(0) + (0)(1) + (-2)(8) = 0 + 0 - 16 = -16.$$

The reader should check the remaining elements.

The product operation for matrices is not commutative. Indeed, if A is 2×3 and B is 3×4, then AB may be found, but BA is undefined since the number of columns of B is different from the number of rows of A. Even if AB and BA are both defined, it is often true that these products are different. This is illustrated in the next example, along with the fact that the product of two nonzero matrices may equal a zero matrix.

Example 1 If $A = \begin{bmatrix} 2 & 2 \\ -1 & -1 \end{bmatrix}$ and $B = \begin{bmatrix} 1 & 2 \\ 1 & 2 \end{bmatrix}$, show that $AB \neq BA$.

Solution Using the definition of product we obtain

$$AB = \begin{bmatrix} 4 & 8 \\ -2 & -4 \end{bmatrix} \quad \text{and} \quad BA = \begin{bmatrix} 0 & 0 \\ 0 & 0 \end{bmatrix}.$$

Hence, $AB \neq BA$. Note that BA is a zero matrix. ∎

It is possible to show that matrix multiplication is associative in the sense that

$$A(BC) = (AB)C$$

provided that the indicated products are defined. That will be the case when A is $m \times n$, B is $n \times p$, and C is $p \times q$. The Distributive Properties also hold if the matrices involved have the proper number of rows and columns. Thus, if A_1 and A_2 are $m \times n$ matrices, and if B_1 and B_2 are $n \times p$ matrices, then

$$A_1(B_1 + B_2) = A_1 B_1 + A_1 B_2$$

and

$$(A_1 + A_2)B_1 = A_1 B_1 + A_2 B_1.$$

As a special case, if all matrices are square, of order n, then the Associative and Distributive Properties are true. We shall not give proofs for these theorems.

Throughout the remainder of this section we shall concentrate on square matrices. The symbol I_n will be used to denote the square matrix of order n which has 1 in each position on the main diagonal and 0 elsewhere. For example,

$$I_2 = \begin{bmatrix} 1 & 0 \\ 0 & 1 \end{bmatrix}, \quad I_3 = \begin{bmatrix} 1 & 0 & 0 \\ 0 & 1 & 0 \\ 0 & 0 & 1 \end{bmatrix},$$

etc. It can be shown that if A is any square matrix of order n, then

$$AI_n = A = I_nA.$$

For that reason I_n is called an **identity matrix of order n**. To illustrate, if $A = (a_{ij})$ is of order 2, then a direct calculation shows that

$$\begin{bmatrix} a_{11} & a_{12} \\ a_{21} & a_{22} \end{bmatrix} \begin{bmatrix} 1 & 0 \\ 0 & 1 \end{bmatrix} = \begin{bmatrix} a_{11} & a_{12} \\ a_{21} & a_{22} \end{bmatrix} = \begin{bmatrix} 1 & 0 \\ 0 & 1 \end{bmatrix} \begin{bmatrix} a_{11} & a_{12} \\ a_{21} & a_{22} \end{bmatrix}.$$

Some, but not all, $n \times n$ matrices A have an **inverse** in the sense that there is a matrix B such that $AB = I_n = BA$. If A has an inverse we denote it by A^{-1} and write

$$AA^{-1} = I_n = A^{-1}A.$$

The symbol A^{-1} is read "A inverse." In matrix theory it is *not* acceptable to use the symbol $1/A$ in place of A^{-1}.

Let us describe a technique for finding the inverse of a square matrix A, whenever it exists. We shall not attempt to justify the following procedure, since that would require concepts which are beyond the scope of this text. Given the $n \times n$ matrix $A = (a_{ij})$, we begin by forming the $n \times (2n)$ matrix:

$$\begin{bmatrix} a_{11} & a_{12} & \cdots & a_{1n} & 1 & 0 & 0 & \cdots & 0 \\ a_{21} & a_{22} & \cdots & a_{2n} & 0 & 1 & 0 & \cdots & 0 \\ \cdot & \cdot & \cdots & \cdot & \cdot & \cdot & \cdot & \cdots & \cdot \\ \cdot & \cdot & \cdots & \cdot & \cdot & \cdot & \cdot & \cdots & \cdot \\ \cdot & \cdot & \cdots & \cdot & \cdot & \cdot & \cdot & \cdots & \cdot \\ a_{n1} & a_{n2} & \cdots & a_{nn} & 0 & 0 & 0 & \cdots & 1 \end{bmatrix}$$

where the $n \times n$ identity matrix I_n appears "to the right" of the matrix A, as indicated. We next apply a succession of elementary row transformations until we arrive at a matrix of the form

$$\begin{bmatrix} 1 & 0 & 0 & \cdots & 0 & b_{11} & b_{12} & \cdots & b_{1n} \\ 0 & 1 & 0 & \cdots & 0 & b_{21} & b_{22} & \cdots & b_{2n} \\ \cdot & \cdot & \cdot & \cdots & \cdot & \cdot & \cdot & \cdots & \cdot \\ \cdot & \cdot & \cdot & \cdots & \cdot & \cdot & \cdot & \cdots & \cdot \\ \cdot & \cdot & \cdot & \cdots & \cdot & \cdot & \cdot & \cdots & \cdot \\ 0 & 0 & 0 & \cdots & 1 & b_{n1} & b_{n2} & \cdots & b_{nn} \end{bmatrix}$$

where the identity matrix I_n appears "to the left" of the $n \times n$ matrix (b_{ij}). It can be shown that (b_{ij}) is the desired inverse A^{-1}.

Example 2 Find A^{-1} if $A = \begin{bmatrix} 3 & 5 \\ 1 & 4 \end{bmatrix}$.

Solution We begin with the matrix

$$\begin{bmatrix} 3 & 5 & 1 & 0 \\ 1 & 4 & 0 & 1 \end{bmatrix}.$$

We next perform elementary row transformations until the identity matrix I_2 appears on the left, as follows:

$$\begin{bmatrix} 3 & 5 & 1 & 0 \\ 1 & 4 & 0 & 1 \end{bmatrix} \rightarrow \begin{bmatrix} 1 & 4 & 0 & 1 \\ 3 & 5 & 1 & 0 \end{bmatrix} \quad \text{Interchange rows 1 and 2.}$$

$$\rightarrow \begin{bmatrix} 1 & 4 & 0 & 1 \\ 0 & -7 & 1 & -3 \end{bmatrix} \quad \begin{array}{l} \text{Add to the second row } -3 \\ \text{times the first row.} \end{array}$$

$$\rightarrow \begin{bmatrix} 1 & 4 & 0 & 1 \\ 0 & 1 & -\frac{1}{7} & \frac{3}{7} \end{bmatrix} \quad \begin{array}{l} \text{Multiply the second row by} \\ -1/7. \end{array}$$

$$\rightarrow \begin{bmatrix} 1 & 0 & \frac{4}{7} & -\frac{5}{7} \\ 0 & 1 & -\frac{1}{7} & \frac{3}{7} \end{bmatrix} \quad \begin{array}{l} \text{Add to the first row } -4 \text{ times} \\ \text{the second row.} \end{array}$$

According to the previous discussion,

$$A^{-1} = \begin{bmatrix} \frac{4}{7} & -\frac{5}{7} \\ -\frac{1}{7} & \frac{3}{7} \end{bmatrix}.$$

To check our work the reader should verify that

$$\begin{bmatrix} 3 & 5 \\ 1 & 4 \end{bmatrix} \begin{bmatrix} \frac{4}{7} & -\frac{5}{7} \\ -\frac{1}{7} & \frac{3}{7} \end{bmatrix} = \begin{bmatrix} 1 & 0 \\ 0 & 1 \end{bmatrix} = \begin{bmatrix} \frac{4}{7} & -\frac{5}{7} \\ -\frac{1}{7} & \frac{3}{7} \end{bmatrix} \begin{bmatrix} 3 & 5 \\ 1 & 4 \end{bmatrix}. \quad \blacksquare$$

Example 3 Find A^{-1} if $A = \begin{bmatrix} -1 & 3 & 1 \\ 2 & 5 & 0 \\ 3 & 1 & -2 \end{bmatrix}$.

Solution As in Example 2, we begin with

$$\begin{bmatrix} -1 & 3 & 1 & 1 & 0 & 0 \\ 2 & 5 & 0 & 0 & 1 & 0 \\ 3 & 1 & -2 & 0 & 0 & 1 \end{bmatrix}$$

and apply elementary row operations until I_3 appears on the left side of the matrix. Thus,

$$\begin{bmatrix} -1 & 3 & 1 & 1 & 0 & 0 \\ 2 & 5 & 0 & 0 & 1 & 0 \\ 3 & 1 & -2 & 0 & 0 & 1 \end{bmatrix}$$

$$\rightarrow \begin{bmatrix} 1 & -3 & -1 & -1 & 0 & 0 \\ 2 & 5 & 0 & 0 & 1 & 0 \\ 3 & 1 & -2 & 0 & 0 & 1 \end{bmatrix}$$ Multiply the first row by -1.

$$\rightarrow \begin{bmatrix} 1 & -3 & -1 & -1 & 0 & 0 \\ 0 & 11 & 2 & 2 & 1 & 0 \\ 0 & 10 & 1 & 3 & 0 & 1 \end{bmatrix}$$ Add to the second row -2 times the first row and then add to the third row -3 times the first row.

$$\rightarrow \begin{bmatrix} 1 & -3 & -1 & -1 & 0 & 0 \\ 0 & 1 & 1 & -1 & 1 & -1 \\ 0 & 10 & 1 & 3 & 0 & 1 \end{bmatrix}$$ Add to the second row -1 times the third row.

$$\rightarrow \begin{bmatrix} 1 & 0 & 2 & -4 & 3 & -3 \\ 0 & 1 & 1 & -1 & 1 & -1 \\ 0 & 0 & -9 & 13 & -10 & 11 \end{bmatrix}$$ Add to the first row 3 times the second row and then add to the third row -10 times the second row.

$$\rightarrow \begin{bmatrix} 1 & 0 & 2 & -4 & 3 & -3 \\ 0 & 1 & 1 & -1 & 1 & -1 \\ 0 & 0 & 1 & -\frac{13}{9} & \frac{10}{9} & -\frac{11}{9} \end{bmatrix}$$ Multiply the third row by $-1/9$.

$$\rightarrow \begin{bmatrix} 1 & 0 & 0 & -\frac{10}{9} & \frac{7}{9} & -\frac{5}{9} \\ 0 & 1 & 0 & \frac{4}{9} & -\frac{1}{9} & \frac{2}{9} \\ 0 & 0 & 1 & -\frac{13}{9} & \frac{10}{9} & -\frac{11}{9} \end{bmatrix}$$ Add to the first row -2 times the third row and then add to the second row -1 times the third row.

Consequently,

$$A^{-1} = \begin{bmatrix} -\frac{10}{9} & \frac{7}{9} & -\frac{5}{9} \\ \frac{4}{9} & -\frac{1}{9} & \frac{2}{9} \\ -\frac{13}{9} & \frac{10}{9} & -\frac{11}{9} \end{bmatrix} = \frac{1}{9} \begin{bmatrix} -10 & 7 & -5 \\ 4 & -1 & 2 \\ -13 & 10 & -11 \end{bmatrix}.$$

The reader should check the fact that

$$AA^{-1} = I_3 = A^{-1}A. \qquad \blacksquare$$

There are many uses for inverses of matrices. One application concerns solutions of systems of linear equations. To illustrate, let us consider the case of two linear equations in two unknowns:

$$\begin{cases} a_{11}x + a_{12}y = k_1 \\ a_{21}x + a_{22}y = k_2. \end{cases}$$

We may express the system by means of matrices as follows:

$$\begin{bmatrix} a_{11}x + a_{12}y \\ a_{21}x + a_{22}y \end{bmatrix} = \begin{bmatrix} k_1 \\ k_2 \end{bmatrix}.$$

If we let

$$A = \begin{bmatrix} a_{11} & a_{12} \\ a_{21} & a_{22} \end{bmatrix}, \quad X = \begin{bmatrix} x \\ y \end{bmatrix}, \quad \text{and} \quad B = \begin{bmatrix} k_1 \\ k_2 \end{bmatrix},$$

this may be written in the form

$$AX = B.$$

If A^{-1} exists, then multiplying both sides of the last equation by A^{-1} gives us $A^{-1}AX = A^{-1}B$. Since $A^{-1}A = I_n$ and $I_nX = X$, this leads to

$$X = A^{-1}B$$

from which the solution (x, y) may be found. This technique may be extended to systems of n linear equations in n unknowns.

Example 4 Solve the following system of equations:

$$\begin{cases} -x + 3y + z = 1 \\ 2x + 5y \quad\quad = 3 \\ 3x + y - 2z = -2. \end{cases}$$

Solution If we let

$$A = \begin{bmatrix} -1 & 3 & 1 \\ 2 & 5 & 0 \\ 3 & 1 & -2 \end{bmatrix}, \quad X = \begin{bmatrix} x \\ y \\ z \end{bmatrix}, \quad \text{and} \quad B = \begin{bmatrix} 1 \\ 3 \\ -2 \end{bmatrix},$$

then, as in the preceding discussion, the given system may be written in terms of matrices as $AX = B$. This implies that $X = A^{-1}B$. The matrix A^{-1} was found in Example 3. Substituting for X, A^{-1}, and B in the last equation gives us

$$\begin{bmatrix} x \\ y \\ z \end{bmatrix} = \frac{1}{9} \begin{bmatrix} -10 & 7 & -5 \\ 4 & -1 & 2 \\ -13 & 10 & -11 \end{bmatrix} \begin{bmatrix} 1 \\ 3 \\ -2 \end{bmatrix} = \frac{1}{9} \begin{bmatrix} 21 \\ -3 \\ 39 \end{bmatrix} = \begin{bmatrix} \frac{7}{3} \\ -\frac{1}{3} \\ \frac{13}{3} \end{bmatrix}.$$

It follows that $x = \frac{7}{3}$, $y = -\frac{1}{3}$, and $z = \frac{13}{3}$. Hence, the ordered triple $(\frac{7}{3}, -\frac{1}{3}, \frac{13}{3})$ is the solution of the given system. ∎

The method of solution employed in Example 4 is beneficial only if A^{-1} is known, or if many systems with the same coefficient matrix are to be considered.

The preferred technique for solving a system of linear equations is still either the method of elimination or the matrix method discussed in Section 8.4.

EXERCISES 8.5

In Exercises 1–8 find $A + B$, $A - B$, $2A$, and $-3B$.

1 $A = \begin{bmatrix} 5 & -2 \\ 1 & 3 \end{bmatrix}$ $B = \begin{bmatrix} 4 & 1 \\ -3 & 2 \end{bmatrix}$

2 $A = \begin{bmatrix} 3 & 0 \\ -1 & 2 \end{bmatrix}$, $B = \begin{bmatrix} 3 & -4 \\ 1 & 1 \end{bmatrix}$

3 $A = \begin{bmatrix} 6 & -1 \\ 2 & 0 \\ -3 & 4 \end{bmatrix}$, $B = \begin{bmatrix} 3 & 1 \\ -1 & 5 \\ 6 & 0 \end{bmatrix}$

4 $A = \begin{bmatrix} 0 & -2 & 7 \\ 5 & 4 & -3 \end{bmatrix}$, $B = \begin{bmatrix} 8 & 4 & 0 \\ 0 & 1 & 4 \end{bmatrix}$

5 $A = \begin{bmatrix} 4 & -3 & 2 \end{bmatrix}$, $B = \begin{bmatrix} 7 & 0 & -5 \end{bmatrix}$

6 $A = \begin{bmatrix} 7 \\ -16 \end{bmatrix}$, $B = \begin{bmatrix} -11 \\ 9 \end{bmatrix}$

7 $A = \begin{bmatrix} 0 & 4 & 0 & 3 \\ 1 & 2 & 0 & -5 \end{bmatrix}$,

   $B = \begin{bmatrix} -3 & 0 & 1 & 3 \\ 2 & 0 & 7 & -2 \end{bmatrix}$

8 $A = \begin{bmatrix} -7 \end{bmatrix}$, $B = \begin{bmatrix} 9 \end{bmatrix}$

In Exercises 9–18 find AB and BA.

9 $A = \begin{bmatrix} 2 & 6 \\ 3 & -4 \end{bmatrix}$, $B = \begin{bmatrix} 5 & -2 \\ 1 & 7 \end{bmatrix}$

10 $A = \begin{bmatrix} 4 & -2 \\ -2 & 1 \end{bmatrix}$, $B = \begin{bmatrix} 2 & 1 \\ 4 & 2 \end{bmatrix}$

11 $A = \begin{bmatrix} 3 & 0 & -1 \\ 0 & 4 & 2 \\ 5 & -3 & 1 \end{bmatrix}$, $B = \begin{bmatrix} 1 & -5 & 0 \\ 4 & 1 & -2 \\ 0 & -1 & 3 \end{bmatrix}$

12 $A = \begin{bmatrix} 5 & 0 & 0 \\ 0 & -3 & 0 \\ 0 & 0 & 2 \end{bmatrix}$, $B = \begin{bmatrix} 3 & 0 & 0 \\ 0 & 4 & 0 \\ 0 & 0 & -2 \end{bmatrix}$

13 $A = \begin{bmatrix} 4 & -3 & 1 \\ -5 & 2 & 2 \end{bmatrix}$, $B = \begin{bmatrix} 2 & 1 \\ 0 & 1 \\ -4 & 7 \end{bmatrix}$

14 $A = \begin{bmatrix} 2 & 1 & -1 & 0 \\ 3 & -2 & 0 & 5 \\ -2 & 1 & 4 & 2 \end{bmatrix}$,

   $B = \begin{bmatrix} 5 & -3 & 1 \\ 1 & 2 & 0 \\ -1 & 0 & 4 \\ 0 & -2 & 3 \end{bmatrix}$

15 $A = \begin{bmatrix} 1 & 2 & 3 \\ 4 & 5 & 6 \\ 7 & 8 & 9 \end{bmatrix}$, $B = \begin{bmatrix} 1 & 0 & 0 \\ 0 & 1 & 0 \\ 0 & 0 & 1 \end{bmatrix}$

16 $A = \begin{bmatrix} 1 & 2 & 3 \\ 2 & 3 & 1 \\ 3 & 1 & 2 \end{bmatrix}$, $B = \begin{bmatrix} 2 & 0 & 0 \\ 0 & 2 & 0 \\ 0 & 0 & 2 \end{bmatrix}$

17 $A = \begin{bmatrix} -3 & 7 & 2 \end{bmatrix}$, $B = \begin{bmatrix} 1 \\ 4 \\ -5 \end{bmatrix}$

18 $A = \begin{bmatrix} 4 & 8 \end{bmatrix}$, $B = \begin{bmatrix} -3 \\ 2 \end{bmatrix}$

In Exercises 19–22 find AB.

19 $A = \begin{bmatrix} 4 & -2 \\ 0 & 3 \\ -7 & 5 \end{bmatrix}$, $B = \begin{bmatrix} 3 \\ 4 \end{bmatrix}$

20 $A = \begin{bmatrix} 4 \\ -3 \\ 2 \end{bmatrix}$, $B = \begin{bmatrix} 5 & 1 \end{bmatrix}$

21 $A = \begin{bmatrix} 2 & 1 & 0 & -3 \\ -7 & 0 & -2 & 4 \end{bmatrix}$,

   $B = \begin{bmatrix} 4 & -2 & 0 \\ 1 & 1 & -2 \\ 0 & 0 & 5 \\ -3 & -1 & 0 \end{bmatrix}$

22 $A = \begin{bmatrix} 1 & 2 & -3 \\ 4 & -5 & 6 \end{bmatrix}$,

$B = \begin{bmatrix} 1 & -1 & 0 & 2 \\ -2 & 3 & 1 & 0 \\ 0 & 4 & 0 & -3 \end{bmatrix}$

23 If $A = \begin{bmatrix} 1 & 2 \\ 0 & -3 \end{bmatrix}$, and $B = \begin{bmatrix} 2 & -1 \\ 3 & 1 \end{bmatrix}$, show that $(A + B)(A - B) \neq A^2 - B^2$, where $A^2 = AA$ and $B^2 = BB$.

24 If A and B are the matrices of Exercise 23, show that $(A + B)(A + B) \neq A^2 + 2AB + B^2$.

25 If A and B are the matrices of Exercise 23 and $C = \begin{bmatrix} 3 & 1 \\ -2 & 0 \end{bmatrix}$, prove that $A(B + C) = AB + AC$.

26 If A, B, and C are the matrices of Exercise 25 prove that $A(BC) = (AB)C$.

Prove the identities given in Exercises 27–30, where

$$A = \begin{bmatrix} a_{11} & a_{12} \\ a_{21} & a_{22} \end{bmatrix}, \quad B = \begin{bmatrix} b_{11} & b_{12} \\ b_{21} & b_{22} \end{bmatrix},$$

$$C = \begin{bmatrix} c_{11} & c_{12} \\ c_{21} & c_{22} \end{bmatrix}$$

and c, d are real numbers.

27 $c(A + B) = cA + cB$

28 $(c + d)A = cA + dA$

29 $A(B + C) = AB + AC$

30 $A(BC) = (AB)C$

In Exercises 31–42 find the inverse of the given matrix, if it exists.

31 $\begin{bmatrix} 2 & -4 \\ 1 & 3 \end{bmatrix}$

32 $\begin{bmatrix} 3 & 2 \\ 4 & 5 \end{bmatrix}$

33 $\begin{bmatrix} 2 & 4 \\ 4 & 8 \end{bmatrix}$

34 $\begin{bmatrix} 3 & -1 \\ 6 & -2 \end{bmatrix}$

35 $\begin{bmatrix} 3 & -1 & 0 \\ 2 & 2 & 0 \\ 0 & 0 & 4 \end{bmatrix}$

36 $\begin{bmatrix} 3 & 0 & 2 \\ 0 & 1 & 0 \\ -4 & 0 & 2 \end{bmatrix}$

37 $\begin{bmatrix} -2 & 2 & 3 \\ 1 & -1 & 0 \\ 0 & 1 & 4 \end{bmatrix}$

38 $\begin{bmatrix} 1 & 2 & 3 \\ -2 & 1 & 0 \\ 3 & -1 & 1 \end{bmatrix}$

39 $\begin{bmatrix} 2 & 0 & 0 \\ 0 & 4 & 0 \\ 0 & 0 & 6 \end{bmatrix}$

40 $\begin{bmatrix} 1 & 1 & 1 \\ 2 & 2 & 2 \\ 3 & 3 & 3 \end{bmatrix}$

41 $\begin{bmatrix} 1 & -1 & 0 & 1 \\ 0 & 1 & -2 & 0 \\ -1 & 2 & 1 & 2 \\ -2 & 1 & 2 & 0 \end{bmatrix}$

42 $\begin{bmatrix} 1 & 2 & 0 & 1 \\ 0 & -1 & 1 & -2 \\ 0 & 0 & 2 & 0 \\ 0 & 0 & 0 & 1 \end{bmatrix}$

43 State conditions on a and b which guarantee that the matrix $\begin{bmatrix} a & 0 \\ 0 & b \end{bmatrix}$ has an inverse, and find a formula for the inverse when it exists.

44 If $abc \neq 0$, find the inverse of $\begin{bmatrix} a & 0 & 0 \\ 0 & b & 0 \\ 0 & 0 & c \end{bmatrix}$.

45 If $A = \begin{bmatrix} a_{11} & a_{12} & a_{13} \\ a_{21} & a_{22} & a_{23} \\ a_{31} & a_{32} & a_{33} \end{bmatrix}$, prove that $AI_3 = A = I_3A$.

46 Prove that $AI_4 = A = I_4A$ for every square matrix A of order 4.

Solve the systems in Exercises 47–50 by the method of Example 4. (Refer to inverses of matrices found in Exercises 31–38.)

47 $\begin{cases} 2x - 4y = 3 \\ x + 3y = 1 \end{cases}$

48 $\begin{cases} 3x + 2y = -1 \\ 4x + 5y = 1 \end{cases}$

49 $\begin{cases} -2x + 2y + 3z = 1 \\ x - y = 3 \\ y + 4z = -2 \end{cases}$

50 $\begin{cases} x + 2y + 3z = -1 \\ -2x + y = 4 \\ 3x - y + z = 2 \end{cases}$

8.6 DETERMINANTS

Throughout this section and the next it is assumed that all matrices under discussion are *square* matrices. Associated with each such matrix A is a number called the **determinant** of A. Determinants can be used to solve systems of linear equations if the number of equations is the same as the number of variables. In this section we shall state the definition and give some basic properties of determinants. It may be difficult to see exactly why the definitions given below are used. One reason will be pointed out in Section 8.8, where it is shown that our definitions arise naturally when solving systems of linear equations.

The determinant of a square matrix A will be denoted by $|A|$. This notation should not be confused with the symbol used for the absolute value of a real number. To avoid any misunderstanding, the expression det A is used in some mathematics texts instead of $|A|$. We shall define $|A|$ by beginning with the case in which A has order 1 and then by increasing the order a step at a time.

If A is a square matrix of order 1, then A has only one element. Thus, $A = [a_{11}]$ and we define $|A| = a_{11}$. If A is a square matrix of order 2, then we may write

$$A = \begin{bmatrix} a_{11} & a_{12} \\ a_{21} & a_{22} \end{bmatrix}$$

and the determinant of A is defined by

$$|A| = a_{11}a_{22} - a_{21}a_{12}.$$

Another notation for $|A|$ is obtained by replacing the brackets in the symbol for A with vertical bars as follows

Definition

$$|A| = \begin{vmatrix} a_{11} & a_{12} \\ a_{21} & a_{22} \end{vmatrix} = a_{11}a_{22} - a_{21}a_{12}.$$

Example 1 Find $|A|$ if $A = \begin{bmatrix} 2 & -1 \\ 4 & -3 \end{bmatrix}$.

Solution By definition,

$$|A| = \begin{vmatrix} 2 & -1 \\ 4 & -3 \end{vmatrix} = (2)(-3) - (4)(-1) = -6 + 4 = -2. \qquad \blacksquare$$

For matrices of higher order it is convenient to introduce additional terminology as follows.

Definition

If A is a matrix of order 3, then the **minor** M_{ij} of an element a_{ij} is the determinant of the matrix of order 2 obtained by deleting row i and column j of A.

Thus, to determine the minor of an element, we discard the row and column in which the element appears and then find determinant of the resulting matrix. To illustrate, given the general 3×3 matrix

$$A = \begin{bmatrix} a_{11} & a_{12} & a_{13} \\ a_{21} & a_{22} & a_{23} \\ a_{31} & a_{32} & a_{33} \end{bmatrix}$$

we obtain

$$M_{11} = \begin{vmatrix} a_{22} & a_{23} \\ a_{32} & a_{33} \end{vmatrix} = a_{22}a_{33} - a_{32}a_{23}$$

$$M_{12} = \begin{vmatrix} a_{21} & a_{23} \\ a_{31} & a_{33} \end{vmatrix} = a_{21}a_{33} - a_{31}a_{23}$$

$$M_{13} = \begin{vmatrix} a_{21} & a_{22} \\ a_{31} & a_{32} \end{vmatrix} = a_{21}a_{32} - a_{31}a_{22}$$

$$M_{23} = \begin{vmatrix} a_{11} & a_{12} \\ a_{31} & a_{32} \end{vmatrix} = a_{11}a_{32} - a_{31}a_{12}$$

and likewise for the other minors M_{21}, M_{22}, M_{31}, M_{32}, and M_{33}.

We shall also make use of the following concept.

Definition

> The **cofactor** A_{ij} of the element a_{ij} is defined by
>
> $$A_{ij} = (-1)^{i+j} M_{ij}.$$

Thus, to obtain the cofactor of a_{ij}, we find the minor and multiply it by 1 or -1 depending on whether the sum of i and j is even or odd, respectively. An easy way to remember the sign $(-1)^{i+j}$ associated with the cofactor A_{ij} is to consider the following "checkerboard" scheme:

$$\begin{bmatrix} + & - & + \\ - & + & - \\ + & - & + \end{bmatrix}$$

where we may regard the $+$ signs as occurring on red squares and the $-$ signs on black squares.

Example 2 If

$$A = \begin{bmatrix} 1 & -3 & 3 \\ 4 & 2 & 0 \\ -2 & -7 & 5 \end{bmatrix}$$

find M_{11}, M_{21}, M_{22}, A_{11}, A_{21}, and A_{22}.

Solution   By definition

$$M_{11} = \begin{vmatrix} 2 & 0 \\ -7 & 5 \end{vmatrix} = (2)(5) - (-7)(0) = 10$$

$$M_{21} = \begin{vmatrix} -3 & 3 \\ -7 & 5 \end{vmatrix} = (-3)(5) - (-7)(3) = 6$$

$$M_{22} = \begin{vmatrix} 1 & 3 \\ -2 & 5 \end{vmatrix} = (1)(5) - (-2)(3) = 11.$$

All that is necessary to obtain the cofactors is to prefix the corresponding minors with the proper signs. Thus, using the definition of cofactor,

$$A_{11} = (-1)^{1+1}M_{11} = (1)(10) = 10$$
$$A_{21} = (-1)^{2+1}M_{21} = (-1)(6) = -6$$
$$A_{22} = (-1)^{2+2}M_{22} = (1)(11) = 11.$$

The checkerboard scheme could also be used to determine the proper signs.   ■

The determinant $|A|$ of a square matrix of order 3 is defined by

Definition

$$|A| = \begin{vmatrix} a_{11} & a_{12} & a_{13} \\ a_{21} & a_{22} & a_{23} \\ a_{31} & a_{32} & a_{33} \end{vmatrix} = a_{11}A_{11} + a_{12}A_{12} + a_{13}A_{13}.$$

Since $A_{11} = (-1)^{1+1}M_{11} = M_{11}$, $A_{12} = (-1)^{1+2}M_{12} = -M_{12}$, and $A_{13} = (-1)^{1+3}M_{13} = M_{13}$, the preceding definition may also be written

$$|A| = a_{11}M_{11} - a_{12}M_{12} + a_{13}M_{13}.$$

If we substitute for M_{11}, M_{12}, and M_{13}, we obtain the following formula for $|A|$ in terms of the elements of A:

$$|A| = a_{11}a_{22}a_{33} - a_{11}a_{32}a_{23} - a_{12}a_{21}a_{33} + a_{12}a_{31}a_{23} + a_{13}a_{21}a_{32} - a_{13}a_{31}a_{22}.$$

The definition of $|A|$ for a square matrix A of order 3 displays a pattern of multiplying each element in row 1 by its cofactor and then adding to find $|A|$. This is referred to as *expanding $|A|$ by the first row.* By actually carrying out the computations, it is not difficult to show that $|A|$ *can be expanded in similar fashion by using any row or column.* As an illustration, the expansion by the second column is

$$|A| = a_{12}A_{12} + a_{22}A_{22} + a_{32}A_{32}$$

$$= a_{12}\left(-\begin{vmatrix} a_{21} & a_{23} \\ a_{31} & a_{32} \end{vmatrix}\right) + a_{22}\left(+\begin{vmatrix} a_{11} & a_{13} \\ a_{31} & a_{33} \end{vmatrix}\right) + a_{32}\left(-\begin{vmatrix} a_{11} & a_{13} \\ a_{21} & a_{23} \end{vmatrix}\right).$$

Applying the definition to the determinants in parentheses, multiplying as indicated, and rearranging the terms in the sum, we could arrive at the formula for $|A|$ in terms of the elements of A. Similarly, the expansion by the third row is

$$|A| = a_{31}A_{31} + a_{32}A_{32} + a_{33}A_{33}$$

$$= a_{31}\left(+\begin{vmatrix} a_{12} & a_{13} \\ a_{22} & a_{23} \end{vmatrix}\right) + a_{32}\left(-\begin{vmatrix} a_{11} & a_{13} \\ a_{21} & a_{23} \end{vmatrix}\right) + a_{33}\left(+\begin{vmatrix} a_{11} & a_{12} \\ a_{21} & a_{22} \end{vmatrix}\right).$$

Once again it can be shown that this agrees with previous expansions.

Example 3   Find $|A|$ if

$$A = \begin{bmatrix} -1 & 3 & 1 \\ 2 & 5 & 0 \\ 3 & 1 & -2 \end{bmatrix}.$$

Solution   Expanding $|A|$ by the second row gives us

$$|A| = (2)A_{21} + (5)A_{22} + (0)A_{23}.$$

Using the definition of cofactor we have

$$A_{21} = (-1)^3 M_{21} = -\begin{vmatrix} 3 & 1 \\ 1 & -2 \end{vmatrix} = -[(3)(-2) - (1)(1)] = 7$$

$$A_{22} = (-1)^4 M_{22} = \begin{vmatrix} -1 & 1 \\ 3 & -2 \end{vmatrix} = [(-1)(-2) - (3)(1)] = -1.$$

Consequently,

$$|A| = (2)(7) + (5)(-1) + (0)A_{23} = 14 - 5 + 0 = 9. \qquad \blacksquare$$

If A is a matrix of order 4, we define the minor M_{ij} of the element a_{ij} as the determinant of the matrix of order 3 obtained by deleting row i and column j of A. The cofactor A_{ij} is again given by $(-1)^{i+j}M_{ij}$. In a manner analogous to the case for order 3, we define

$$|A| = a_{11}A_{11} + a_{12}A_{12} + a_{13}A_{13} + a_{14}A_{14}.$$

The last formula is called *the expansion of* $|A|$ *by the first row.* In terms of minors, the formula may be written as

$$|A| = a_{11}M_{11} - a_{12}M_{12} + a_{13}M_{13} - a_{14}M_{14}.$$

It can be shown that the same number is obtained if $|A|$ is expanded by any other row or column.

The method of defining determinants of matrices of arbitrary order n should now be apparent. If a_{ij} is an element of a matrix of order $n > 1$, then the **minor** M_{ij} is defined as the determinant of the matrix of order $n - 1$ obtained by deleting row i and column j. The **cofactor** A_{ij} is defined as $(-1)^{i+j}M_{ij}$. The sign $(-1)^{i+j}$ associated with A_{ij} can be remembered by using a checkerboard similar to that used for order 3, extending the rows and columns as far as necessary. We then define the determinant $|A|$ of a matrix A of order n as the expansion by the first row, that is,

$$|A| = a_{11}A_{11} + a_{12}A_{12} + \cdots + a_{1n}A_{1n}$$

or, in terms of minors,

$$|A| = a_{11}M_{11} - a_{12}M_{12} + \cdots + a_{1n}(-1)^{1+n}M_{1n}.$$

As was the case with matrices of small order, the number $|A|$ may be found by using *any* row or column. Specifically, we have the following theorem.

The Expansion Theorem for Determinants

If A is a square matrix of order $n > 1$, then the determinant $|A|$ may be found by multiplying the elements of any row (or column) by their respective cofactors, and adding the resulting products.

The proof of this theorem is difficult and may be found in texts on matrix theory. The theorem is quite useful if many zeros appear in a row or column, as illustrated in the following example.

Example 4 Find $|A|$ if

$$A = \begin{bmatrix} 1 & 0 & 2 & 5 \\ -2 & 1 & 5 & 0 \\ 0 & 0 & -3 & 0 \\ 0 & -1 & 0 & 3 \end{bmatrix}.$$

Solution Note that all but one of the elements in the third row is zero. Hence if we expand $|A|$ by the third row there will be at most one nonzero term. Specifically,

$$|A| = (0)A_{31} + (0)A_{32} + (-3)A_{33} + (0)A_{34} = -3A_{33}$$

where

$$A_{33} = \begin{vmatrix} 1 & 0 & 5 \\ -2 & 1 & 0 \\ 0 & -1 & 3 \end{vmatrix}.$$

Expanding A_{33} by column 1, we obtain

$$A_{33} = (1)\begin{vmatrix} 1 & 0 \\ -1 & 3 \end{vmatrix} + (-2)\left(-\begin{vmatrix} 0 & 5 \\ -1 & 3 \end{vmatrix}\right) + 0\begin{vmatrix} 0 & 5 \\ 1 & 0 \end{vmatrix} = 3 + 10 + 0 = 13.$$

Therefore, $$|A| = -3A_{33} = (-3)(13) = -39. \qquad \blacksquare$$

In general, if all but one element a in some row (or column) of A is zero and if the determinant $|A|$ is expanded by that row (or column), then all terms drop out except the product of the element a with its cofactor. We will make important use of this fact in the next section.

If *every* element in a row (or column) of a matrix A is zero, then upon expanding $|A|$ by that row (or column) we obtain the number 0. This gives us the following result.

Theorem

> If every element of a row (or column) of a square matrix A is zero, then $|A| = 0$.

EXERCISES 8.6

In each of Exercises 1–4 find all the minors and cofactors of the elements in the given matrix.

1. $\begin{bmatrix} 2 & 4 & -1 \\ 0 & 3 & 2 \\ -5 & 7 & 0 \end{bmatrix}$ 2. $\begin{bmatrix} 5 & -2 & 1 \\ 4 & 7 & 0 \\ -3 & 4 & -1 \end{bmatrix}$

3. $\begin{bmatrix} 7 & -1 \\ 5 & 0 \end{bmatrix}$ 4. $\begin{bmatrix} -6 & 4 \\ 3 & 2 \end{bmatrix}$

5–8 Find the determinants of the matrices given in Exercises 1–4.

Find the determinants of the matrices in Exercises 9–20.

9. $\begin{bmatrix} -5 & 4 \\ -3 & 2 \end{bmatrix}$ 10. $\begin{bmatrix} 6 & 4 \\ -3 & 2 \end{bmatrix}$

11. $\begin{bmatrix} a & -a \\ b & -b \end{bmatrix}$ 12. $\begin{bmatrix} c & d \\ -d & c \end{bmatrix}$

13. $\begin{bmatrix} 3 & 1 & -2 \\ 4 & 2 & 5 \\ -6 & 3 & -1 \end{bmatrix}$ 14. $\begin{bmatrix} 2 & -5 & 1 \\ -3 & 1 & 6 \\ 4 & -2 & 3 \end{bmatrix}$

15. $\begin{bmatrix} -5 & 4 & 1 \\ 3 & -2 & 7 \\ 2 & 0 & 6 \end{bmatrix}$ 16. $\begin{bmatrix} 2 & 7 & -3 \\ 1 & 0 & 4 \\ 4 & -1 & -2 \end{bmatrix}$

17. $\begin{bmatrix} 3 & -1 & 2 & 0 \\ 4 & 0 & -3 & 5 \\ 0 & 6 & 0 & 0 \\ 1 & 3 & -4 & 2 \end{bmatrix}$

18. $\begin{bmatrix} 2 & 5 & 1 & 0 \\ -4 & 0 & -3 & 0 \\ 3 & -2 & 1 & 6 \\ -1 & 4 & 2 & 0 \end{bmatrix}$

19. $\begin{bmatrix} 0 & b & 0 & 0 \\ 0 & 0 & c & 0 \\ a & 0 & 0 & 0 \\ 0 & 0 & 0 & d \end{bmatrix}$ 20. $\begin{bmatrix} a & u & v & w \\ 0 & b & x & y \\ 0 & 0 & c & z \\ 0 & 0 & 0 & d \end{bmatrix}$

Verify the identities in Exercises 21–28 by expanding each determinant.

21. $\begin{vmatrix} a & b \\ c & d \end{vmatrix} = - \begin{vmatrix} c & d \\ a & b \end{vmatrix}$

22. $\begin{vmatrix} a & b \\ c & d \end{vmatrix} = - \begin{vmatrix} b & a \\ d & c \end{vmatrix}$

23. $\begin{vmatrix} a & kb \\ c & kd \end{vmatrix} = k \begin{vmatrix} a & b \\ c & d \end{vmatrix}$

24. $\begin{vmatrix} a & b \\ kc & kd \end{vmatrix} = k \begin{vmatrix} a & b \\ c & d \end{vmatrix}$

25. $\begin{vmatrix} a & b \\ c & d \end{vmatrix} = \begin{vmatrix} a & b \\ ka + c & kb + d \end{vmatrix}$

26. $\begin{vmatrix} a & b \\ c & d \end{vmatrix} = \begin{vmatrix} a & ka + b \\ c & kc + d \end{vmatrix}$

27. $\begin{vmatrix} a & b \\ c & d \end{vmatrix} + \begin{vmatrix} a & e \\ c & f \end{vmatrix} = \begin{vmatrix} a & b + e \\ c & d + f \end{vmatrix}$

28. $\begin{vmatrix} a & b \\ c & d \end{vmatrix} + \begin{vmatrix} a & b \\ e & f \end{vmatrix} = \begin{vmatrix} a & b \\ c + e & d + f \end{vmatrix}$

29. Prove that if a square matrix A of order 2 has two identical rows or columns, then $|A| = 0$.

30. Repeat Exercise 29 for a matrix of order 3.

8.7 PROPERTIES OF DETERMINANTS

The method of evaluating a determinant by means of the Expansion Theorem stated in Section 8.6 is not very efficient for matrices of high order. For example, if a determinant of a matrix of order 10 is expanded by any row, a sum of 10 terms is obtained, where each term contains the determinant of a matrix of order 9 (that is, a cofactor of the original matrix). If any of the latter determinants is expanded by a row (or column), a sum of 9 terms is obtained, each containing the determinant of a matrix of order 8. Hence, at this stage there are 90 determinants of matrices of order 8 to evaluate! The process could be continued until only determinants of matrices of order 2 remain. Unless many elements of the original matrix are zero, it is an enormous task to carry out all of the computations.

We shall now consider some rules which make the process of evaluating determinants simpler. These rules are used mainly for introducing zeros into the determinant. They may also be used to change the determinant to echelon form, that is, a form in which the elements below the main diagonal elements a_{ii} are all zero. The transformations on rows stated in the next theorem are the same as the elementary row transformations of a matrix introduced in Section 8.4. However, for determinants we may also employ similar transformations on columns.

Theorem on Row and Column Transformations of a Determinant

Let A be a matrix of order n.

(i)   If a matrix B is obtained from A by interchanging two rows (or columns), then $|B| = -|A|$.

(ii) If B is obtained from A by multiplying every element of one row (or column) of A by a real number k, then $|B| = k|A|$.

(iii) If B is obtained from A by adding to any row (or column) of A, k times another row (or column), where k is any real number, then $|B| = |A|$.

When using this theorem to justify manipulations with determinants, we shall refer to the rows (or columns) of the *determinant* in the obvious way. For example, property (iii) may be phrased: "Adding the product of k times another row (or column) to any row (or column) of a determinant does not affect the value of the determinant."

We shall not give a general proof of the preceding theorem. For the case of matrices of orders 2 or 3 the theorem can be proved by evaluating $|B|$. For example, given a matrix $A = (a_{ij})$ of order 3, suppose that B is the matrix obtained by adding to row 2 the product of k times row 1, that is,

$$B = \begin{bmatrix} a_{11} & a_{12} & a_{13} \\ ka_{11} + a_{21} & ka_{12} + a_{22} & ka_{13} + a_{23} \\ a_{31} & a_{32} & a_{33} \end{bmatrix}.$$

To evaluate $|B|$ we expand by row 2, obtaining

$$|B| = (ka_{11} + a_{21})\left(-\begin{vmatrix} a_{12} & a_{13} \\ a_{32} & a_{33} \end{vmatrix}\right) + (ka_{12} + a_{22})\begin{vmatrix} a_{11} & a_{13} \\ a_{31} & a_{33} \end{vmatrix}$$

$$+ (ka_{13} + a_{23})\left(-\begin{vmatrix} a_{11} & a_{12} \\ a_{31} & a_{32} \end{vmatrix}\right).$$

The determinants which appear in this equation are those associated with the cofactors A_{21}, A_{22}, and A_{23} of the original matrix. Hence,

$$|B| = (ka_{11} + a_{21})A_{21} + (ka_{12} + a_{22})A_{22} + (ka_{13} + a_{23})A_{23}$$

which may also be written in the form

$$|B| = k(a_{11}A_{21} + a_{12}A_{22} + a_{13}A_{23}) + (a_{21}A_{21} + a_{22}A_{22} + a_{23}A_{23}).$$

The second expression in parentheses equals $|A|$. (Why?) By actually carrying out the computations it is possible to show that the first expression in parentheses is zero. Consequently,

$$|B| = 0 + |A| = |A|.$$

Statement (i) of the last theorem is often phrased: "Interchanging two rows (or columns) changes the sign of a determinant." As illustrations we have

$$\begin{vmatrix} 2 & 0 & 1 \\ 6 & 4 & 3 \\ 0 & 3 & 5 \end{vmatrix} = -\begin{vmatrix} 6 & 4 & 3 \\ 2 & 0 & 1 \\ 0 & 3 & 5 \end{vmatrix} \quad \text{and} \quad \begin{vmatrix} 2 & 0 & 1 \\ 6 & 4 & 3 \\ 0 & 3 & 5 \end{vmatrix} = -\begin{vmatrix} 1 & 0 & 2 \\ 3 & 4 & 6 \\ 5 & 3 & 0 \end{vmatrix}$$

where in the first case rows 1 and 2 were interchanged and in the second case we interchanged columns 1 and 3.

As an illustration of part (iii) of the theorem,

$$\begin{vmatrix} 1 & -3 & 4 \\ 2 & -1 & 0 \\ 3 & 1 & 6 \end{vmatrix} = \begin{vmatrix} 1 & -3 & 4 \\ 0 & 5 & -8 \\ 3 & 1 & 6 \end{vmatrix}$$

where we have added to the second row the product of -2 times the first row. This type of manipulation is very important for evaluating determinants of large order as will be shown below in Example 1. In like manner,

$$\begin{vmatrix} 1 & -3 & 4 \\ 2 & -1 & 0 \\ 3 & 1 & 6 \end{vmatrix} = \begin{vmatrix} -5 & -3 & 4 \\ 0 & -1 & 0 \\ 5 & 1 & 6 \end{vmatrix}$$

where we have added to the first column the product of 2 times the second column.

Part (ii) of the theorem will be illustrated in Examples 2 and 3.

Theorem

> If two rows (or columns) of a square matrix A are identical, then $|A| = 0$.

Proof   If B is the matrix obtained from A by interchanging the two identical rows (or columns), then B and A are the same and consequently $|B| = |A|$. However, by (i) of the Theorem on Row and Column Transformations of a Determinant, $|B| = -|A|$ and hence $-|A| = |A|$, which implies that $|A| = 0$. ∎

Example 1   Find $|A|$ if

$$A = \begin{bmatrix} 2 & 3 & 0 & 4 \\ 0 & 5 & -1 & 6 \\ 1 & 0 & -2 & 3 \\ -3 & 2 & 0 & -5 \end{bmatrix}$$

Solution   We plan to use (iii) of the Theorem on Row and Column Transformations to introduce many zeros in some row or column. To do this, it is convenient to work with an element of the matrix which equals 1 or -1, since this enables us to avoid the use of fractions. If no such element appears in the original matrix, it is always possible to introduce the number 1 by using (iii) or (ii) of the theorem. In this example there is no such problem, since 1 appears in row 3 and -1 in row 2. Let us work with the element 1 and introduce zero everywhere else in the first column, as shown below:

$$|A| = \begin{vmatrix} 0 & 3 & 4 & -2 \\ 0 & 5 & -1 & 6 \\ 1 & 0 & -2 & 3 \\ 0 & 2 & -6 & 4 \end{vmatrix}$$

Add to the first row -2 times the third row and then add to the fourth row 3 times the third row.

$$= (1) \begin{vmatrix} 3 & 4 & -2 \\ 5 & -1 & 6 \\ 2 & -6 & 4 \end{vmatrix}$$

Expand by column 1.

$$= \begin{vmatrix} 23 & 4 & 22 \\ 0 & -1 & 0 \\ -28 & -6 & -32 \end{vmatrix}$$

Add to the first column 5 times the second column and then add to the third column 6 times the second column.

$$= (-1) \begin{vmatrix} 23 & 22 \\ -28 & -32 \end{vmatrix}$$

Expand by row 2.

$$= (-1)[(23)(-32) - (-28)(22)]$$ By definition.

$$= 120$$

∎

Part (ii) of the Theorem on Row and Column Transformations is useful for finding factors of determinants. To illustrate, for a determinant of a matrix of order 3, we have the following:

$$\begin{vmatrix} a_{11} & a_{12} & a_{13} \\ ka_{21} & ka_{22} & ka_{23} \\ a_{31} & a_{32} & a_{33} \end{vmatrix} = k \begin{vmatrix} a_{11} & a_{12} & a_{13} \\ a_{21} & a_{22} & a_{23} \\ a_{31} & a_{32} & a_{33} \end{vmatrix}.$$

Similar formulas hold if k is a common factor of the elements of some other row or column. When the theorem is used in this way, we often use the phrase "k is a common factor in the row (or column)."

Example 2 Find $|A|$ if

$$A = \begin{bmatrix} 14 & -6 & 4 \\ 4 & -5 & 12 \\ -21 & 9 & -6 \end{bmatrix}.$$

Solution

$$|A| = 2 \begin{vmatrix} 7 & -3 & 2 \\ 4 & -5 & 12 \\ -21 & 9 & -6 \end{vmatrix} \qquad \text{2 is a common factor in row 1.}$$

$$= (2)(-3) \begin{vmatrix} 7 & -3 & 2 \\ 4 & -5 & 12 \\ 7 & -3 & 2 \end{vmatrix} \qquad \text{-3 is a common factor in row 3.}$$

$$= 0 \qquad \text{Two rows are identical.} \qquad \blacksquare$$

Example 3 Without expanding, show that $a - b$ is a factor of

$$\begin{vmatrix} 1 & 1 & 1 \\ a & b & c \\ a^2 & b^2 & c^2 \end{vmatrix}.$$

Solution

$$\begin{vmatrix} 1 & 1 & 1 \\ a & b & c \\ a^2 & b^2 & c^2 \end{vmatrix} = \begin{vmatrix} 0 & 1 & 1 \\ a - b & b & c \\ a^2 - b^2 & b^2 & c^2 \end{vmatrix} \qquad \text{Add to the first column -1 times the second column.}$$

$$= (a - b) \begin{vmatrix} 0 & 1 & 1 \\ 1 & b & c \\ a + b & b^2 & c^2 \end{vmatrix} \qquad \text{$a - b$ is a common factor of column 1.} \qquad \blacksquare$$

EXERCISES 8.7

Without expanding, explain why the statements in Exercises 1–14 are true.

1 $\begin{vmatrix} 1 & 0 & 1 \\ 0 & 1 & 1 \\ 1 & 1 & 0 \end{vmatrix} = - \begin{vmatrix} 1 & 0 & 1 \\ 1 & 1 & 0 \\ 0 & 1 & 1 \end{vmatrix}$

2 $\begin{vmatrix} 1 & 0 & 1 \\ 0 & 1 & 1 \\ 1 & 1 & 0 \end{vmatrix} = - \begin{vmatrix} 1 & 1 & 0 \\ 0 & 1 & 1 \\ 1 & 0 & 1 \end{vmatrix}$

3 $\begin{vmatrix} 1 & 0 & 1 \\ 2 & 1 & 0 \\ 1 & 1 & 2 \end{vmatrix} = \begin{vmatrix} 1 & 0 & 1 \\ 2 & 1 & 0 \\ 0 & 1 & 1 \end{vmatrix}$

4 $\begin{vmatrix} 1 & 1 & 2 \\ 1 & 0 & 1 \\ 2 & 1 & 1 \end{vmatrix} = \begin{vmatrix} 0 & 1 & 1 \\ 1 & 0 & 1 \\ 2 & 1 & 1 \end{vmatrix}$

5 $\begin{vmatrix} 2 & 4 & 2 \\ 1 & 2 & 4 \\ 2 & 6 & 4 \end{vmatrix} = 4 \begin{vmatrix} 1 & 2 & 1 \\ 1 & 2 & 4 \\ 1 & 3 & 2 \end{vmatrix}$

6 $\begin{vmatrix} 2 & 1 & 6 \\ 4 & 3 & 3 \\ 2 & 1 & 3 \end{vmatrix} = 6 \begin{vmatrix} 1 & 1 & 2 \\ 2 & 3 & 1 \\ 1 & 1 & 1 \end{vmatrix}$

7 $\begin{vmatrix} 1 & -1 & 2 \\ 1 & 2 & -1 \\ 1 & -1 & 2 \end{vmatrix} = 0$

8 $\begin{vmatrix} 1 & -1 & 1 \\ 0 & 1 & 0 \\ -1 & 0 & -1 \end{vmatrix} = 0$

9 $\begin{vmatrix} 1 & 5 \\ -3 & 2 \end{vmatrix} = - \begin{vmatrix} 1 & 5 \\ 3 & -2 \end{vmatrix}$

10 $\begin{vmatrix} 2 & -2 \\ 1 & 1 \end{vmatrix} = - \begin{vmatrix} -2 & 2 \\ 1 & 1 \end{vmatrix}$

11 $\begin{vmatrix} 0 & 0 & 1 \\ 1 & 0 & 0 \\ 0 & 0 & 1 \end{vmatrix} = 0$ **12** $\begin{vmatrix} 1 & 0 & 1 \\ 0 & 0 & 0 \\ 1 & 1 & 0 \end{vmatrix} = 0$

13 $\begin{vmatrix} 1 & -1 & -2 \\ -1 & 2 & 1 \\ 0 & 1 & 1 \end{vmatrix} = \begin{vmatrix} 1 & -1 & 0 \\ -1 & 2 & -1 \\ 0 & 1 & 1 \end{vmatrix}$

14 $\begin{vmatrix} a & 0 & 0 \\ 0 & b & 0 \\ 0 & 0 & c \end{vmatrix} = - \begin{vmatrix} 0 & 0 & a \\ 0 & b & 0 \\ c & 0 & 0 \end{vmatrix}$

In each of Exercises 15–24 find the determinant of the matrix after introducing zeros, as in Example 1.

15 $\begin{bmatrix} 3 & 1 & 0 \\ -2 & 0 & 1 \\ 1 & 3 & -1 \end{bmatrix}$ **16** $\begin{bmatrix} -3 & 0 & 4 \\ 1 & 2 & 0 \\ 4 & 1 & -1 \end{bmatrix}$

17 $\begin{bmatrix} 5 & 4 & 3 \\ -3 & 2 & 1 \\ 0 & 7 & -2 \end{bmatrix}$ **18** $\begin{bmatrix} 0 & 2 & -6 \\ 5 & 1 & -3 \\ 6 & -2 & 5 \end{bmatrix}$

19 $\begin{bmatrix} 2 & 2 & -3 \\ 3 & 6 & 9 \\ -2 & 5 & 4 \end{bmatrix}$ **20** $\begin{bmatrix} 3 & 8 & 5 \\ 5 & 3 & -6 \\ 2 & 4 & -2 \end{bmatrix}$

21 $\begin{bmatrix} 3 & 1 & -2 & 2 \\ 2 & 0 & 1 & 4 \\ 0 & 1 & 3 & 5 \\ -1 & 2 & 0 & -3 \end{bmatrix}$

22 $\begin{bmatrix} 3 & 2 & 0 & 4 \\ -2 & 0 & 5 & 0 \\ 4 & -3 & 1 & 6 \\ 2 & -1 & 2 & 0 \end{bmatrix}$

23 $\begin{bmatrix} 2 & -2 & 0 & 0 & -3 \\ 3 & 0 & 3 & 2 & -1 \\ 0 & 1 & -2 & 0 & 2 \\ -1 & 2 & 0 & 3 & 0 \\ 0 & 4 & 1 & 0 & 0 \end{bmatrix}$

24 $\begin{bmatrix} 2 & 0 & -1 & 0 & 2 \\ 1 & 3 & 0 & 0 & 1 \\ 0 & 4 & 3 & 0 & -1 \\ -1 & 2 & 0 & -2 & 0 \\ 0 & 1 & 5 & 0 & -4 \end{bmatrix}$

25 Prove that

$$\begin{vmatrix} 1 & 1 & 1 \\ a & b & c \\ a^2 & b^2 & c^2 \end{vmatrix} = (a - b)(b - c)(c - a).$$

(*Hint*: See Example 3.)

26 Prove that

$$\begin{vmatrix} 1 & 1 & 1 \\ a & b & c \\ a^3 & b^3 & c^3 \end{vmatrix} = (a-b)(b-c)(c-a)(a+b+c).$$

27 If A is a matrix of order 4 of the form

$$A = \begin{bmatrix} a_{11} & a_{12} & a_{13} & a_{14} \\ 0 & a_{22} & a_{23} & a_{24} \\ 0 & 0 & a_{33} & a_{34} \\ 0 & 0 & 0 & a_{44} \end{bmatrix}$$

show that $|A| = a_{11}a_{22}a_{33}a_{44}$.

28 If

$$A = \begin{bmatrix} a & b & 0 & 0 \\ c & d & 0 & 0 \\ 0 & 0 & e & f \\ 0 & 0 & g & h \end{bmatrix}$$

prove that

$$|A| = \begin{vmatrix} a & b \\ c & d \end{vmatrix} \begin{vmatrix} e & f \\ g & h \end{vmatrix}.$$

29 If $A = (a_{ij})$ and $B = (b_{ij})$ are arbitrary square matrices of order 2, prove that $|AB| = |A|\,|B|$.

30 If $A = (a_{ij})$ is a square matrix of order n and k is any real number, prove that $|kA| = k^n|A|$. (*Hint*: Use (ii) of the Theorem on Row and Column Transformations of a Determinant.)

8.8 CRAMER'S RULE

Determinants arise naturally in the study of solutions of systems of linear equations. To illustrate, let us consider the following case of two general linear equations in two unknowns:

$$\begin{cases} a_{11}x + a_{12}y = k_1 \\ a_{21}x + a_{22}y = k_2 \end{cases}$$

where at least one nonzero coefficient appears in each equation. We may as well assume that $a_{11} \ne 0$, for otherwise $a_{12} \ne 0$ and we could regard y as the "first" variable instead of x. We shall use the matrix method to obtain the matrix of an equivalent system. Changing the matrix of the system by means of elementary row transformations we obtain

$$\begin{bmatrix} a_{11} & a_{12} & k_1 \\ a_{21} & a_{22} & k_2 \end{bmatrix}$$

$$\rightarrow \begin{bmatrix} a_{11} & a_{12} & k_1 \\ 0 & a_{22} - \left(\dfrac{a_{12}a_{21}}{a_{11}}\right) & k_2 - \left(\dfrac{a_{21}k_1}{a_{11}}\right) \end{bmatrix} \quad \begin{array}{l}\text{Add to the second row} \\ -a_{21}/a_{11} \text{ times the first} \\ \text{row.}\end{array}$$

$$\rightarrow \begin{bmatrix} a_{11} & a_{12} & k_1 \\ 0 & (a_{11}a_{22} - a_{12}a_{21}) & (a_{11}k_2 - a_{21}k_1) \end{bmatrix} \quad \text{Multiply row 2 by } a_{11}.$$

Thus, the given system is equivalent to

$$\begin{cases} a_{11}x + a_{12}y = k_1 \\ (a_{11}a_{22} - a_{12}a_{21})y = a_{11}k_2 - a_{21}k_1. \end{cases}$$

Notice that the numbers in the second equation may be written in determinant form as follows:

$$\begin{cases} a_{11}x + a_{12}y = k_1 \\ \begin{vmatrix} a_{11} & a_{12} \\ a_{21} & a_{22} \end{vmatrix} y = \begin{vmatrix} a_{11} & k_1 \\ a_{21} & k_2 \end{vmatrix}. \end{cases}$$

The following results now follow from the discussion in Section 8.2.

(a) If

$$\begin{vmatrix} a_{11} & a_{12} \\ a_{21} & a_{22} \end{vmatrix} = 0 \quad \text{and} \quad \begin{vmatrix} a_{11} & k_1 \\ a_{21} & k_2 \end{vmatrix} = 0$$

then the equations are dependent.

(b) If

$$\begin{vmatrix} a_{11} & a_{12} \\ a_{21} & a_{22} \end{vmatrix} = 0 \quad \text{and} \quad \begin{vmatrix} a_{11} & k_1 \\ a_{21} & k_2 \end{vmatrix} \neq 0$$

then the equations are inconsistent.

(c) If

$$\begin{vmatrix} a_{11} & a_{12} \\ a_{21} & a_{22} \end{vmatrix} \neq 0$$

then the equations are consistent.

If (c) occurs, then we can solve the second equation for y, obtaining

$$y = \frac{\begin{vmatrix} a_{11} & k_1 \\ a_{21} & k_2 \end{vmatrix}}{\begin{vmatrix} a_{11} & a_{12} \\ a_{21} & a_{22} \end{vmatrix}}$$

The corresponding value for x may be found by substituting for y in the first equation. It can be shown that this leads to

$$x = \frac{\begin{vmatrix} k_1 & a_{12} \\ k_2 & a_{22} \end{vmatrix}}{\begin{vmatrix} a_{11} & a_{12} \\ a_{21} & a_{22} \end{vmatrix}}$$

This proves that *if the determinant of the coefficient matrix of a system of two linear equations in two variables is not zero, then the system has a unique solution.* The last two formulas for x and y as quotients of certain determinants constitute what is known as **Cramer's Rule**.

There is an easy way to remember Cramer's Rule. Let

$$D = \begin{bmatrix} a_{11} & a_{12} \\ a_{21} & a_{22} \end{bmatrix}$$

be the coefficient matrix of the system and let D_x denote the matrix obtained from D by replacing the coefficients a_{11}, a_{21} of x by the numbers k_1, k_2, respectively. Similarly, let D_y denote the matrix obtained from D by replacing the coefficients a_{12}, a_{22} of y by the numbers k_1, k_2, respectively. Thus,

$$D_x = \begin{bmatrix} k_1 & a_{12} \\ k_2 & a_{22} \end{bmatrix}, \quad D_y = \begin{bmatrix} a_{11} & k_1 \\ a_{21} & k_2 \end{bmatrix}.$$

According to the previous discussion, if $|D| \neq 0$, the solution (x, y) is given by

Cramer's Rule

$$x = \frac{|D_x|}{|D|}, \quad y = \frac{|D_y|}{|D|}.$$

Example 1 Use Cramer's Rule to solve the system

$$\begin{cases} 2x - 3y = -4 \\ 5x + 7y = 1. \end{cases}$$

Solution The determinant of the coefficient matrix is

$$|D| = \begin{vmatrix} 2 & -3 \\ 5 & 7 \end{vmatrix} = 29.$$

Using the notation introduced previously,

$$|D_x| = \begin{vmatrix} -4 & -3 \\ 1 & 7 \end{vmatrix} = -25, \quad |D_y| = \begin{vmatrix} 2 & -4 \\ 5 & 1 \end{vmatrix} = 22.$$

Hence, $$x = \frac{|D_x|}{|D|} = \frac{-25}{29}, \quad y = \frac{|D_y|}{|D|} = \frac{22}{29}.$$

Thus, the system has the unique solution $(-25/29, 22/29)$. ∎

Let us briefly consider the case of a general homogeneous system of two linear equations in x and y; that is,

$$\begin{cases} a_{11}x + a_{12}y = 0 \\ a_{21}x + a_{22}y = 0. \end{cases}$$

In this event the determinants $|D_x|$ and $|D_y|$ in Cramer's Rule are both zero. (Why?) Consequently, if

$$|D| = \begin{vmatrix} a_{11} & a_{12} \\ a_{21} & a_{22} \end{vmatrix} \neq 0$$

then $x = 0/|D| = 0$ and $y = 0/|D| = 0$; that is, the only solution is the trivial one $(0, 0)$. This proves that *if the given system of homogeneous equations has a nontrivial solution, then*

$$\begin{vmatrix} a_{11} & a_{12} \\ a_{21} & a_{22} \end{vmatrix} = 0.$$

It can be shown, conversely, that if the determinant of the coefficient matrix is zero, then a homogeneous system has a nontrivial solution.

The previous discussion can be extended to systems of n linear equations in n variables. It is possible to show that such a system has a unique solution if and only if the determinant of the coefficient matrix is different from zero. If the system is homogeneous, then nontrivial solutions exist if and only if the determinant of the coefficient matrix is zero.

Cramer's Rule can be extended to systems of n linear equations in n variables x_1, $x_2, \ldots, x_n$, where each equation is written in the form

$$a_1 x_1 + a_2 x_2 + \cdots + a_n x_n = a.$$

To solve such a system, let D denote the coefficient matrix and let D_{x_i} denote the matrix obtained by replacing the coefficients of x_i in D by the column of numbers $k_1, \ldots, k_n$, which appears to the right of the equals signs in the system. It can be shown that if $|D| \neq 0$, then the system has a unique solution given by the following:

Cramer's Rule (General Form)

$$x_1 = \frac{|D_{x_1}|}{|D|}, \quad x_2 = \frac{|D_{x_2}|}{|D|}, \ldots, \quad x_n = \frac{|D_{x_n}|}{|D|}.$$

It can be shown that if $|D| = 0$, the equations are dependent or inconsistent, depending on whether all the $|D_{x_i}|$ are zero or at least one of them is not zero.

Example 2   Use Cramer's Rule to solve the system

$$\begin{cases} x \quad\;\; - 2z = 3 \\ \quad -y + 3z = 1 \\ 2x \quad\;\; + 5z = 0. \end{cases}$$

Solution   We shall merely list the various determinants which are used, leaving the reader to check the answers:

$$|D| = \begin{vmatrix} 1 & 0 & -2 \\ 0 & -1 & 3 \\ 2 & 0 & 5 \end{vmatrix} = -9, \quad |D_x| = \begin{vmatrix} 3 & 0 & -2 \\ 1 & -1 & 3 \\ 0 & 0 & 5 \end{vmatrix} = -15,$$

$$|D_y| = \begin{vmatrix} 1 & 3 & -2 \\ 0 & 1 & 3 \\ 2 & 0 & 5 \end{vmatrix} = 27, \quad |D_z| = \begin{vmatrix} 1 & 0 & 3 \\ 0 & -1 & 1 \\ 2 & 0 & 0 \end{vmatrix} = 6.$$

By Cramer's Rule, the solution is

$$x = \frac{|D_x|}{|D|} = \frac{-15}{-9} = \frac{5}{3}, \quad y = \frac{|D_y|}{|D|} = \frac{27}{-9} = -3, \quad z = \frac{|D_z|}{|D|} = \frac{6}{-9} = -\frac{2}{3}. \;\blacksquare$$

Cramer's Rule is an inefficient method to apply if there are a large number of equations, since many determinants of matrices of high order must be evaluated. Note also that Cramer's Rule cannot be used directly if $|D| = 0$ or if the number of equations is not the same as the number of variables. In general, the method of elimination or the matrix method is far superior to Cramer's Rule.

EXERCISES 8.8

1–18 Use Cramer's Rule to solve the systems in Exercises 1–18 of Section 8.2.

19–26 Use Cramer's Rule to solve the systems in Exercises 1–8 of Section 8.3.

27–30 Use Cramer's Rule to solve the systems in Exercises 17–20 of Section 8.3.

8.9 SYSTEMS OF INEQUALITIES

Our previous work with inequalities was restricted to inequalities in one variable. The notion of inequalities in several variables can be developed in a manner similar to our work with equations in several variables. For example, expressions of the form

$$3x + y < 5y^2 + 1$$
$$2x^2 \geq 4 - 3y$$

are called **inequalities in x and y**. As with equations, a **solution** of an inequality in x and y is defined as an ordered pair (a, b) which produces a true statement when a and b are substituted for x and y, respectively. The **graph of an inequality** is the graph of the totality of solutions. Two inequalities are **equivalent** if they have exactly the same solutions. An inequality in x and y can be simplified by adding an expression in x and y to both sides or by multiplying both sides by such an expression, provided we are careful about signs. Similar definitions and remarks hold for inequalities in more than two variables. We shall restrict our discussion, however, to the case of inequalities in two variables.

Example 1 Find the solutions and sketch the graph of the inequality

$$3x - 3 < 5x - y.$$

Solution Adding the expression $y + 3 - 3x$ to both sides, we obtain the equivalent inequality $y < 2x + 3$. Hence, the solutions consist of all ordered pairs (x, y) such that $y < 2x + 3$. It is convenient to denote the solutions as follows:

$$\{(x, y) : y < 2x + 3\}.$$

There is a close relationship between the graph of the inequality $y < 2x + 3$ and the graph of the equation $y = 2x + 3$. The graph of the equation is the straight line sketched in Figure 8.5. For each real number a, the point on the line with abscissa a has coordinates $(a, 2a + 3)$. A point $P(a, b)$ belongs to the graph of the *inequality* if and only if $b < 2a + 3$; that is, if and only if the point $P(a, b)$ lies directly below the point with coordinates $(a, 2a + 3)$ as shown in Figure 8.5. It follows that the graph of the inequality $y < 2x + 3$ consists of all

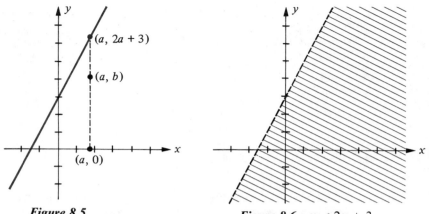

Figure 8.5   **Figure 8.6** $y < 2x + 3$

points in the plane which lie below the line $y = 2x + 3$. In Figure 8.6 we have shaded a portion of the graph. Dashes used for the line indicate that it is not part of the graph of the inequality. ∎

A region of the type shown in Figure 8.6 is called a **half-plane**. More precisely, when the line is *not* included, we refer to such a region as an **open half-plane**. If the line *is* included, as would be the case for the graph of the inequality $y \leq 2x + 3$, then the region is called a **closed half-plane**.

By an argument similar to that used in Example 1, it can be shown that the graph of the inequality $y > 2x + 3$ is the open half-plane which lies *above* the line with equation $y = 2x + 3$.

If an inequality involves only polynomials of the first degree in x and y, as was the case in Example 1, it is called a **linear inequality**.

The procedure used in Example 1 can be generalized to inequalities of the form $y < f(x)$, where f is a function. Specifically, the following can be established.

Theorem

> If f is a function, then the graph of the inequality $y < f(x)$ is the set of points which lie *below* the graph of the equation $y = f(x)$. Similarly, the graph of $y > f(x)$ is the set of points which lie *above* the graph of $y = f(x)$.

Example 2 Find the solutions and sketch the graph of the inequality

$$x(x + 1) - 2y > 3(x - y).$$

Solution   The given inequality is equivalent to

$$x^2 + x - 2y > 3x - 3y.$$

Adding $3y - x^2 - x$ to both sides we obtain

$$y > 2x - x^2.$$

Hence, the solutions are $\{(x, y): y > 2x - x^2\}$. To find the graph, we begin by sketching the graph of $y = 2x - x^2$ (a parabola) with dashes as illustrated in Figure 8.7. Using the preceding theorem, the graph is the region above the parabola, as indicated by the shaded portion of the figure.

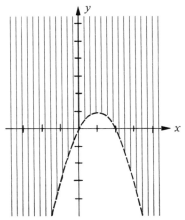

Figure 8.7 $y > 2x - x^2$

The following result can also be proved.

Theorem

> If g is a function, then the graph of the inequality $x < g(y)$ is the set of points to the *left* of the graph of the equation $x = g(y)$. Similarly, the graph of $x > g(y)$ is the set of points to the *right* of the graph of $x = g(y)$.

Example 3 Sketch the graph of $x \geq y^2$.

Solution The graph of the equation $x = y^2$ is a parabola. By the preceding theorem, the graph of the inequality consists of all points on the parabola together with the points in the region to the right of the parabola (see Figure 8.8).

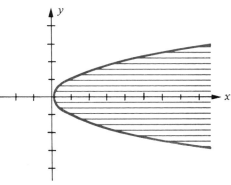

Figure 8.8 $x \geq y^2$

It is sometimes necessary to work simultaneously with several inequalities in two variables. In this case we refer to the given inequalities as a **system of inequalities**. The **solutions of a system** of inequalities are, by definition, the common solutions of all the inequalities in the system. It should be clear how to define **equivalent systems** and the **graph of a system** of inequalities. The following examples illustrate a method for solving systems of inequalities.

Example 4   Find the solutions and sketch the graph of the system

$$\begin{cases} x + y \le 4 \\ 2x - y \le 4. \end{cases}$$

Solution   The given system is equivalent to

$$\begin{cases} y \le 4 - x \\ y \ge 2x - 4. \end{cases}$$

We begin by sketching the graphs of the lines $y = 4 - x$ and $y = 2x - 4$. The lines intersect at the point $(\frac{8}{3}, \frac{4}{3})$ shown in Figure 8.9. The graph of $y \le 4 - x$ includes the points on the graph of $y = 4 - x$ together with the points which lie below this line. The graph of $y \ge 2x - 4$ includes the points on the graph of $y = 2x - 4$ together with the points which lie above this line. A portion of each of these regions is shown in Figure 8.9. The graph of the system consists of the points that are in *both* regions. This corresponds to the cross-hatched region shown in Figure 8.9.

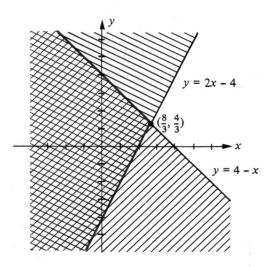

Figure 8.9

Example 5 Sketch the graph of the system

$$\begin{cases} x + y \le 4 \\ 2x - y \le 4 \\ \quad\quad x \ge 0 \\ \quad\quad y \ge 0. \end{cases}$$

Solution The first two inequalities are the same as those considered in Example 4 and hence the points on the graph of the present system must lie within the region shown in Figure 8.9. In addition, the third and fourth inequalities in the system tell us that the points must lie in the first quadrant. This gives us the region shown in Figure 8.10.

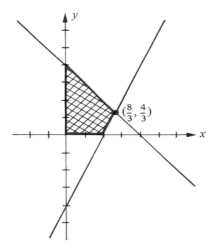

Figure 8.10   ■

Example 6 Sketch the graph of the system

$$\begin{cases} \quad\quad x^2 + y^2 \le 1 \\ (x - 1)^2 + y^2 \le 1. \end{cases}$$

Solution The graph of the equation $x^2 + y^2 = 1$ is a unit circle with center at the origin, and the graph of $(x - 1)^2 + y^2 = 1$ is a unit circle with center at the point $C(1, 0)$. To find the points of intersection of the two circles, let us solve the equation $x^2 + y^2 = 1$ for y^2, obtaining $y^2 = 1 - x^2$. Substituting for y^2 in $(x - 1)^2 + y^2 = 1$ leads to the following equations:

$$(x - 1)^2 + (1 - x^2) = 1$$
$$x^2 - 2x + 1 + 1 - x^2 = 1$$
$$-2x = -1$$
$$x = 1/2.$$

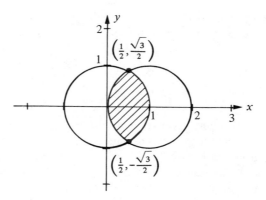

Figure 8.11

The corresponding values for y are given by

$$y^2 = 1 - x^2 = 1 - (1/2)^2 = 3/4$$

and hence $y = \pm\sqrt{3}/2$. Thus, the points of intersection are $(1/2, \sqrt{3}/2)$ and $(1/2, -\sqrt{3}/2)$ as shown in Figure 8.11. By the distance formula, the graphs of the given inequalities are the regions within and on the two circles. The graph of the system consists of the points common to both regions, as indicated by the shaded portion of the figure. ∎

EXERCISES 8.9

In each of Exercises 1–10 find the solutions and sketch the graph of the given inequality.

1 $3x - 2y < 6$

2 $4x + 3y < 12$

3 $2x + 3y \geq 2y + 1$

4 $2x - y > 3$

5 $y + 2 < x^2$

6 $y^2 - x \leq 0$

7 $x^2 + 1 \leq y$

8 $y - x^3 < 1$

9 $yx^2 \geq 1$

10 $x^2 + 4 \geq y$

In each of Exercises 11–24 sketch the graph of the given system.

11 $\begin{cases} 3x + y < 3 \\ 4 - y < 2x \end{cases}$

12 $\begin{cases} y + 2 < 2x \\ y - x > 4 \end{cases}$

13 $\begin{cases} y - x < 0 \\ 2x + 5y < 10 \end{cases}$

14 $\begin{cases} 2y - x \leq 4 \\ 3y + 2x < 6 \end{cases}$

15 $\begin{cases} 3x + y \leq 6 \\ y - 2x \geq 1 \\ \quad x \geq -2 \\ \quad y \leq 4 \end{cases}$

16 $\begin{cases} 3x - 4y \geq 12 \\ x - 2y \leq 2 \\ \quad x \geq 9 \\ \quad y \leq 5 \end{cases}$

17 $\begin{cases} x^2 + y^2 \leq 4 \\ x + y \geq 1 \end{cases}$

18 $\begin{cases} x^2 + y^2 > 1 \\ x^2 + y^2 < 4 \end{cases}$

19 $\begin{cases} x^2 \leq 1 - y \\ x \geq 1 + y \end{cases}$

20 $\begin{cases} x - y^2 < 0 \\ x + y^2 > 0 \end{cases}$

21 $\begin{cases} y < 3^x \\ y > 2^x \\ x \geq 0 \end{cases}$

22 $\begin{cases} y \geq \log x \\ y - x \leq 1 \\ x \geq 1 \end{cases}$

23 $\begin{cases} y \leq \log x \\ y + x \geq 1 \\ x \leq 10 \end{cases}$

24 $\begin{cases} y \leq 3^{-x} \\ y \geq 2^{-x} \\ y < 9 \end{cases}$

25 The manager of a baseball team wishes to buy bats and balls, costing \$3.50 and \$2.50, respectively. If the maximum amount available is \$40 and if at least two balls and three bats are required, find a system of inequalities describing all the possibilities and sketch the graph.

26 An office worker wishes to purchase some 40-cent postage stamps and also some 50-cent stamps, totaling not more than \$35. Moreover, it is desired to have at least twice as many 40-cent stamps as 50-cent stamps and more than ten 40-cent stamps. Find a system of inequalities describing all the possibilities and sketch the graph.

27 A store sells two brands of television sets. Customer demand indicates that it is necessary to stock at least twice as many sets of brand A as of brand B. It is also necessary to have on hand at least 20 of brand A and 10 of brand B. If there is room for not more than 100 sets in the store, find a system of inequalities describing all possibilities and sketch the graph.

28 An auditorium contains 600 seats. For a certain event it is planned to charge \$4.00 for certain seats and \$3.00 for others. At least 225 tickets are to be sold for \$3.00, and total sales of more than \$2,000 is desired. Find a system of inequalities describing all possibilities and sketch the graph.

29 A woman wishes to invest \$15,000 in two different savings accounts. She also wants to have at least \$2,000 in each account, with the amount in one account being at least three times that in the other. Find a system of inequalities describing all possibilities and sketch the graph.

30 The manager of a college bookstore stocks two types of notebooks, the first wholesaling for 55 cents and the second for 85 cents. If the maximum amount to be spent is \$600 and if an inventory of at least 300 of the 85-cent variety and 400 of the 55-cent variety is desired, find a system of inequalities describing all possibilities and sketch the graph.

8.10 LINEAR PROGRAMMING

In applications, problems sometimes arise which require finding solutions of systems of inequalities. A typical problem is that of finding maximum and minimum values of certain expressions involving variables which are subject to various constraints. If all the expressions and inequalities are linear in the variables, then a technique called **linear programming** may be used to help solve such problems. This technique has become very important in businesses where decisions must be made concerning the best use of stock, parts, or manufacturing processes. Usually the objective of management is to maximize profit or to minimize cost. Since there are often many choices, it may be extremely difficult to arrive at a correct decision. A mathematical theory such as that afforded by linear programming can simplify the task considerably. The logical development of the theorems and techniques which are needed would take us beyond the objectives of this text. We shall, therefore, limit ourselves to several examples.

Example 1 A manufacturer of a certain product has two warehouses, W_1 and W_2. There are 80 units of his product stored at W_1 and 70 units at W_2. Two customers, A and B, order 35 units and 60 units, respectively. The shipping cost from each warehouse to A and B is determined according to the following table. How should the order be filled so as to minimize the total shipping cost?

Warehouse	Customer	Shipping cost per unit
W_1	A	$ 8
W_1	B	12
W_2	A	10
W_2	B	13

Solution   If we let x denote the number of units to be sent to A from W_1, then $35 - x$ units must be sent from W_2 to A. Similarly, if y denotes the number of units to be sent from W_1 to B, then $60 - y$ units must be sent from W_2 to B. We wish to determine values for x and y which make the total shipping costs minimal. We first note that, since x and y are between 35 and 60, respectively, the pair (x, y) must be a solution of the following system of inequalities:

$$0 \leq x \leq 35, \quad 0 \leq y \leq 60.$$

The graph of this system is the rectangular region shown in Figure 8.12.

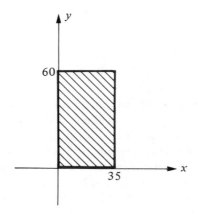

Figure 8.12

There are further constraints on x and y which make it possible to reduce the size of the region above. Since the total number of units shipped from W_1 cannot exceed 80 and the total shipped from W_2 cannot exceed 70, the pair (x, y) must also be a solution of the system

$$\begin{cases} x + y \leq 80 \\ (35 - x) + (60 - y) \leq 70. \end{cases}$$

This system is equivalent to

$$\begin{cases} x + y \leq 80 \\ x + y \geq 25. \end{cases}$$

The graph of the last system is the region between the parallel lines $x + y = 80$ and $x + y = 25$ (see Figure 8.13). Since the pair (x, y) that we seek must be a solution of this system and also the system $0 \le x \le 35$, $0 \le y \le 60$, the corresponding point must lie in the region shown in Figure 8.14.

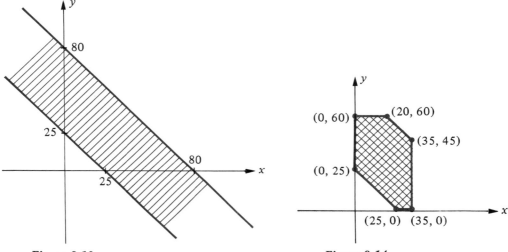

Figure 8.13   **Figure 8.14**

Now let C denote the total cost (in dollars) of shipping the merchandise to A and B. We see from the table that the cost of shipping the 35 units to A is $8x + 10(35 - x)$, whereas the cost of shipping the 60 units to B is $12y + 13(60 - y)$. Hence, the total cost is

$$C = (8x + 350 - 10x) + (12y + 780 - 13y)$$

or $$C = 1130 - 2x - y.$$

For each point (x, y) of the region shown in Figure 8.14 we obtain a value for C. For example, at $(20, 40)$

$$C = 1130 - 40 - 40 = 1050;$$

and at $(10, 50)$, $$C = 1130 - 20 - 50 = 1060.$$

Since x and y must be integers, there is only a finite number of possible values for C. By checking each possibility, we could find the pair (x, y) which produces the smallest cost. However, since there is a very large number of pairs, the task of checking each one would be very tedious. This is where the theory developed in linear programming is helpful. It can be shown that if we are interested in the value C of a linear expression $ax + by + c$, where each pair (x, y) is a solution of a system of linear inequalities and hence corresponds to a point which is in the intersection of half-planes, then C takes on its maximum and minimum values at a point of intersection of the lines which determine the half-planes. This means that, in order to determine the minimum (or maximum) value of C, we need only check the points $(0, 25)$, $(0, 60)$, $(20, 60)$, $(35, 45)$, $(35, 0)$, and $(25, 0)$ shown in Figure 8.14. The values are arranged in tabular form below.

Point	$1130 - 2x - y = C$
$(0, 25)$	$1130 - 2(0) - 25 = 1105$
$(0, 60)$	$1130 - 2(0) - 60 = 1070$
$(20, 60)$	$1130 - 2(20) - 60 = 1030$
$(35, 45)$	$1130 - 2(35) - 45 = 1015$
$(35, 0)$	$1130 - 2(35) - 0 = 1060$
$(25, 0)$	$1130 - 2(25) - 0 = 1080$

According to our remarks, the minimal shipping cost \$1,015 occurs if $x = 35$ and $y = 45$. This means that the manufacturer should ship all of the units to A from W_1. In addition, the manufacturer should ship 45 units to B from W_1 and 15 units to B from W_2. Note that the *maximum* shipping cost will occur when $x = 0$ and $y = 25$, that is, when all 35 units are shipped to A from W_2 and when B receives 25 units from W_1 and 35 units from W_2. ∎

The preceding example illustrates how linear programming can be used to minimize the cost in a certain situation. The next example has to do with maximization of profit.

Example 2 A firm manufactures two products X and Y. For each product it is necessary to use three different machines A, B, and C. In order to manufacture one unit of product X, machine A must be used for 3 hours, machine B for 1 hour, and machine C for 1 hour. To manufacture one unit of product Y requires 2 hours on A, 2 hours on B, and 1 hour on C. The profit on product X is \$500 per unit and the profit on product Y is \$350 per unit. Machine A is available for a total of 24 hours per day; however, B can only be used for 16 hours and C for 9 hours. If the machines are available when needed (subject to the noted total hour restrictions), determine the number of units of each product that should be manufactured each day in order to maximize the profit.

Solution   The following table summarizes the data given in the statement of the problem.

Machine	Hours required for 1 unit of X	Hours required for one unit of Y	Hours available
A	3	2	24
B	1	2	16
C	1	1	9

Let x and y denote the number of units of products X and Y, respectively, to be produced per day. Since each unit of product X requires 3 hours on machine A,

x units require $3x$ hours. Similarly, since each unit of product Y requires 2 hours on A, y units require $2y$ hours. Hence, the total number of hours per day that machine A must be used is $3x + 2y$. Since A can be used for at most 24 hours per day, we have

$$3x + 2y \le 24.$$

Using the same type of reasoning on rows two and three of the table we see that

$$x + 2y \le 16$$
$$x + y \le 9.$$

This system of three linear inequalities, together with the obvious inequalities

$$x \ge 0, \quad y \ge 0$$

states, in mathematical form, the restraints which occur in the manufacturing process. The graph of the preceding system of five linear inequalities is sketched in Figure 8.15. The points shown in the figure are found by solving systems of linear equations. Specifically, $(6, 3)$ is a solution of the system $3x + 2y = 24$, $x + y = 9$, and $(2, 7)$ is a solution of the system $x + 2y = 16$, $x + y = 9$.

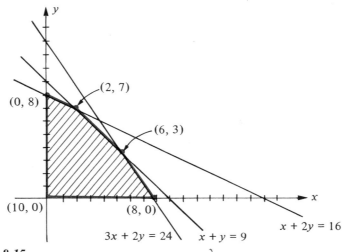

Figure 8.15

Since the production of each unit of product X results in a profit of \$500, and each unit of product Y yields a profit of \$350, the profit P obtained by producing x units of X together with y units of Y is given by

$$P = 500x + 350y.$$

As pointed out in the solution of Example 1, the maximum and minimum values of P occur at certain of the points shown in Figure 8.15. The values of P at all the points are given in the following table.

(x, y)	$500x + 350y = P$
$(0, 8)$	$500(0) + 350(8) = 2800$
$(2, 7)$	$500(2) + 350(7) = 3450$
$(6, 3)$	$500(6) + 350(3) = 4050$
$(8, 0)$	$500(8) + 350(0) = 4000$
$(0, 0)$	$500(0) + 350(0) = 0$

We see from the table that a maximum profit of $4,050 occurs if there is a daily production of 6 units of product X and 3 units of product Y. ∎

The two illustrations we have given in this section are elementary problems in linear programming which can be solved by rather crude methods. The much more complicated problems that occur in practice are usually solved by employing matrix techniques which are adapted for solutions by computers.

EXERCISES 8.10

1 A manufacturer of tennis rackets makes a profit of $15 on each Set Point racket and $8 on each Double Fault racket. To meet dealer demand, daily production of Double Faults should be between 30 and 80, whereas the number of Set Points should be between 10 and 30. In order to maintain high quality, the total number of rackets produced should not exceed 80 per day. How many of each type should be manufactured daily to maximize the profit?

2 A manufacturer of CB radios makes a profit of $25 on a deluxe model and $30 on a standard model. The company wishes to produce at least 80 deluxe models and at least 100 standard models per day. To maintain high quality, the daily production should not exceed 200 radios. How many of each type should be produced daily in order to maximize the profit?

3 Two substances S and T each contain two types of ingredients I and G. One pound of S contains 2 ounces of I and 4 ounces of G. One pound of T contains 2 ounces of I and 6 ounces of G. It is desired to combine quantities of the two substances to obtain a mixture which contains at least 9 ounces of I and 20 ounces of G. If the cost of S is $3.00 per pound and the cost of T is $4.00 per pound, how much of each substance should be used to keep the cost to a minimum?

4 A stationery company makes two types of notebooks. Type M sells for $1.25 and type N sells for $0.90. It costs the company $1.00 to produce one type M notebook and $0.75 to produce one of type N. The company has the facilities to manufacture between 2,000 and 3,000 of type M and between 3,000 and 6,000 of type N, but not more than 7,000 altogether. How many notebooks of each type should be manufactured to maximize the difference between the selling prices and the costs of production?

5 In Example 1 of this section, if the shipping costs are $12 per unit from W_1 to A, $10 per unit from W_2 to A, $16 per unit from W_1 to B, and $12 per unit from W_2 to B, determine how the order should be filled so as to minimize shipping costs.

6 A coffee company purchases mixed lots of coffee beans and then grades them into premium, regular and unusable beans. The company needs at least 280 tons of premium-grade and 200 tons of regular-grade coffee beans. The company can purchase ungraded coffee from two suppliers A and B in any amount desired. Samples from the two suppliers contain the following percentages of premium, regular, and unusable beans:

Supplier	Premium	Regular	Unusable
A	20%	50%	30%
B	40%	20%	40%

If A charges $125 per ton and B charges $200 per ton, how much should the company purchase from each supplier to fulfill its needs at minimum cost?

7 A farmer has 100 acres available for planting two crops A and B. The seed for crop A costs \$4 per acre and the seed for crop B costs \$6 per acre. The total cost of labor will amount to \$20 per acre for crop A and \$10 per acre for crop B. The expected income from crop A is \$110 per acre, whereas from B it is \$150 per acre. If the farmer does not wish to spend more than \$480 for seed and \$1,400 for labor, how many acres of each crop should be planted in order to obtain the maximum profit?

8 A firm manufactures two products A and B. For each product it is necessary to employ two different machines X and Y. To manufacture product A, machine X must be used for $\frac{1}{2}$ hour and machine Y for 1 hour. To manufacture product B, machine X must be used for 2 hours and machine Y for 2 hours. The profit on product A is \$20 per unit and the profit on B is \$50 per unit. If machine X can be used for 8 hours per day and machine Y for 12 hours per day, determine how many units of each product should be manufactured each day in order to maximize the profit.

9 Three substances X, Y, and Z each contain four ingredients A, B, C, and D. The percentage of

each ingredient and the cost in cents per ounce of substances are given in the following table.

Substance	A	B	C	D	Cost/ Ounce
X	20%	10%	25%	45%	25¢
Y	20%	40%	15%	25%	35¢
Z	10%	20%	25%	45%	50¢

If the cost is to be minimum, how many ounces of each substance should be combined in order to obtain a mixture of 20 ounces containing at least 14% A, 16% B, and 20% C? What combination would make the cost greatest?

10 A man plans to operate a stand at a one-day fair in which he will sell bags of peanuts and bags of candy. He has \$100 available to purchase his stock, which will cost 10¢ per bag of peanuts and 20¢ per bag of candy. He intends to sell the peanuts at 15¢ and the candy at 26¢ per bag. His stand can accomodate up to 500 bags of peanuts and 400 bags of candy. From past experience he knows that he will sell no more than a total of 700 bags. Find the number of bags of each that he should have available in order to maximize his profit. What is the maximum profit?

8.11 REVIEW

Concepts

Define or discuss each of the following.

1 System of equations

2 Solution of a system of equations

3 Equivalent systems of equations

4 System of linear equations

5 Homogeneous system of linear equations

6 An $m \times n$ matrix

7 A square matrix of order n.

8 The coefficient matrix of a system of linear equations; the augmented matrix

9 Elementary row transformations

10 The sum and product of two matrices

11 Zero matrix **12** Identity matrix

13 Inverse of a matrix **14** Minor

15 Cofactor **16** Determinant

17 Properties of determinants

18 Cramer's Rule

19 System of inequalities

20 Linear programming

Exercises

Find the solutions of the systems of equations in Exercises 1–16.

1 $\begin{cases} 2x - 3y = 4 \\ 5x + 4y = 1 \end{cases}$

2 $\begin{cases} x - 3y = 4 \\ -2x + 6y = 2 \end{cases}$

3 $\begin{cases} y + 4 = x^2 \\ 2x + y = -1 \end{cases}$

4 $\begin{cases} x^2 + y^2 = 25 \\ x - y = 7 \end{cases}$

5 $\begin{cases} 9x^2 + 16y^2 = 140 \\ x^2 - 4y^2 = 4 \end{cases}$

6 $\begin{cases} 2x = y^2 + 3z \\ x = y^2 + z - 1 \\ x^2 = xz \end{cases}$

7 $\begin{cases} \dfrac{1}{x} + \dfrac{3}{y} = 7 \\ \dfrac{4}{x} - \dfrac{2}{y} = 1 \end{cases}$

8 $\begin{cases} 2^x + 3^{y+1} = 10 \\ 2^{x+1} - 3^y = 5 \end{cases}$

9 $\begin{cases} 3x + y - 2z = -1 \\ 2x - 3y + z = 4 \\ 4x + 5y - z = -2 \end{cases}$

10 $\begin{cases} x + 3y = 0 \\ y - 5z = 3 \\ 2x + z = -1 \end{cases}$

11 $\begin{cases} 4x - 3y - z = 0 \\ x - y - z = 0 \\ 3x - y + 3z = 0 \end{cases}$

12 $\begin{cases} 2x + y - z = 0 \\ x - 2y + z = 0 \\ 3x + 3y + 2z = 0 \end{cases}$

13 $\begin{cases} 4x + 2y - z = 1 \\ 3x + 2y + 4z = 2 \end{cases}$

14 $\begin{cases} 2x + y = 6 \\ x - 3y = 17 \\ 3x + 2y = 7 \end{cases}$

15 $\begin{cases} \dfrac{4}{x} + \dfrac{1}{y} + \dfrac{2}{z} = 4 \\ \dfrac{2}{x} + \dfrac{3}{y} - \dfrac{1}{z} = 1 \\ \dfrac{1}{x} + \dfrac{1}{y} + \dfrac{1}{z} = 4 \end{cases}$

16 $\begin{cases} 2x - y + 3z - w = -3 \\ 3x + 2y - z + w = 13 \\ x - 3y + z - 2w = -4 \\ -x + y + 4z + 3w = 0 \end{cases}$

Find the solutions and sketch the graphs of the systems in Exercises 17–20.

17 $\begin{cases} x^2 + y^2 < 16 \\ y - x^2 > 0 \end{cases}$

18 $\begin{cases} y - x \leq 0 \\ y + x \geq 2 \\ x \leq 5 \end{cases}$

19 $\begin{cases} x - 2y \leq 2 \\ y - 3x \leq 4 \\ 2x + y \leq 4 \end{cases}$

20 $\begin{cases} x^2 - y < 0 \\ y - 2x < 5 \\ xy < 0 \end{cases}$

Find the determinants of the matrices in Exercises 21–30

21 $[-6]$

22 $\begin{bmatrix} 3 & 4 \\ -6 & -5 \end{bmatrix}$

23 $\begin{bmatrix} 3 & -4 \\ 6 & 8 \end{bmatrix}$

24 $\begin{bmatrix} 0 & 4 & -3 \\ 2 & 0 & 4 \\ -5 & 1 & 0 \end{bmatrix}$

25 $\begin{bmatrix} 2 & -3 & 5 \\ -4 & 1 & 3 \\ 3 & 2 & -1 \end{bmatrix}$

26 $\begin{bmatrix} 3 & 1 & -2 \\ -5 & 2 & -4 \\ 7 & 3 & -6 \end{bmatrix}$

27 $\begin{bmatrix} 5 & 0 & 0 & 0 \\ 6 & -3 & 0 & 0 \\ 1 & 4 & -4 & 0 \\ 7 & 2 & 3 & 2 \end{bmatrix}$

28 $\begin{bmatrix} 1 & 2 & 0 & 3 & 1 \\ -2 & -1 & 4 & 1 & 2 \\ 3 & 0 & -1 & 0 & -1 \\ 2 & -3 & 2 & -4 & 2 \\ -1 & 1 & 0 & 1 & 3 \end{bmatrix}$

29 $\begin{bmatrix} 2 & 0 & 1 & 0 & -1 \\ 0 & 1 & 0 & 1 & 2 \\ 2 & -2 & 1 & -2 & 0 \\ 0 & 0 & -2 & 0 & 1 \\ 1 & -1 & 0 & -1 & 0 \end{bmatrix}$

30 $\begin{bmatrix} 1 & 2 & 0 & 0 & 0 \\ 3 & 4 & 0 & 0 & 0 \\ 0 & 0 & 1 & 2 & 3 \\ 0 & 0 & 2 & -1 & 1 \\ 0 & 0 & 1 & 3 & -1 \end{bmatrix}$

31 Find the determinant of the $n \times n$ matrix (a_{ij}), where $a_{ij} = 0$ if $i \neq j$.

32 Without expanding show that

$$\begin{vmatrix} 1 & a & b+c \\ 1 & b & a+c \\ 1 & c & a+b \end{vmatrix} = 0.$$

Find the inverses of the matrices in Exercises 33–36.

33 $\begin{bmatrix} 5 & -4 \\ -3 & 2 \end{bmatrix}$

34 $\begin{bmatrix} 2 & -1 & 0 \\ 1 & 4 & 2 \\ 3 & -2 & 1 \end{bmatrix}$

35 $\begin{bmatrix} 3 & -1 & 0 & 0 \\ 1 & 2 & 0 & 0 \\ 0 & 0 & -1 & -2 \\ 0 & 0 & 5 & 3 \end{bmatrix}$

36 $\begin{bmatrix} 2 & 0 & 0 & 0 \\ 0 & 3 & 0 & 0 \\ 0 & 0 & 4 & 0 \\ 0 & 0 & 0 & 5 \end{bmatrix}$

In Exercises 37–46 express as a single matrix.

37 $\begin{bmatrix} 2 & -1 & 0 \\ 3 & 0 & -2 \end{bmatrix} \begin{bmatrix} 2 & -1 & 3 \\ 0 & 3 & 0 \\ 1 & 4 & 2 \end{bmatrix}$

38 $\begin{bmatrix} 4 & 2 \\ 5 & -3 \end{bmatrix} \begin{bmatrix} 3 \\ 7 \end{bmatrix}$

39 $\begin{bmatrix} 2 & 0 \\ 1 & 4 \\ -2 & 3 \end{bmatrix} \begin{bmatrix} 0 & 2 & -3 \\ 4 & 5 & 1 \end{bmatrix}$

40 $\begin{bmatrix} 0 & -2 & 3 \\ 4 & 1 & 2 \end{bmatrix} \begin{bmatrix} 2 & 0 \\ 3 & 8 \\ 2 & -7 \end{bmatrix}$

41 $2\begin{bmatrix} 0 & -1 & -4 \\ 3 & 2 & 1 \end{bmatrix} - 3\begin{bmatrix} 4 & -2 & 1 \\ 0 & 5 & -1 \end{bmatrix}$

42 $\begin{bmatrix} 1 & 3 \\ 2 & 4 \end{bmatrix} \begin{bmatrix} a & 0 \\ 0 & a \end{bmatrix}$

43 $\begin{bmatrix} a & 0 \\ 0 & b \end{bmatrix} \begin{bmatrix} 1 & 3 \\ 2 & 4 \end{bmatrix}$

44 $\begin{bmatrix} 3 & 2 \\ 0 & 0 \end{bmatrix} \begin{bmatrix} -2 & 0 \\ 3 & 0 \end{bmatrix}$

45 $\begin{bmatrix} 1 & 2 \\ 3 & 4 \end{bmatrix} \left\{ \begin{bmatrix} 2 & -4 \\ 3 & 7 \end{bmatrix} + \begin{bmatrix} 1 & 5 \\ -2 & -3 \end{bmatrix} \right\}$

46 $\begin{bmatrix} 3 & 2 & 5 \\ -3 & 4 & 7 \\ 6 & 5 & 1 \end{bmatrix} \begin{bmatrix} 3 & 2 & 5 \\ -3 & 4 & 7 \\ 6 & 5 & 1 \end{bmatrix}^{-1}$

Verify Exercises 47 and 48 without expanding the determinants.

47 $\begin{vmatrix} 2 & 4 & -6 \\ 1 & 4 & 3 \\ 2 & 2 & 0 \end{vmatrix} = 12\begin{vmatrix} 1 & 1 & -1 \\ 1 & 2 & 1 \\ 2 & 1 & 0 \end{vmatrix}$

48 $\begin{vmatrix} a & b & c \\ d & e & f \\ g & h & k \end{vmatrix} = \begin{vmatrix} d & e & f \\ g & h & k \\ a & b & c \end{vmatrix}$

49 Suppose that $A = (a_{ij})$ is a square matrix of order n such that $a_{ij} = 0$ if $i < j$. Prove that

$$|A| = a_{11}a_{22} \cdots a_{nn}.$$

50 If $A = (a_{ij})$ is any 2×2 matrix such that $|A| \neq 0$, prove that A has an inverse and find a general formula for A^{-1}

9

COMPLEX NUMBERS

Complex numbers *were invented to help solve equations of the form* $f(x) = 0$, *where* $f(x)$ *is any polynomial. In this chapter we shall define complex numbers and consider some of their basic properties.*

9.1 DEFINITION OF COMPLEX NUMBERS

Although real numbers are adequate for many mathematical and scientific problems, there is a serious defect in the system when it comes to solving some equations. Indeed, since the square of a real number cannot be negative, an equation such as $x^2 = -5$ has no solutions if x is restricted to $\mathbb{R}$. For many applications we need a mathematical system in which such equations have solutions. Fortunately, it is possible to construct such a system: the system of *complex numbers* defined in this section.

Let us consider the problem of inventing a new mathematical system $\mathbb{C}$ which contains the real number system $\mathbb{R}$ and which can be used to solve equations such as $x^2 = -5$. Since $\mathbb{C}$ is to be used to find solutions of equations, it must possess *operations*, that is, rules which may be applied to every pair of its elements to obtain another element. As a matter of fact, we would like to define operations called *addition* and *multiplication* in such a way that if we restrict the elements to the subset $\mathbb{R}$, then the operations behave in the same way as addition and multiplication of real numbers. Since we are extending the notions of addition and multiplication to the set $\mathbb{C}$, we shall continue to use the symbols $+$ and $\cdot$ for those operations.

In order to gain some insight into the construction of $\mathbb{C}$, let us begin by taking an intuitive approach. If we want equations of the form $x^2 = -k$ to have solutions when k is a positive real number, then in particular when $k = 1$ it is necessary for $\mathbb{C}$ to contain some element i such that $i^2 = -1$. If b is in $\mathbb{R}$, then b is also in $\mathbb{C}$, and since $\mathbb{C}$ is to be closed relative to multiplication, bi must be in $\mathbb{C}$. Moreover, if a is in $\mathbb{R}$

and if $\mathbb{C}$ is to be closed relative to addition, then $a + bi$ is in $\mathbb{C}$. Thus, $\mathbb{C}$ contains elements of the form $a + bi$, $c + di$, etc., where a, b, c, and d are real numbers and $i^2 = -1$. If we want the Commutative, Associative, and Distributive Properties to be valid, then these elements must add as follows:

$$(a + bi) + (c + di) = (a + c) + (bi + di)$$

or

(A)
$$(a + bi) + (c + di) = (a + c) + (b + d)i.$$

Similarly, if $i^2 = -1$, the following manipulations should be valid:

$$\begin{aligned}
(a + bi)(c + di) &= (a + bi)c + (a + bi)(di) \\
&= ac + (bi)c + a(di) + (bi)(di) \\
&= ac + (bc)i + (ad)i + (bd)(i^2) \\
&= ac + (bd)(-1) + (ad)i + (bc)i \\
&= (ac - bd) + (ad + bc)i.
\end{aligned}$$

To summarize, the following rule for multiplication must hold in $\mathbb{C}$:

(M)
$$(a + bi)(c + di) = (ac - bd) + (ad + bc)i.$$

The preceding discussion indicates the manner in which elements behave if a system of the required type is to exist. Moreover, our discussion provides a key to the actual construction of $\mathbb{C}$. Thus, we begin by *defining* a **complex number** as any symbol of the form $a + bi$, where a and b are real numbers. The real number a is called the **real part** of the complex number and bi is called the **imaginary part**. At the outset the letter i is given no specific meaning and the $+$ sign which appears in $a + bi$ is not to be interpreted as the symbol for addition, but only as part of the notation for a complex number. As above, $\mathbb{C}$ will denote the set of all complex numbers. Two complex numbers $a + bi$ and $c + di$ are said to be **equal**, and we write

$$a + bi = c + di \quad \text{if and only if} \quad a = c \text{ and } b = d.$$

Next we *define* addition and multiplication of complex numbers by means of formulas (A) and (M). It should be observed that the $+$ sign is used in three different ways in (A). First, by our previous remarks, it is part of the symbol for a complex number. Second, it is used to denote addition of the complex numbers $a + bi$ and $c + di$. Third, it is the addition sign for real numbers, as in the expressions $a + c$ and $b + d$. The need for remembering this threefold use of $+$ will disappear after we agree on the notational conventions which follow.

Let us consider the subset $\mathbb{R}'$ of $\mathbb{C}$ consisting of all complex numbers of the form $a + 0i$, where a is a real number. By associating a with $a + 0i$, we obtain a one-to-one correspondence between the set $\mathbb{R}$ and $\mathbb{R}'$. Applying (A) and (M) to the elements of

$\mathbb{R}'$ (by letting $b = d = 0$), we obtain

$$(a + 0i) + (c + 0i) = (a + c) + 0i$$
$$(a + 0i)(c + 0i) = ac + 0i.$$

This shows that, in order to add (or multiply) two elements of $\mathbb{R}'$, we merely add (or multiply) the real parts, *disregarding* the imaginary parts. Hence, as far as properties of addition and multiplication are concerned, the only difference between $\mathbb{R}$ and $\mathbb{R}'$ is the notation for the elements. Accordingly, we shall use the symbol a in place of $a + 0i$. For example, an element such as 3 of $\mathbb{R}$ (or $\mathbb{C}$) is considered the same as the element $3 + 0i$ of $\mathbb{C}$. It is also convenient to abbreviate the complex number $0 + bi$ by the symbol bi. Applying (A) gives us

$$(a + 0i) + (0 + bi) = (a + 0) + (0 + b)i = a + bi.$$

This indicates that $a + bi$ may be thought of as the sum of two complex numbers a and bi (that is, $a + 0i$ and $0 + bi$). With these agreements on notation, all the $+$ signs in (A) may be regarded as addition of complex numbers.

Example 1   Express each of the following in the form $a + bi$, where a and b are real numbers.

(a) $(3 + 4i) + (2 + 5i)$ (b) $(3 + 4i)(2 + 5i)$ (c) $(3 + 4i)^2$

Solutions

(a)   Applying (A),

$$(3 + 4i) + (2 + 5i) = (3 + 2) + (4 + 5)i = 5 + 9i.$$

(b)   Using (M),

$$(3 + 4i)(2 + 5i) = (3 \cdot 2 - 4 \cdot 5) + (3 \cdot 5 + 4 \cdot 2)i = -14 + 23i.$$

(c)   Exponents are defined in $\mathbb{C}$ exactly as they are in $\mathbb{R}$. Thus,

$$(3 + 4i)^2 = (3 + 4i)(3 + 4i)$$
$$= (3 \cdot 3 - 4 \cdot 4) + (3 \cdot 4 + 4 \cdot 3)i$$
$$= -7 + 24i. \qquad \blacksquare$$

It is not difficult to show that addition and multiplication of complex numbers are both commutative and associative. The distributive property is also true. The identity element relative to addition is 0 (or, equivalently, $0 + 0i$), since

$$(a + bi) + 0 = (a + bi) + (0 + 0i)$$
$$= (a + 0) + (b + 0)i$$
$$= a + bi.$$

As usual, we refer to 0 as **zero** or the **zero element**. It follows from (M) that the *product* of zero and any complex number is zero. We may also use (M) to prove that 1 (that is, $1 + 0i$) is the identity element relative to multiplication.

If $(-a) + (-b)i$ is added to $a + bi$, we obtain 0. This implies that $(-a) + (-b)i$ is the additive inverse of $a + bi$, that is,

$$-(a + bi) = (-a) + (-b)i.$$

We shall postpone the discussion of multiplicative inverses until the next section.

Subtraction of complex numbers is defined using additive inverses as follows:

$$(a + bi) - (c + di) = (a + bi) + [-(c + di)].$$

Since $-(c + di) = (-c) + (-d)i$, it follows that

$$(a + bi) - (c + di) = (a - c) + (b - d)i.$$

The special case with $b = c = 0$ gives us

$$(a + 0i) - (0 + di) = (a - 0) + (0 - d)i$$

which may be written in the form

$$a - di = a + (-d)i.$$

The preceding formula is useful when the real number associated with the imaginary part of a complex number is negative.

If c, d, and k are real numbers, then, by (M) and our agreement on notation,

$$k(c + di) = (k + 0i)(c + di) = (kc - 0d) + (kd + 0c)i$$

that is,

$$k(c + di) = kc + (kd)i.$$

One illustration of this formula is

$$3(5 + 2i) = 15 + 6i.$$

The special case with $k = -1$ gives us

$$(-1)(c + di) = (-c) + (-d)i = -(c + di).$$

Hence, as in $\mathbb{R}$, the additive inverse of a complex number may be found by multiplying it by -1.

The complex number $0 + 1i$ (or, equivalently, $1i$) will be denoted by i. Observe that

$$b(0 + 1i) = (b \cdot 0) + (b \cdot 1)i = 0 + bi = bi.$$

Thus, the symbol bi which has been used throughout this section may be regarded as the *product* of b and i.

Finally, using (M) with $a = c = 0$ and $b = d = 1$, we obtain

$$i^2 = (0 + 1i)(0 + 1i)$$
$$= (0 \cdot 0 - 1 \cdot 1) + (0 \cdot 1 + 1 \cdot 0)i$$
$$= -1 + 0i.$$

This gives us the following important rule:

$$\boxed{i^2 = -1.}$$

If we collect all of the formulas and remarks made in this section it becomes evident that, when working with complex numbers, *we may treat all symbols just as though they represented real numbers with exactly one exception: wherever the symbol i^2 appears it may be replaced by* -1. Consequently, manipulations can be carried out without referring to (A) or (M), which is what we had in mind from the very beginning of our discussion! We shall use this technique in the solution of the next example. If, as in Example 2, we are asked to write an expression in the form $a + bi$, we shall also accept the form $a - di$ since we have seen that it equals $a + (-d)i$.

Example 2   Write each of the following in the form $a + bi$.
   (a) $4(2 + 5i) - (3 - 4i)$ (b) $(4 - 3i)(2 + i)$
   (c) $i(3 - 2i)^2$ (d) i^{51}

Solutions
   (a) $4(2 + 5i) - (3 - 4i) = 8 + 20i - 3 + 4i = 5 + 24i$
   (b) $(4 - 3i)(2 + i) = 8 - 6i + 4i - 3i^2 = 11 - 2i$
   (c) $i(3 - 2i)^2 = i(9 - 12i + 4i^2) = i(5 - 12i) = 5i - 12i^2 = 12 + 5i$
   (d)   Taking successive powers of i, we obtain $i^1 = i$, $i^2 = -1$, $i^3 = -i$, $i^4 = 1$, and then the cycle starts over: $i^5 = i$, $i^6 = i^2 = -1$, etc. In particular, $i^{51} = i^{48}i^3 = (i^4)^{12}i^3 = (1)^{12}i^3 = i^3 = -i$.   ∎

Finally, it should be pointed out that there is another way to define complex numbers. Observe that each symbol $a + bi$ determines a unique ordered pair (a, b) of real numbers. Conversely, every ordered pair (a, b) can be used to obtain a symbol $a + bi$. In this way we obtain a one-to-one correspondence between the symbols $a + bi$ and ordered pairs (a, b). The correspondence suggests using ordered pairs of real numbers to define the system $\mathbb{C}$. Formulas (A) and (M) can then be used to motivate definitions for addition and multiplication. Specifically, we define $\mathbb{C}$ as the set of all ordered pairs of real numbers subject to the following two laws:

$$(a, b) + (c, d) = (a + c, b + d)$$
$$(a, b)(c, d) = (ac - bd, ad + bc).$$

Notice the manner in which (A) and (M) were used to help formulate the definition. We merely replaced symbols such as $a + bi$ by (a, b) and translated the rules accordingly. By a suitable change in notation we can obtain the $a + bi$ form for complex numbers introduced in this section.

EXERCISES 9.1

In each of Exercises 1–36 write the expression in the form $a + bi$.

1 $(3 + 2i) + (-5 + 4i)$

2 $(8 - 5i) + (2 - 3i)$

3 $(-4 + 5i) + (2 - i)$

4 $(5 + 7i) + (-8 - 4i)$

5 $(16 + 10i) - (9 + 15i)$

6 $(2 - 6i) - (7 + 2i)$

7 $-(-2 + 7i) + (-6 + 6i)$

8 $-(5 - 3i) - (-3 - 4i)$

9 $7 - (3 - 7i)$ **10** $-9 + (5 + 9i)$

11 $5i - (6 + 2i)$ **12** $(10 + 7i) - 12i$

13 $(4 + 3i)(-1 + 2i)$ **14** $(3 - 6i)(2 + i)$

15 $(-7 + i)(-3 + i)$ **16** $(5 + 2i)(5 - 2i)$

17 $(3 + 4i)(3 - 4i)$ **18** $7i(13 + 8i)$

19 $-9i(4 - 8i)$ **20** $(6i)(-2i)$

21 $4(8 - 11i)$ **22** $-3(-6 + 12i)$

23 $(-3i)(5i)$ **24** $(1 - i)(1 + i)$

25 $(\sqrt{7} + \sqrt{3}i)(\sqrt{7} - \sqrt{3}i)$

26 $(-7 + 3i)^2$

27 $(3 + 2i)^2$ **28** $4i(2 + 5i)^2$

29 $i(3 - 2i)(5 + i)$ **30** $(1 + i)^4$

31 $\left(-\dfrac{1}{2} - \dfrac{\sqrt{3}}{2}i\right)^3$ **32** $\left(-\dfrac{1}{2} + \dfrac{\sqrt{3}}{2}i\right)^3$

33 i^{42} **34** i^{23}

35 i^{157} **36** $(-i)^{50}$

In Exercises 37–40 solve for x and y, if x and y are real.

37 $5x + 6i = -8 + 2yi$

38 $7 - 4yi = 9x + 3i$

39 $i(2x - 4y) = 4x + 2 + 3yi$

40 $(2x + y) + (3x - 4y)i = (x - 2) + (2y - 5)i$

9.2 CONJUGATES AND INVERSES

The complex number $a - bi$ is called the **conjugate** of the complex number $a + bi$. Since

$$(a + bi)(a - bi) = a^2 + b^2$$

we see that the product of a complex number and its conjugate is a real number. If $a^2 + b^2 \neq 0$, then multiplying both sides of the last equation by $1/(a^2 + b^2)$ and rearranging terms on the left-hand side gives us

$$\left(\frac{1}{a^2 + b^2}\right)(a - bi)(a + bi) = 1.$$

Hence, if $a + bi \neq 0$, then the complex number $a + bi$ has a multiplicative inverse denoted by $(a + bi)^{-1}$, or $1/(a + bi)$, where

$$\frac{1}{a + bi} = \left(\frac{1}{a^2 + b^2}\right)(a - bi).$$

If $c + di \neq 0$, we define the **quotient**

$$\frac{a + bi}{c + di}$$

to be the product of $a + bi$ and $1/(c + di)$. We can write this quotient in the form $u + vi$, where u and v are real numbers, by multiplying numerator and denominator by the conjugate $c - di$ of the denominator as follows:

$$\frac{a + bi}{c + di} = \frac{a + bi}{c + di} \cdot \frac{c - di}{c - di}$$

$$= \frac{(ac + bd) + (bc - ad)i}{c^2 + d^2}$$

$$= \left(\frac{ac + bd}{c^2 + d^2}\right) + \left(\frac{bc - ad}{c^2 + d^2}\right)i.$$

The above technique may also be used to find the multiplicative inverse of $(a + bi)$. Specifically, we multiply numerator and denominator of $1/(a + bi)$ by $a - bi$ as follows:

$$\frac{1}{a + bi} = \frac{1}{a + bi} \cdot \frac{a - bi}{a - bi}$$

$$= \frac{a - bi}{a^2 + b^2} = \frac{1}{a^2 + b^2}(a - bi).$$

Example 1   Express each of the following in the form $a + bi$.

(a) $\dfrac{1}{9 + 2i}$ (b) $\dfrac{7 - i}{3 - 5i}$

Solutions

(a)
$$\frac{1}{9 + 2i} = \frac{1}{9 + 2i} \cdot \frac{9 - 2i}{9 - 2i} = \frac{9 - 2i}{81 + 4} = \frac{9}{85} - \frac{2}{85}i$$

(b)
$$\frac{7 - i}{3 - 5i} = \frac{7 - i}{3 - 5i} \cdot \frac{3 + 5i}{3 + 5i}$$

$$= \frac{21 - 3i + 35i - 5i^2}{9 - 25i^2}$$

$$= \frac{26 + 32i}{34} = \frac{13}{17} + \frac{16}{17}i$$ ∎

Conjugates of complex numbers have several interesting and useful properties. To simplify the notation, if $z = a + bi$ is a complex number, then its conjugate will be denoted by $\bar{z}$, that is, $\bar{z} = a - bi$.

Theorem on Conjugates

> If z and w are complex numbers, then
>
> (i) $\overline{z + w} = \overline{z} + \overline{w}$
>
> (ii) $\overline{z \cdot w} = \overline{z} \cdot \overline{w}$
>
> (iii) $\overline{z^n} = (\overline{z})^n$, for every positive integer n
>
> (iv) $\overline{z} = z$ if and only if z is real.

Proof

Let $z = a + bi$ and $w = c + di$, where a, b, c, and d are real numbers. Since $z + w = (a + c) + (b + d)i$, we have, by the definition of conjugate and properties of addition of complex numbers,

$$\overline{z + w} = (a + c) - (b + d)i$$
$$= (a - bi) + (c - di)$$
$$= \overline{z} + \overline{w}.$$

That proves (i).

Multiplication gives us $z \cdot w = (ac - bd) + (ad + bc)i$ and hence the conjugate is

$$\overline{z \cdot w} = (ac - bd) - (ad + bc)i = (a - bi)(c - di) = \overline{z} \cdot \overline{w},$$

which proves (ii).

If we set $w = z$ in (ii), then $\overline{z \cdot z} = \overline{z} \cdot \overline{z}$, that is, $\overline{z^2} = \overline{z}^2$. We may then write $\overline{z^3} = \overline{z^2 \cdot z} = \overline{z^2} \cdot \overline{z} = \overline{z}^2 \cdot \overline{z} = \overline{z}^3$. Continuing in this manner, it appears that $\overline{z^n} = \overline{z}^n$ for all positive integers n. A complete proof of (iii) requires the method of mathematical induction discussed in Chapter Eleven.

Finally, let us prove (iv). If $z = a + bi$ is real, then $b = 0$ and $\overline{z} = a - 0i = a + 0i = z$. Conversely, if $\overline{z} = z$, then $a - bi = a + bi$ and $-b = b$. It follows that $b = 0$; that is, z is real. ∎

It is not difficult to extend (i) and (ii) of the preceding theorem to more than two complex numbers. For example, if z, w, and u are complex numbers, then applying (i) twice we have

$$\overline{(z + w) + u} = \overline{z + w} + \overline{u} = \overline{z} + \overline{w} + \overline{u}.$$

The analogous result holds for more than three complex numbers. This fact may be stated: "The conjugate of a sum of complex numbers equals the sum of the conjugates." A similar result is true for products.

Real numbers may be represented geometrically by means of points on a coordinate line. We can also obtain geometric representations for complex numbers by using points in a coordinate plane. Specifically, each complex number $a + bi$ determines a unique ordered pair (a, b). The corresponding point $P(a, b)$ in a coordinate plane is called the **geometric representation of** $a + bi$. To emphasize that we are assigning complex numbers to points in a plane, the point $P(a, b)$ will be labeled $a + bi$. A coordinate plane with a complex number assigned to each point is referred to as the **complex plane** instead of the *xy*-plane. Also, according to this scheme, the *x*-axis is called the **real axis** and the *y*-axis the **imaginary axis**. In Figure 9.1 we have indicated the geometric representations of several complex numbers. Note that, to obtain the point corresponding to the conjugate $a - bi$ of any complex number $a + bi$, we simply reflect through the real axis.

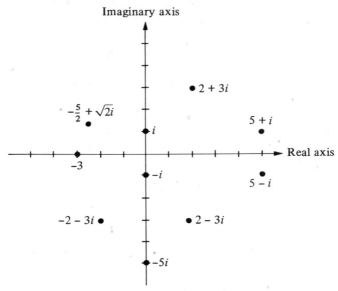

Figure 9.1

In Chapter One we defined the concept of the absolute value $|a|$ of a real number a and noted that geometrically, a is the distance between the origin and the point on a coordinate line which corresponds to a. It is natural, therefore, to interpret the absolute value $a + bi$ of a complex number as the distance $\sqrt{a^2 + b^2}$ between the origin of a complex plane and the point (a, b) corresponding to $a + bi$, as in the following definition.

Definition

> The **absolute value** of a complex number $a + bi$ is denoted by $|a + bi|$ and is defined to be the nonnegative real number $\sqrt{a^2 + b^2}$.

Example 2   Find   (a) $|2 - 6i|$;   (b) $|3i|$.

Solutions   Using the definition of absolute value we obtain
(a) $|2 - 6i| = \sqrt{4 + 36} = \sqrt{40} = 2\sqrt{10}$;
(b) $|3i| = \sqrt{0 + 9} = 3$.   ∎

It is worth noting that the points which correspond to all of the complex numbers having a fixed absolute value lie on a circle with center at the origin in the complex plane. For example, the points corresponding to the complex numbers z with $|z| = 1$ lie on a unit circle.

EXERCISES 9.2

3 -27

In each of Exercises 1–28 express the given number in the form $a + bi$.

1 $\dfrac{1}{3 + 2i}$

2 $\dfrac{1}{5 + 8i}$

3 $\dfrac{7}{5 - 6i}$

4 $\dfrac{-3}{2 - 5i}$

5 $\dfrac{4 - 3i}{2 + 4i}$

6 $\dfrac{4 + 3i}{-1 + 2i}$

7 $\dfrac{6 + 4i}{1 - 5i}$

8 $\dfrac{7 - 6i}{-5 - i}$

9 $\dfrac{21 - 7i}{i}$

10 $\dfrac{10 + 9i}{-3i}$

11 $\dfrac{2 - 3i}{1 + i} + \dfrac{7 + 4i}{3 + 5i}$

12 $\dfrac{6 - 2i}{3 + i} - \dfrac{3 - 7i}{i}$

13 $8 + \dfrac{4 - i}{1 + 2i}$

14 $\dfrac{1}{10 - i} + 5i$

15 $\dfrac{1}{(1 + i)^3}$

16 $\dfrac{(1 - i)^3}{1 + i}$

17 $\dfrac{4 - i^2}{i - 2}$

18 $\dfrac{4i^2 - 25}{5 + 2i}$

19 $\left(\dfrac{1}{5i}\right)^3$

20 $\dfrac{1}{(3 + 2i)^2}$

21 $|3 - 4i|$

22 $|5 + 8i|$

23 $|-6 - 7i|$

24 $|1 - i|$

25 $|8i|$

26 $|i^7|$

27 $|i^{500}|$

28 $|-15i|$

Represent the complex numbers in Exercises 29–38 geometrically.

29 $4 + 2i$

30 $-5 + 3i$

31 $3 - 5i$

32 $-2 - 6i$

33 $-(3 - 6i)$

34 $(1 + 2i)^2$

35 $2i(2 + 3i)$

36 $(-3i)(2 - i)$

37 $(1 + i)^2$

38 $4(-1 + 2i)$

In each of Exercises 39–42 find all complex numbers z that satisfy the given equation, and express z in the form $a + bi$.

39 $5z + 3i = 2iz + 4$

40 $2iz - 6 = 9i + z$

41 $(z - 2i)^2 = (z + 3i)^2$

42 $z(z + 4i) = (z + i)(z - 3i)$

If $z = a + bi$ and $w = c + di$, verify the identities in Exercises 43–50.

43 $\bar{\bar{z}} = z$

44 $\overline{z - w} = \bar{z} - w$

45 $|z| = \sqrt{z\bar{z}}$

46 $|-z| = |z|$

47 $|\bar{z}| = |z|$

48 $|z| = 0$ if and only if $z = 0$

49 $|zw| = |z|\,|w|$

50 $|z/w| = |z|/|w|, \quad w \neq 0$

9.3 COMPLEX ROOTS OF EQUATIONS

It is easy to see that if p is any positive real number, then the equation $x^2 = -p$ has solutions in $\mathbb{C}$. As a matter of fact, one solution is $\sqrt{p}i$, since

$$(\sqrt{p}i)^2 = (\sqrt{p})^2 i^2 = p(-1) = -p.$$

Similarly, $-\sqrt{p}i$ is also a solution. Moreover, they are the only solutions, for if a complex number z is a solution, then $z^2 + p = 0$ and hence

$$(z + \sqrt{p}i)(z - \sqrt{p}i) = 0.$$

This implies that either $z = -\sqrt{p}i$ or $z = \sqrt{p}i$.

The next definition is motivated by the fact that $(\sqrt{p}i)^2 = -p$.

Definition

> If p is a positive real number, then the **principal square root** of $-p$ is denoted by $\sqrt{-p}$ and is defined to be the complex number $\sqrt{p}i$.

As illustrations, we have

$$\sqrt{-9} = \sqrt{9}i = 3i, \quad \sqrt{-5} = \sqrt{5}i, \quad \sqrt{-1} = \sqrt{1}i = i.$$

Care must be taken in using the radical sign if the radicand is negative. For example, the formula $\sqrt{a}\sqrt{b} = \sqrt{ab}$ which holds for positive real numbers is not true when both a and b are negative. To illustrate,

$$\sqrt{-3}\sqrt{-3} = (\sqrt{3}i)(\sqrt{3}i) = (\sqrt{3})^2 i^2 = 3(-1) = -3$$

whereas,

$$\sqrt{(-3)(-3)} = \sqrt{9} = 3.$$

Hence,

$$\sqrt{-3}\sqrt{-3} \neq \sqrt{(-3)(-3)}.$$

However, if only *one* of a or b is negative, then it can be shown that $\sqrt{a}\sqrt{b} = \sqrt{ab}$. In general, we shall not apply laws of radicals if radicands are negative. Instead, we shall change the form of radicals before performing any operations, as illustrated in the next example.

Example 1   Express $(5 - \sqrt{-3})(-1 + \sqrt{-4})$ in the form $a + bi$.

Solution

$$\begin{aligned}(5 - \sqrt{-3})(-1 + \sqrt{-4}) &= (5 - \sqrt{3}i)(-1 + 2i) \\ &= -5 - 2\sqrt{3}i^2 + 10i + \sqrt{3}i \\ &= (-5 + 2\sqrt{3}) + (10 + \sqrt{3})i \end{aligned}$$ ∎

In Section 2.3 we proved that if a, b and c are real numbers such that $b^2 - 4ac \geq 0$ and $a \neq 0$, then the solutions of the quadratic equation $ax^2 + bx + c = 0$ are

$$\frac{-b + \sqrt{b^2 - 4ac}}{2a} \quad \text{and} \quad \frac{-b - \sqrt{b^2 - 4ac}}{2a}.$$

We may now extend this fact to the case where $b^2 - 4ac < 0$. Indeed, the same manipulations used to obtain the quadratic formula, together with the developments in this chapter, show that if $b^2 - 4ac < 0$, then the solutions of $ax^2 + bx + c = 0$ are the two *complex* numbers given above. Notice that the solutions are conjugates of one another. In terms of quadratic functions, given $f(x) = ax^2 + bx + c$, where $a \neq 0$, if $b^2 - 4ac = 0$, then f has one real zero; if $b^2 - 4ac > 0$, then f has two distinct real zeros; and if $b^2 - 4ac < 0$, then f has two distinct conjugate complex zeros.

Example 2   Find the solutions of the equation $5x^2 + 2x + 1 = 0$.

Solution   By the Quadratic Formula we have

$$x = \frac{-2 \pm \sqrt{4 - 20}}{10} = \frac{-2 \pm \sqrt{-16}}{10} = \frac{-2 \pm 4i}{10}.$$

Dividing numerator and denominator by 2, we see that the solutions of the equation are $-\frac{1}{5} + (\frac{2}{5})i$ and $-\frac{1}{5} - (\frac{2}{5})i$. ∎

Example 3 Find the roots of the equation $x^3 - 1 = 0$.

Solution   The given equation may be written as

$$(x - 1)(x^2 + x + 1) = 0.$$

Setting each factor equal to zero and solving the resulting equations, we obtain the solutions

$$1, \quad \frac{-1 \pm \sqrt{1 - 4}}{2}$$

which may be written as

$$1, \quad -\frac{1}{2} + \frac{\sqrt{3}}{2} i, \quad -\frac{1}{2} - \frac{\sqrt{3}}{2} i.$$ ∎

The three roots of $x^3 - 1 = 0$ are called the **cube roots of unity**. It can be shown that if n is any positive integer, then the equation $x^n - 1 = 0$ has n distinct complex roots. They are called the **nth roots of unity**.

It is easy to form a quadratic equation having complex roots z and w. We merely write

$$(x - z)(x - w) = 0.$$

The most important situation occurs if $w = \bar{z}$, the conjugate of z. In this case the equation has the form

$$(x - z)(x - \bar{z}) = 0$$

or equivalently $x^2 - (z + \bar{z})x + z\bar{z} = 0.$

If $z = a + bi$, then $\bar{z} = a - bi$ and hence

$$z + \bar{z} = 2a \quad \text{and} \quad z\bar{z} = a^2 + b^2;$$

that is, the quadratic equation has real coefficients.

Example 4 Find a quadratic equation which has roots $3 + 2i$ and $3 - 2i$.

Solution   By the preceding remarks the equation is given by

$$(x - 3 - 2i)(x - 3 + 2i) = 0.$$

This simplifies to $x^2 - 6x + 13 = 0$. ∎

Quadratic equations with complex coefficients may also be considered. It can be shown that the solutions are again given by the quadratic formula. Since, in this case, $b^2 - 4ac$ may be complex, the solution of such an equation may involve finding the square root of a complex number. We shall not discuss the general theory of roots of complex numbers here.

EXERCISES 9.3

In each of Exercises 1–12 express the given number in the form $a + bi$.

1 $\sqrt{-5}\sqrt{-5}$ **2** $(-\sqrt{5})(\sqrt{-5})$

3 $(3 + \sqrt{-16}) - (6 - \sqrt{-9})$

4 $(4 - \sqrt{-4}) + (3 - \sqrt{-25})$

5 $(7 + \sqrt{-5})(7 - \sqrt{-5})$

6 $(\sqrt{-9} + 3)(\sqrt{-16} - 5)$

7 $\sqrt{-2}(\sqrt{2} + \sqrt{-2})$

8 $\sqrt{-3}(3 - \sqrt{-12})$ **9** $(\sqrt{-10})^3$

10 $\sqrt{(-10)^3}$ **11** $\dfrac{5 + \sqrt{-4}}{1 + \sqrt{-9}}$

12 $\dfrac{1}{6 + \sqrt{-3}}$

Find the solutions of the equations in Exercises 13–26.

13 $x^2 - 3x + 10 = 0$ **14** $x^2 - 5x + 20 = 0$

15 $x^2 + 2x + 5 = 0$ **16** $x^2 + 3x + 6 = 0$

17 $4x^2 + x + 3 = 0$ **18** $-3x^2 + x - 5 = 0$

19 $x^3 - 125 = 0$ **20** $x^3 + 27 = 0$

21 $x^6 - 64 = 0$ **22** $x^4 = 81$

23 $4x^4 + 25x^2 + 36 = 0$

24 $27x^4 + 21x^2 + 4 = 0$

25 $x^3 + 3x^2 + 4x = 0$

26 $8x^3 - 12x^2 + 2x - 3 = 0$

In each of Exercises 27–38 find a quadratic equation with the given roots.

27 $4 + i, \quad 4 - i$

28 $1 + 3i, \quad 1 - 3i$

29 $-3 + 2i, \quad -3 - 2i$

30 $\frac{1}{2} + \frac{1}{3}i, \quad \frac{1}{2} - \frac{1}{3}i$

31 $-5i, \quad 5i$

32 $\sqrt{2}i, \quad -\sqrt{2}i$

33 $10, \quad i$ **34** $6, \quad -6i$

35 $i, \quad 1/i$ **36** $i, \quad i^3$

37 $2i, \quad 3i$ **38** $1 + i, \quad 2 + 2i$

39 Prove that if z is any complex number, then $z + \bar{z}$ is a real number. Is $z - \bar{z}$ necessarily a real number?

40 Prove that the sum and product of the roots of the equation $ax^2 + bx + c = 0$ are $-b/a$ and c/a, respectively.

9.4 TRIGONOMETRIC FORM FOR COMPLEX NUMBERS

The geometric representation of a complex number as a point in a coordinate plane leads to a method of representation that involves trigonometric functions. Let us consider a nonzero complex number $z = a + bi$ and its geometric representation

$P(a, b)$. Let θ be any angle in standard position whose terminal side lies on the segment OP and let $r = |z|$, that is, $r = d(O, P) = \sqrt{a^2 + b^2}$. An illustration of this situation is shown in Figure 9.2.

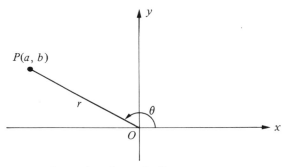

Figure 9.2 $z = a + bi = r(\cos \theta + i \sin \theta)$

Since $a = r \cos \theta$, $b = r \sin \theta$, and $z = a + bi$, we obtain

$$z = (r \cos \theta) + (r \sin \theta)i$$

which may be written in the form

$$z = r(\cos \theta + i \sin \theta).$$

It is important to remember that the number r is the absolute value of z. If $z = 0$, then $r = 0$, and hence the preceding form may be used to represent the complex number 0 where θ is *any* angle. We call $r(\cos \theta + i \sin \theta)$ the **trigonometric form** or **polar form** for the complex number $z = a + bi$. The trigonometric form for z is not unique since there are an unlimited number of different choices for the angle θ. When the trigonometric form is used, the absolute value r of z is often referred to as the **modulus** of z and the angle θ associated with z is called the **argument** (or **amplitude**) of z.

Example 1   Express each of the following complex numbers in trigonometric form.
   (a) $-4 + 4i$ (b) $2\sqrt{3} - 2i$ (c) $2 + 7i$

Solutions   The three complex numbers are represented geometrically in Figure 9.3. Using the trigonometric form gives us the following.

(a)
$$-4 + 4i = 4\sqrt{2}\left[\cos \frac{3\pi}{4} + i \sin \frac{3\pi}{4}\right]$$

(b)
$$2\sqrt{3} - 2i = 4\left[\cos \left(-\frac{\pi}{6}\right) + i \sin \left(-\frac{\pi}{6}\right)\right]$$

(c)
$$2 + 7i = \sqrt{53}\left[\cos \left(\arctan \frac{7}{2}\right) + i \sin \left(\arctan \frac{7}{2}\right)\right]$$

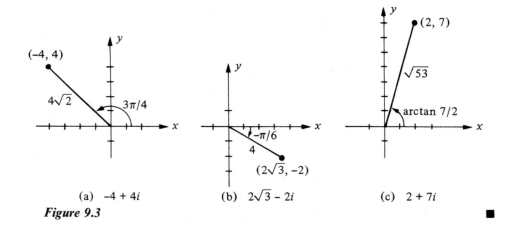

(a) $-4 + 4i$ (b) $2\sqrt{3} - 2i$ (c) $2 + 7i$

Figure 9.3   ∎

If complex numbers are expressed in trigonometric form, then multiplications and divisions may be carried out in the simple manner indicated by the next theorem.

Theorem

> If the trigonometric forms for two complex numbers z_1 and z_2 are given by
>
> $$z_1 = r_1(\cos\theta_1 + i\sin\theta_1) \quad \text{and} \quad z_2 = r_2(\cos\theta_2 + i\sin\theta_2)$$
>
> then
>
> (i) $z_1 z_2 = r_1 r_2[\cos(\theta_1 + \theta_2) + i\sin(\theta_1 + \theta_2)]$
>
> (ii) $\dfrac{z_1}{z_2} = \dfrac{r_1}{r_2}[\cos(\theta_1 - \theta_2) + i\sin(\theta_1 - \theta_2)], \quad z_2 \neq 0.$

Proof

We shall prove (i) and leave (ii) as an exercise. Thus,

$$z_1 z_2 = r_1(\cos\theta_1 + i\sin\theta_1)\cdot r_2(\cos\theta_2 + i\sin\theta_2)$$
$$= r_1 r_2\{[\cos\theta_1\cos\theta_2 - \sin\theta_1\sin\theta_2] + i[\sin\theta_1\cos\theta_2 + \cos\theta_1\sin\theta_2]\}.$$

Applying the addition formulas for $\cos(\theta_1 + \theta_2)$ and $\sin(\theta_1 + \theta_2)$ gives us (i).

∎

Part (i) of the preceding theorem states that the modulus of a product of two complex numbers is the product of their moduli, and the argument is the sum of their arguments. An analogous statement can be made for (ii).

Example 2   Use trigonometric forms to find $z_1 z_2$ and z_1/z_2 if $z_1 = 2\sqrt{3} - 2i$ and $z_2 = -1 + \sqrt{3}i$. Check by using the methods of Sections 9.1 and 9.2.

Solution   From Example 1 we have

$$z_1 = 2\sqrt{3} - 2i = 4\left[\cos\left(-\frac{\pi}{6}\right) + i\sin\left(-\frac{\pi}{6}\right)\right].$$

Similarly, the reader may verify that

$$z_2 = -1 + \sqrt{3}i = 2\left[\cos\frac{2\pi}{3} + i\sin\frac{2\pi}{3}\right].$$

Applying (i) of the preceding theorem, we have

$$z_1 z_2 = 8\left[\cos\left(-\frac{\pi}{6} + \frac{2\pi}{3}\right) + i\sin\left(-\frac{\pi}{6} + \frac{2\pi}{3}\right)\right]$$

$$= 8\left[\cos\frac{\pi}{2} + i\sin\frac{\pi}{2}\right]$$

$$= 8i.$$

As a check, using the methods of Section 9.1, we have

$$z_1 z_2 = (2\sqrt{3} - 2i)(-1 + \sqrt{3}i)$$

$$= (-2\sqrt{3} + 2\sqrt{3}) + (2 + 6)i$$

$$= 8i$$

which is in agreement with our previous answer.

Applying (ii) of the preceding theorem we obtain

$$\frac{z_1}{z_2} = 2\left[\cos\left(-\frac{\pi}{6} - \frac{2\pi}{3}\right) + i\sin\left(-\frac{\pi}{6} - \frac{2\pi}{3}\right)\right]$$

$$= 2\left[\cos\left(-\frac{5\pi}{6}\right) + i\sin\left(-\frac{5\pi}{6}\right)\right]$$

$$= 2\left[-\frac{\sqrt{3}}{2} + i\left(-\frac{1}{2}\right)\right]$$

$$= -\sqrt{3} - i.$$

Using the methods of Section 9.2 gives us

$$\frac{z_1}{z_2} = \frac{2\sqrt{3} - 2i}{-1 + \sqrt{3}i} \cdot \frac{-1 - \sqrt{3}i}{-1 - \sqrt{3}i}$$

$$= \frac{(-2\sqrt{3} - 2\sqrt{3}) + (2 - 6)i}{4}$$

$$= -\sqrt{3} - i.$$ ∎

EXERCISES 9.4 $3-27$

Change the complex numbers in Exercises 1–20 to trigonometric form.

1   $1 - i$ 2 $\sqrt{3} + i$

3   $-4\sqrt{3} + 4i$ 4 $-2 - 2i$

5   $-20i$ 6 15

7   $-5(1 + \sqrt{3}i)$ 8 $-6i$

9   -7 10 $2i(1 - \sqrt{3}i)$

11   $2 + i$ 12 $4 - 3i$

13   $-4 - 4i$ 14 $-10 + 10i$

15   $6i$ 16 -5

17   $-4\sqrt{3} + 4i$ 18 $-5 - 5\sqrt{3}i$

19   12 20 0

In Exercises 21–28 find z_1z_2 and z_1/z_2 by changing to trigonometric form and then using the last theorem

of this section. Check by using the methods of Section 9.2.

21   $z_1 = -1 + i$, $z_2 = 1 + i$

22   $z_1 = \sqrt{3} - i$, $z_2 = -\sqrt{3} - i$

23   $z_1 = -2 - 2\sqrt{3}i$, $z_2 = 5i$

24   $z_1 = -5 + 5i$, $z_2 = -3i$

25   $z_1 = -10$, $z_2 = -4$

26   $z_1 = 2i$, $z_2 = -3i$

27   $z_1 = 4$, $z_2 = 2 - i$

28   $z_1 = -3$, $z_2 = 5 + 2i$

29   Extend (i) of the last theorem in this section to the case of three complex numbers. Generalize to n complex numbers.

30   Prove (ii) of the last theorem in this section.

9.5   DE MOIVRE'S THEOREM AND nTH ROOTS OF COMPLEX NUMBERS

If z is a complex number and n is a positive integer, then a complex number w is called an ***n*th root** of z if $w^n = z$. In this section we shall show that every nonzero complex number has n distinct nth roots. Since $\mathbb{R}$ is in $\mathbb{C}$, it will also follow that every nonzero real number has n distinct nth roots. If a is a positive real number and $n = 2$, then we already know that the roots are $\sqrt{a}$ and $-\sqrt{a}$.

If, in the last theorem of Section 9.4, we let both z_1 and z_2 equal $z = r(\cos\theta + i\sin\theta)$, we obtain

$$z^2 = r^2(\cos 2\theta + i\sin 2\theta).$$

If we now apply the same theorem to z and z^2, then

$$z^2 \cdot z = (r^2 \cdot r)[\cos(2\theta + \theta) + i\sin(2\theta + \theta)]$$

or $$z^3 = r^3(\cos 3\theta + i\sin 3\theta).$$

Next, applying the theorem to z^3 and z produces

$$z^4 = r^4(\cos 4\theta + i\sin 4\theta).$$

In general, we have the following result, which is named after the French mathematician Abraham De Moivre (1667–1754).

De Moivre's Theorem

For every integer n,

$$[r(\cos\theta + i\sin\theta)]^n = r^n[\cos n\theta + i\sin n\theta].$$

A complete proof of the theorem may be given with the aid of the method of mathematical induction discussed in Chapter Eleven.

Example 1 Find $(1 + i)^{20}$.

Solution Introducing the trigonometric form,

$$1 + i = \sqrt{2}\left(\cos\frac{\pi}{4} + i\sin\frac{\pi}{4}\right).$$

Next, by De Moivre's Theorem,

$$(1 + i)^{20} = (2^{1/2})^{20}\left[\cos 20\left(\frac{\pi}{4}\right) + i\sin 20\left(\frac{\pi}{4}\right)\right]$$

$$= 2^{10}(\cos 5\pi + i\sin 5\pi)$$

$$= -1024. \qquad\blacksquare$$

If a nonzero complex number z has an nth root w, then $w^n = z$. If the trigonometric forms for w and z are

$$w = s(\cos\alpha + i\sin\alpha) \quad \text{and} \quad z = r(\cos\theta + i\sin\theta),$$

then, applying De Moivre's Theorem to $w^n = z$, we get

$$s^n[\cos n\alpha + i\sin n\alpha] = r(\cos\theta + i\sin\theta).$$

If two complex numbers are equal, then so are their absolute values. Consequently, $s^n = r$, and since s and r are nonnegative, $s = \sqrt[n]{r}$. Substituting s^n for r in the last displayed equation and dividing both sides by s^n, we obtain

$$\cos n\alpha + i\sin n\alpha = \cos\theta + i\sin\theta.$$

It follows that

$$\cos n\alpha = \cos\theta \quad \text{and} \quad \sin n\alpha = \sin\theta.$$

Since both the sine and cosine functions have period 2π, the last two equations are true if and only if $n\alpha$ and θ differ by a multiple of 2π. Thus, for some integer k,

$$n\alpha = \theta + 2\pi k$$

and hence,

$$\alpha = \frac{\theta + 2\pi k}{n}.$$

Substituting in the trigonometric form for w, we obtain the formula

$$w = \sqrt[n]{r}\left[\cos\left(\frac{\theta + 2\pi k}{n}\right) + i\sin\left(\frac{\theta + 2\pi k}{n}\right)\right].$$

If we substitute $k = 0, 1, \ldots, n - 1$ successively, there result n distinct values for w and hence n distinct nth roots of z. No other value of k will produce a new nth root. For example, if $k = n$, we obtain the angle $(\theta + 2\pi n)/n$, or $\theta/n + 2\pi$, which gives us the same nth root as $k = 0$. Similarly, $k = n + 1$ yields the same nth root as $k = 1$, etc. The same is true for negative values of k. Our discussion also shows that the numbers given by the formula are the only possible nth roots of z. We have proved the following theorem.

Theorem on nth Roots

If $z = r(\cos\theta + i\sin\theta)$ is any nonzero complex number and if n is any positive integer, then z has precisely n distinct nth roots. Moreover, the roots are given by

$$\sqrt[n]{r}\left[\cos\left(\frac{\theta + 2\pi k}{n}\right) + i\sin\left(\frac{\theta + 2\pi k}{n}\right)\right]$$

where $k = 0, 1, \ldots, n - 1$.

Note that the nth roots of z all have modulus $\sqrt[n]{r}$ and hence they lie on a circle of radius $\sqrt[n]{r}$ with center at O. Moreover, they are equispaced on this circle since the difference in the arguments of successive nth roots is $2\pi/n$.

It is sometimes convenient to use degree measure for θ. In this event the formula in the preceding theorem becomes

$$\sqrt[n]{r}\left[\cos\left(\frac{\theta + k \cdot 360°}{n}\right) + i\sin\left(\frac{\theta + k \cdot 360°}{n}\right)\right]$$

where $k = 0, 1, \ldots, n - 1$.

Example 2   Find the four fourth roots of $-8(1 + i\sqrt{3})$.

Solution   The geometric representation of the given number is shown in Figure 9.4.

Introducing trigonometric form, we have

$$-8(1 + i\sqrt{3}) = 16[\cos 240° + i\sin 240°].$$

By the preceding theorem, with $n = 4$, the fourth roots are given by

$$2\left[\cos\left(\frac{240° + k \cdot 360°}{4}\right) + i\sin\left(\frac{240° + k \cdot 360°}{4}\right)\right]$$

where $k = 0, 1, 2,$ and 3. The formula may be rewritten as

$$2[\cos(60° + k \cdot 90°) + i\sin(60° + k \cdot 90°)].$$

Substituting 0, 1, 2, and 3 for k yields the following fourth roots:

$$2(\cos 60° + i\sin 60°) = 1 + \sqrt{3}i$$
$$2(\cos 150° + i\sin 150°) = -\sqrt{3} + i$$
$$2(\cos 240° + i\sin 240°) = -1 - \sqrt{3}i$$
$$2(\cos 330° + i\sin 330°) = \sqrt{3} - i.$$

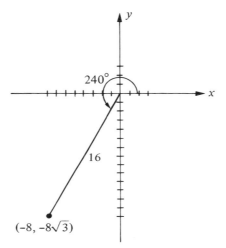

Figure 9.4

Example 3   Find the six sixth roots of -1.

Solution   Writing $-1 = 1(\cos \pi + i\sin \pi)$ and using the Theorem on nth Roots with $n = 6$, we find that the sixth roots of -1 are given by

$$\cos\left(\frac{\pi + 2\pi k}{6}\right) + i\sin\left(\frac{\pi + 2\pi k}{6}\right)$$

or

$$\cos\left(\frac{\pi}{6} + \frac{\pi}{3}k\right) + i\sin\left(\frac{\pi}{6} + \frac{\pi}{3}k\right).$$

Substituting 0, 1, 2, 3, 4, and 5 for k gives us the sixth roots of -1, namely,

$$\cos \pi/6 + i\sin \pi/6 = \sqrt{3}/2 + (1/2)i$$
$$\cos \pi/2 + i\sin \pi/2 = i$$
$$\cos 5\pi/6 + i\sin 5\pi/6 = -\sqrt{3}/2 + (1/2)i$$
$$\cos 7\pi/6 + i\sin 7\pi/6 = -\sqrt{3}/2 - (1/2)i$$
$$\cos 3\pi/2 + i\sin 3\pi/2 = -i$$
$$\cos 11\pi/6 + i\sin 11\pi/6 = \sqrt{3}/2 - (1/2)i.$$

The special case in which $z = 1$ is of particular interest. The n distinct nth roots of 1 are called the **nth roots of unity**.

Example 4 Find the three third roots of unity.

Solution   Writing $1 = \cos 0 + i \sin 0$ and using the Theorem on nth Roots with $n = 3$, we obtain the three roots

$$\cos \frac{2\pi k}{3} + i \sin \frac{2\pi k}{3}$$

where $k = 0, 1$, and 2. Substituting for k gives us

$$\cos 0 + i \sin 0 = 1$$

$$\cos \frac{2\pi}{3} + i \sin \frac{2\pi}{3} = -\frac{1}{2} + \left(\frac{\sqrt{3}}{2}\right)i$$

$$\cos \frac{4\pi}{3} + i \sin \frac{4\pi}{3} = -\frac{1}{2} - \left(\frac{\sqrt{3}}{2}\right)i.$$

∎

Compare this solution with Example 3 of Section 9.3.

EXERCISES 9.5

In Exercises 1–12 use De Moivre's Theorem to express the given numbers in the form $a + bi$, where a and b are real numbers.

1 $(3 + 3i)^5$

2 $(1 + i)^{12}$

3 $(1 - i)^{10}$

4 $(-1 + i)^8$

5 $(1 - \sqrt{3}i)^3$

6 $(1 - \sqrt{3}i)^5$

7 $\left(-\dfrac{\sqrt{2}}{2} + \dfrac{\sqrt{2}}{2}i\right)^{15}$

8 $\left(\dfrac{\sqrt{2}}{2} + \dfrac{\sqrt{2}}{2}i\right)^{25}$

9 $\left(-\dfrac{\sqrt{3}}{2} - \dfrac{1}{2}i\right)^{20}$

10 $\left(-\dfrac{\sqrt{3}}{2} - \dfrac{1}{2}i\right)^{50}$

11 $(\sqrt{3} + i)^7$

12 $(-2 - 2i)^{10}$

13 Find the two square roots of $1 + \sqrt{3}i$.

14 Find the two square roots of $-9i$.

15 Find the four fourth roots of $-1 - \sqrt{3}i$.

16 Find the four fourth roots of $-8 + 8\sqrt{3}i$.

17 Find the three cube roots of $-27i$.

18 Find the three cube roots of $64i$.

19 Find the six sixth roots of unity.

20 Find the eight eighth roots of unity.

21 Find the five fifth roots of $1 + i$.

22 Find the five fifth roots of $-\sqrt{3} - i$.

Find the solutions of the equations in Exercises 23–30.

23 $x^4 - 16 = 0$

24 $x^6 - 64 = 0$

25 $x^6 + 64 = 0$

26 $x^5 + 1 = 0$

27 $x^3 + 8i = 0$

28 $x^3 - 64i = 0$

29 $x^5 - 243 = 0$

30 $x^4 + 81 = 0$

9.6 REVIEW

Concepts

Define or discuss the following.

1 The system of complex numbers

2 The conjugate of a complex number

3 The absolute value of a complex number

4 The nth roots of unity

5 Geometric representation of a complex number

6 The complex plane

7 Trigonometric form for complex numbers

8 De Moivre's Theorem

Exercises

Express each number in Exercises 1–20 in the form $a + bi$ where a and b are real numbers.

1 $(7 + 5i) + (-8 + 3i)$

2 $(-4 + 7i) - (3 - 3i)$

3 $(4 + 2i)(-5 + 4i)$

4 $(-8 + 2i)(-7 + 5i)$

5 $8i(7 - 5i)$

6 $(-5 + 6i)(-5 - 6i)$

7 $(9 + 4i)(9 - 4i)$ 8 $(3 + 8i)^2$

9 $i(6 + i)(2 - 3i)$ 10 $(6 - 3i)/(2 + 7i)$

11 $(4 + 6i)/(5 - i)$ 12 $(20 - 8i)/4i$

13 $1/(10 - 8i)$ 14 $3/(-2i)$

15 $|6 - 10i|$ 16 $|-11 + 10i|$

17 $-5(3 + i) - (4 + 2i)$

18 $(2 + i)^4$

19 $(1 + i)^{-1} - (1 - i)^{-1}$

20 i^{99}

Find the solutions of the equations in Exercises 21–24.

21 $5x^2 - 2x + 3 = 0$ 22 $3x^2 + x + 6 = 0$

23 $6x^4 + 29x^2 + 28 = 0$

24 $x^4 + x^2 + 1 = 0$

In each of Exercises 25–28 find a quadratic equation with the given roots.

25 $7 + i, \quad 7 - i$ 26 $12i, \quad -12i$

27 $-1 - 4i, \quad -1 + 4i$ 28 $1 + i, \quad 3i$

Change the complex numbers in Exercises 29–34 to trigonometric form.

29 $-10 + 10i$ 30 $2 - 2\sqrt{3}i$

31 -17 32 $-12i$

33 $-5\sqrt{3} - 5i$ 34 $4 + 5i$

In each of Exercises 35–38 use De Moivre's Theorem to write the given number in the form $a + bi$, where a and b are real numbers.

35 $(-\sqrt{3} + i)^9$ 36 $\left(\dfrac{\sqrt{2}}{2} - \dfrac{\sqrt{2}}{2}i\right)^{30}$

37 $(3 - 3i)^5$ 38 $(2 + 2\sqrt{3}i)^{10}$

39 Find the three cube roots of -27.

40 Find the solutions of the equation $x^5 - 32 = 0$.

10

ZEROS OF POLYNOMIALS

The concept of polynomial was introduced in Chapter One. At that time we were interested primarily in basic manipulations such as adding, multiplying, and factoring. In this chapter we shall explore the theory of polynomials more deeply and discuss equations which involve polynomials of any degree.

10.1 PROPERTIES OF DIVISION

Every polynomial in a variable x, with real coefficients, can be written in the form

$$a_n x^n + a_{n-1} x^{n-1} + \cdots + a_1 x + a_0,$$

where n is a nonnegative integer and the coefficients $a_0, a_1, \ldots, a_n$ are real numbers. Recall that if $a_n \neq 0$ the polynomial has **degree n**. If all the coefficients of a polynomial are zero it is called the **zero polynomial** and is denoted by 0. It is customary not to assign a degree to the zero polynomial. If we allow the coefficients a_i to be complex numbers we obtain a polynomial in x with *complex* coefficients.

Throughout this chapter, whenever we consider arbitrary polynomials, we will assume that the coefficients a_i are either real or complex numbers. When the term "number" is used, we mean either a real or complex number. Symbols such as $f(x)$ and $g(x)$ will be used to denote polynomials in x. This is a natural notation, since each polynomial determines a function f, where the functional value $f(c)$ is obtained by substituting c for x in the given polynomial. If $f(c) = 0$, then c is called a **zero**, or **root**, of the polynomial $f(x)$.

If a polynomial $g(x)$ is a factor of a polynomial $f(x)$, we say that $f(x)$ is **divisible** by $g(x)$. For example, $x^2 - 25$ is divisible by $x - 5$ and by $x + 5$. A division process may also be introduced if $g(x)$ is *not* a factor of $f(x)$. The process is similar to that used for integers. For example, the integer 24 has positive factors 1, 2, 3, 4, 6, 8, 12, and 24; that is, 24 is *divisible* by those numbers. A different situation exists if we divide 24 by nonfactors such as 5, 7, 15, and 32. In such cases a process called *long division* is used which yields a *quotient* and *remainder*. Let us consider the process from elementary arithmetic where, for example, the work involved in dividing 4126 by 23 can be arranged as follows:

$$
\begin{array}{r}
179 \\
23 \overline{\smash{)}4126} \\
23 \\
\hline
182 \\
161 \\
\hline
216 \\
207 \\
\hline
9
\end{array}
$$

The number 179 is called the *quotient* and 9 the *remainder*. To complete the terminology, 23 is called the *divisor* and 4126 the *dividend*. The remainder should always be less than the divisor for, otherwise, the quotient can be increased. The preceding result is sometimes written as follows:

$$\frac{4126}{23} = 179 + \frac{9}{23}.$$

Multiplying by 23 gives us

$$4126 = (23)(179) + 9.$$

This form is very useful for theoretical purposes and can be generalized to arbitrary integers. Specifically, it can be shown that if a and b are integers with $b > 0$, then there exist unique integers q and r such that

$$a = bq + r$$

where $0 \le r < b$. The integer q is called the **quotient** and r the **remainder** in the division of a by b.

A similar discussion can be given for polynomials. For example, the polynomial $x^4 - 16$ is divisible by $x^2 - 4$, $x^2 + 4$, $x + 2$, and $x - 2$; but $x^2 + 3x + 1$ is not a factor of $x^4 - 16$. However, by another process called *long division*, we write

$$
\begin{array}{r}
x^2 - 3x + 8 \\
x^2 + 3x + 1 \overline{\smash{)}x^4 \qquad\qquad - 16} \\
x^4 + 3x^3 + x^2 \\
\hline
-3x^3 - x^2 \\
-3x^3 - 9x^2 - 3x \\
\hline
8x^2 + 3x - 16 \\
8x^2 + 24x + 8 \\
\hline
-21x - 24
\end{array}
$$

which yields the quotient $x^2 - 3x + 8$ and the remainder $-21x - 24$. In this division we proceed as indicated until we arrive at a polynomial (the remainder) which is either 0 or has smaller degree than the divisor. We shall assume familiarity with this process and not attempt to justify it. As with integers, the result of this division process is often written as follows:

$$\frac{x^4 - 16}{x^2 + 3x + 1} = (x^2 - 3x + 8) + \left(\frac{-21x - 24}{x^2 + 3x + 1}\right).$$

Multiplying by $x^2 + 3x + 1$,

$$x^4 - 16 = (x^2 + 3x + 1)(x^2 - 3x + 8) + (-21x - 24).$$

This has the same general form $a = bq + r$ that was given for integers.

The preceding example illustrates the following theorem, which we state without proof.

Division Algorithm for Polynomials

> If $f(x)$ and $g(x)$ are polynomials and if $g(x) \neq 0$, then there exist unique polynomials $q(x)$ and $r(x)$ such that
>
> $$f(x) = g(x)q(x) + r(x)$$
>
> where either $r(x) = 0$ or the degree of $r(x)$ is less than the degree of $g(x)$. The polynomial $q(x)$ is called the **quotient** and $r(x)$ the **remainder** in the division of $f(x)$ by $g(x)$.

An interesting special case occurs if $f(x)$ is divided by a linear polynomial of the form $x - c$, where c is a number. If $x - c$ is a factor of $f(x)$, then

$$f(x) = (x - c)q(x)$$

for some polynomial $q(x)$; that is, the remainder $r(x)$ is 0. If $x - c$ is not a factor of $f(x)$, then the degree of the remainder $r(x)$ is less than the degree of the divisor $x - c$, and hence $r(x)$ must have degree 0. This, in turn, means that the remainder is a nonzero number. Consequently, in all cases we have

$$f(x) = (x - c)q(x) + d$$

where d is some number (possibly $d = 0$). If c is substituted for x in $f(x) = (x - c)q(x) + d$, we obtain

$$f(c) = (c - c)q(c) + d$$

which reduces to $f(c) = d$, that is,

$$f(x) = (x - c)q(x) + f(c).$$

We have proved the following theorem.

Remainder Theorem

> If a polynomial $f(x)$ is divided by $x - c$, then the remainder is $f(c)$.

Example 1 If $f(x) = x^3 - 3x^2 + x + 5$, use the Remainder Theorem to find $f(2)$.

Solution According to the Remainder Theorem, $f(2)$ is the remainder when $f(x)$ is divided by $x - 2$. By long division,

$$
\begin{array}{r}
x^2 - x - 1 \\
x - 2 \overline{\smash{)}\, x^3 - 3x^2 + x + 5} \\
\underline{x^3 - 2x^2} \\
-x^2 + x \\
\underline{-x^2 + 2x} \\
-x + 5 \\
\underline{-x + 2} \\
3
\end{array}
$$

and hence $f(2) = 3$. To check our work we have, by direct substitution, $f(2) = 2^3 - 3(2)^2 + 2 + 5 = 3$. ∎

Factor Theorem

> A polynomial $f(x)$ has a factor $x - c$ if and only if $f(c) = 0$.

Proof By the Remainder Theorem, $f(x) = (x - c)q(x) + f(c)$ for some quotient $q(x)$. If $f(c) = 0$, then $f(x) = (x - c)q(x)$; that is, $x - c$ is a factor of $f(x)$. Conversely, if $x - c$ is a factor, then the remainder upon division of $f(x)$ by $x - c$ must be 0, and hence, by the Remainder Theorem, $f(c) = 0$. ∎

The Factor Theorem is useful for finding factors of polynomials, as illustrated in the next example.

Example 2 Show that $x - 2$ is a factor of the polynomial

$$f(x) = x^3 - 4x^2 + 3x + 2.$$

Solution Since $f(2) = 8 - 16 + 6 + 2 = 0$, it follows from the Factor Theorem that $x - 2$ is a factor of $f(x)$. Of course, another method of solution would be to divide $f(x)$ by $x - 2$ and show that the remainder is 0. The quotient in the division would be another factor of $f(x)$. ∎

Example 3 Find a polynomial $f(x)$ of degree 3 that has zeros 2, -1, and 3.

Solution By the Factor Theorem, $f(x)$ has factors $x - 2$, $x + 1$, and $x - 3$. We may then write

$$f(x) = a(x - 2)(x + 1)(x - 3)$$

where any nonzero value may be assigned to a. If we let $a = 1$ and multiply, we obtain

$$f(x) = x^3 - 4x^2 + x + 6.$$ ∎

EXERCISES 10.1

In Exercises 1–6 find the quotient $q(x)$ and the remainder $r(x)$ if $f(x)$ is divided by $g(x)$.

1 $f(x) = x^4 + 3x^3 - 2x + 5, \quad g(x) = x^2 + 2x - 4$

2 $f(x) = 4x^3 - x^2 + x - 3, \quad g(x) = x^2 - 5x$

3 $f(x) = 5x^3 - 2x, \quad g(x) = 2x^2 + 1$

4 $f(x) = 3x^4 - x^3 - x^2 + 3x + 4, \quad g(x) = 2x^3 - x + 4$

5 $f(x) = 7x^3 - 5x + 2, \quad g(x) = 2x^4 - 3x^2 + 9$

6 $f(x) = 10x - 4, \quad g(x) = 8x^2 - 5x + 17$

In Exercises 7–12 use the Remainder Theorem to find $f(c)$.

7 $f(x) = 2x^3 - x^2 - 5x + 3, \quad c = 4$

8 $f(x) = 4x^3 - 3x^2 + 7x + 10, \quad c = 3$

9 $f(x) = x^4 + 5x^3 - x^2 + 5, \quad c = -2$

10 $f(x) = x^4 - 7x^2 + 2x - 8, \quad c = -3$

11 $f(x) = x^6 - 3x^4 + 4, \quad c = \sqrt{2}$

12 $f(x) = x^5 - x^4 + x^3 - x^2 + x - 1, \quad c = i$

13 Determine k so that $f(x) = x^3 + kx^2 - kx + 10$ is divisible by $x + 3$.

14 Determine all values of k such that $f(x) = k^2x^3 - 4kx - 3$ is divisible by $x - 1$.

15 Use the Factor Theorem to show that $x - 2$ is a factor of $f(x) = x^4 - 3x^3 - 2x^2 + 5x + 6$.

16 Show that $x + 2$ is a factor of $f(x) = x^{12} - 4096$.

17 Prove that $f(x) = 3x^4 + x^2 + 5$ has no factor of the form $x - c$ where c is a real number.

18 Find the remainder if the polynomial $3x^{100} + 5x^{85} - 4x^{38} + 2x^{17} - 6$ is divided by $x + 1$.

19 Use the Factor Theorem to prove that $x - y$ is a factor of $x^n - y^n$, for all positive integers n. If n is even, show that $x + y$ is also a factor of $x^n - y^n$.

20 If n is an odd positive integer, prove that $x + y$ is a factor of $x^n + y^n$.

21 Prove that a polynomial $f(x)$ has a zero if and only if $f(x)$ has a first-degree polynomial as a factor.

22 Comment on the following "proof" that if $f(x)$ is divided by $x - c$, then the remainder is $f(c)$.

$$\begin{array}{r} f \\ x - c \overline{\smash{)}\, f(x)} \\ \underline{f(x) - f(c)} \\ f(c) \end{array}$$

10.2 SYNTHETIC DIVISION

When applying the Remainder Theorem (Section 10.1), it is necessary to divide by polynomials of the form $x - c$. The process referred to as *synthetic division* simplifies the work if divisors are of that form. We shall illustrate the process by means of examples.

If the polynomial $3x^4 - 8x^3 + 9x + 5$ is divided by $x - 2$, we obtain

$$
\begin{array}{r}
3x^3 - 2x^2 - 4x + 1 \\
x - 2 \overline{\big)\, 3x^4 - 8x^3 + 0x^2 + 9x + 5} \\
\underline{3x^4 - 6x^3} \\
-2x^3 + 0x^2 \\
\underline{-2x^3 + 4x^2} \\
-4x^2 + 9x \\
\underline{-4x^2 + 8x} \\
x + 5 \\
\underline{x - 2} \\
7
\end{array}
$$

where the term $0x^2$ has been inserted in the dividend so that *all* powers of x are accounted for. Since this technique of long division seems to involve a great deal of labor for so simple a problem, we look for a means of simplifying the notation. After arranging the terms which involve like powers of x in vertical columns, it is seen that the repeated expressions $3x^4$, $-2x^3$, $-4x^2$, and x may be deleted without too much chance of confusion. Also, it appears unnecessary to "bring down" the terms $0x^2$, $9x$, and 5 from the dividend as indicated. With the elimination of those repetitions, our work takes on this form:

$$
\begin{array}{r}
3x^3 - 2x^2 - 4x + 1 \\
x - 2 \overline{\big)\, 3x^4 - 8x^3 + 0x^2 + 9x + 5} \\
\underline{-6x^3} \\
-2x^3 \\
\underline{4x^2} \\
-4x^2 \\
\underline{8x} \\
x \\
\underline{-2} \\
7
\end{array}
$$

If we take care to keep like powers of x under one another, and if we account for missing terms by means of zero coefficients, then some labor can be saved by omitting the symbol x. Doing this in the preceding display, we obtain the following:

$$
\begin{array}{r}
3 \quad -2 \quad -4 \quad 1 \\
1 - 2 \overline{\big)\, 3 \quad -8 \quad 0 \quad 9 \quad 5} \\
\underline{-6} \\
-2 \\
\underline{4} \\
-4 \\
\underline{8} \\
1 \\
\underline{-2} \\
7
\end{array}
$$

Since the divisor is a polynomial of the form $x - c$, the two coefficients in the far left position are always $1 - c$. With this in mind, we shall discard the coefficient 1. Moreover, to make our notation more compact, let us move the numbers up in the following way:

$$
\begin{array}{r}
3 \quad -2 \quad -4 \quad1 \\
\underline{-2\,\big|\,3 \quad -8 \quad0 \quad9 \quad5} \\
-6 \quad4 \quad8 \quad -2 \\
\hline
-2 \quad -4 \quad1 \quad7
\end{array}
$$

If we now insert the leading coefficient 3 in the first position of the last row, the first four numbers of that row are the coefficients 3, -2, -4, and 1 of the quotient, and the final number 7 is the remainder. Since there is no need to write the coefficients of the quotient two times, we discard the first row in our scheme, obtaining

$$
\begin{array}{r}
-2\,\big|\,3 \quad -8 \quad0 \quad9 \quad5 \\
-6 \quad4 \quad8 \quad -2 \\
\hline
3 \quad -2 \quad -4 \quad1 \quad7
\end{array}
$$

where the top line has also been deleted since there is no longer any need for it.

There is a simple way of interpreting the last display. Note that every number in the second row can be obtained by multiplying the number in the third row of the *preceding* column by -2. Moreover, each number in the third row can be found by subtracting the number above it in the second row from the corresponding number in the first row. This suggests a method for carrying out this procedure without actually thinking of the division process. After arranging the terms of the polynomial in decreasing powers of x, we write the coefficients in a row, supplying 0 for any missing term. Next we write $-c$ (in the previous case, -2) to the left of this row. Next we bring down the leading coefficient 3 to the third row. Then we multiply that number by -2 to obtain the first number, -6, in the second row. We subtract -6 from -8 to obtain the second number, -2, in the third row, and then we multiply by $-c$ (in our case, -2) to obtain the second number, 4, in the second row. Again we subtract to get the third number, -4, in the third row. This process is continued until the final number in the third row (the remainder) is obtained.

It is possible to avoid subtractions if the number c is used in place of $-c$ in the far left position of the first row. In this event, when the process is used, the signs of the elements in the second row are changed, and hence to find elements of the third row, we *add* the number above it in the second row to the corresponding number in the first row. With this change we obtain

$$
\begin{array}{r}
2\,\big|\,3 \quad -8 \quad0 \quad9 \quad5 \\
6 \quad -4 \quad -8 \quad2 \\
\hline
3 \quad -2 \quad -4 \quad1 \quad7
\end{array}
$$

The latter scheme is called the process of **synthetic division**.

Example 1 Use synthetic division to find the quotient and remainder if $2x^4 + 5x^3 - 2x - 8$ is divided by $x + 3$.

Solution   Since we are to divide by $x + 3$, the c in the expression $x - c$ is -3. Hence, the synthetic division takes this form:

$$
\begin{array}{r|rrrrr}
-3 & 2 & 5 & 0 & -2 & -8 \\
   &   & -6 & 3 & -9 & 33 \\
\hline
   & 2 & -1 & 3 & -11 & 25
\end{array}
$$

The first four numbers in the third row are the coefficients of the quotient and the last number is the remainder. Hence, the quotient is $2x^3 - x^2 + 3x - 11$ and the remainder is 25. ■

Synthetic division can be used to find values of polynomial functions, as illustrated in the next example.

Example 2   If $f(x) = 3x^5 - 38x^3 + 5x^2 - 1$, use synthetic division to find $f(4)$.

Solution   By the Remainder Theorem, $f(4)$ is the remainder when $f(x)$ is divided by $x - 4$. Dividing synthetically, we obtain

$$
\begin{array}{r|rrrrrr}
4 & 3 & 0 & -38 & 5 & 0 & -1 \\
  &   & 12 & 48 & 40 & 180 & 720 \\
\hline
  & 3 & 12 & 10 & 45 & 180 & 719
\end{array}
$$

Consequently, $f(4) = 719$. ■

Synthetic division may be employed to help find zeros of polynomials. By the method illustrated in the preceding example, $f(c) = 0$ if and only if the remainder in the synthetic division by $x - c$ is 0.

Example 3   Show that -11 is a zero of the polynomial

$$f(x) = x^3 + 8x^2 - 29x + 44.$$

Solution   Dividing synthetically by $x + 11$ gives us

$$
\begin{array}{r|rrrr}
-11 & 1 & 8 & -29 & 44 \\
    &   & -11 & 33 & -44 \\
\hline
    & 1 & -3 & 4 & 0
\end{array}
$$

Thus, $f(-11) = 0$. ■

The preceeding example shows that -11 is a solution of the equation $x^3 + 8x^2 - 29x + 44 = 0$. In the next section we shall use synthetic division in this way to find solutions of equations.

EXERCISES 10.2

In each of Exercises 1–10 use synthetic division to find the quotient and remainder if the first polynomial is divided by the second.

1 $2x^3 - 3x^2 + 4x - 5$, $x - 2$

2 $3x^3 - 4x^2 - x + 8$, $x + 4$

3 $x^3 - 8x - 5$, $x + 3$

4 $5x^3 - 6x^2 + 15$, $x - 4$

5 $3x^5 + 6x^2 + 7$, $x + 2$

6 $-2x^4 + 10x - 3$, $x - 3$

7 $4x^4 - 5x^2 + 1$, $x - \frac{1}{2}$

8 $9x^3 - 6x^2 + 3x - 4$, $x - \frac{1}{3}$

9 $x^n - 1$, $x - 1$, where n is any positive integer

10 $x^n + 1$, $x + 1$, where n is any positive integer

Use synthetic division to solve Exercises 11–16.

11 If $f(x) = x^4 - 4x^3 + x^2 - 3x - 5$, find $f(2)$ and $f(-2)$.

12 If $f(x) = 0.3x^3 + 0.04x - 0.034$, find $f(0.2)$ and $f(-0.2)$.

13 If $f(x) = x^6 - x^5 + x^4 - x^3 + x^2 - x + 1$, find $f(4)$.

14 If $f(x) = 8x^5 - 3x^2 + 7$, find $f(1/2)$.

15 If $f(x) = x^2 + 3x - 5$, find $f(2 + \sqrt{3})$.

16 If $f(x) = x^3 - 3x^2 - 8$, find $f(1 + \sqrt{2})$.

In Exercises 17–20 use synthetic division to show that c is a zero of $f(x)$.

17 $f(x) = 3x^4 + 8x^3 - 2x^2 - 10x + 4$, $c = -2$

18 $f(x) = 4x^3 - 9x^2 - 8x - 3$, $c = 3$

19 $f(x) = 4x^3 - 6x^2 + 8x - 3$, $c = \frac{1}{2}$

20 $f(x) = 27x^4 - 9x^3 + 3x^2 + 6x + 1$, $c = -\frac{1}{3}$

10.3   THE NUMBER OF ZEROS OF A POLYNOMIAL

The Factor Theorem (Section 10.1) indicates that there is a close relationship between the zeros of a polynomial $f(x)$ and the factors of $f(x)$. Indeed, $f(c) = 0$ if and only if $x - c$ is a factor. Unfortunately, except in special cases, zeros of polynomials are very difficult to find. For example, given the polynomial $f(x) = x^5 - 3x^4 + 4x^3 + 4x - 10$, there are no obvious zeros. Moreover, there is no device such as the quadratic formula which can be used to produce the zeros. In spite of the practical difficulty of determining zeros of polynomials, it is possible to make some headway concerning the *theory* of such zeros. The results in this section form the basis for work in what is known as **The Theory of Equations**.

Fundamental Theorem of Algebra

> If $f(x)$ is a polynomial of positive degree, then $f(x)$ has at least one complex zero.

The usual proof of this theorem requires results from the field of mathematics called *functions of a complex variable*. In turn, a prerequisite for studying the latter field is a strong background in calculus. The first proof of the Fundamental Theorem of Algebra was given by the German mathematician Carl Friedrich Gauss (1777–1855), who is considered by many to be the greatest mathematician of all time.

As a special case, if all the coefficients of $f(x)$ are real, then $f(x)$ has at least one complex zero. We should also remark that if $a + bi$ is a complex zero of a polynomial, it may happen that $b = 0$, in which case we refer to the number as a **real zero**. If the Fundamental Theorem is combined with the Factor Theorem, the following useful corollary is obtained.

Corollary

> Every polynomial of positive degree has a factor of the form $x - c$, where c is a real or complex number.

This corollary enables us, at least in theory, to express every polynomial $f(x)$ of positive degree as a product of polynomials of degree 1. If $f(x)$ has a degree $n > 0$, then applying the corollary,

$$f(x) = (x - c_1)f_1(x)$$

where c_1 is some number and $f_1(x)$ is a polynomial of degree $n - 1$. If $n - 1 > 0$, we may apply the corollary again, obtaining

$$f_1(x) = (x - c_2)f_2(x)$$

where c_2 is some number and $f_2(x)$ is a polynomial of degree $n - 2$. Hence,

$$f(x) = (x - c_1)(x - c_2)f_2(x).$$

Continuing this process, after n steps we arrive at a polynomial $f_n(x)$ of degree 0. Thus, $f_n(x) = a$ for some nonzero number a and we may write

$$f(x) = a(x - c_1)(x - c_2) \cdots (x - c_n)$$

where each c_i is a zero of $f(x)$. Evidently, the leading coefficient of the polynomial on the right in the last equation is a. It follows that a is the leading coefficient of $f(x)$. We have proved the following theorem.

Theorem

> If $f(x)$ is a polynomial of degree $n > 0$, then there exist n numbers $c_1, c_2, \ldots, c_n$ such that
> $$f(x) = a(x - c_1)(x - c_2) \cdots (x - c_n)$$
> where a is the leading coefficient of $f(x)$. Each number c_i is a zero of $f(x)$.

Corollary

> A polynomial of degree $n > 0$ has at most n different zeros.

Proof

We shall give an indirect proof. Suppose $f(x)$ has *more* than n different zeros. Let us choose $n + 1$ of these zeros and label them $c_1, c_2, \ldots, c_n$, and c. We may use the c_i to obtain the factorization indicated in the statement of the theorem. Substituting c for x and using the fact that $f(c) = 0$, we obtain

$$0 = a(c - c_1)(c - c_2) \cdots (c - c_n).$$

However, each factor on the right side is different from zero because $c \neq c_i$ for every i. Since the product of nonzero numbers cannot equal zero, we have a contradiction. ∎

Example 1 Find a polynomial $f(x)$ of degree 3 with zeros 2, -1, and 3, which has the value 5 at $x = 1$.

Solution   By the Factor Theorem, $f(x)$ has factors $x - 2$, $x + 1$, and $x - 3$. No other factors of degree 1 exist, since, by the Factor Theorem another linear factor $x - c$ would produce a fourth zero of $f(x)$ in violation of the last corollary. Hence, $f(x)$ has the form

$$f(x) = a(x - 2)(x + 1)(x - 3)$$

for some number a. If $f(x)$ has the value 5 at $x = 1$, then $f(1) = 5$, that is,

$$a(1 - 2)(1 + 1)(1 - 3) = 5, \quad \text{or} \quad 4a = 5.$$

Consequently, $a = 5/4$ and

$$f(x) = \frac{5}{4}(x - 2)(x + 1)(x - 3).$$

Multiplying the four factors we obtain

$$f(x) = \frac{5}{4}x^3 - 5x^2 + \frac{5}{4}x + \frac{15}{2}. \qquad \blacksquare$$

The numbers $c_1, c_2, \ldots, c_n$ in the last theorem are not necessarily all different. To illustrate, the polynomial $f(x) = x^3 + x^2 - 5x + 3$ has the factorization

$$f(x) = (x + 3)(x - 1)(x - 1).$$

If a factor $x - c$ occurs m times in the factorization, then c is called a **zero of multiplicity** m of $f(x)$. In the preceding illustration 1 is a zero of multiplicity 2, and -3 is a zero of multiplicity 1.

As another illustration, if

$$f(x) = (x - 2)(x - 4)^3(x + 1)^2$$

then $f(x)$ has degree 6 and possesses three distinct zeros 2, 4, and -1, where 2 has multiplicity 1, 4 has multiplicity 3, and -1 has multiplicity 2.

If a zero of multiplicity m is counted as m zeros, then the last theorem tells us that a polynomial $f(x)$ of degree $n > 0$ has *at least* n zeros (not necessarily all different). Combining this with the fact that $f(x)$ has at *most* n zeros, we obtain the next result.

Theorem

> If $f(x)$ is a polynomial of degree $n > 0$ and if a zero of multiplicity m is counted m times, then $f(x)$ has precisely n zeros.

Example 2 Express $f(x) = x^5 - 4x^4 + 13x^3$ as a product of linear factors and list the five zeros of $f(x)$.

Solution   We begin by writing

$$f(x) = x^3(x^2 - 4x + 13).$$

By the Quadratic Formula, the zeros of the polynomial $x^2 - 4x + 13$ are given by

$$\frac{4 \pm \sqrt{16 - 52}}{2} = \frac{4 \pm \sqrt{-36}}{2} = \frac{4 \pm 6i}{2} = 2 \pm 3i.$$

Hence, by the Factor Theorem, $x^2 - 4x + 13$ has factors $x - (2 + 3i)$ and $x - (2 - 3i)$, and we obtain the desired factorization

$$f(x) = x \cdot x \cdot x \cdot (x - 2 - 3i)(x - 2 + 3i).$$

Since $x - 0$ occurs as a factor three times, the number 0 is a zero of multiplicity three, and the five zeros of $f(x)$ are 0, 0, 0, $2 + 3i$, and $2 - 3i$. ∎

The next theorem may be used to obtain information about the real zeros of a polynomial $f(x)$ with real coefficients. In the statement of the theorem it is assumed that the terms of $f(x)$ are arranged in order of decreasing powers of x, and that terms with zero coefficients are deleted. In this case we say there is a **variation of sign** in $f(x)$ if two consecutive coefficients have opposite signs. To illustrate, the polynomial

$$f(x) = 2x^5 - 7x^4 + 3x^2 + 6x - 5$$

has three variations in sign, since there is one variation from $2x^5$ to $-7x^4$, a second from $-7x^4$ to $3x^2$, and a third from $6x$ to -5.

The theorem also refers to the variations of sign in $f(-x)$. Using the previous illustration, note that

$$\begin{aligned} f(-x) &= 2(-x)^5 - 7(-x)^4 + 3(-x)^2 + 6(-x) - 5 \\ &= -2x^5 - 7x^4 + 3x^2 - 6x - 5. \end{aligned}$$

Hence, there are two variations of sign in $f(-x)$, one from $-7x^4$ to $3x^2$, and a second from $3x^2$ to $-6x$.

Descartes' Rule of Signs

If $f(x)$ is a polynomial with real cofficients, then

(i)   the number of positive real solutions of the equation $f(x) = 0$ either is equal to the number of variations of sign in $f(x)$ or is less than that number by an even integer.

(ii) the number of negative real solutions of the equation $f(x) = 0$ either is equal to the number of variations of sign in $f(-x)$ or is less than that number by an even integer.

The proof of Descartes' Rule is rather technical and will not be given in this text.

Example 3   Discuss the number of possible positive and negative real solutions of the equation

$$2x^5 - 7x^4 + 3x^2 + 6x - 5 = 0.$$

Solution   The polynomial $f(x)$ on the left side of the equation is the same as that given in the illustration preceding the statement of Descartes' Rule. Since there are three variations of sign in $f(x)$, the equation has either three positive real solutions or one positive real solution.

Since $f(-x) = -2x^5 - 7x^4 + 3x^2 - 6x - 5$ has two variations of sign, the given equation has either two negative solutions or no negative solutions.

The following table summarizes the various possibilities that can occur.

Number of positive solutions	1	1	3	3
Number of negative solutions	0	2	0	2
Number of complex solutions	4	2	2	0

■

Example 4   Discuss the nature of the roots of the equation

$$3x^5 + 4x^3 + 2x - 5 = 0.$$

Solution   The polynomial $f(x)$ on the left side of the equation has one variation of sign and hence, by (i) of Descartes' Rule of Signs, there is precisely one positive real root. Since

$$\begin{aligned} f(-x) &= 3(-x)^5 + 4(-x)^3 + 2(-x) - 5 \\ &= -3x^5 - 4x^3 - 2x - 5 \end{aligned}$$

there are no variations of sign in $f(-x)$ and hence, by (ii) of Descartes' Rule of Signs, there are no negative real roots. Thus the equation has one real root and four complex roots. Moreover, since $f(0) = -5$ and $f(1) = 4$, the real root is between 0 and 1. (Why?)   ■

When applying Descartes' Rule, roots of multiplicity k should be counted as k roots. For example, given $x^2 - 2x + 1 = 0$, the indicated polynomial has two variations of sign and hence there are either two positive real roots or none. The factored form of the equation is $(x - 1)^2 = 0$, and hence the equation has the root 1 of multiplicity 2.

We shall conclude this section with a discussion of *bounds* for the real solutions of an equation $f(x) = 0$, where $f(x)$ is a polynomial with real coefficients. By definition, a real number b is an **upper bound** for the solutions if no solution is greater than b. A real number a is a **lower bound** for the solutions if no solution is less than a. Thus, if r is a real solution of $f(x) = 0$, then $a \le r \le b$. Note that upper and lower bounds are not unique, since any number greater than b is also an upper bound, and any number less than a is a lower bound.

We may use synthetic division to find upper and lower bounds for the solutions of $f(x) = 0$. Recall that if $f(x)$ is divided synthetically by $x - c$, then the third row which appears in the division process consists of the coefficients of the quotient $q(x)$ together with the remainder $f(c)$. The following theorem indicates how this third row may be used to find upper and lower bounds for the real solutions.

**Bounds
for Real
Zeros of
Polynomials**

Suppose $f(x)$ is a polynomial with real coefficients and positive leading coefficient, and let $f(x)$ be divided synthetically by $x - c$.

(i)   If $c > 0$, and if all numbers in the third row of the division process are either positive or zero, then c is an upper bound for the real solutions of $f(x) = 0$.

(ii) If $c < 0$, and if the numbers in the third row of the division process are alternatively positive or negative (where a 0 in the third row is to be considered either positive or negative), then c is a lower bound for the real solutions of $f(x) = 0$.

A general proof of this theorem can be patterned after the solution given in the next example.

Example 5   Find upper and lower bounds for the real solutions of the equation

$$2x^3 + 5x^2 - 8x - 7 = 0.$$

Solution   If we divide synthetically by $x - 1$ and $x - 2$, we obtain

```
1 | 2    5   -8   -7         2 | 2    5   -8   -7
        2    7   -1                  4   18   20
    ―――――――――――――――――             ―――――――――――――――――
      2    7   -1   -8              2    9   10   13
```

Since all numbers in the third row of the synthetic division by $x - 2$ are positive, it follows from (i) of the preceding theorem that 2 is an upper bound for the solutions of the given equation. This fact is also evident if we express the division by $x - 2$ in the Division Algorithm form

$$2x^3 + 5x^2 - 8x - 7 = (x - 2)(2x^2 + 9x + 10) + 13.$$

If $x > 2$, then it is obvious that the right side of the equation is positive, and hence is not zero. Consequently, $2x^3 + 5x^2 - 8x - 7$ is not zero if $x > 2$.

Dividing synthetically by $x - (-3)$ and $x - (-4)$ gives us

```
-3 | 2    5   -8   -7        -4 | 2    5   -8   -7
        -6    3   15                 -8   12  -16
    ―――――――――――――――――             ―――――――――――――――――
      2   -1   -5    8              2   -3    4  -23
```

Since the numbers in the third row of the synthetic division by $x - (-4)$ are alternately positive or negative, it follows from (ii) of the preceding theorem that -4 is a lower bound for the solutions. This can also be proved by expressing the division by $x + 4$ in the form

$$2x^3 + 5x^2 - 8x - 7 = (x + 4)(2x^2 - 3x + 4) - 23.$$

If $x < -4$, then the right side of this equation is negative (Why?) and therefore, is not zero. Hence, $2x^3 + 5x^2 - 8x - 7$ is not zero if $x < -4$.   ∎

EXERCISES 10.3

In each of Exercises 1–6, find a polynomial $f(x)$ of degree 3 with the indicated zeros and satisfying the given conditions.

1 $5, -2, -3$; $f(2) = 4$

2 $4, 1, -6$; $f(5) = 2$

3 $2, 3, 1$; $f(0) = 12$

4 $\sqrt{2}, \pi, 0$; $f(0) = 0$

5 $2 + i, 2 - i, -4$; $f(1) = 3$

6 $1 + 2i, 1 - 2i, 5$; $f(-2) = 1$

7 Find a polynomial of degree 4 such that both -2 and 3 are zeros of multiplicity 2.

8 Find a polynomial of degree 5 such that -2 is a zero of multiplicity 3 and 4 is a zero of multiplicity 2.

9 Find a polynomial $f(x)$ of degree 8 such that 2 is a zero of multiplicity 3, 0 is a zero of multiplicity 5, and $f(3) = 54$.

10 Find a polynomial $f(x)$ of degree 7 such that 1 is a zero of multiplicity 2, -1 is a zero of multiplicity 2, 0 is a zero of multiplicity 3, and $f(2) = 36$.

In Exercises 11–18 find the zeros of the polynomials and state the multiplicity of each zero.

11 $f(x) = (x + 4)^3(3x - 4)$

12 $f(x) = (x - 5)^2(4x + 7)^3$

13 $f(x) = 2x^5 - 8x^4 - 10x^3$

14 $f(x) = (4x^2 - 5)^2$

15 $f(x) = (9x^2 - 25)^4(x^2 + 16)$

16 $f(x) = (2x^2 + 13x - 7)^3$

17 $f(x) = (x^2 + x - 2)^2(x^2 - 4)$

18 $f(x) = 4x^6 + x^4$

19 Show that -3 is a zero of multiplicity 2 of the polynomial $f(x) = x^4 + 7x^3 + 13x^2 - 3x - 18$ and express $f(x)$ as a product of linear factors.

20 Show that 4 is a zero of multiplicity 2 of the polynomial $f(x) = x^4 - 9x^3 + 22x^2 - 32$ and express $f(x)$ as a product of linear factors.

21 Show that 1 is a zero of multiplicity 5 of $f(x) = x^6 - 4x^5 + 5x^4 - 5x^2 + 4x - 1$ and express $f(x)$ as a product of linear factors.

22 Show that -1 is a zero of multiplicity 4 of $f(x) = x^5 + x^4 - 6x^3 - 14x^2 - 11x - 3$ and express $f(x)$ as a product of linear factors.

In each of Exercises 23–30 use Descartes' Rule of Signs to determine the number of possible positive and negative solutions of the given equation.

23 $4x^3 - 6x^2 + x - 3 = 0$

24 $5x^3 - 6x - 4 = 0$

25 $4x^3 + 2x^2 + 1 = 0$

26 $3x^3 - 4x^2 + 3x = 0$

27 $3x^4 + 2x^3 - 4x + 2 = 0$

28 $2x^4 - x^3 + x^2 - 3x + 4 = 0$

29 $x^5 + 4x^4 + 3x^3 - 4x + 2 = 0$

30 $2x^6 + 5x^5 + 2x^2 - 3x + 4 = 0$

In each of Exercises 31–36 find the smallest and largest integers that are upper and lower bounds, respectively, for the real solutions of the given equation.

31 $x^3 - 4x^2 - 5x + 7 = 0$

32 $2x^3 - 5x^2 + 4x - 8 = 0$

33 $x^4 - x^3 - 2x^2 + 3x + 6 = 0$

34 $2x^4 - 9x^3 - 8x - 10 = 0$

35 $2x^5 - 13x^3 + 2x - 5 = 0$

36 $3x^5 + 2x^4 - x^3 - 8x^2 - 7 = 0$

37 Let $f(x)$ and $g(x)$ be polynomials of degree not greater than n, where n is a positive integer. Show that if $f(x)$ and $g(x)$ are equal in value for more than n distinct values of x, then $f(x)$ and $g(x)$ are identical, that is, coefficients of like powers are the same. (*Hint*: Write

$$f(x) = a_n x^n + a_{n-1}x^{n-1} + \cdots + a_1 x + a_0$$
$$g(x) = b_n x^n + b_{n-1}x^{n-1} + \cdots + b_1 x + b_0$$

and consider $h(x) = f(x) - g(x) = (a_n - b_n)x^n + \cdots + (a_0 - b_0)$. Then show that $h(x)$ has more than n distinct zeros and conclude that $a_i = b_i$ for all i.)

10.4 COMPLEX AND RATIONAL ZEROS OF POLYNOMIALS

In this section we shall concentrate on polynomials with real coefficients. An interesting fact about such polynomials is illustrated in Example 2 of the preceding section, where the two complex zeros of $x^5 - 4x^4 + 13x^3$ were conjugates of one another. That is no accident, since the following general result is true.

Theorem

> If $f(x)$ is a polynomial of degree $n > 0$, with real coefficients, and if z is a complex zero of $f(x)$, then the conjugate $\bar{z}$ is also a zero of $f(x)$.

Proof   We may write

$$f(x) = a_n x^n + a_{n-1} x^{n-1} + \cdots + a_1 x + a_0$$

where the a_i are real numbers and $a_n \neq 0$. If $f(z) = 0$, then

$$a_n z^n + a_{n-1} z^{n-1} + \cdots + a_1 z + a_0 = 0.$$

If two complex numbers are equal, then so are their conjugates. Consequently, the conjugates of each side of the latter equation are equal, that is,

$$\overline{a_n z^n + a_{n-1} z^{n-1} + \cdots + a_1 z + a_0} = \bar{0} = 0.$$

(The fact that $\bar{0} = 0$ follows from (iv) of the Theorem on Conjugates in Section 9.2.) As pointed out in the preceding chapter, the conjugate of a sum of complex numbers equals the sum of the conjugates, and therefore,

$$\overline{a_n z^n} + \overline{a_{n-1} z^{n-1}} + \cdots + \overline{a_1 z} + \overline{a_0} = 0.$$

Since, for each i,

$$\overline{a_i z^i} = \overline{a_i} \cdot \overline{z^i} = \overline{a_i} \cdot \overline{z}^i = a_i \bar{z}^i$$

we obtain

$$a_n \bar{z}^n + a_{n-1} \bar{z}^{n-1} + \cdots + a_1 \bar{z} + a_0 = 0.$$

The last equation states that $f(\bar{z}) = 0$, which completes the proof.   ∎

Example 1   Find a polynomial $f(x)$ of degree 4 that has real coefficients and zeros $2 + i$ and $-3i$.

Solution   By the preceding theorem, $f(x)$ must also have zeros $2 - i$ and $3i$, and hence by the Factor Theorem (Section 10.1), $f(x)$ has factors $x - (2 + i)$, $x - (2 - i)$, $x - (-3i)$, and $x - (3i)$. Multiplying those factors gives us a polynomial of the required type. Thus,

$$\begin{aligned} f(x) &= [x - (2 + i)][x - (2 - i)](x - 3i)(x + 3i) \\ &= (x^2 - 4x + 5)(x^2 + 9) \\ &= x^4 - 4x^3 + 14x^2 - 36x + 45. \end{aligned}$$ ∎

If a polynomial with real coefficients is factored as on page 463, some of the factors $x - c_i$ may have a complex coefficient c_i. However, it is always possible to obtain a factorization into polynomials with real coefficients, as stated in the next theorem.

Theorem

> Every polynomial with real coefficients and positive degree n can be expressed as a product of linear and quadratic polynomials with real coefficients, where the quadratic factors have no real zeros.

Proof

Since $f(x)$ has precisely n complex zeros $c_1, c_2, \ldots, c_n$, we may write

$$f(x) = a(x - c_1)(x - c_2) \cdots (x - c_n)$$

where a is the leading coefficient of $f(x)$. Of course, some of the c_i may be real; that is, their imaginary parts may be zero. In such cases we obtain the linear factors referred to in the statement of the theorem. If a zero c_i is not real, then by the preceding theorem the conjugate $\overline{c_i}$ is also a zero of $f(x)$ and hence must be one of the numbers $c_1, c_2, \ldots, c_n$. This implies that both $x - c_i$ and $x - \overline{c_i}$ appear in the factorization of $f(x)$. If those factors are multiplied, we obtain

$$(x - c_i)(x - \overline{c_i}) = x^2 - (c_i + \overline{c_i})x + c_i\overline{c_i}$$

which has *real* coefficients, since $c_i + \overline{c_i}$ and $c_i\overline{c_i}$ are real numbers. Thus, the complex zeros of $f(x)$ and their conjugates give rise to quadratic polynomials which are irreducible over $\mathbb{R}$. This completes the proof. ∎

Example 2   Express $x^4 - 2x^2 - 3$ (a) as a product of linear polynomials, and (b) as a product of linear and quadratic polynomials with real coefficients which are irreducible over $\mathbb{R}$.

Solutions

(a)   The zeros of the given polynomial can be found by solving the equation $x^4 - 2x^2 - 3 = 0$, which may be regarded as quadratic in x^2. Solving for x^2 by means of the quadratic formula, we obtain

$$x^2 = \frac{2 \pm \sqrt{4 + 12}}{2} = \frac{2 \pm 4}{2}$$

or $x^2 = 3$ and $x^2 = -1$. Hence, the zeros are $\sqrt{3}$, $-\sqrt{3}$, i, and $-i$, and we obtain the factorization

$$x^4 - 2x^3 - 3 = (x - \sqrt{3})(x + \sqrt{3})(x - i)(x + i).$$

(b)   Multiplying the last two factors in the preceding factorization gives us

$$x^4 - 2x^2 - 3 = (x - \sqrt{3})(x + \sqrt{3})(x^2 + 1)$$

which is of the form stated in the preceding theorem.

The solution of this example could also have been obtained by factoring the original expression without first finding the zeros. Thus,

$$x^4 - 2x^2 - 3 = (x^2 - 3)(x^2 + 1)$$
$$= (x + \sqrt{3})(x - \sqrt{3})(x + i)(x - i).$$ ∎

We have already pointed out that it is generally very difficult to find zeros of polynomials of high degree. However, if all the coefficients are integers or rational numbers, there is a method for finding the *rational* zeros, if they exist. The method is a consequence of the following theorem.

Theorem on Rational Zeros

> Suppose that $f(x) = a_n x^n + a_{n-1} x^{n-1} + \cdots + a_1 x + a_0$ is a polynomial with integer coefficients. If c/d is a rational zero of $f(x)$, where c and d have no common prime factors and $c > 0$, then c is a factor of a_0 and d is a factor of a_n.

Proof

Let us show that c is a factor of a_0. If $c = 1$, the theorem follows at once, since 1 is a factor of *any* number. Now suppose that $c \neq 1$. In this case $c/d \neq 1$, for if $c/d = 1$, we obtain $c = d$, and since c and d have no prime factor in common, this implies that $c = d = 1$, a contradiction. Hence, in the following discussion we have $c \neq 1$ and $c \neq d$.

Since $f(c/d) = 0$,

$$a_n(c^n/d^n) + a_{n-1}(c^{n-1}/d^{n-1}) + \cdots + a_1(c/d) + a_0 = 0.$$

Multiplying by d^n and then adding $-a_0 d^n$ to both sides, we obtain

$$a_n c^n + a_{n-1} c^{n-1} d + \cdots + a_1 c d^{n-1} = -a_0 d^n$$

or $\qquad c(a_n c^{n-1} + a_{n-1} c^{n-2} d + \cdots + a_1 d^{n-1}) = -a_0 d^n.$

This shows that c is a factor of the integer $a_0 d^n$. If c is factored into primes, say $c = p_1 p_2 \cdots p_k$, then each prime p_i is also a factor of $a_0 d^n$. However, by hypothesis, none of the p_i is a factor of d. This implies that each p_i is a factor of a_0, that is, c is a factor of a_0. A similar argument may be used to prove that d is a factor of a_n. ∎

The technique of using the preceding theorem for finding rational solutions of equations with integral coefficients is illustrated in the following example.

Example 3 Find all rational solutions of the equation

$$3x^4 + 14x^3 + 14x^2 - 8x - 8 = 0.$$

Solution The problem is equivalent to finding the rational zeros of the indicated polynomial. According to the last theorem, if c/d is a rational zero and $c > 0$, then c is a divisor of -8 and d is a divisor of 3. Hence, the possible choices for c are 1, 2, 4, and 8, and the choices for d are ± 1 and ± 3. Consequently, any rational roots are included among the numbers ± 1, ± 2, ± 4, ± 8, $\pm \frac{1}{3}$, $\pm \frac{2}{3}$, $\pm \frac{4}{3}$, and $\pm \frac{8}{3}$. The number of possibilities can be reduced by finding upper and lower bounds for the real solutions; however, we shall not do so here. It is necessary to check to see which, if any, are zeros. The method of synthetic division is recommended for carrying out this task. We find that

$$
\begin{array}{r|rrrrr}
-2 & 3 & 14 & 14 & -8 & -8 \\
   &   & -6 & -16 & 4 & 8 \\
\hline
   & 3 & 8 & -2 & -4 & 0
\end{array}
$$

which shows that -2 is a zero. Moreover, the synthetic division provides the coefficients of the quotient in the division of the polynomial by $x + 2$. Hence, we have the following factorization of the given polynomial:

$$(x + 2)(3x^3 + 8x^2 - 2x - 4).$$

The remaining solutions of the equation must be zeros of the second factor, and therefore we may use the latter polynomial to check for solutions. Dividing by $x + \frac{2}{3}$ synthetically gives us

$$
\begin{array}{r|rrrr}
-\frac{2}{3} & 3 & 8 & -2 & -4 \\
             &   & -2 & -4 & 4 \\
\hline
             & 3 & 6 & -6 & 0
\end{array}
$$

and therefore $-\frac{2}{3}$ is a zero.

The remaining zeros are solutions of the equation $3x^2 + 6x - 6 = 0$, or equivalently, $x^2 + 2x - 2 = 0$. By the Quadratic Formula, this equation has solutions

$$\frac{-2 \pm \sqrt{4 - 4(-2)}}{2} = \frac{-2 \pm \sqrt{12}}{2} = \frac{-2 \pm 2\sqrt{3}}{2}.$$

Hence, the given polynomial has two rational roots, -2 and $-\frac{2}{3}$, and two irrational roots, $-1 + \sqrt{3}$ and $-1 - \sqrt{3}$. ∎

The Theorem on Rational Zeros may also be applied to equations with rational coefficients. We merely multiply both sides of the equation by the least common denominator of all the coefficients to obtain an equation with integral coefficients and then proceed as above.

Example 4   Find all rational solutions of the equation

$$\left(\tfrac{2}{3}\right)x^4 + \left(\tfrac{1}{2}\right)x^3 - \left(\tfrac{5}{4}\right)x^2 - x - \left(\tfrac{1}{6}\right) = 0.$$

Solution Multiplying both sides of the equation by 12 produces the equivalent equation

$$8x^4 + 6x^3 - 15x^2 - 12x - 2 = 0.$$

If c/d is a rational solution, then the choices for c are 1 and 2 and the choices for d are ± 1, ± 2, ± 4, and ± 8. Hence, the only possible rational roots are ± 1, ± 2, $\pm \frac{1}{2}$, $\pm \frac{1}{4}$, and $\pm \frac{1}{8}$. By trial we have

$$
\begin{array}{r|rrrrr}
-\frac{1}{2} & 8 & 6 & -15 & -12 & -2 \\
 & & -4 & -1 & 8 & 2 \\
\hline
 & 8 & 2 & -16 & -4 & 0
\end{array}
$$

and hence, $-\frac{1}{2}$ is a solution. Using synthetic division on the coefficients of the quotient, we obtain

$$
\begin{array}{r|rrrr}
-\frac{1}{4} & 8 & 2 & -16 & -4 \\
 & & -2 & 0 & 4 \\
\hline
 & 8 & 0 & -16 & 0
\end{array}
$$

and consequently, $-\frac{1}{4}$ is a solution. The last synthetic division gave us the quotient $8x^2 - 16$. Setting this equal to zero and solving, we obtain $x^2 = 2$, or $x = \pm\sqrt{2}$. Thus, the given equation has rational solutions $-\frac{1}{2}$, $-\frac{1}{4}$ and irrational solutions $\sqrt{2}$, $-\sqrt{2}$. ∎

The discussion in this section gives no practical information about finding the irrational zeros of polynomials. The examples we have worked are not typical of problems encountered in applications. Indeed, it is not unusual for a polynomial with rational coefficients to have *no* rational zeros. Except in the simplest cases, the best that can be accomplished is to find decimal approximations to the irrational zeros. There exist methods which may be used to approximate irrational zeros to any degree of accuracy (see, for example, Calculator Exercises 4.2). A standard way to approximate zeros is to use a technique studied in calculus called Newton's Method. In practice, computers have taken over the task of approximating irrational solutions of equations.

EXERCISES 10.4

In each of Exercises 1–8, find a polynomial with real coefficients which has the given zeros and degree.

1 $4 + i$; degree 2

2 $3 - 3i$; degree 2

3 $4, 3 - 2i$; degree 3

4 $-2, 1 + 5i$; degree 3

5 $1 + 4i, 2 - i$; degree 4

6 $3, 0, 8 - 7i$; degree 4

7 $5i, 1 + i, 0$; degree 5

8 $i, 2i, 3i$; degree 6

9 Does there exist a polynomial of degree 3 with real coefficients having zeros 1, -1, and i? Justify your answer.

10   The complex number i is a zero of the polynomial $f(x) = x^3 - ix^2 + 2ix + 2$; however, the conjugate $-i$ of i is not a zero. Why doesn't this contradict the first theorem of this section?

In Exercises 11–28 find all solutions of the given equations.

11 $x^3 - x^2 - 10x - 8 = 0$

12 $x^3 + x^2 - 14x - 24 = 0$

13 $2x^3 - 3x^2 - 17x + 30 = 0$

14 $12x^3 + 8x^2 - 3x - 2 = 0$

15 $x^4 + 3x^3 - 30x^2 - 6x + 56 = 0$

16 $3x^5 - 10x^4 - 6x^3 + 24x^2 + 11x - 6 = 0$

17 $2x^3 - 7x^2 - 10x + 24 = 0$

18 $2x^3 - 3x^2 - 8x - 3 = 0$

19 $6x^3 + 19x^2 + x - 6 = 0$

20 $6x^3 + 5x^2 - 17x - 6 = 0$

21 $8x^3 + 18x^2 + 45x + 27 = 0$

22 $3x^3 - x^2 + 11x - 20 = 0$

23 $x^4 - x^3 - 9x^2 - 3x - 36 = 0$

24 $3x^4 + 16x^3 + 28x^2 + 31x + 30 = 0$

25 $9x^4 + 15x^3 - 20x^2 - 20x + 16 = 0$

26 $15x^4 + 4x^3 + 11x^2 + 4x - 4 = 0$

27 $4x^5 + 12x^4 - 41x^3 - 99x^2 + 10x + 24 = 0$

28 $4x^5 + 24x^4 - 13x^3 - 174x^2 + 9x + 270 = 0$

Show that the equations in Exercises 29–36 have no rational roots.

29 $x^3 + 3x^2 - 4x + 6 = 0$

30 $3x^3 - 4x + 7x + 5 = 0$

31 $x^5 - 3x^3 + 4x^2 + x - 2 = 0$

32 $2x^5 + 3x^3 + 7 = 0$

33 $3x^4 + 7x^3 + 3x^2 - 8x - 10 = 0$

34 $2x^4 - 3x^3 + 6x^2 - 24x + 5 = 0$

35 $8x^4 + 16x^3 - 26x^2 - 12x + 15 = 0$

36 $5x^5 + 2x^4 + x^3 - 10x^2 - 4x - 2 = 0$

37   If n is an odd positive integer, prove that a polynomial of degree n with real coefficients has at least one real zero.

38   Show that the first theorem of this section is not necessarily true if $f(x)$ has complex coefficients.

39   Complete the proof of the Theorem on Rational Zeros by showing that d is a factor of a_n.

40   If a polynomial of the form $x^n + a_{n-1}x^{n-1} + \cdots + a_1 x + a_0$, where each a_i is an integer, has a rational root r, show that r is an integer and is a factor of a_0.

10.5   REVIEW

Concepts

Define or discuss each of the following.

1   The division algorithm for polynomials

2   The Remainder Theorem

3   The Factor Theorem

4   Synthetic division

5   The Fundamental Theorem of Algebra

6   The multiplicity of a zero of a polynomial

7   The number of zeros of a polynomial

8   Descartes' Rule of Signs

9   Upper and lower bounds for zeros of polynomials

10   Rational zeros of polynomials

Exercises

In each of Exercises 1–4 find the quotient and re-mainder if $f(x)$ is divided by $g(x)$.

1 $f(x) = 3x^5 - 4x^3 + x + 5$, $g(x) = x^3 - 2x + 7$

2 $f(x) = 7x^2 + 3x - 10$, $g(x) = x^3 - x^2 + 10$

3 $f(x) = 9x + 4$, $g(x) = 2x - 5$

4 $f(x) = 4x^3 - x^2 + 2x - 1$, $g(x) = x^2$

5 If $f(x) = -4x^4 + 3x^3 - 5x^2 + 7x - 10$, use the Remainder Theorem to find $f(-2)$.

6 Use the Remainder Theorem to prove that $x - 3$ is a factor of $2x^4 - 5x^3 - 4x^2 + 9$.

In Exercises 7 and 8 use synthetic division to find the quotient and remainder if $f(x)$ is divided by $g(x)$.

7 $f(x) = 6x^5 - 4x^2 + 8$, $g(x) = x + 2$

8 $f(x) = 2x^3 + 5x^2 - 2x + 1$, $g(x) = x - \sqrt{2}$

In Exercises 9 and 10 find polynomials with real coefficients which have the indicated zeros, degrees, and satisfy the given conditions.

9 $-3 + 5i$, -1; degree 3; $f(1) = 4$

10 $1 - i$, 3, 0; degree 4; $f(2) = -1$

11 Find a polynomial of degree 5 such that -3 is a zero of multiplicity 2 and 0 is a zero of multi-plicity 5.

12 Show that 2 is a zero of multiplicity 3 of the poly-nomial $x^5 - 4x^4 - 3x^3 + 34x^2 - 52x + 24$ and express this polynomial as a product of linear factors.

In Exercises 13 and 14 find the zeros of the poly-nomials and state the multiplicity of each zero.

13 $(x^2 - 2x + 1)^2(x^2 + 2x - 3)$

14 $x^6 + 2x^4 + x^2$

For each equation in Exercises 15 and 16, (a) use Descartes' Rule of Signs to determine the number of possible positive and negative solutions. (b) find the smallest and largest integers that are upper and lower bounds, respectively, of the real solutions.

15 $2x^4 - 4x^3 + 2x^2 - 5x - 7 = 0$

16 $x^5 - 4x^3 + 6x^2 + x + 4 = 0$

17 Prove that $7x^6 + 2x^4 + 3x^2 + 10$ has no real roots.

Find all solutions of the equations in Exercises 18–20.

18 $x^4 + 9x^3 + 31x^2 + 49x + 30 = 0$

19 $16x^3 - 20x^2 - 8x + 3 = 0$

20 $x^4 - 7x^2 + 6 = 0$

11

SEQUENCES, SERIES, AND PROBABILITY

The method of proof called mathematical induction *considered in the first section of this chapter is very important in all branches of mathematics. In particular, in Section 11.2 it is used to prove the* Binomial Theorem. *Section 11.3 contains a discussion of* sequences *and* summation notation. *Of special interest are the* arithmetic *and* geometric *sequences considered in Sections 11.4 and 11.5. Sections 11.6 and 11.7 contain a discussion of counting processes that arise frequently in mathematics and in everyday life. Among these are the notions of* permutations *and* combinations. *The final section contains an introduction to probability.*

11.1 MATHEMATICAL INDUCTION

If n is a positive integer, let P_n denote the statement

$$(xy)^n = x^n y^n$$

where x and y are real numbers. Thus, P_1 represents the statement $(xy)^1 = x^1 y^1$, P_2 denotes $(xy)^2 = x^2 y^2$, P_3 is $(xy)^3 = x^3 y^3$, etc. It is easy to show that P_1, P_2, and P_3 are *true* statements. However, since the set of positive integers is infinite, it is impossible to check the validity of P_n for every positive integer n. In order to give a proof, the method of mathematical induction is required. This method is based on the following fundamental axiom.

Axiom of Mathematical Induction

Suppose a set S of positive integers has the following two properties:

(i) S contains the integer 1.

(ii) Whenever S contains a positive integer k, S also contains $k + 1$.

Then S contains every positive integer.

If S is a set of positive integers satisfying property (ii), then whenever S contains an arbitrary positive integer k, it must also contain the next positive integer, $k + 1$. If S also satisfies property (i), then S contains 1 and hence by (ii), S contains $1 + 1$, or 2. Applying (ii) again, we see that S contains $2 + 1$, or 3. Once again, S must contain $3 + 1$, or 4. If we continue in this manner, it can be argued that if n is any *specific* positive integer, then n is in S, since we can proceed a step at a time, eventually reaching n Although this argument does not *prove* the axiom, it certainly makes it plausible.

We shall use the preceding axiom to establish the following fundamental principle.

Principle of Mathematical Induction

If with each positive integer n there is associated a statement P_n, then all the statements P_n are true provided the following two conditions hold:

(i) P_1 is true.
(ii) Whenever k is a positive integer such that P_k is true, then P_{k+1} is also true.

Proof

Assume that conditions (i) and (ii) of the Principle hold, and let S denote the set of all positive integers n such that P_n is true. By assumption, P_1 is true, and consequently, 1 is in S. Thus, S satisfies property (i) of the Axiom of Mathematical Induction. Whenever S contains a positive integer k, then by the definition of S, P_k is true and hence from condition (ii) of the Principle, P_{k+1} is also true. This means that S contains $k + 1$. We have shown that whenever S contains a positive integer k, then S also contains $k + 1$. Consequently, property (ii) of the Axiom of Mathematical Induction is true, and hence S contains every positive integer; that is, P_n is true for every positive integer n. ■

There are other variations of the Principle of Mathematical Induction. One is stated later in this section. In most of our work the statement P_n will usually be given in the form of an equation involving the arbitrary positive integer n, as in our illustration $(xy)^n = x^n y^n$.

When applying the Principle of Mathematical Induction always follow these two steps:

Step (i)   Prove that P_1 is true.
Step (ii)   Assume that P_k is true and prove that P_{k+1} is true.

Step (ii) is usually the most confusing for the beginning student. We do not *prove* that P_k is true (except for $k = 1$). Instead, we show that *if* P_k is true, then the statement P_{k+1} is true. That is all that is necessary according to the Principle of Mathematical Induction. The assumption that P_k is true is referred to as the **induction hypothesis**.

Many interesting formulas about positive integers can be established by using mathematical induction, two of which are illustrated in Examples 1 and 2. Others appear in the Exercises.

Example 1 Prove that for every positive integer n, the sum of the first n positive integers is $n(n + 1)/2$.

Solution If n is any positive integer, let P_n denote the statement

$$1 + 2 + 3 + \cdots + n = \frac{n(n + 1)}{2}$$

where, by convention, when $n \le 4$, the left side is adjusted so that there are precisely n terms in the sum. The following are some special cases of P_n: If $n = 2$, then P_2 is

$$1 + 2 = \frac{2(2 + 1)}{2}, \quad \text{or} \quad 3 = 3.$$

If $n = 3$, then P_3 is

$$1 + 2 + 3 = \frac{3(3 + 1)}{2}, \quad \text{or} \quad 6 = 6.$$

If $n = 5$, then P_5 is

$$1 + 2 + 3 + 4 + 5 = \frac{5(5 + 1)}{2}, \quad \text{or} \quad 15 = 15.$$

Although it is instructive to check the validity of P_n for several values of n as we did above it, it unnecessary to do so. We need only apply the two step process outlined prior to this example. Thus, we proceed as follows.

Step (i). If we substitute $n = 1$ in P_n, then, by convention, the left side collapses to 1 and the right side is $\frac{1(1 + 1)}{2}$, which also equals 1. This proves that P_1 is true.

Step (ii). Assume that P_k is true. Thus, the induction hypothesis is

$$1 + 2 + 3 + \cdots + k = \frac{k(k + 1)}{2}.$$

Our goal is to prove that P_{k+1} is true, that is,

$$1 + 2 + 3 + \cdots + (k + 1) = \frac{(k + 1)[(k + 1) + 1]}{2}.$$

By the induction hypothesis we already have a formula for the sum of the first k positive integers. Hence, a formula for the sum of the first $k + 1$ positive integers may be found simply by adding $(k + 1)$ to both sides. Doing so and simplifying, we obtain

$$1 + 2 + 3 + \cdots + k + (k + 1) = \frac{k(k + 1)}{2} + (k + 1)$$

$$= \frac{k(k + 1) + 2(k + 1)}{2}$$

$$= \frac{k^2 + 3k + 2}{2}$$

$$= \frac{(k + 1)(k + 2)}{2}$$

$$= \frac{(k + 1)[(k + 1) + 1]}{2}.$$

We have shown that P_{k+1} is true and, therefore, the proof by mathematical induction is complete. ∎

Example 2   Prove that, for each positive integer n,

$$1^2 + 3^2 + \cdots + (2n - 1)^2 = \frac{n(2n - 1)(2n + 1)}{3}.$$

Solution   For each positive integer n, let P_n denote the given statement. Note that this is a formula for the sum of the squares of the first n odd positive integers. We again follow the two-step procedure used in Example 1.

Step (i).   Substituting 1 for n in P_n, we obtain

$$1^2 = \frac{(1)(2 - 1)(2 + 1)}{3} = \frac{3}{3} = 1$$

which shows that P_1 is true.

Step (ii).   Assume that P_k is true. Thus, the induction hypothesis is

$$1^2 + 3^2 + \cdots + (2k - 1)^2 = \frac{k(2k - 1)(2k + 1)}{3}.$$

We wish to prove that P_{k+1} is true, that is,

$$1^2 + 3^2 + \cdots + [2(k + 1) - 1]^2$$
$$= \frac{(k + 1)[2(k + 1) - 1][2(k + 1) + 1]}{3}.$$

This equation for P_{k+1} simplifies to

$$1^2 + 3^2 + \cdots + (2k + 1)^2 = \frac{(k + 1)(2k + 1)(2k + 3)}{3}.$$

Observe that the second from the last term on the left-hand side of this equation is $(2k - 1)^2$. (Why?) In a manner similar to the solution of Example 1, we may obtain the left side of P_{k+1} by adding $(2k + 1)^2$ to both sides of the equation stated in the induction hypothesis. This gives us

$$1^2 + 3^2 + \cdots + (2k - 1)^2 + (2k + 1)^2$$
$$= \frac{k(2k - 1)(2k + 1)}{3} + (2k + 1)^2.$$

We leave it as an exercise for the reader to show that the right side of the preceding equation may be written in the form of the right side of P_{k+1}. This proves that P_{k+1} is true, and hence P_n is true for every n.   ∎

The Laws of Exponents can be proved by mathematical induction. We shall use the following definition of exponents.

Definition of Exponents

If x is any real number, then

(i) $x^1 = x$.

(ii)   whenever k is a positive integer for which x^k is defined, let $x^{k+1} = x^k \cdot x$.

A definition of this type is called a **recursive definition**. In general, if a concept is defined for every positive integer n in such a way that the case corresponding to $n = 1$ is given, and if it is also stated how any case after the first is obtained from the preceding one, then the definition is a recursive definition. For example, by (i) of the definition we have $x^1 = x$. Next, applying (ii) of the definition we obtain

$$x^2 = x^{1+1} = x^1 \cdot x = x \cdot x.$$

Since x^2 is now defined, we may employ (ii) again (with $k = 2$), obtaining

$$x^3 = x^{2+1} = x^2 \cdot x = (x \cdot x) \cdot x.$$

This defines x^3, and hence (ii) of the definition may be used again to obtain x^4. Thus,

$$x^4 = x^{3+1} = x^3 \cdot x = [(x \cdot x) \cdot x] \cdot x.$$

Observe that this agrees with the formulation of x^n as a product of x by itself n times. It can be shown by mathematical induction that x^n is defined for every positive integer n.

Example 3   If x is a real number, prove that $x^m \cdot x^n = x^{m+n}$ for all positive integers m and n.

Solution   Let m be an arbitrary positive integer. For each positive integer n, let P_n denote the statement

$$x^m \cdot x^n = x^{m+n}.$$

We shall employ the two-step process used in previous examples to prove that P_n is true for every positive integer n.

Step (i).   To show that P_1 is true we may use (i) and (ii) of the Definition of Exponents as follows:

$$\begin{aligned} x^m \cdot x^1 &= x^m \cdot x \\ &= x^{m+1} \end{aligned}$$

which is the statement P_n with $n = 1$. Hence, P_1 is true.

Step (ii).   Assume that P_k is true. Thus, the induction hypothesis is

$$x^m \cdot x^k = x^{m+k}.$$

We wish to prove that P_{k+1} is true, that is,

$$x^m \cdot x^{k+1} = x^{m+(k+1)}.$$

The proof may be arranged as follows, where reasons are stated to the right of each step.

$$\begin{aligned} x^m \cdot x^{k+1} &= x^m \cdot (x^k \cdot x) &&\text{((ii) of the definition of exponents)} \\ &= (x^m \cdot x^k) \cdot x &&\text{(associative law in } \mathbb{R}) \\ &= x^{m+k} \cdot x &&\text{(induction hypothesis)} \\ &= x^{(m+k)+1} &&\text{((ii) of the definition of exponents)} \\ &= x^{m+(k+1)} &&\text{(associative law for integers)}. \end{aligned}$$

This completes the proof.   ∎

Consider a positive integer j and suppose that with each integer $n \geq j$ there is associated a statement P_n. For example, if $j = 6$, then the statements are numbered $P_6, P_7, P_8, \ldots$. The principle of mathematical induction may be extended to cover this situation. Just as before, two steps are used. Specifically, to prove that the statements S_n are true for $n \geq j$, we use the following two steps.

Extended Principle of Mathematical Induction

(i′)   Prove that S_j is true.

(ii′)   Assume that S_k is true for $k \geq j$ and prove that S_{k+1} is true.

Example 4   Let a be a nonzero real number such that $a > -1$. Prove that $(1 + a)^n > 1 + na$ for every integer $n \geq 2$.

Solution   For each positive integer n, let P_n denote the inequality $(1 + a)^n > 1 + na$. Note that P_1 is *false*, since $(1 + a)^1 = 1 + (1)(a)$. However, we can show that P_n is true for $n \geq 2$ by using the Extended Principle with $j = 2$.

Step (i').   We first note that $(1 + a)^2 = 1 + 2a + a^2$. Since $a \neq 0$, we have $a^2 > 0$ and therefore $1 + 2a + a^2 > 1 + 2a$. This gives us $(1 + a)^2 > 1 + 2a$, and hence P_2 is true.

Step (ii').   Assume that P_k is true. Thus, the induction hypothesis is

$$(1 + a)^k > 1 + ka.$$

We wish to show that P_{k+1} is true, that is,

$$(1 + a)^{k+1} > 1 + (k + 1)a.$$

Since $a > -1$, we have $a + 1 > 0$, and hence multiplying both sides of the induction hypothesis by $1 + a$ will not change the inequality sign. Consequently,

$$(1 + a)^k(1 + a) > (1 + ka)(1 + a)$$

which may be rewritten as

$$(1 + a)^{k+1} > 1 + ka + a + ka^2$$

or as

$$(1 + a)^{k+1} > 1 + (k + 1)a + ka^2.$$

Since $ka^2 > 0$, we have

$$1 + (k + 1)a + ka^2 > 1 + (k + 1)a$$

and therefore,

$$(1 + a)^{k+1} > 1 + (k + 1)a.$$

Thus, P_{k+1} is true and the proof is complete.   ∎

EXERCISES 11.1

In each of Exercises 1–18, prove that the given formula is true for every positive integer n.

1 $2 + 4 + 6 + \cdots + 2n = n(n + 1)$

2 $1 + 4 + 7 + \cdots + (3n - 2) = \dfrac{n(3n - 1)}{2}$

3 $1 + 3 + 5 + \cdots + (2n - 1) = n^2$

4 $3 + 9 + 15 + \cdots + (6n - 3) = 3n^2$

5 $2 + 7 + 12 + \cdots + (5n - 3) = \dfrac{n}{2}(5n - 1)$

6 $2 + 6 + 18 + \cdots + 2 \cdot 3^{n-1} = 3^n - 1$

7 $1 + 2 \cdot 2 + 3 \cdot 2^2 + 4 \cdot 2^3 + \cdots + n \cdot 2^{n-1}$
   $= 1 + (n - 1) \cdot 2^n$

8 $(-1)^1 + (-1)^2 + (-1)^3 + \cdots + (-1)^n$
   $= \dfrac{(-1)^n - 1}{2}$

9 $1^2 + 2^2 + 3^2 + \cdots + n^2 = \dfrac{n(n + 1)(2n + 1)}{6}$

10 $1^3 + 2^3 + 3^3 + \cdots + n^3 = \left[\dfrac{n(n + 1)}{2}\right]^2$

11 $\dfrac{1}{1 \cdot 2} + \dfrac{1}{2 \cdot 3} + \dfrac{1}{3 \cdot 4} + \cdots + \dfrac{1}{n(n + 1)} = \dfrac{n}{n + 1}$

12 $\dfrac{1}{1 \cdot 2 \cdot 3} + \dfrac{1}{2 \cdot 3 \cdot 4} + \dfrac{1}{3 \cdot 4 \cdot 5} + \cdots$

$+ \dfrac{1}{n(n+1)(n+2)} = \dfrac{n(n+3)}{4(n+1)(n+2)}$

13 $3 + 3^2 + 3^3 + \cdots + 3^n = \frac{3}{2}(3^n - 1)$

14 $1^3 + 3^3 + 5^3 + \cdots + (2n-1)^3 = n^2(2n^2 - 1)$

15 $n < 2^n$

16 $1 + 2n \leq 3^n$

17 $1 + 2 + 3 + \cdots + n < \frac{1}{8}(2n+1)^2$

18 If $0 < a < b$, then $\left(\dfrac{a}{b}\right)^{n+1} < \left(\dfrac{a}{b}\right)^{n}$.

Prove that the statements in Exercises 19–22 are true for every positive integer n.

19 3 is a factor of $n^3 - n + 3$.

20 2 is a factor of $n^2 + n$.

21 4 is a factor of $5^n - 1$.

22 9 is a factor of $10^{n+1} + 3 \cdot 10^n + 5$.

23 Use mathematical induction to prove that if a is any real number greater than 1, then $a^n > 1$ for every positive integer n.

24 If $a \neq 1$, prove that

$$1 + a + a^2 + \cdots + a^{n-1} = \dfrac{a^n - 1}{a - 1}$$

for every positive integer n.

25 If a and b are real numbers, use mathematical induction to prove that $(ab)^n = a^n b^n$ for every positive integer n.

26 If a is a real number, prove that $(a^m)^n = a^{mn}$ for all positive integers m and n.

27 Use mathematical induction to prove that $a - b$ is a factor of $a^n - b^n$ for every positive integer n. (*Hint*: $a^{k+1} - b^{k+1} = a^k(a - b) + (a^k - b^k)b$.)

28 Prove that $a + b$ is a factor of $a^{2n-1} + b^{2n-1}$ for every positive integer n.

29 If z is a complex number and $\bar{z}$ is its conjugate, prove that $\overline{z^n} = \bar{z}^n$ for every positive integer n.

30 Prove that, for every positive integer n, if $z_1, z_2, \ldots, z_n$ are complex numbers, then $\overline{z_1 z_2 \cdots z_n} = \bar{z}_1 \bar{z}_2 \cdots \bar{z}_n$.

31 Prove that

$$\log(a_1 a_2 \cdots a_n) = \log a_1 + \log a_2 + \cdots + \log a_n$$

for all $n \geq 2$, where each a_i is a positive real number.

32 Prove the **Generalized Distributive Law**

$$a(b_1 + b_2 + \cdots + b_n) = ab_1 + ab_2 + \cdots + ab_n$$

for all $n \geq 2$, where a and each b_i are real numbers.

33 Prove that

$$a + ar + ar^2 + \cdots + ar^{n-1} = \dfrac{a(1 - r^n)}{1 - r}$$

where n is any positive integer and a and r are real numbers with $r \neq 1$.

34 Prove that

$$a + (a + d) + (a + 2d) + \cdots + [a + (n-1)d]$$
$$= (n/2)[2a + (n-1)d]$$

where n is any positive integer and a and d are real numbers.

35 If a and b are real numbers and n is any positive integer, prove that

$$(a - b)(a^{n-1} + a^{n-2}b + \cdots + ab^{n-2} + b^{n-1})$$
$$= a^n - b^n.$$

36 Use mathematical induction to prove De Moivre's Theorem:

$$[r(\cos\theta + i\sin\theta)]^n = r^n[\cos n\theta + i\sin n\theta]$$

for every positive integer n.

37 Prove that, for every positive integer $n \geq 3$, the sum of the interior angles of a polygon of n sides is $(n - 2) \cdot 180°$.

38 Prove, by mathematical induction, that if a finite set S consists of n elements, then the number of subsets of S is 2^n.

Prove that the formulas in Exercises 39 and 40 are true for every positive integer n.

39 $\sin(\theta + n\pi) = (-1)^n \sin\theta$

40 $\cos(\theta + n\pi) = (-1)^n \cos\theta$

11.2 THE BINOMIAL THEOREM

It is often necessary to work with expressions of the form $(a + b)^n$, where a and b are mathematical expressions of some type and n is a large positive integer. There exists a general formula for *expanding* $(a + b)^n$, that is, for expressing it as a sum. The theorem which gives us the formula is called the **Binomial Theorem**. In order to obtain the formula let us first consider cases where n is a small positive integer. By examining the patterns which emerge, we shall make an educated guess about the nature of the general formula. Finally, we shall prove, by means of mathematical induction, that our guess is correct.

If we actually perform the multiplications, the following expansions of $(a + b)^n$ are obtained for the cases $n = 2, 3, 4,$ and 5:

$$(a + b)^2 = a^2 + 2ab + b^2$$
$$(a + b)^3 = a^3 + 3a^2b + 3ab^2 + b^3$$
$$(a + b)^4 = a^4 + 4a^3b + 6a^2b^2 + 4ab^3 + b^4$$
$$(a + b)^5 = a^5 + 5a^4b + 10a^3b^2 + 10a^2b^3 + 5ab^4 + b^5.$$

Let us now make some observations regarding the above expansions of $(a + b)^n$. We see that there are always $n + 1$ terms, the first being a^n and the last b^n. Each intermediate term contains a product of the form $a^i b^j$, where $i + j = n$. Moreover, as we move from one term to the next, the exponent associated with a decreases by 1 whereas the exponent associated with b increases by 1. The pattern for the coefficients is rather interesting. If we start at one end of the expansion and consider successive terms, the coefficients match those obtained by starting at the other end of the expansion and proceeding in the reverse direction.

It requires some ingenuity to find a formula for the general term in the expansion. The second term of the preceding special cases for $(a + b)^n$ is always $na^{n-1}b$. The third term is seen to be

$$\frac{n(n - 1)}{2} a^{n-2}b^2.$$

If there are more than three terms, the fourth term is

$$\frac{n(n - 1)(n - 2)}{3 \cdot 2} a^{n-3}b^3.$$

It appears that, for each term, *if we take the product of the coefficient and the exponent of a and then divide by the number of the term, we obtain the coefficient of the next term* in the expansion. For example, applying this rule to the general fourth term displayed above gives us the following fifth term (if it exists):

$$\frac{n(n - 1)(n - 2)(n - 3)}{4 \cdot 3 \cdot 2} a^{n-4}b^4.$$

The denominators in these terms may be abbreviated by employing the **factorial notation**. If n is any positive integer, then the symbol $n!$ (read "n factorial") is defined by

$$n! = n(n-1)(n-2)\cdots 1$$

where there are n factors on the right side. As special cases we have

$$1! = 1, \quad 2! = 2 \cdot 1, \quad 3! = 3 \cdot 2 \cdot 1 = 6, \quad 4! = 4 \cdot 3 \cdot 2 \cdot 1 = 24, \quad 5! = 5 \cdot 4 \cdot 3 \cdot 2 \cdot 1 = 120.$$

To ensure that certain formulas will be true for all *nonnegative* integers, we define $0! = 1$. The factorial notation may also be defined recursively by writing $1! = 1$ and, for any positive integer k, $(k+1)! = (k+1)k!$.

If we compare the coefficients with the exponents in the general fourth and fifth terms of $(a+b)^n$, we are led to believe that, for a positive integer r, the $(r+1)$st term in the expansion of $(a+b)^n$ is given by

$$\frac{n(n-1)(n-2)\cdots(n-r+1)}{r!}a^{n-r}b^r.$$

Note that, if $r = n$, then the coefficient reduces to $n!/n!$ and we obtain b^n, the last term in the expansion. If $r = n - 1$, then we obtain nab^{n-1}, which is the second from the last term. Hence, we *conjecture* that the following formula is true for every positive integer n and all real or complex numbers a and b.

**The Binomial
Theorem**

$$(a+b)^n = a^n + na^{n-1}b + \frac{n(n-1)}{2!}a^{n-2}b^2 + \cdots$$

$$+ \frac{n(n-1)(n-2)\cdots(n-r+1)}{r!}a^{n-r}b^r$$

$$+ \cdots + nab^{n-1} + b^n$$

Proof

We shall use mathematical induction as follows. For each positive integer n, let P_n denote the statement given in the Binomial Theorem.

Step (i). If $n = 1$, the statement reduces to $(a+b)^1 = a^1 + b^1$. Consequently, P_1 is true.

Step (ii). Assume that P_k is true. Thus, the induction hypothesis is

$$(a+b)^k = a^k + ka^{k-1}b + \frac{k(k-1)}{2!}a^{k-2}b^2 + \cdots$$

$$+ \frac{k(k-1)(k-2)\cdots(k-r+2)}{(r-1)!}a^{k-r+1}b^{r-1}$$

$$+ \frac{k(k-1)(k-2)\cdots(k-r+1)}{r!}a^{k-r}b^r$$

$$+ \cdots + kab^{k-1} + b^k$$

where we have shown both the rth and the $(r+1)$st terms in the expansion.

If we multiply both sides of the last equation by $(a + b)$, we obtain

$$(a + b)^{k+1} = \left[a^{k+1} + ka^kb + \frac{k(k-1)}{2!} a^{k-1}b^2 + \cdots \right.$$

$$+ \frac{k(k-1)\cdots(k-r+1)}{r!} a^{k-r+1}b^r + \cdots + ab^k \Bigg]$$

$$+ \left[a^kb + ka^{k-1}b^2 + \cdots + \frac{k(k-1)\cdots(k-r+2)}{(r-1)!} a^{k-r+1}b^r \right.$$

$$+ \cdots + kab^k + b^{k+1} \Bigg]$$

where the terms in the first pair of brackets result from multiplying the right side of the induction hypothesis by a and the terms in the second pair of brackets result from multiplying by b. Rearranging and combining terms, we have

$$(a + b)^{k+1} = a^{k+1} + (k+1)a^kb + \left[\frac{k(k-1)}{2!} + k \right] a^{k-1}b^2 + \cdots$$

$$+ \left[\frac{k(k-1)\cdots(k-r+1)}{r!} + \frac{k(k-1)\cdots(k-r+2)}{(r-1)!} \right] a^{k-r+1}b^r$$

$$+ \cdots + (1+k)ab^k + b^{k+1}.$$

It is left to the reader to show that if the coefficients are simplified, then we obtain statement P_n with $k + 1$ substituted for n. Thus, P_{k+1} is true and, therefore, P_n holds for every positive integer n. ∎

Example 1   Find the binomial expansion of $(2x + 3y^2)^4$.

Solution   Using the Binomial Theorem with $a = 2x$, $b = 3y^2$, and $n = 4$, we obtain

$$(2x + 3y^2)^4 = (2x)^4 + 4(2x)^3(3y^2) + \frac{4 \cdot 3}{2!} (2x)^2(3y^2)^2$$

$$+ \frac{4 \cdot 3 \cdot 2}{3!} (2x)(3y^2)^3 + \frac{4 \cdot 3 \cdot 2 \cdot 1}{4!} (3y^2)^4.$$

This simplifies to

$$(2x + 3y^2)^4 = 16x^4 + 96x^3y^2 + 216x^2y^4 + 216xy^6 + 81y^8. \quad ∎$$

The symbol $\binom{n}{r}$ is often used to denote the coefficient of $a^{n-r}b^r$ in the Binomial Theorem, that is,

$$\binom{n}{r} = \frac{n(n-1)\cdots(n-r+1)}{r!}.$$

An equivalent formula is

$$\binom{n}{r} = \frac{n!}{r!(n-r)!}$$

where r may be assigned any integral value between 0 and n. If we substitute n for r in the first of these formulas we obtain

$$\binom{n}{n} = \frac{n!}{n!} = 1.$$

It is also convenient to *define*

$$\binom{n}{0} = 1.$$

The formulas for $\binom{n}{r}$ may then be used if r is any integer such that $0 \le r \le n$. Employing this notation gives us the following.

The Binomial Theorem (Alternate Form)

$$(a+b)^n = \binom{n}{0}a^n b^0 + \binom{n}{1}a^{n-1}b + \binom{n}{2}a^{n-2}b^2 + \cdots + \binom{n}{r}a^{n-r}b^r$$
$$+ \cdots + \binom{n}{n-1}a^{n-(n-1)}b^{n-1} + \binom{n}{n}a^0 b^n$$

The numbers $\binom{n}{r}$ are called **binomial coefficients**. As a special case, if $n = 4$,

$$(a+b)^4 = \binom{4}{0}a^4 b^0 + \binom{4}{1}a^3 b + \binom{4}{2}a^2 b^2 + \binom{4}{3}ab^3 + \binom{4}{4}a^0 b^4.$$

It is easy to show that this reduces to the expansion of $(a+b)^4$ given at the beginning of this section.

The next example illustrates the fact that if one of a or b is negative, then the terms of the expansion are alternately positive and negative.

Example 2 Expand $\left(\frac{1}{x} - 2\sqrt{x}\right)^5$.

Solution Letting $a = 1/x$, $b = -2\sqrt{x}$, and $n = 5$ in the Binomial Theorem, we obtain

$$\left(\frac{1}{x} - 2\sqrt{x}\right)^5 = \left(\frac{1}{x}\right)^5 + 5\left(\frac{1}{x}\right)^4(-2\sqrt{x}) + \frac{5\cdot4}{2!}\left(\frac{1}{x}\right)^3(-2\sqrt{x})^2$$
$$+ \frac{5\cdot4\cdot3}{3!}\left(\frac{1}{x}\right)^2(-2\sqrt{x})^3 + \frac{5\cdot4\cdot3\cdot2}{4!}\left(\frac{1}{x}\right)(-2\sqrt{x})^4$$
$$+ \frac{5\cdot4\cdot3\cdot2\cdot1}{5!}(-2\sqrt{x})^5.$$

This simplifies to

$$\left(\frac{1}{x} - 2\sqrt{x}\right)^5 = \frac{1}{x^5} - \frac{10}{x^{7/2}} + \frac{40}{x^2} - \frac{80}{x^{1/2}} + 80x - 32x^{5/2}. \qquad \blacksquare$$

For certain problems it is only required to find a specific term in the expansion of $(a + b)^n$. To work such problems we first find the exponent r that is to be assigned to b. Notice that, by the Binomial Theorem, *the exponent of b is always one less than the number of the term.* Once r is found, the exponent of a is $n - r$. Referring to the form of the $(r + 1)$st term, we see that the coefficient of the term involving $a^{n-r}b^r$ is of the form $p/r!$, *where p is the product of r factors and where the first factor is n and each factor is one less than the preceding factor.*

Example 3 Find the fifth term in the expansion of $(4x^2 + 3/y)^{13}$.

Solution   We let $a = 4x^2$ and $b = 3/y$. The exponent of b in the fifth term is 4 and hence the exponent of a is 9. From the discussion of the preceding paragraph we obtain

$$\frac{13 \cdot 12 \cdot 11 \cdot 10}{4!} (4x^2)^9 (3/y)^4. \qquad \blacksquare$$

Finally, there is an interesting triangular array of numbers, called **Pascal's Triangle**, which can be used to obtain the binomial coefficients. The numbers are arranged as follows.

$$
\begin{array}{ccccccccccccc}
 & & & & & & 1 & & & & & & \\
 & & & & & 1 & & 1 & & & & & \\
 & & & & 1 & & 2 & & 1 & & & & \\
 & & & 1 & & 3 & & 3 & & 1 & & & \\
 & & 1 & & 4 & & 6 & & 4 & & 1 & & \\
 & 1 & & 5 & & 10 & & 10 & & 5 & & 1 & \\
1 & & 6 & & 15 & & 20 & & 15 & & 6 & & 1 \\
\end{array}
$$

The numbers in the second row are the coefficients in the expansion of $(a + b)^1$; those in the third row are the coefficients determined by $(a + b)^2$; those in the fourth row are obtained from $(a + b)^3$, and so on. Each number in the array which is different from 1 can be found by adding the two numbers in the previous row which appear above and immediately to the left and right of the number.

EXERCISES 11.2

In each of Exercises 1–12, expand and simplify the given expression.

1 $(a + b)^6$

2 $(a + b)^7$

3 $(a - b)^8$

4 $(a - b)^9$

5 $(3x - 5y)^4$

6 $(2t - s)^5$

7 $(u^2 + 4v)^5$

8 $(\frac{1}{2}c + d^3)^4$

9 $(r^{-2} - 2r)^6$

10 $(x^{1/2} - y^{-1/2})^6$

11 $(1 + x)^{10}$

12 $(1 - x)^{10}$

13 Find the first four terms in the binomial expansion of $(3c^{2/5} + c^{4/5})^{25}$.

14 Find the first three terms and the last three terms in the binomial expansion of $(x^3 + 5x^{-2})^{20}$.

15 Find the last two terms in the expansion of $(4b^{-1} - 3b)^{15}$.

16 Find the last three terms in the expansion of $(s - 2t^3)^{12}$.

Solve Exercises 17–28 without expanding completely.

17 Find the fifth term in the expansion of $(3a^2 + \sqrt{b})^9$.

18 Find the sixth term in the expansion of $\left(\frac{2}{c} + \frac{c^2}{3}\right)^7$.

19 Find the seventh term in the expansion of $(\frac{1}{2}u - 2v)^{10}$.

20 Find the fourth term in the expansion of $(2x^3 - y^2)^6$.

21 Find the middle term in the expansion of $(x^{1/3} + y^{1/3})^{12}$.

22 Find the two middle terms in the expansion of $(rs + t)^7$.

23 Find the term which does not contain x in the expansion of $\left(6x - \frac{1}{2x}\right)^{10}$.

24 Find the term involving x^8 in the expansion of $(y + 3x^2)^6$.

25 Find the term containing y^6 in the expansion of $(x - 2y^3)^4$.

26 Find the term containing b^9 in the expansion of $(5a + 2b^3)^4$.

27 Find the term containing c^3 in the expansion of $(\sqrt{c} + \sqrt{d})^{10}$.

28 In the expansion of $(xy - 2y^{-3})^8$ find the term which does not contain y.

29 Use the first four terms in the binomial expansion of $(1 + 0.02)^{10}$ to approximate $(1.02)^{10}$. Also approximate $(1.02)^{10}$ by using logarithms and then compare answers.

30 Use the first four terms in the binomial expansion of $(1 - 0.01)^4$ to approximate $(0.99)^4$. Compare with the answer obtained by approximating $(0.99)^4$ through the use of logarithms.

11.3 INFINITE SEQUENCES AND SUMMATION NOTATION

A function f from a set X to a set Y is a correspondence that associates with each element x of X a unique element $f(x)$ of Y. Up to now the domain X has usually been an interval of real numbers. In this section we shall consider a different class of functions.

Definition

An **infinite sequence** is a function whose domain is the set of positive integers.

For convenience we sometimes refer to infinite sequences merely as *sequences*. In this book the range of an infinite sequence will be a set of real numbers.

If f is an infinite sequence, then to each positive integer n there corresponds a real number $f(n)$. These numbers in the range of f may be represented by writing

$$f(1), f(2), f(3), \ldots, f(n), \ldots$$

where the dots at the end indicate that the sequence does not terminate. The number $f(1)$ is called the **first term** of the sequence, $f(2)$ the **second term** and, in general, $f(n)$ the **nth term** of the sequence. It is customary to use a subscript notation instead of the functional notation and write these numbers as

$$a_1, a_2, a_3, \ldots, a_n, \ldots$$

where it is understood that for each positive integer n, the symbol a_n denotes the real number $f(n)$. In this way we obtain an infinite collection of real numbers which is *ordered* in the sense that there is a first number, a second number, a forty-fifth number, and so on. Although sequences are functions, an ordered collection of the type displayed above will also be referred to as an infinite sequence. If we wish to convert the collection to a function f we let $f(n) = a_n$ for all positive integers n.

From the definition of equality of functions we see that a sequence

$$a_1, a_2, a_3, \ldots, a_n, \ldots$$

is **equal** to a sequence $$b_1, b_2, b_3, \ldots, b_n, \ldots$$

if and only if $a_i = b_i$ for every positive integer i. Infinite sequences are often defined by stating a formula for the nth term, as in the following example.

Example 1   List the first four terms and the tenth term of the sequence whose nth term is as follows.

(a) $a_n = \dfrac{n}{n+1}$ (b) $a_n = 2 + (0.1)^n$

(c) $a_n = (-1)^{n+1}\dfrac{n^2}{3n-1}$ (d) $a_n = 4$

Solutions   To find the first four terms we substitute, successively, $n = 1, 2, 3,$ and 4 in the formula for a_n. The tenth term is found by substituting 10 for n. Doing this and simplifying gives us the following:

	First four terms	Tenth term
(a)	$\frac{1}{2}, \frac{2}{3}, \frac{3}{4}, \frac{4}{5}$	$\frac{10}{11}$
(b)	2.1, 2.01, 2.001, 2.0001	2.0000000001
(c)	$\frac{1}{2}, -\frac{4}{5}, \frac{9}{8}, -\frac{16}{11}$	$-\frac{100}{29}$
(d)	4, 4, 4, 4	4

It is not essential that a formula for a_n be given. Indeed, sometimes that is impossible, as illustrated in the next example.

Example 2   List the first seven terms of the sequence whose nth term a_n is the nth positive prime number.

Solution   The first seven primes are 2, 3, 5, 7, 11, 13, and 17. Hence, $a_1 = 2, a_2 = 3$, $a_3 = 5$, $a_4 = 7$, $a_5 = 11$, $a_6 = 13$, and $a_7 = 17$. No one has found a formula which yields the nth prime number. Our sequence is perfectly legitimate, however, since a_n is uniquely determined for every positive integer n.   ∎

An infinite sequence is sometimes defined by stating a recursive definition for the terms, as in the following example.

Example 3   Find the first four terms and the nth term of the infinite sequence defined by

$$a_1 = 3 \quad \text{and} \quad a_{k+1} = 2a_k, \quad \text{for} \quad k \geq 1.$$

Solution   The sequence is defined recursively since the first term is given and, moreover, whenever a term a_k of the sequence is known, then the next term a_{k+1} can be found. Thus,

$$a_1 = 3$$
$$a_2 = 2a_1 = 2 \cdot 3 = 6$$
$$a_3 = 2a_2 = 2 \cdot 2 \cdot 3 = 2^2 \cdot 3 = 12$$
$$a_4 = 2a_3 = 2 \cdot 2 \cdot 2 \cdot 3 = 2^3 \cdot 3 = 24.$$

We have written the terms as products, so as to gain some insight into the nature of the nth term. Continuing, we obtain $a_5 = 2^4 \cdot 3$ and $a_6 = 2^5 \cdot 3$; and it appears that

$$a_n = 2^{n-1} \cdot 3$$

for every positive integer n. We shall prove that this guess is correct by means of mathematical induction. If we let P_n denote the statement $a_n = 2^{n-1} \cdot 3$, then P_1 is true since $a_1 = 2^0 \cdot 3 = 3$. Next, *assume* that P_k is true, that is, $a_k = 2^{k-1} \cdot 3$. We then have

$$
\begin{aligned}
a_{k+1} &= 2a_k & &\text{(definition of } a_{k+1}) \\
&= 2 \cdot 2^{k-1} \cdot 3 & &\text{(induction hypothesis)} \\
&= 2^k \cdot 3 & &\text{(a law of exponents)} \\
&= 2^{(k+1)-1} \cdot 3 & &\text{(Why?)}
\end{aligned}
$$

which shows that P_{k+1} is true. Hence, $a_n = 2^{n-1} \cdot 3$ for every positive integer n.   ∎

It is important to observe that if only the first few terms of an infinite sequence are known, then it is impossible to predict additional terms. For example, if we were given 3, 6, 9, . . . and asked to find the fourth term, we could not proceed without further information. The infinite sequence with nth term

$$a_n = 3n + (1 - n)^3(2 - n)^2(3 - n)$$

has for its first four terms 3, 6, 9, and 120. It is possible to describe sequences where the first three terms are 3, 6, and 9 and the fourth term is *any* given number. This shows that when we work with infinite sequences it is essential to have specific information about the nth term or to know a general scheme for obtaining each term from the preceding one.

It is often desirable to find the sum of many terms of an infinite sequence. For ease in expressing such sums we use the **summation notation** described below.

Given an infinite sequence

$$a_1, a_2, a_3, \ldots, a_n, \ldots$$

the symbol $\sum_{i=1}^{m} a_i$ represents the sum of the first m terms, that is,

Summation Notation

$$\sum_{i=1}^{m} a_i = a_1 + a_2 + a_3 + \cdots + a_m.$$

The Greek capital letter $\sum$ (sigma) indicates a sum and the symbol a_i represents the ith term. The letter i is called the **index of summation** or the **summation variable**, and the numbers 1 and m indicate the extreme values of the summation variable.

Example 4   Find $\sum_{i=1}^{4} i^2(i - 3)$.

Solution   In this case, $a_i = i^2(i - 3)$. To find the indicated sum we merely substitute, in succession, the integers 1, 2, 3, and 4 for i and add the resulting terms. Thus,

$$\sum_{i=1}^{4} i^2(i - 3) = 1^2(1 - 3) + 2^2(2 - 3) + 3^2(3 - 3) + 4^2(4 - 3)$$

$$= (-2) + (-4) + 0 + 16 = 10. \qquad \blacksquare$$

The letter used for the summation variable is arbitrary. To illustrate, if we use j for the summation variable, then

$$\sum_{j=1}^{m} a_j = a_1 + a_2 + a_3 + \cdots + a_m$$

which is the same as $\sum_{i=1}^{m} a_i$. Other symbols can be used similarly. As a numerical example, the sum in Example 4 can be written as

$$\sum_{k=1}^{4} k^2(k - 3).$$

If n is a positive integer, then the sum of the first n terms of an infinite sequence will be denoted by S_n. For example, given $a_1, a_2, a_3, \ldots, a_n, \ldots$,

$$S_1 = a_1$$
$$S_2 = a_1 + a_2$$
$$S_3 = a_1 + a_2 + a_3$$
$$S_4 = a_1 + a_2 + a_3 + a_4$$

and, in general,

$$S_n = \sum_{i=1}^{n} a_i = a_1 + a_2 + \cdots + a_n.$$

The number S_n is called the **nth partial sum** of the sequence $a_1, a_2, a_3, \ldots, a_n, \ldots$, and the infinite sequence

$$S_1, S_2, S_3, \ldots, S_n, \ldots$$

is called a **sequence of partial sums**. Sequences of partial sums are very important in calculus, where the concept of *infinite series* is introduced. We shall discuss some special types of infinite series in Section 11.5.

Example 5 Find the first four terms and the nth term of the sequence of partial sums associated with the sequence $1, 2, 3, \ldots, n, \ldots$ of positive integers.

Solution The first four terms of the partial sum sequence are

$$S_1 = 1$$
$$S_2 = 1 + 2 = 3$$
$$S_3 = 1 + 2 + 3 = 6$$
$$S_4 = 1 + 2 + 3 + 4 = 10.$$

From Example 1 of Section 11.1 we see that

$$S_n = 1 + 2 + 3 + \cdots + n = \frac{n(n+1)}{2}. \qquad \blacksquare$$

If a_i is the same for all positive integers i, say $a_i = c$, where c is a real number, then

$$\sum_{i=1}^{n} a_i = a_1 + a_2 + a_3 + \cdots + a_n$$
$$= c + c + c + \cdots + c$$
$$= nc$$

that is,

$$\sum_{i=1}^{n} c = nc.$$

This formula can also be established by mathematical induction (see Exercise 60).

The domain of the summation variable does not have to begin at 1. For example, the following is self-explanatory:

$$\sum_{i=4}^{8} a_i = a_4 + a_5 + a_6 + a_7 + a_8.$$

As another variation, if the first term of an infinite sequence is a_0, as in

$$a_0, a_1, a_2, \ldots, a_n, \ldots$$

then sums of the form

$$\sum_{i=0}^{n} a_i = a_0 + a_1 + a_2 + \cdots + a_n$$

may be considered. Note that this is the sum of the first $n + 1$ terms of the given sequence.

Example 6   Find $\displaystyle\sum_{i=0}^{3} \frac{2^i}{(i + 1)}$.

Solution

$$\sum_{i=0}^{3} \frac{2^i}{(i + 1)} = \frac{2^0}{(0 + 1)} + \frac{2^1}{(1 + 1)} + \frac{2^2}{(2 + 1)} + \frac{2^3}{(3 + 1)}$$

$$= 1 + 1 + \frac{4}{3} + 2$$

$$= \frac{16}{3}.$$ ∎

Summation notation can be used to denote polynomials compactly. For example, in place of

$$f(x) = a_0 + a_1 x + a_2 x^2 + \cdots + a_n x^n$$

we may write

$$f(x) = \sum_{i=0}^{n} a_i x^i.$$

As another illustration, the rather cumbersome formula for the Binomial Theorem can be written as

$$(a + b)^n = \sum_{r=0}^{n} \binom{n}{r} a^{n-r} b^r.$$

The following theorem concerning sums has many uses in advanced courses in mathematics.

Theorem on Sums	If $a_1, a_2, \ldots, a_n, \ldots$ and $b_1, b_2, \ldots, b_n, \ldots$ are infinite sequences, then for every positive integer n,

$$\text{(i)} \quad \sum_{i=1}^{n} (a_i + b_i) = \sum_{i=1}^{n} a_i + \sum_{i=1}^{n} b_i$$

$$\text{(ii)} \quad \sum_{i=1}^{n} (a_i - b_i) = \sum_{i=1}^{n} a_i - \sum_{i=1}^{n} b_i$$

$$\text{(iii)} \quad \sum_{i=1}^{n} ca_i = c\left(\sum_{i=1}^{n} a_i\right), \quad \text{for every number } c.$$

Proof

Although the theorem can be proved by mathematical induction, we shall use an argument that makes the truth of the formulas transparent. We begin as follows:

$$\sum_{i=1}^{n} (a_i + b_i) = (a_1 + b_1) + (a_2 + b_2) + (a_3 + b_3) + \cdots + (a_n + b_n).$$

Using the Commutative and Associative Properties many times, we may rearrange the terms on the right to produce

$$\sum_{i=1}^{n} (a_i + b_i) = (a_1 + a_2 + a_3 + \cdots + a_n) + (b_1 + b_2 + b_3 + \cdots + b_n).$$

Expressing the right side in summation notation gives us formula (i).

For formula (iii) we have

$$\sum_{i=1}^{n} (ca_i) = ca_1 + ca_2 + ca_3 + \cdots + ca_n$$

$$= c(a_1 + a_2 + a_3 + \cdots + a_n)$$

$$= c\left(\sum_{i=1}^{n} a_i\right).$$

The proof of (ii) is left as an exercise. ■

EXERCISES 11.3

In Exercises 1–20, find the first five terms and the eighth term of the sequence which has the given nth term.

1 $a_n = 12 - 3n$

2 $a_n = \dfrac{3}{5n - 2}$

3 $a_n = \dfrac{3n - 2}{n^2 + 1}$

4 $a_n = 10 + \dfrac{1}{n}$

5 $a_n = 9$

6 $a_n = (n - 1)(n - 2)(n - 3)$

7 $a_n = 2 + (-0.1)^n$

8 $a_n = 4 + (0.1)^n$

9 $a_n = (-1)^{n-1}\dfrac{n + 7}{2n}$

10 $a_n = (-1)^n\dfrac{6 - 2n}{\sqrt{n + 1}}$

11 $a_n = 1 + (-1)^{n+1}$

12 $a_n = (-1)^{n+1} + (0.1)^{n-1}$

13 $a_n = \dfrac{2^n}{n^2 + 2}$ **14** $a_n = \sqrt{2}$

15 a_n is the square of the nth prime.

16 a_n is the nth prime greater than 50.

17 a_n is the largest prime less than $10n$.

18 a_n is the smallest prime greater than $15n$.

19 a_n is the number of decimal places in $(0.1)^n$.

20 a_n is the number of positive integers less than n^3.

Find the first five terms of the infinite sequences defined recursively in Exercises 21–28.

21 $a_1 = 2, \quad a_{k+1} = 3a_k - 5$

22 $a_1 = 5, \quad a_{k+1} = 7 - 2a_k$

23 $a_1 = -3, \quad a_{k+1} = a_k^2$

24 $a_1 = 128, \quad a_{k+1} = a_k/4$

25 $a_1 = 5, \quad a_{k+1} = ka_k$

26 $a_1 = 3, \quad a_{k+1} = 1/a_k$

27 $a_1 = 2, \quad a_{k+1} = (a_k)^k$

28 $a_1 = 2, \quad a_{k+1} = (a_k)^{1/k}$

29 A test question lists the first four terms of a sequence as 2, 4, 6, and 8 and asks for the fifth term. Show that the fifth term can be any real number a by finding the nth term of a sequence which has for its first five terms, 2, 4, 6, 8, and a.

30 The number of bacteria in a certain culture doubles every day. If the initial number of bacteria is 500, how many are present after one day? Two days? Three days? Find a formula for the number of bacteria present after n days.

Find the number given in each of Exercises 31–46.

31 $\displaystyle\sum_{k=1}^{5} (2k - 7)$ **32** $\displaystyle\sum_{k=1}^{6} (10 - 3k)$

33 $\displaystyle\sum_{k=1}^{4} (k^2 - 5)$ **34** $\displaystyle\sum_{k=1}^{10} [1 + (-1)^k]$

35 $\displaystyle\sum_{k=0}^{5} k(k - 2)$ **36** $\displaystyle\sum_{k=0}^{4} (k - 1)(k - 3)$

37 $\displaystyle\sum_{k=3}^{6} \frac{k - 5}{k - 1}$ **38** $\displaystyle\sum_{k=1}^{6} \frac{3}{k + 1}$

39 $\displaystyle\sum_{k=1}^{5} (-3)^{k-1}$ **40** $\displaystyle\sum_{k=0}^{4} 3(2^k)$

41 $\displaystyle\sum_{k=1}^{100} 100$ **42** $\displaystyle\sum_{k=1}^{1000} 5$

43 $\displaystyle\sum_{k=1}^{n} (k^2 + 3k + 5)$ (*Hint*: Use the Theorem on Sums to write the sum as $\sum_{k=1}^{n} k^2 + 3\sum_{k=1}^{n} k + \sum_{k=1}^{n} 5$. Next employ Exercise 9 and Example 1 of Section 11.1, together with the formula for $\sum_{k=1}^{n} c$.

44 $\displaystyle\sum_{k=1}^{n} (3k^2 - 2k + 1)$ **45** $\displaystyle\sum_{k=1}^{n} (2k - 3)^2$

46 $\displaystyle\sum_{k=1}^{n} (k^3 + 2k^2 - k + 4)$ (*Hint*: See Exercise 10 of Section 11.1).

Express the sums in Exercises 47–56 in terms of summation notation.

47 $1 + 5 + 9 + 13 + 17$

48 $2 + 5 + 8 + 11 + 14$

49 $\dfrac{1}{2} + \dfrac{2}{5} + \dfrac{3}{8} + \dfrac{4}{11}$

50 $\dfrac{1}{4} + \dfrac{2}{9} + \dfrac{3}{14} + \dfrac{4}{19}$

51 $1 - \dfrac{x^2}{2} + \dfrac{x^4}{4} - \dfrac{x^6}{6} + \cdots + (-1)^{n+1} \dfrac{x^{2n}}{2n}$

52 $2 - 4 + 8 - 16 + 32 - 64$

53 $1 - \dfrac{1}{2} + \dfrac{1}{3} - \dfrac{1}{4} + \dfrac{1}{5} - \dfrac{1}{6} + \dfrac{1}{7}$

54 $1 + x + \dfrac{x^2}{2} + \dfrac{x^3}{3} + \cdots + \dfrac{x^n}{n}$

55 $\dfrac{1}{1 \cdot 2} + \dfrac{1}{2 \cdot 3} + \dfrac{1}{3 \cdot 4} + \cdots + \dfrac{1}{99 \cdot 100}$

56 $\dfrac{1}{1 \cdot 2 \cdot 3} + \dfrac{1}{2 \cdot 3 \cdot 4} + \dfrac{1}{3 \cdot 4 \cdot 5} + \cdots + \dfrac{1}{98 \cdot 99 \cdot 100}$

57 Prove (ii) of the Theorem on Sums.

58 Extend (i) of the Theorem on Sums to $\sum_{i=1}^{n} (a_i + b_i + c_i)$.

59 Prove the Theorem on Sums by mathematical induction.

60 Prove $\sum_{i=1}^{n} c = nc$ by mathematical induction.

11.4 ARITHMETIC SEQUENCES

In this section and the next we shall concentrate on two special types of sequences. The first type may be defined as follows.

Definition

> An **arithmetic sequence** is a sequence such that successive terms differ by the same real number.

Arithmetic sequences are also called **arithmetic progressions**. By definition, a sequence

$$a_1, a_2, a_3, \ldots, a_n, \ldots$$

is arithmetic if and only if there is a real number d such that

$$a_{k+1} - a_k = d$$

for every positive integer k. The number d is called the **common difference** associated with the arithmetic sequence.

Example 1   Show that the sequence

$$1, 4, 7, 10, \ldots, 3n - 2, \ldots$$

is arithmetic and find the common difference.

Solution   Since $a_n = 3n - 2$, it follows that for every positive integer k,

$$a_{k+1} - a_k = [3(k + 1) - 2] - (3k - 2)$$
$$= 3k + 3 - 2 - 3k + 2 = 3.$$

Hence, by definition, the given sequence is arithmetic with common difference 3.
∎

Given an arithmetic sequence, we know that

$$a_{k+1} = a_k + d$$

for every positive integer k. This provides a recursive formula for obtaining successive terms. Beginning with any real number a_1, we can generate an arithmetic sequence with common difference d simply by adding d to a_1, then to $a_1 + d$, and so on, obtaining

$$a_1, \quad a_1 + d, \quad a_1 + 2d, \quad a_1 + 3d, \quad a_1 + 4d, \ldots.$$

It is evident that the nth term a_n of this sequence is given by

***n*th Term of an Arithmetic Sequence**

$$a_n = a_1 + (n - 1)d.$$

This formula can be proved by mathematical induction.

Example 2   Find the fifteenth term of the arithmetic sequence whose first three terms are 20, 16.5, and 13.

Solution   The common difference is -3.5. (Why?) Substituting $a_1 = 20, d = -3.5$, and $n = 15$ in the formula $a_n = a_1 + (n-1)d$,

$$a_{15} = 20 + (15-1)(-3.5) = 20 - 49 = -29. \qquad \blacksquare$$

Example 3   If the fourth term of an arithmetic sequence is 5 and the ninth term is 20, find the sixth term.

Solution   Substituting $n = 4$ and $n = 9$ in the formula $a_n = a_1 + (n-1)d$, and using the fact that $a_4 = 5$ and $a_9 = 20$, we obtain the following system of linear equations in the variables a_1 and d:

$$\begin{cases} 5 = a_1 + (4-1)d \\ 20 = a_1 + (9-1)d. \end{cases}$$

This system has the unique solution $d = 3$ and $a_1 = -4$. (Verify!) Substitution in the formula $a_n = a_1 + (n-1)d$ gives us

$$a_6 = (-4) + (6-1)(3) = 11. \qquad \blacksquare$$

Theorem

If $a_1, a_2, \ldots, a_n, \ldots$ is an arithmetic sequence with common difference d, then the nth partial sum S_n is given by both

$$S_n = \frac{n}{2}\left[2a_1 + (n-1)d\right] \quad \text{and} \quad S_n = \frac{n}{2}(a_1 + a_n).$$

Proof

We may write

$$S_n = a_1 + a_2 + a_3 + \cdots + a_n$$
$$= a_1 + (a_1 + d) + (a_1 + 2d) + \cdots + [a_1 + (n-1)d].$$

Employing the Commutative and Associative Properties many times we obtain

$$S_n = (a_1 + a_1 + a_1 + \cdots + a_1) + [d + 2d + \cdots + (n-1)d]$$

where a_1 appears n times within the first parentheses. It follows that

$$S_n = na_1 + d[1 + 2 + \cdots + (n-1)].$$

The expression within brackets is the sum of the first $n-1$ positive integers. From Example 1 of Section 11.1 (with $n-1$ in place of n), that sum equals

$$\frac{(n-1)n}{2}.$$

Substituting in the last equation for S_n and factoring gives us

$$S_n = na_1 + d\frac{n(n-1)}{2}$$

$$= \frac{n}{2}\left[2a_1 + (n-1)d\right].$$

Since $a_n = a_1 + (n-1)d$, we get

$$S_n = \frac{n}{2}(a_1 + a_n).$$ ∎

Example 4 Find the sum of all the even integers from 2 through 100.

Solution The given problem is equivalent to finding the sum of the first fifty terms of the arithmetic sequence $2, 4, 6, \ldots, 2n, \ldots$. Substituting $n = 50$, $a_1 = 2$, and $a_{50} = 100$ in the second formula of the preceding theorem,

$$S_{50} = \frac{50}{2}(2 + 100) = 2550.$$

In order to check our work we use the first formula given in the theorem. Thus,

$$S_{50} = \frac{50}{2}[2 \cdot 2 + (50 - 1)2] = 25[4 + 98] = 2550.$$ ∎

The **arithmetic mean** of two numbers a and b is defined as $(a + b)/2$. This is also called the **average** of a and b. Note that

$$a, \quad \frac{a+b}{2}, \quad b$$

is an arithmetic sequence. This concept may be generalized as follows. If $c_1, c_2, \ldots, c_k$ are real numbers such that

$$a, c_1, c_2, \ldots, c_k, b$$

is an arithmetic sequence, then $c_1, c_2, \ldots, c_k$ are called the **k arithmetic means** of the numbers a and b. The process of determining such numbers is referred to as *inserting k arithmetic means between a and b.*

Example 5 Insert three arithmetic means between 2 and 9.

Solution We wish to find three numbers c_1, c_2, and c_3 such that $2, c_1, c_2, c_3, 9$ is an arithmetic sequence. The common difference d may be found by using the formula $a_n = a_1 + (n-1)d$ with $n = 5$, $a_5 = 9$, and $a_1 = 2$. This gives us

$$9 = 2 + (5-1)d, \quad \text{or} \quad d = \tfrac{7}{4}.$$

The three arithmetic means are

$$c_1 = a_1 + d = 2 + \tfrac{7}{4} = \tfrac{15}{4}$$
$$c_2 = c_1 + d = \tfrac{15}{4} + \tfrac{7}{4} = \tfrac{22}{4} = \tfrac{11}{2}$$
$$c_3 = c_2 + d = \tfrac{11}{2} + \tfrac{7}{4} = \tfrac{29}{4}. \qquad \blacksquare$$

EXERCISES 11.4

In each of Exercises 1–8 find the fifth term, the tenth term, and the nth term of the given arithmetic sequence.

1 $2, 6, 10, 14, \ldots$

2 $16, 13, 10, 7, \ldots$

3 $3, 2.7, 2.4, 2.1, \ldots$

4 $-6, -4.5, -3, -1.5, \ldots$

5 $-7, -3.9, -0.8, 2.3, \ldots$

6 $x - 8, x - 3, x + 2, x + 7, \ldots$

7 $\log 3, \log 9, \log 27, \log 81, \ldots$

8 $\log 1000, \log 100, \log 10, 0, \ldots$

9 Find the twelfth term of the arithmetic sequence whose first two terms are 9.1 and 7.5.

10 Find the eleventh term of the arithmetic sequence whose first two terms are $2 + \sqrt{2}$ and 3.

11 The sixth and seventh terms of an arithmetic sequence are 2.7 and 5.2. Find the first term.

12 Given an arithmetic sequence with $a_3 = 7$ and $a_{20} = 43$, find a_{15}.

In Each of Exercises 13–16 find the sum S_n of the arithmetic sequence which satisfies the given conditions.

13 $a_1 = 40, \quad d = -3, \quad n = 30$

14 $a_1 = 5 \quad d = 0.1, \quad n = 40$

15 $a_1 = -9, \quad a_{10} = 15, \quad n = 10$

16 $a_7 = 7/3, \quad d = -2/3, \quad n = 15$

Find the sums in Exercises 17–20.

17 $\displaystyle\sum_{k=1}^{20} (3k - 5)$

18 $\displaystyle\sum_{k=1}^{12} (7 - 4k)$

19 $\displaystyle\sum_{k=1}^{18} \left(\tfrac{1}{2}k + 7\right)$

20 $\displaystyle\sum_{k=1}^{10} \left(\tfrac{1}{4}k + 3\right)$

21 How many integers between 32 and 395 are divisible by 6? Find their sum.

22 How many negative integers greater than -500 are divisible by 33? Find their sum.

23 How many terms are in an arithmetic sequence with first term -2, common difference $\tfrac{1}{4}$, and sum 21?

24 How many terms are in an arithmetic sequence with sixth term -3, common difference 0.2, and sum -33?

25 Insert five arithmetic means between 2 and 10.

26 Insert three arithmetic means between 3 and -5.

27 A pile of logs has 24 logs in the first layer, 23 in the second, 22 in the third, and so on. The last layer contains 10 logs. Find the total number of logs in the pile.

28 A seating section in a certain athletic stadium has 30 seats in the first row, 32 seats in the second, 34 in the third, and so on, until the tenth row is reached, after which there are 10 more rows, each containing 50 seats. Find the total number of seats in the section.

29 A man wishes to construct a ladder with nine rungs which diminish uniformly from 24 inches at the base to 18 inches at the top. Determine the lengths of the seven intermediate rungs.

30 A boy on a bicycle coasts down a hill, covering 4 feet the first second and in each succeeding second 5 feet more than in the preceding second. If he reaches the bottom of the hill in 11 seconds, find the total distance traveled.

501 11.5 GEOMETRIC SEQUENCES

11.5 GEOMETRIC SEQUENCES

Another important type of infinite sequence is defined as follows.

Definition

> A sequence
> $$a_1, a_2, \ldots, a_n, \ldots$$
> is a **geometric sequence** if and only if there is a real number $r \neq 0$ such that
> $$\frac{a_{k+1}}{a_k} = r$$
> for every positive integer k.

Geometric sequences are also called **geometric progressions**. The number r in the definition is called the **common ratio** associated with the geometric sequence. Thus $a_1, a_2, \ldots, a_n, \ldots$ is geometric (with common ratio r) if and only if

$$a_{k+1} = a_k r$$

for every positive integer k. As with arithmetic sequences, this provides a recursive method for obtaining terms. Beginning with any nonzero real number a_1, we *multiply* by the number r successively, obtaining

$$a_1, a_1 r, a_1 r^2, a_1 r^3, \ldots$$

It appears that the nth term a_n of this sequence is given by

nth Term of a Geometric Sequence

$$a_n = a_1 r^{n-1}.$$

This can be proved by mathematical induction.

Example 1 Find the first five terms and the tenth term of the geometric sequence having first term 3 and common ratio $-1/2$.

Solution If we let $a_1 = 3$ and $r = -1/2$, then the first five terms are,

$$3, \quad -\frac{3}{2}, \quad \frac{3}{4}, \quad -\frac{3}{8}, \quad \frac{3}{16}.$$

Using the formula $a_n = a_1 r^{n-1}$ with $n = 10$, we obtain

$$a_{10} = 3\left(-\frac{1}{2}\right)^9 = -\frac{3}{512}.$$ ∎

Example 2 If the third term of a geometric sequence is 5 and the sixth term is -40, find the eighth term.

Solution   We are given $a_3 = 5$ and $a_6 = -40$. Substituting $n = 3$ and $n = 6$ in the formula $a_n = a_1 r^{n-1}$ leads to the following system of equations:

$$\begin{cases} 5 = a_1 r^2 \\ -40 = a_1 r^5. \end{cases}$$

Since $r \neq 0$, the first equation is equivalent to $a_1 = 5/r^2$. Substituting for a_1 in the second equation we obtain

$$-40 = \left(\frac{5}{r^2}\right) \cdot r^5 = 5r^3.$$

Hence, $r^3 = -8$ and $r = -2$. If we now substitute -2 for r in the equation $5 = a_1 r^2$, we obtain $a_1 = 5/4$. Finally, using $a_n = a_1 r^{n-1}$ with $n = 8$ gives us

$$a_8 = \left(\frac{5}{4}\right)(-2)^7 = -160. \qquad \blacksquare$$

Let us find a formula for S_n, the nth partial sum of a geometric sequence. If the first term is a_1 and the common ratio is r, then

$$S_n = a_1 + a_1 r + a_1 r^2 + \cdots + a_1 r^{n-2} + a_1 r^{n-1}.$$

If $r = 1$, this gives us $S_n = na_1$. Next suppose that $r \neq 1$. Multiplying both sides of the general formula for S_n by r, we obtain

$$rS_n = a_1 r + a_1 r^2 + a_1 r^3 + \cdots + a_1 r^{n-1} + a_1 r^n.$$

If we subtract the preceding equation from that for S_n, then many terms on the right side drop out, leaving

$$S_n - rS_n = a_1 - a_1 r^n \quad \text{or} \quad (1 - r)S_n = a_1(1 - r^n).$$

Since $r \neq 1$, we have $1 - r \neq 0$, and dividing both sides by $1 - r$ gives us

$$S_n = \frac{a_1(1 - r^n)}{1 - r}.$$

We have proved the following theorem.

Theorem

> The nth partial sum of a geometric sequence with first term a_1 and common ratio $r \neq 1$ is
> $$S_n = a_1 \frac{(1 - r^n)}{1 - r}.$$

Example 3   Find the sum of the first five terms of the geometric sequence which begins as follows: $1, 0.3, 0.09, 0.027, \ldots$.

Solution   We let $a_1 = 1$, $r = 0.3$, and $n = 5$ in the formula for S_n, obtaining

$$S_5 = 1\left(\frac{1 - (0.3)^5}{1 - 0.3}\right)$$

which reduces to $S_5 = 1.4251$.   ■

Example 4   A man wishes to save money by setting aside 1 cent the first day, 2 cents the second day, 4 cents the third day and so on, doubling the amount each day. If this is continued, how much must be set aside on the fifteenth day? Assuming he does not run out of money, what is the total amount saved at the end of 30 days?

Solution   The amount (in cents) set aside on successive days forms a geometric sequence

$$1, 2, 4, 8, \ldots$$

with first term 1 and common ratio 2. The amount needed for the fifteenth day is found by using $a_n = a_1 r^{n-1}$ with $a_1 = 1$ and $n = 15$. This gives us $1 \cdot 2^{14}$, or \$163.84. To find the total amount set aside after 30 days, we use the formula for S_n with $n = 30$. Thus,

$$S_{30} = 1\frac{(1 - 2^{30})}{1 - 2},$$

which simplifies to \$10,737,418.23.   ■

Given the geometric series with first term a_1 and common ratio $r \neq 1$, we may write the formula for S_n of the last theorem in the form

$$S_n = \frac{a_1}{1 - r} - \frac{a_1}{1 - r}\, r^n.$$

Although we shall not do it here, it can be shown that if $|r| < 1$, then r^n *approaches the number 0 as n increases* without bound; that is, we can make r^n as close as we wish to 0 by taking n sufficiently large. It follows that S_n approaches $a_1/(1 - r)$ as n increases without bound. Using the notation introduced in our work with rational functions in Chapter Four, this may be expressed symbolically as

$$S_n \to \frac{a_1}{1 - r} \quad \text{as} \quad n \to \infty.$$

The number $a_1/(1 - r)$ is called *the sum of the infinite geometric series*

$$a_1 + a_1 r + a_1 r^2 + \cdots + a_1 r^{n-1} + \cdots.$$

This gives us the following result.

Theorem

> If $|r| < 1$, then the infinite geometric series
> $$a_1 + a_1 r + a_1 r^2 + \cdots + a_1 r^{n-1} + \cdots$$
> has the sum $a_1/(1 - r)$.

Intuitively, the preceding theorem implies that if we add more and more terms of the indicated infinite geometric series, the sums get closer and closer to $a_1/(1-r)$. The next example illustrates how the theorem can be used to show that every real number represented by a repeating decimal is rational.

Example 5 Find the rational number which corresponds to the infinite repeating decimal $5.4\overline{27}$, where the bar means that the block of digits underneath is to be repeated indefinitely.

Solution From the decimal expression $5.4272727\ldots$ we obtain the infinite series

$$5.4 + 0.027 + 0.00027 + 0.0000027 + \cdots.$$

The part of the expression after the first term is

$$0.027 + 0.00027 + 0.0000027 + \cdots$$

which has the form given in the last theorem with $a_1 = 0.027$ and $r = 0.01$. Hence, the sum of this infinite geometric series is

$$\frac{0.027}{1 - 0.01} = \frac{0.027}{0.99} = \frac{27}{990} = \frac{3}{110}.$$

Thus, it appears that the desired number is $5.4 + 3/110$, or $597/110$. A check by long division shows that $597/110$ does equal the given repeating decimal. ∎

In general, given any infinite sequence $a_1, a_2, \ldots, a_n, \ldots$, an expression of the form

$$a_1 + a_2 + \cdots + a_n + \cdots$$

is called an **infinite series**, or simply a **series**. In summation notation, this series is denoted by either

$$\sum_{n=1}^{\infty} a_n \quad \text{or} \quad \sum a_n$$

where it is understood that the summation variable in the last sum is n. Each number a_i is called a **term** of the series and a_n is called the **nth term**. Since only finite sums may be added algebraically, it is necessary to *define* what is meant by an infinite sum. A natural way to do so is to use the sequence of partial sums

$$S_1, S_2, \ldots, S_n, \ldots.$$

If there is a number S such that $S_n \to S$ as $n \to \infty$, then, as in our discussion of infinite geometric series, we call S the **sum** of the series and write

$$S = a_1 + a_2 + \cdots + a_n + \cdots.$$

Thus, in Example 5 we may write

$$\frac{597}{110} = 5.4 + 0.027 + 0.00027 + 0.0000027 + \cdots.$$

The next example provides an illustration of a non-geometric infinite series. Since a complete discussion is beyond the scope of this text, the solution is presented in an intuitive manner.

Example 6 Show that the infinite series

$$\frac{1}{1 \cdot 2} + \frac{1}{2 \cdot 3} + \frac{1}{3 \cdot 4} + \cdots + \frac{1}{n(n+1)} + \cdots$$

has a sum

Solution The nth term $a_n = 1/n(n+1)$ has the partial fraction decomposition

$$a_n = \frac{1}{n(n+1)}$$

$$= \frac{1}{n} - \frac{1}{n+1}.$$

Consequently, the nth partial sum of the series may be written

$$S_n = a_1 + a_2 + a_3 + \cdots + a_n$$

$$= \left(1 - \frac{1}{2}\right) + \left(\frac{1}{2} - \frac{1}{3}\right) + \left(\frac{1}{3} - \frac{1}{4}\right) + \cdots + \left(\frac{1}{n} - \frac{1}{n+1}\right)$$

$$= 1 - \frac{1}{n+1}.$$

Since $1/(n+1) \to 0$ as $n \to \infty$, it follows that $S_n \to 1$, and we may write

$$1 = \frac{1}{1 \cdot 2} + \frac{1}{2 \cdot 3} + \frac{1}{3 \cdot 4} + \cdots + \frac{1}{n(n+1)} + \cdots.$$

This means that, as we add more and more terms of the series, the sums get closer and closer to 1. ■

If the terms of an infinite sequence are alternately positive and negative, and if we consider the expression

$$a_1 + (-a_2) + a_3 + (-a_4) + \cdots + [(-1)^{n+1}a_n] + \cdots$$

where all the a_i are positive real numbers, then this expression is referred to as an *alternating infinite series*, and we write it in the form

$$a_1 - a_2 + a_3 - a_4 + \cdots + (-1)^{n+1}a_n + \cdots.$$

Illustrations of alternating infinite series can be obtained by using infinite geometric series with negative common ratio.

Infinite series have many applications in mathematics and the sciences. The reader is referred to texts on calculus for a rigorous treatment of this important concept.

EXERCISES 11.5

In each of Exercises 1–12 find the fifth term, the eighth term, and the nth term of the given geometric sequence.

1 $8, 4, 2, 1, \ldots$

2 $4, 1.2, 0.36, 0.108, \ldots$

3 $300, -30, 3, -0.3, \ldots$

4 $1, -\sqrt{3}, 3, -\sqrt{27}, \ldots$

5 $5, 25, 125, 625, \ldots$

6 $2, 6, 18, 54, \ldots$

7 $4, -6, 9, -13.5, \ldots$

8 $162, -54, 18, -6, \ldots$

9 $1, -x^2, x^4, -x^6, \ldots$

10 $1, -\dfrac{x}{3}, \dfrac{x^2}{9}, -\dfrac{x^3}{27}, \ldots$

11 $2, 2^{x+1}, 2^{2x+1}, 2^{3x+1}, \ldots$

12 $10, 10^{2x-1}, 10^{4x-3}, 10^{6x-5}, \ldots$

13 Find the sixth term of the geometric sequence whose first two terms are 4 and 6.

14 Find the seventh term of the geometric sequence that has 2 and $-\sqrt{2}$ for its second and third terms, respectively.

15 In a certain geometric sequence $a_5 = 1/16$ and $r = 3/2$. Find a_1 and S_5.

16 Given a geometric sequence in which $a_4 = 4$ and $a_7 = 12$, find r and a_{10}.

Find the sum in each of Exercises 17–20.

17 $\displaystyle\sum_{k=1}^{10} 3^k$

18 $\displaystyle\sum_{k=1}^{9} (-\sqrt{5})^k$

19 $\displaystyle\sum_{k=0}^{9} (-\tfrac{1}{2})^{k+1}$

20 $\displaystyle\sum_{k=1}^{7} (3^{-k})$

21 A vacuum pump removes one-half of the air in a container at each stroke. After 10 strokes, what percentage of the original amount of air remains in the container?

22 The yearly depreciation of a certain machine is 25% of its value at the beginning of the year. If the original cost of the machine is $20,000, find its value after 6 years.

23 A culture of bacteria increases 20% every hour. If the original culture contains 10,000 bacteria, find a formula for the number of bacteria present after t hours. How many bacteria are in the culture at the end of 10 hours?

24 If an amount P of money is deposited in a savings account which pays interest at a rate of r percent per year compounded quarterly and if the principal and accumulated interest are left in the account, find a formula for the total amount in the account after n years.

Find the sums of the infinite geometric series in Exercises 25–30, whenever they exist.

25 $1 - \dfrac{1}{2} + \dfrac{1}{4} - \dfrac{1}{8} + \cdots$

26 $2 + \dfrac{2}{3} + \dfrac{2}{9} + \dfrac{2}{27} + \cdots$

27 $1.5 + 0.015 + 0.00015 + \cdots$

28 $1 - 0.1 + 0.01 - 0.001 + \cdots$

29 $\sqrt{2} - 2 + \sqrt{8} - 4 + \cdots$

30 $250 - 100 + 40 - 16 + \cdots$

In each of Exercises 31–38 find the rational number represented by the given repeating decimal.

31 $0.\overline{23}$

32 $0.0\overline{71}$

33 $2.4\overline{17}$

34 $10.\overline{55}$

35 $5.\overline{146}$

36 $3.2\overline{394}$

37 $1.6\overline{124}$

38 $123.61\overline{83}$

39 A rubber ball is dropped from a height of 10 meters. If it rebounds approximately one-half the distance after each fall, use an infinite geometric series to approximate the total distance the ball travels before coming to rest.

40 The bob of a pendulum swings through an arc 24 cm long on its first swing. If each successive swing is approximately five-sixths the length of the preceding swing, use an infinite geometric series to approximate the total distance it travels before coming to rest.

11.6 PERMUTATIONS

Suppose that four teams are involved in a tournament in which first, second, third, and fourth places will be determined. For identification purposes, we label the teams a, b, c, and d. Let us find the number of different ways that first and second place can be decided. It is convenient to use a **tree diagram** as in Figure 11.1. Beginning at the word "start," the four possibilities for first place are listed. From each of these an arrow points to a possible second-place finisher. In this problem the second-place position is occupied by a team other than the one in the first-place position and hence *three* arrows are drawn from each first position. The final standings list the possible outcomes, from left to right. These are found by following all the different paths (*branches* of the tree) which lead from the word "start" to the second-place team. The total number of outcomes is 12, which we note is the product of the number of choices for first place and the number of choices for second place (after the first has been determined).

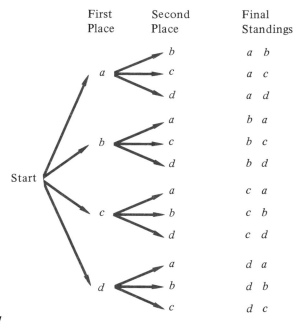

First Place	Second Place	Final Standings
a	b	a b
	c	a c
	d	a d
b	a	b a
	c	b c
	d	b d
c	a	c a
	b	c b
	d	c d
d	a	d a
	b	d b
	c	d c

Figure 11.1

Let us now find the total number of ways that first, second, third, and fourth positions can be filled. To sketch a tree diagram we may begin by drawing arrows from the word "start" to each possible first-place finisher a, b, c, or d. Then we draw arrows from those to possible second-place finishers, as was done in Figure 11.1. Next, from each second-place position we draw arrows indicating the possible third-place positions. Finally, we draw arrows to the fourth-place team. If we consider only the case where team a finishes in first place, we have the diagram shown in

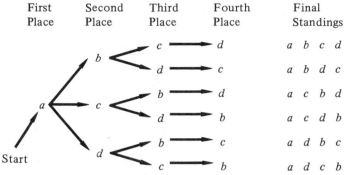

Figure 11.2

Figure 11.2. Note that there are six possible final standings in which team *a* occupies first place. In a complete tree diagram there would also be three other branches of this type corresponding to first place for *b*, *c*, and *d*, respectively. A complete diagram would display the following 24 possibilities for the final standings:

$$abcd, \quad abdc, \quad acbd, \quad acdb, \quad adbc, \quad adcb,$$
$$bacd, \quad badc, \quad bcad, \quad bcda, \quad bdac, \quad bdca,$$
$$cabd, \quad cadb, \quad cbad, \quad cbda, \quad cdab, \quad cdba,$$
$$dabc, \quad dacb, \quad dbac, \quad dbca, \quad dcab, \quad dcba.$$

Note that the number 24 is the product of the number of ways (4) that first place may occur, the number of ways (3) that second place may occur (after first place has been determined), the number of possible outcomes (2) for third place (after second place has been decided), and the number of ways (1) that fourth place can occur (after the first three places have been taken).

The discussion above illustrates the following general rule, which we accept as a basic axiom of counting.

Fundamental Counting Principle	Let $E_1, E_2, \ldots, E_k$ be a sequence of k events. If for each i, E_i can occur in m_i ways, then the total number of ways the events may take place is the product $m_1 m_2 \cdots m_k$.

Returning to the first illustration above, we let E_1 represent the determination of the first-place team, so that $m_1 = 4$. If E_2 denotes the determination of the second-place team, then $m_2 = 3$. Hence, the number of outcomes for the sequence E_1, E_2 is $4 \cdot 3 = 12$, which is the same as that found by means of the tree diagram. If we proceed to E_3, the determination of the third-place team, then $m_3 = 2$, and hence, $m_1 m_2 m_3 = 24$. Finally, if E_1, E_2, and E_3 have occurred, there is only one possible outcome for E_4. Thus, $m_4 = 1$ and $m_1 m_2 m_3 m_4 = 24$.

Instead of teams, let us now regard *a*, *b*, *c*, and *d* merely as symbols and consider the various *orderings* or *arrangements* that may be assigned to these symbols, taking them either two at a time, three at a time, or four at a time. By abstracting in this way we may apply our methods to other similar situations. For example, the problem of determining the number of two-digit numbers that can be formed from the digits 1, 2, 3, and 4 so that no digit occurs twice in any number is essentially the same as our illustration of listing first and second place in the tournament. Incidentally, the arrangements discussed above are called **arrangements without repetitions**, since a symbol may not be used twice in the same arrangement. In Example 1 below we consider arrangements in which repetitions *are* allowed.

Previously we defined ordered pairs and ordered triples. Similarly, an **ordered 4-tuple** is a set containing four elements x_1, x_2, x_3, x_4 in which an ordering has been specified, so that one of the elements may be referred to as the *first element*, another as the *second element*, and so on. The symbol (x_1, x_2, x_3, x_4) is used for the ordered 4-tuple having first element x_1, second element x_2, third element x_3, and fourth element x_4. In general, for any positive integer r, we speak of the **ordered r-tuple**

$$(x_1, x_2, \ldots, x_r)$$

as a set of elements in which x_1 is designated as the first element, x_2 as the second element, and so on.

Example 1   How many ordered triples can be obtained by using the letters a, b, c, and d? How many ordered 4-tuples can be obtained? How many r-tuples?

Solution   To answer the first question we must determine the number of symbols of the form (x_1, x_2, x_3) that can be obtained by using only the letters a, b, c, and d. This is not the same as listing first, second, and third place as in our previous illustration, since we have not ruled out the possibility of repetitions. For example, (a, b, a), (a, a, b), and (b, a, a) are different triples of the desired type. If for $i = 1, 2, 3$, we let E_i represent the determination of x_i in the triple (x_1, x_2, x_3), then, since repetitions are allowed, there are four possibilities for each of E_1, E_2, and E_3. Hence, by the Fundamental Counting Principle, the total number of ordered triples is $4 \cdot 4 \cdot 4$, or 64. Similarly, the number of possible 4-tuples (x_1, x_2, x_3, x_4) is $4 \cdot 4 \cdot 4 \cdot 4 = 256$. Evidently, the number of r-tuples is the product $4 \cdot 4 \cdot \cdots \cdot 4$, where 4 appears as a factor r times. That product equals 4^r. ∎

Example 2   A class consists of 60 girls and 40 boys. In how many ways can a president, vice-president, treasurer, and secretary be chosen if the treasurer must be a girl, the secretary must be a boy, and a student may not hold more than one office?

Solution   Let E_1 represent the choice of treasurer, E_2 the choice of secretary, and E_3 and E_4 the choice for president and vice-president, respectively. As in the Fundamental Counting Principle, we let m_i denote the number of different ways E_i can occur, for $i = 1, 2, 3$, and 4. It follows that $m_1 = 60$, $m_2 = 40$, $m_3 = 98$, and $m_4 = 97$. (Why?) By the Fundamental Counting Principle, the total number of possibilities is $60 \cdot 40 \cdot 98 \cdot 97 = 22,814,400$. ∎

The preceding example illustrates the fact that, if some of the events E_i are specialized in any way, then in applying the Fundamental Counting Principle, those special E_i should be performed first in the sequence.

Let n be a positive integer and let S be a set consisting of n distinct elements. When we speak about sets we are generally not concerned about the order or arrangement of the elements. In the remainder of this section, however, the arrangement of the elements will be our main concern. In the following definition, S denotes a set containing n elements, and r is a positive integer such that $r \leq n$.

Definition

> A **permutation** of r elements of a set S is an arrangement, without repetitions, of r elements of S.

We shall use the symbol $_nP_r$ to denote the number of different permutations of r elements that can be obtained from a set containing n elements. It is also common to refer to a permutation of r elements of a set containing n elements as a **permutation of n elements taken r at a time**. As a special case, $_nP_n$ denotes the number of arrangements of n elements of S, that is, $_nP_n$ is the number of ways of arranging *all* the elements of S.

In our first illustration involving the four teams a, b, c, and d, we had $_4P_2 = 12$, since there were 12 different ways of arranging the four teams in groups of two. It was also shown that the number of ways to arrange all the elements a, b, c, and d is 24. In the present notation this would be written $_4P_4 = 24$.

It is not difficult to find a general formula for $_nP_r$. If S denotes a set containing n elements, then the problem of determining $_nP_r$ is equivalent to determining the number of different r-tuples $(x_1, x_2, \ldots, x_r)$, where each x_i is an element of S and no element of S appears twice in the same r-tuple. We shall find this number by means of the Fundamental Counting Principle. For each $i = 1, 2, \ldots, r$, let E_i represent the determination of the element x_i and let m_i be the number of different ways of performing E_i. We wish to apply the sequence $E_1, E_2, \ldots, E_r$. There are n possible choices for x_1 and, consequently, we have $m_1 = n$. Since repetitions are not allowed, there are $n - 1$ choices for x_2, so that $m_2 = n - 1$. Continuing in this manner, we successively obtain $m_3 = n - 2$, $m_4 = n - 3$, and ultimately $m_r = n - (r - 1)$, or $m_r = n - r + 1$. Hence, by the Fundamental Counting Principle, we have

$$_nP_r = n(n - 1)(n - 2) \cdots (n - r + 1)$$

where there are r factors on the right side of the equation. As illustrations, if we consider the special cases $r = 1, 2, 3$, and 4, then $n - r + 1$ equals n, $n - 1$, $n - 2$, and $n - 3$, respectively, and using the formula for $_nP_r$ gives us

$$_nP_1 = n$$
$$_nP_2 = n(n - 1)$$
$$_nP_3 = n(n - 1)(n - 2)$$
$$_nP_4 = n(n - 1)(n - 2)(n - 3).$$

Example 3   Find $_nP_r$ if   (a) $n = 5, r = 2$;   (b) $n = 6, r = 4$;   (c) $n = 5, r = 5$.

Solutions   Using the formula for $_nP_r$,

(a) $_5P_2 = 5 \cdot 4 = 20$

(b) $_6P_4 = 6 \cdot 5 \cdot 4 \cdot 3 = 360$

(c) $_5P_5 = 5 \cdot 4 \cdot 3 \cdot 2 \cdot 1 = 120.$ ■

Example 4   A baseball team consists of nine players. Find the number of ways of arranging the first four positions in the batting order, if the pitcher is excluded.

Solution   Using the formula for $_nP_r$ with $n = 8$ and $r = 4$,

$$_8P_4 = 8 \cdot 7 \cdot 6 \cdot 5 = 1680.$$ ■

If $r = n$ in $_nP_r$ we obtain the number of different arrangements of *all* the elements of S. In this case, $n - r + 1 = n - n + 1 = 1$, and hence,

$$_nP_n = n(n - 1)(n - 2) \cdots 3 \cdot 2 \cdot 1 = n!$$

Consequently, $_nP_n$ is the product of the first n positive integers. We may also obtain a form for $_nP_r$ which involves the factorial notation. If r and n are positive integers with $r \leq n$, then

$$\frac{n!}{(n - r)!} = \frac{n(n - 1)(n - 2) \cdots (n - r + 1) \cdot [(n - r)!]}{(n - r)!}$$

$$= n(n - 1)(n - 2) \cdots (n - r + 1).$$

Comparison with the formula for $_nP_r$ gives us

$$\boxed{_nP_r = \frac{n!}{(n - r)!}.}$$

EXERCISES 11.6

Solve Exercises 1–14 by using tree diagrams or the Fundamental Counting Principle.

1 How many three-digit numbers can be formed from the digits 1, 2, 3, 4, and 5 if
(a) repetitions are not allowed?
(b) repetitions are allowed?

2 Repeat Exercise 1 for four-digit numbers.

3 How many numbers can be formed from the digits 1, 2, 3, and 4 if repetitions are not allowed? (*Note*: 42 and 231 are examples of such numbers.)

4 Work Exercise 3 if repetitions are allowed.

5 If eight basketball teams take part in a tournament, find the number of different ways that first, second, and third place can be decided, if ties are not allowed.

6 Repeat Exercise 5 for twelve teams.

7 A girl has four skirts and six blouses. How many different skirt–blouse combinations can she wear?

8 If the girl in Exercise 7 also has three sweaters,

in how many different ways can she combine the three articles of clothing?

9 In a certain state, automobile license plates start with one letter of the alphabet followed by five numerals, using the digits 0, 1, 2, . . . , 9. Find how many different license plates are possible if:
   (a) the first digit following the letter cannot be 0,
   (b) the first letter cannot be *O* or *I* and the first digit cannot be 0.

10 Two dice are tossed, one after the other. In how many different ways can they fall? List the number of different ways the sum of the dots can equal:
   (a) three   (b) five
   (c) seven   (d) nine
   (e) eleven.

11 A row of six seats in a classroom is to be filled by selecting individuals from a group of ten students. In how many different ways can the seats be occupied? If there are six boys and four girls in the group and if boys and girls are to be alternated, find the number of different seating arrangements.

12 A student in a certain school may take mathematics at 8, 10, 11, or 2 o'clock; English at 9, 10, 1, or 2; and History at 8, 11, 2, or 3. Find the number of different ways in which the student can schedule the three courses.

13 In how many different ways can a test consisting of ten true-or-false questions be answered?

14 A test consists of six multiple-choice questions, and the number of choices for each question is five. In how many different ways can the test be answered?

In each of Exercises 15–22, find the given number.

15 $_7P_3$ 16 $_8P_5$

17 $_9P_6$ 18 $_5P_3$

19 $_5P_5$ 20 $_4P_4$

21 $_6P_1$ 22 $_5P_1$

Use permutations to solve Exercises 23–30.

23 In how many different ways can eight people be seated in a row?

24 In how many different ways can ten books be arranged on a shelf?

25 A signal man has six different flags. How many different signals can be sent by placing three flags, one above the other, on a flag pole?

26 In how many different ways can five books be selected from a set of twelve different books?

27 How many three-digit numbers can be formed from the digits 2, 4, 6, 8, and 9 if
   (a) repetitions are not allowed?
   (b) repetitions are allowed?

28 There are 24 letters in the Greek alphabet. How many fraternities may be specified by choosing three Greek letters if:
   (a) repetitions are not allowed?
   (b) repetitions are allowed?

29 How many seven-digit phone numbers can be formed from the digits 0, 1, 2, 3, . . . , 9 if the first digit may not be 0?

30 After selecting nine players for a baseball game, the manager of the team arranges the batting order so that the pitcher bats last and the best hitter bats fourth. In how many different ways can the remainder of the batting order be arranged?

11.7 DISTINGUISHABLE PERMUTATIONS AND COMBINATIONS

Certain problems involve finding different arrangements of objects, some of which are indistinguishable. For example, suppose we are given five discs of the same size, where three are black, one is white, and one is red. Let us find the number of ways they can be arranged in a row so that different color arrangements are obtained. If the discs were all different, then the number of arrangements would be 5!, or 120. However, since some of the discs have the same appearance, we cannot obtain 120 different arrangements. To clarify this point, let us write

$$B \quad B \quad B \quad W \quad R$$

for the arrangement having black discs in the first three positions in the row, the white disc in the fourth position, and the red disc in the fifth position. Now the first three discs can be arranged in 3!, or 6, different ways, but these arrangements cannot be distinguished from one another because the first three discs look alike. We say that those 3! permutations are **nondistinguishable**. Similarly, given any other arrangement, say

$$B \quad R \quad B \quad W \quad B,$$

there are 3! different ways of arranging the three black discs, but again each such arrangement is nondistinguishable from the others. Let us call two arrangements of the five objects **distinguishable permutations** if one arrangement cannot be obtained from the other by rearranging the like objects. Thus, B B B W R and B R B W B are distinguishable permutations. Let n denote the number of distinguishable permutations. Since with each such arrangement there corresponds 3! nondistinguishable permutations, we must have $3!n = 5!$, the number of permutations of five *different* objects. Hence, $n = 5!/3! = 5 \cdot 4 = 20$. By the same type of reasoning we can obtain the following extension of the previous problem:

> If r objects in a collection of n objects are alike, and if the remaining objects are different from each other and also from the r objects, then the number of distinguishable permutations of the n objects is $n!/r!$.

The preceding result can be generalized to the case in which there are several subcollections of indistinguishable objects. For example, consider eight discs, four of which are black, three white, and one red. In this case, with each arrangement, such as

$$B \quad W \quad B \quad W \quad B \quad W \quad B \quad R,$$

there are 4! arrangements of the black discs and 3! arrangements of the white discs which have no effect on the color arrangement. Hence there are 4!3! arrangements of discs that can be made which will not produce distinguishable permutations. If we let n denote the number of *distinguishable* permutations of the objects, then it follows that $4!3!n = 8!$, since 8! is the number of permutations we would obtain if the objects were all different. This gives us

$$n = \frac{8!}{4!3!} = \frac{8 \cdot 7 \cdot 6 \cdot 5}{3!} \cdot \frac{4!}{4!} = 280.$$

The following general result can be proved.

Theorem on Distinguishable Permutations

Consider a collection of n objects in which n_1 are alike, n_2 are alike of another kind, . . . , n_k are alike of a further kind, and let

$$n = n_1 + n_2 + \cdots + n_k.$$

Then the number of distinguishable permutations of the n objects is

$$\frac{n!}{n_1!n_2! \cdots n_k!}.$$

Example 1 Find the number of distinguishable permutations of the letters in the word "MISSISSIPPI."

Solution   In this example we are given a collection of eleven objects in which four are of one kind (the letter S), four are of another kind (I), two are of a third kind (P), and one is of a fourth kind (M). Hence, by the preceding theorem, the number of distinguishable permutations is

$$\frac{11!}{4!4!2!1!}$$

which simplifies to 34,650. ∎

When we work with permutations our concern is with the orderings or arrangements of elements. Let us now ignore the order or arrangement of elements and consider the following question: Given a set containing n distinct elements, in how many ways can a subset of r elements be chosen, where $r \leq n$? Before answering, let us state a definition.

Definition

A **combination** of r elements of a set S is a subset of S which contains r distinct elements.

We shall use the symbol ${}_nC_r$ to denote the number of combinations of r elements that can be obtained from a set containing n elements. The phrase "a combination of n elements taken r at a time" is often used to describe ${}_nC_r$.

If S contains n elements, then, to find ${}_nC_r$, we must find the total number of subsets of the form

$$\{x_1, x_2, \ldots, x_r\}$$

where the x_i are different elements of S. Since the elements of the above subset can be arranged in $r!$ different ways, each such subset produces $r!$ different r-tuples. Hence, the total number of different r-tuples is $r!{}_nC_r$. However, in the previous section we found that the number of r-tuples is

$$_nP_r = \frac{n!}{(n-r)!}.$$

This implies that

$$r!{}_nC_r = \frac{n!}{(n-r)!}$$

and hence,

$$\boxed{{}_nC_r = \frac{n!}{(n-r)!\,r!}.}$$

Example 2 A baseball squad has six outfielders, seven infielders, five pitchers, and two catchers. In how many different ways can a team of nine players be chosen?

Solution The number of ways of choosing three outfielders from the six candidates is

$${}_6C_3 = \frac{6!}{3!(6-3)!} = \frac{6!}{3!\,3!} = 20.$$

The number of ways of choosing the four infielders is

$${}_7C_4 = \frac{7!}{4!(7-4)!} = \frac{7 \cdot 6 \cdot 5}{3!} = 35.$$

There are five ways of choosing a pitcher and two choices for the catcher. It follows from the Fundamental Counting Principle that the total number of ways to choose a team is $20 \cdot 35 \cdot 5 \cdot 2 = 7000$. ∎

It is worth noting that if $r = n$, then the right side of the formula for ${}_nC_r$ becomes

$$\frac{n!}{(n-n)!\,n!} = \frac{n!}{0!\,n!} = 1.$$

Moreover, ${}_nC_n$ is the number of subsets consisting of n elements that can be obtained from a set of n elements. This is also equal to 1. Hence, the formula for ${}_nC_r$ is true for $r = n$.

It is convenient to assign a meaning to ${}_nC_r$ when $r = 0$. If the formula is to be true in this case, then

$${}_nC_0 = \frac{n!}{n!\,0!} = 1.$$

Hence, we *define* ${}_nC_0 = 1$, which is the same as ${}_nC_n$. Finally, for consistency we also *define* ${}_0C_0 = 1$. Thus, ${}_nC_r$ has meaning for all nonnegative integers n and r with $r \le n$.

As a final observation, note that

$${}_nC_r = \frac{n!}{(n-r)!\,r!} = \frac{n(n-1)\cdots(n-r+1)}{r!}.$$

This shows that ${}_nC_r$ *is identical with the binomial coefficient* $\binom{n}{r}$ (see Section 11.2). An interesting application of that fact is given in the following example.

Example 3 If a set S contains n elements, find the number of distinct subsets of S.

Solution   Let r be any nonnegative integer such that $r \leq n$. From our previous work the number of subsets of S which contain r elements is $_nC_r$, or $\binom{n}{r}$

Hence, to find the total number of subsets, we form the sum

$$\binom{n}{0} + \binom{n}{1} + \binom{n}{2} + \binom{n}{3} + \cdots + \binom{n}{n}.$$

This is precisely the binomial expansion of $(1 + 1)^n$. Thus, there are 2^n subsets of a set of n elements. In particular, a set of 3 elements has 2^3, or 8, different subsets. A set of 4 elements has 2^4, or 16, subsets. A set of 10 elements has 2^{10}, or 1024, subsets. ∎

EXERCISES 11.7

In Exercises 1–8 find the given number.

1 $_7C_3$

2 $_8C_4$

3 $_9C_8$

4 $_6C_2$

5 $_nC_{n-1}$

6 $_nC_1$

7 $_7C_0$

8 $_5C_5$

9 If five black, three red, two white, and two green discs are to be arranged in a row, find the number of possible color arrangements for the discs.

10 Work Exercise 9 if there are three black, three red, three white, and three green discs.

11 Find the number of distinguishable permutations of the letters in the work "BOOKKEEPER."

12 Find the number of distinguishable permutations of the letters in the word "MOON." List all of the permutations.

13 Ten boys wish to play a basketball game. In how many different ways can two teams consisting of five players each be chosen?

14 A student may answer any six of ten questions in an examination. In how many ways can a choice be made? How many choices are possible if the first two questions must be answered?

15 How many straight lines are determined by eight points if no three of the points are collinear? How many triangles are determined?

16 A committee of five persons is to be chosen from a group of twelve men and eight women. If the committee is to consist of three men and two women, determine the number of ways of selecting the committee.

17 A student has five mathematics books, four history books, and eight fiction books. In how many ways can they be arranged on a shelf if books in the same category are kept next to one another?

18 A basketball squad consists of twelve individuals. Disregarding positions, in how many ways can a team of five be selected? If the center of a team must be selected from two specific individuals on the squad and the other four members of the team from the remaining ten players, find the number of different teams possible.

19 A football squad consists of three centers, ten linemen who can play either guard or tackle, three quarterbacks, six halfbacks, four ends, and four fullbacks. In how many ways can a team consisting of one center, two guards, two tackles, two ends, two halfbacks, a quarterback, and a fullback be selected?

20 In how many different ways can seven keys be arranged on a key ring if the keys can slide completely around the ring?

11.8 PROBABILITY

If two dice are tossed, what are the chances of rolling a 7? If a person is dealt five cards from a standard deck of 52 playing cards, what is the likelihood of obtaining three aces? In the 17th century, similar questions about games of chance led to the study of what is now called *probability*. Since that time, the theory of probability has grown extensively. It is now used to predict outcomes of a large variety of situations which arise in the natural and social sciences.

Any chance process, such as flipping a coin, rolling a die, being dealt a card from a deck, determining whether or not a manufactured item is defective, or finding the blood pressure of an arbitrary individual, will be called an **experiment**. A result of an experiment is called an **outcome**. We shall restrict our discussion to experiments for which outcomes are **equally likely**. This means, for example, that if a coin is flipped, it is assumed that the possibility of obtaining a head is the same as that of obtaining a tail. Similarly, if a die is tossed, we assume that the die is *fair*, in the sense that there is an equal chance of obtaining either a 1, 2, 3, 4, 5, or 6. The set S of all possible outcomes of an experiment is called the **sample space** of the experiment. Thus, if the experiment consists of flipping a coin, and we let H or T denote the outcome of obtaining a head or tail, respectively, then the sample space S may be denoted by

$$S = \{H, T\}.$$

If a fair die is tossed, then the set S of all possible outcomes is given by

$$S = \{1, 2, 3, 4, 5, 6\}.$$

The following definition expresses, in mathematical terms, the notion of obtaining *particular* outcomes of an experiment.

Definition

> Let S be the sample space of an experiment. Any subset of S is called an **event** associated with the experiment.

Let us consider the experiment of tossing a single die, so that $S = \{1, 2, 3, 4, 5, 6\}$. If $E = \{4\}$, then the event associated with the experiment consists of the outcome of obtaining a 4 on the toss. Different events may be associated with the same experiment. For example, if we let $E = \{1, 3, 5\}$, then this event consists of obtaining an odd number on a toss of the die.

As another illustration, suppose the experiment consists of flipping two coins, one after the other. If we let HH denote the outcome of two heads appearing, HT that of a head appearing on the first coin and a tail on the second, etc., then the sample space S of the experiment may be denoted by

$$S = \{HH, HT, TH, TT\}.$$

If we let

$$E = \{HT, TH\}$$

then the event consists of a head appearing on one of the coins and a tail on the other.

Next we shall define what is meant by the *probability* of an event. Throughout our discussion it is assumed that the sample space S of an experiment contains only a finite number of elements. If E is an event, the symbols $n(E)$ and $n(S)$ will denote the number of elements in E and S, respectively.

Definition

Let S be the sample space of an experiment and E an event. The **probability** $P(E)$ of E is given by

$$P(E) = \frac{n(E)}{n(S)}.$$

Since E is a subset of S, it follows that $0 \le n(E) \le n(S)$ and hence,

$$0 \le P(E) \le 1.$$

Note that $P(E) = 1$ if $E = S$, whereas $P(E) = 0$ if E contains no elements.

The next example provides several illustrations of the preceding definition if E contains exactly one element.

Example 1

(a)   If a coin is flipped, find the probability that a head will turn up.

(b)   If a fair die is tossed, find the probability of obtaining a 4.

(c)   If two coins are flipped, find the probability that both coins turn up heads.

Solutions   For each experiment we shall list sets S and E, and then use the definition to find $P(E)$.

(a) $S = \{H, T\}, \quad E = \{H\}, \quad P(E) = \dfrac{n(E)}{n(S)} = \dfrac{1}{2}$

(b) $S = \{1, 2, 3, 4, 5, 6\}, \quad E = \{4\}, \quad P(E) = \dfrac{n(E)}{n(S)} = \dfrac{1}{6}$

(c) $S = \{HH, HT, TH, TT\}, \quad E = \{HH\}, \quad P(E) = \dfrac{n(E)}{n(S)} = \dfrac{1}{4}$ ∎

In (a) of Example 1 we found that the probability of obtaining a head on a flip of a coin is 1/2. We take this to mean that if a coin is flipped many times, the number of heads that turn up should be approximately one-half the total number of flips. Thus, for 100 flips, a head should turn up approximately 50 times. Of course, it is unlikely that this number will be *exactly* 50. A probability of 1/2 implies that if we let the number of flips increase, then the number of heads that turn up *approaches* 1/2 the total number of flips. Similar remarks can be made for (b) and (c) of Example 1.

In the next two examples we consider experiments in which an event contains more than one element.

Example 2   If two dice are tossed, what is the probability of rolling   (a) 7?   (b) 9?

Solutions

(a)   Let us refer to one die as "the first die" and the other as "the second die." We shall use ordered pairs to represent outcomes as follows: $(2, 4)$ will denote the outcome of obtaining a 2 on the first die and a 4 on the second; $(5, 3)$ represents a 5 on the first die and a 3 on the second; etc. Since there are six different possibilities for the first number of the ordered pair and, with each of these, six possibilities for the second number, the total number of ordered pairs is 36. Hence, if S is the sample space, then $n(S) = 36$. The event E corresponding to rolling a 7 is given by

$$E = \{(1, 6), (2, 5), (3, 4), (4, 3), (5, 2), (6, 1)\}$$

and consequently,

$$P(E) = \frac{n(E)}{n(S)} = \frac{6}{36} = \frac{1}{6}.$$

(b)   If E is the event corresponding to rolling a 9, then

$$E = \{(3, 6), (4, 5), (5, 4), (6, 3)\}$$

and $$P(E) = \frac{n(E)}{n(S)} = \frac{4}{36} = \frac{1}{9}.$$ ■

In the next example (and in the exercises), when it is stated that several cards are drawn from a deck, we mean that each card is removed from the deck and is *not* replaced before the next card is drawn.

Example 3   If five cards are drawn from a standard deck of 52 playing cards, find the probability that all five cards are hearts.

Solution   The sample space S of the experiment is the set of all possible five-card hands that can be formed from the 52 cards in the deck. It follows from our work in Section 11.7 that $n(S) = {}_{52}C_5$.

Since there are 13 cards in the heart suit, the number of different ways of obtaining a hand which contains five hearts is ${}_{13}C_5$. Hence, if E represents this event, then

$$P(E) = \frac{n(E)}{n(S)} = \frac{{}_{13}C_5}{{}_{52}C_5}$$

$$= \frac{13!}{5!8!} \div \frac{52!}{5!47!} = \frac{13!47!}{8!52!}.$$

It is left to the reader to check that

$$P(E) = \frac{1,287}{2,598,960} \approx 0.0005$$

or
$$P(E) \approx \frac{5}{10,000} \approx \frac{1}{2,000}.$$

This implies that if the experiment is performed many times, then a five card heart hand should be drawn approximately once every 2,000 times. ∎

Suppose S is the sample space of an experiment, and E_1 and E_2 are two events associated with the experiment. Suppose further that E_1 and E_2 have no elements in common, that is, E_1 and E_2 are *disjoint* sets. If $E = E_1 \cup E_2$, then

$$n(E) = n(E_1 \cup E_2) = n(E_1) + n(E_2)$$

and hence

$$P(E) = \frac{n(E_1) + n(E_2)}{n(S)} = \frac{n(E_1)}{n(S)} + \frac{n(E_2)}{n(S)}$$

that is,

$$P(E) = P(E_1) + P(E_2).$$

Thus, the probability of E is the sum of the probabilities of E_1 and E_2.

The next theorem states that this result can be extended to any number of events $E_1, E_2, \ldots, E_k$ which are *mutually disjoint*, in the sense that, if $i \neq j$, then E_i and E_j have no elements in common.

Theorem

Suppose $E_1, E_2, \ldots, E_k$ are mutually disjoint events associated with the same experiment. If

$$E = E_1 \cup E_2 \cup \cdots \cup E_k,$$

then
$$P(E) = P(E_1) + P(E_2) + \cdots + P(E_k).$$

The theorem may be proved by mathematical induction.

Example 4 If two dice are tossed, find the probability of rolling either a 7 or a 9.

Solution   Let E_1 denote the event of rolling a 7, and E_2 that of rolling a 9. We wish to find the probability of the event $E = E_1 \cup E_2$. From Example 3, $P(E_1) = 1/6$ and $P(E_2) = 1/9$. Hence, by the last theorem,

$$P(E) = P(E_1) + P(E_2)$$

$$= \frac{1}{6} + \frac{1}{9} = \frac{3+2}{18} = \frac{5}{18}.$$ ∎

The results of this section provide only an introduction to the theory of probability. There are many other types of problems which may be considered. For example, given two events E_1 and E_2, we have not discussed how to find $P(E_1 \cup E_2)$ if E_1 and E_2 have elements in common, or how to find $P(E_2)$ *after* E_1 has occurred. Such questions are left for independent investigation or for future courses.

EXERCISES 11.8

1 A single card is drawn from a standard 52-card deck. Find the probability that the card is:
  (a) a king
  (b) a king or a queen
  (c) a king, a queen, or a jack

2 A single card is drawn from a 52-card deck. Find the probability that the card is:
  (a) a heart
  (b) a heart or a diamond
  (c) a heart, a diamond, or a club

3 If a single die is tossed, find the probability of obtaining
  (a) a 4   (b) a 6
  (c) either a 4 or a 6

4 An urn contains 5 red balls, 6 green balls, and 4 white balls. If a single ball is drawn, find the probability that it is:
  (a) red   (b) green
  (c) either red or white

5 If two dice are tossed, find the probability of rolling:
  (a) 11   (b) 8
  (c) either 11 or 8

6 If two dice are tossed, find the probability that the sum of the dots is greater than 5.

7 If three dice are tossed, find the probability that the sum of the dots is 5.

8 If three dice are tossed, find the probability that a 6 turns up on exactly one die.

9 If three coins are flipped, find the probability that exactly two heads turn up.

10 If four coins are flipped, find the probability of obtaining two heads and two tails.

In Exercises 11–16, suppose 5 cards are drawn from a 52-card deck. Find the probability of obtaining the indicated cards.

11 four of a kind (i.e., four aces, four kings, etc.)

12 three aces and two kings

13 four diamonds and one spade

14 five face cards

15 a flush (five cards, all of the same suit)

16 a royal flush (an ace, king, queen, jack, and 10 of the same suit)

17 A true-or-false test contains eight questions. If a student guesses the answer for each question, find the probability that:
  (a) 8 answers are correct.
  (b) 7 answers are correct and 1 is incorrect.
  (c) 6 answers are correct and 2 are incorrect.
  (d) at least 6 answers are correct.

18 A 6-member committee is to be chosen by drawing names of individuals from a hat. If the hat contains the names of 8 men and 14 women, find the probability that the committee will consist of 3 men and 3 women.

19 (a) If S is the sample space of an experiment and E is an event, let E' denote the elements of S which are not in E. Prove that $P(E') = 1 - P(E)$.
  (b) If 5 cards are drawn from a 52-card deck, use (a) to find the probability of obtaining at least one ace.

20 If 5 cards are drawn from a 52-card deck, use (a) of Exercise 19 to find the probability of obtaining at least one heart.

11.9 REVIEW

Concepts

Define or discuss each of the following.

1 The axiom of mathematical induction

2 The principle of mathematical induction

3 The Binomial Theorem

4 Binomial coefficients

5 Infinite sequence

6 Summation notation

7 The nth partial sum of an infinite sequence

8 Arithmetic sequence

9 Arithmetic mean of two numbers

10 Geometric sequence

11 Infinite geometric series

12 Infinite series

13 Probability of an event

Exercises

Prove that the statements in Exercises 1–5 are true for every positive integer n.

1 $2 + 5 + 8 + \cdots + (3n - 1) = \dfrac{n(3n + 1)}{2}$

2 $2^2 + 4^2 + 6^2 + \cdots + (2n)^2$

$= \dfrac{2n(2n + 1)(n + 1)}{3}$

3 $\dfrac{1}{1 \cdot 3} + \dfrac{1}{3 \cdot 5} + \dfrac{1}{5 \cdot 7} + \cdots + \dfrac{1}{(2n - 1)(2n + 1)}$

$= \dfrac{n}{2n + 1}$

4 $1 \cdot 2 + 2 \cdot 3 + 3 \cdot 4 + \cdots + n(n + 1)$

$= \dfrac{n(n + 1)(n + 2)}{3}$

5 3 is a factor of $n^3 + 2n$.

6 Prove that $2^n > n^2$ for every positive integer $n \geq 5$.

7 Expand and simplify $(x^2 - 3y)^6$.

8 Find the first four terms in the binomial expansion of $(a^{2/5} + 2a^{-3/5})^{20}$.

9 Find the sixth term in the expansion of $(b^3 - \frac{1}{2}c^2)^9$.

10 In the expansion of $(2c^3 + 5c^{-2})^{10}$ find the term which does not contain c.

In each of Exercises 11–14 find the first four terms

and the seventh term of the sequence that has the given nth term.

11 $a_n = \dfrac{5n}{3 - 2n^2}$

12 $a_n = (-1)^{n+1} - (0.1)^n$

13 $a_n = 1 + (-\frac{1}{2})^{n-1}$

14 $a_n = \dfrac{2^n}{(n + 1)(n + 2)(n + 3)}$

Find the first five terms of the infinite sequence defined recursively in each of Exercises 15–18.

15 $a_1 = 10, \quad a_{k+1} = 1 + (1/a_k)$

16 $a_1 = 2, \quad a_{k+1} = a_k!$

17 $a_1 = 9, \quad a_{k+1} = \sqrt{a_k}$

18 $a_1 = 1, \quad a_{k+1} = (1 + a_k)^{-1}$

Find the number represented by the sum in each of Exercises 19–22.

19 $\displaystyle\sum_{k=1}^{5} (k^2 + 4)$

20 $\displaystyle\sum_{k=2}^{6} \dfrac{2k - 8}{k - 1}$

21 $\displaystyle\sum_{k=1}^{100} 10$

22 $\displaystyle\sum_{i=1}^{4} (2^i - 10)$

In Exercises 23–26 use summation notation to represent the sums.

23 $3 + 6 + 9 + 12 + 15$

24 $2 + 4 + 8 + 16 + 32 + 64 + 128$

25 $100 - 95 + 90 - 85 + 80$

26 $a_0 + a_4 x^4 + a_8 x^8 + \cdots + a_{100} x^{100}$

27   Find the tenth term and the sum of the first ten terms of the arithmetic sequence whose first two terms are $4 + \sqrt{3}$ and 3.

28   Find the sum of the first eight terms of an arithmetic sequence in which the fourth term is 9 and the common difference is -5.

29   The fifth and thirteenth terms of an arithmetic sequence are 5 and 77, respectively. Find the first term and the tenth term.

30   Insert four arithmetic means between 20 and -10.

31   Find the tenth term of the geometric sequence whose first two terms are $1/8$ and $1/4$.

32   If a geometric sequence has 3 and -0.3 as its third and fourth terms, find the eighth term.

33   Find a positive number c such that 4, c, 8 are successive terms of a geometric sequence.

34   In a certain geometric sequence the eighth term is 100 and the common ratio is $-3/2$. Find the first term.

Find the sums in Exercises 35–38.

35 $\displaystyle\sum_{k=1}^{20} (3k - 5)$ **36** $\displaystyle\sum_{k=1}^{10} (6 - \tfrac{1}{2}k)$

37 $\displaystyle\sum_{k=1}^{10} (2^k - \tfrac{1}{2})$ **38** $\displaystyle\sum_{k=1}^{8} (\tfrac{1}{2} - 2^k)$

39   Find the sum of the infinite geometric series

$$1 - \frac{2}{5} + \frac{4}{25} - \frac{8}{125} + \cdots$$

40   Find the rational number whose decimal representation is $6.\overline{274}$.

41   In how many ways can thirteen cards be selected from a deck of 52 cards? In how many ways can thirteen cards be selected if one wishes to obtain five spades, three hearts, three clubs, and two diamonds?

42   How many four-digit numbers can be formed from the digits 1, 2, 3, 4, 5, and 6 if
(a)   repetitions are not allowed?
(b)   repetitions are allowed?

43   If a student must answer eight of twelve questions on an examination, in how many ways can a choice be made? How many choices are possible if the first three questions must be answered?

44   If six black, five red, four white, and two green discs are to be arranged in a row, find the number of possible color arrangements.

45   If 1,000 tickets are sold for a raffle, find the probability of winning if an individual purchases:
(a)   1 ticket   (b)   10 tickets
(c)   50 tickets.

46   If four coins are flipped, find the probability of obtaining 1 head and 3 tails.

47   A quiz consists of six true-or-false questions, and at least four correct answers are required for a passing grade. If a student guesses at each answer, what is the probability of   (a) passing?
(b) failing?

48   If a single die is tossed and then a card is drawn from a 52-card deck, what is the probability of obtaining:
(a)   a 6 on the die and the king of hearts?
(b)   *either* a 6 on the die or the king of hearts?

Table 1. Common Logarithms

N	0	1	2	3	4	5	6	7	8	9
1.0	.0000	.0043	.0086	.0128	.0170	.0212	.0253	.0294	.0334	.0374
1.1	.0414	.0453	.0492	.0531	.0569	.0607	.0645	.0682	.0719	.0755
1.2	.0792	.0828	.0864	.0899	.0934	.0969	.1004	.1038	.1072	.1106
1.3	.1139	.1173	.1206	.1239	.1271	.1303	.1335	.1367	.1399	.1430
1.4	.1461	.1492	.1523	.1553	.1584	.1614	.1644	.1673	.1703	.1732
1.5	.1761	.1790	.1818	.1847	.1875	.1903	.1931	.1959	.1987	.2014
1.6	.2041	.2068	.2095	.2122	.2148	.2175	.2201	.2227	.2253	.2279
1.7	.2304	.2330	.2355	.2380	.2405	.2430	.2455	.2480	.2504	.2529
1.8	.2553	.2577	.2601	.2625	.2648	.2672	.2695	.2718	.2742	.2765
1.9	.2788	.2810	.2833	.2856	.2878	.2900	.2923	.2945	.2967	.2989
2.0	.3010	.3032	.3054	.3075	.3096	.3118	.3139	.3160	.3181	.3201
2.1	.3222	.3243	.3263	.3284	.3304	.3324	.3345	.3365	.3385	.3404
2.2	.3424	.3444	.3464	.3483	.3502	.3522	.3541	.3560	.3579	.3598
2.3	.3617	.3636	.3655	.3674	.3692	.3711	.3729	.3747	.3766	.3784
2.4	.3802	.3820	.3838	.3856	.3874	.3892	.3909	.3927	.3945	.3962
2.5	.3979	.3997	.4014	.4031	.4048	.4065	.4082	.4099	.4116	.4133
2.6	.4150	.4166	.4183	.4200	.4216	.4232	.4249	.4265	.4281	.4298
2.7	.4314	.4330	.4346	.4362	.4378	.4393	.4409	.4425	.4440	.4456
2.8	.4472	.4487	.4502	.4518	.4533	.4548	.4564	.4579	.4594	.4609
2.9	.4624	.4639	.4654	.4669	.4683	.4698	.4713	.4728	.4742	.4757
3.0	.4771	.4786	.4800	.4814	.4829	.4843	.4857	.4871	.4886	.4900
3.1	.4914	.4928	.4942	.4955	.4969	.4983	.4997	.5011	.5024	.5038
3.2	.5051	.5065	.5079	.5092	.5105	.5119	.5132	.5145	.5159	.5172
3.3	.5185	.5198	.5211	.5224	.5237	.5250	.5263	.5276	.5289	.5302
3.4	.5315	.5328	.5340	.5353	.5366	.5378	.5391	.5403	.5416	.5428
3.5	.5441	.5453	.5465	.5478	.5490	.5502	.5514	.5527	.5539	.5551
3.6	.5563	.5575	.5587	.5599	.5611	.5623	.5635	.5647	.5658	.5670
3.7	.5682	.5694	.5705	.5717	.5729	.5740	.5752	.5763	.5775	.5786
3.8	.5798	.5809	.5821	.5832	.5843	.5855	.5866	.5877	.5888	.5899
3.9	.5911	.5922	.5933	.5944	.5955	.5966	.5977	.5988	.5999	.6010
4.0	.6021	.6031	.6042	.6053	.6064	.6075	.6085	.6096	.6107	.6117
4.1	.6128	.6138	.6149	.6160	.6170	.6180	.6191	.6201	.6212	.6222
4.2	.6232	.6243	.6253	.6263	.6274	.6284	.6294	.6304	.6314	.6325
4.3	.6335	.6345	.6355	.6365	.6375	.6385	.6395	.6405	.6415	.6425
4.4	.6435	.6444	.6454	.6464	.6474	.6484	.6493	.6503	.6513	.6522
4.5	.6532	.6542	.6551	.6561	.6571	.6580	.6590	.6599	.6609	.6618
4.6	.6628	.6637	.6646	.6656	.6665	.6675	.6684	.6693	.6702	.6712
4.7	.6721	.6730	.6739	.6749	.6758	.6767	.6776	.6785	.6794	.6803
4.8	.6812	.6821	.6830	.6839	.6848	.6857	.6866	.6875	.6884	.6893
4.9	.6902	.6911	.6920	.6928	.6937	.6946	.6955	.6964	.6972	.6981
5.0	.6990	.6998	.7007	.7016	.7024	.7033	.7042	.7050	.7059	.7067
5.1	.7076	.7084	.7093	.7101	.7110	.7118	.7126	.7135	.7143	.7152
5.2	.7160	.7168	.7177	.7185	.7193	.7202	.7210	.7218	.7226	.7235
5.3	.7243	.7251	.7259	.7267	.7275	.7284	.7292	.7300	.7308	.7316
5.4	.7324	.7332	.7340	.7348	.7356	.7364	.7372	.7380	.7388	.7396
5.5	.7404	.7412	.7419	.7427	.7435	.7443	.7451	.7459	.7466	.7474
5.6	.7482	.7490	.7497	.7505	.7513	.7520	.7528	.7536	.7543	.7551
5.7	.7559	.7566	.7574	.7582	.7589	.7597	.7604	.7612	.7619	.7627
5.8	.7634	.7642	.7649	.7657	.7664	.7672	.7679	.7686	.7694	.7701
5.9	.7709	.7716	.7723	.7731	.7738	.7745	.7752	.7760	.7767	.7774
6.0	.7782	.7789	.7796	.7803	.7810	.7818	.7825	.7832	.7839	.7846
6.1	.7853	.7860	.7868	.7875	.7882	.7889	.7896	.7903	.7910	.7917
6.2	.7924	.7931	.7938	.7945	.7952	.7959	.7966	.7973	.7980	.7987
6.3	.7993	.8000	.8007	.8014	.8021	.8028	.8035	.8041	.8048	.8055
6.4	.8062	.8069	.8075	.8082	.8089	.8096	.8102	.8109	.8116	.8122
6.5	.8129	.8136	.8142	.8149	.8156	.8162	.8169	.8176	.8182	.8189
6.6	.8195	.8202	.8209	.8215	.8222	.8228	.8235	.8241	.8248	.8254
6.7	.8261	.8267	.8274	.8280	.8287	.8293	.8299	.8306	.8312	.8319
6.8	.8325	.8331	.8338	.8344	.8351	.8357	.8363	.8370	.8376	.8382
6.9	.8388	.8395	.8401	.8407	.8414	.8420	.8426	.8432	.8439	.8445
7.0	.8451	.8457	.8463	.8470	.8476	.8482	.8488	.8494	.8500	.8506
7.1	.8513	.8519	.8525	.8531	.8537	.8543	.8549	.8555	.8561	.8567
7.2	.8573	.8579	.8585	.8591	.8597	.8603	.8609	.8615	.8621	.8627
7.3	.8633	.8639	.8645	.8651	.8657	.8663	.8669	.8675	.8681	.8686
7.4	.8692	.8698	.8704	.8710	.8716	.8722	.8727	.8733	.8739	.8745
7.5	.8751	.8756	.8762	.8768	.8774	.8779	.8785	.8791	.8797	.8802
7.6	.8808	.8814	.8820	.8825	.8831	.8837	.8842	.8848	.8854	.8859
7.7	.8865	.8871	.8876	.8882	.8887	.8893	.8899	.8904	.8910	.8915
7.8	.8921	.8927	.8932	.8938	.8943	.8949	.8954	.8960	.8965	.8971
7.9	.8976	.8982	.8987	.8993	.8998	.9004	.9009	.9015	.9020	.9025
8.0	.9031	.9036	.9042	.9047	.9053	.9058	.9063	.9069	.9074	.9079
8.1	.9085	.9090	.9096	.9101	.9106	.9112	.9117	.9122	.9128	.9133
8.2	.9138	.9143	.9149	.9154	.9159	.9165	.9170	.9175	.9180	.9186
8.3	.9191	.9196	.9201	.9206	.9212	.9217	.9222	.9227	.9232	.9238
8.4	.9243	.9248	.9253	.9258	.9263	.9269	.9274	.9279	.9284	.9289
8.5	.9294	.9299	.9304	.9309	.9315	.9320	.9325	.9330	.9335	.9340
8.6	.9345	.9350	.9355	.9360	.9365	.9370	.9375	.9380	.9385	.9390
8.7	.9395	.9400	.9405	.9410	.9415	.9420	.9425	.9430	.9435	.9440
8.8	.9445	.9450	.9455	.9460	.9465	.9469	.9474	.9479	.9484	.9489
8.9	.9494	.9499	.9504	.9509	.9513	.9518	.9523	.9528	.9533	.9538
9.0	.9542	.9547	.9552	.9557	.9562	.9566	.9571	.9576	.9581	.9586
9.1	.9590	.9595	.9600	.9605	.9609	.9614	.9619	.9624	.9628	.9633
9.2	.9638	.9643	.9647	.9652	.9657	.9661	.9666	.9671	.9675	.9680
9.3	.9685	.9689	.9694	.9699	.9703	.9708	.9713	.9717	.9722	.9727
9.4	.9731	.9736	.9741	.9745	.9750	.9754	.9759	.9763	.9768	.9773
9.5	.9777	.9782	.9786	.9791	.9795	.9800	.9805	.9809	.9814	.9818
9.6	.9823	.9827	.9832	.9836	.9841	.9845	.9850	.9854	.9859	.9863
9.7	.9868	.9872	.9877	.9881	.9886	.9890	.9894	.9899	.9903	.9908
9.8	.9912	.9917	.9921	.9926	.9930	.9934	.9939	.9943	.9948	.9952
9.9	.9956	.9961	.9965	.9969	.9974	.9978	.9983	.9987	.9991	.9996

TABLE 2. VALUES OF THE TRIGONOMETRIC FUNCTIONS

Table 2. Values of the Trigonometric Functions

t	degrees	sin t	cos t	tan t	cot t	sec t	csc t	degrees	t
.0000	0°00'	.0000	1.0000	.0000	—	1.000	—	90°00'	1.5708
.0029	10	.0029	1.0000	.0029	343.8	1.000	343.8	50	1.5679
.0058	20	.0058	1.0000	.0058	171.9	1.000	171.9	40	1.5650
.0087	30	.0087	1.0000	.0087	114.6	1.000	114.6	30	1.5621
.0116	40	.0116	.9999	.0116	85.94	1.000	85.95	20	1.5592
.0145	50	.0145	.9999	.0145	68.75	1.000	68.76	10	1.5563
.0175	1°00'	.0175	.9998	.0175	57.29	1.000	57.30	89°00'	1.5533
.0204	10	.0204	.9998	.0204	49.10	1.000	49.11	50	1.5504
.0233	20	.0233	.9997	.0233	42.96	1.000	42.98	40	1.5475
.0262	30	.0262	.9997	.0262	38.19	1.000	38.20	30	1.5446
.0291	40	.0291	.9996	.0291	34.37	1.000	34.38	20	1.5417
.0320	50	.0320	.9995	.0320	31.24	1.001	31.26	10	1.5388
.0349	2°00'	.0349	.9994	.0349	28.64	1.001	28.65	88°00'	1.5359
.0378	10	.0378	.9993	.0378	26.43	1.001	26.45	50	1.5330
.0407	20	.0407	.9992	.0407	24.54	1.001	24.56	40	1.5301
.0436	30	.0436	.9990	.0437	22.90	1.001	22.93	30	1.5272
.0465	40	.0465	.9989	.0466	21.47	1.001	21.49	20	1.5243
.0495	50	.0494	.9988	.0495	20.21	1.001	20.23	10	1.5213
.0524	3°00'	.0523	.9986	.0524	19.08	1.001	19.11	87°00'	1.5184
.0553	10	.0552	.9985	.0553	18.07	1.002	18.10	50	1.5155
.0582	20	.0581	.9983	.0582	17.17	1.002	17.20	40	1.5126
.0611	30	.0610	.9981	.0612	16.35	1.002	16.38	30	1.5097
.0640	40	.0640	.9980	.0641	15.60	1.002	15.64	20	1.5068
.0669	50	.0669	.9978	.0670	14.92	1.002	14.96	10	1.5039
.0698	4°00'	.0698	.9976	.0699	14.30	1.002	14.34	86°00'	1.5010
.0727	10	.0727	.9974	.0729	13.73	1.003	13.76	50	1.4981
.0756	20	.0756	.9971	.0758	13.20	1.003	13.23	40	1.4952
.0785	30	.0785	.9969	.0787	12.71	1.003	12.75	30	1.4923
.0814	40	.0814	.9967	.0816	12.25	1.003	12.29	20	1.4893
.0844	50	.0843	.9964	.0846	11.83	1.004	11.87	10	1.4864
.0873	5°00'	.0872	.9962	.0875	11.43	1.004	11.47	85°00'	1.4835
.0902	10	.0901	.9959	.0904	11.06	1.004	11.10	50	1.4806
.0931	20	.0929	.9957	.0934	10.71	1.004	10.76	40	1.4777
.0960	30	.0958	.9954	.0963	10.39	1.005	10.43	30	1.4748
.0989	40	.0987	.9951	.0992	10.08	1.005	10.13	20	1.4719
.1018	50	.1016	.9948	.1022	9.788	1.005	9.839	10	1.4690
.1047	6°00'	.1045	.9945	.1051	9.514	1.006	9.567	84°00'	1.4661
.1076	10	.1074	.9942	.1080	9.255	1.006	9.309	50	1.4632
.1105	20	.1103	.9939	.1110	9.010	1.006	9.065	40	1.4603
.1134	30	.1132	.9936	.1139	8.777	1.006	8.834	30	1.4573
.1164	40	.1161	.9932	.1169	8.556	1.007	8.614	20	1.4544
.1193	50	.1190	.9929	.1198	8.345	1.007	8.405	10	1.4515
.1222	7°00'	.1219	.9925	.1228	8.144	1.008	8.206	83°00'	1.4486
		cos t	sin t	cot t	tan t	csc t	sec t	degrees	t

t	degrees	sin t	cos t	tan t	cot t	sec t	csc t	degrees	t
.1222	7°00'	.1219	.9925	.1228	8.144	1.008	8.206	83°00'	1.4486
.1251	10	.1248	.9922	.1257	7.953	1.008	8.016	50	1.4457
.1280	20	.1276	.9918	.1287	7.770	1.008	7.834	40	1.4428
.1309	30	.1305	.9914	.1317	7.596	1.009	7.661	30	1.4399
.1338	40	.1334	.9911	.1346	7.429	1.009	7.496	20	1.4370
.1367	50	.1363	.9907	.1376	7.269	1.009	7.337	10	1.4341
.1396	8°00'	.1392	.9903	.1405	7.115	1.010	7.185	82°00'	1.4312
.1425	10	.1421	.9899	.1435	6.968	1.010	7.040	50	1.4283
.1454	20	.1449	.9894	.1465	6.827	1.011	6.900	40	1.4254
.1484	30	.1478	.9890	.1495	6.691	1.011	6.765	30	1.4224
.1513	40	.1507	.9886	.1524	6.561	1.012	6.636	20	1.4195
.1542	50	.1536	.9881	.1554	6.435	1.012	6.512	10	1.4166
.1571	9°00'	.1564	.9877	.1584	6.314	1.012	6.392	81°00'	1.4137
.1600	10	.1593	.9872	.1614	6.197	1.013	6.277	50	1.4108
.1629	20	.1622	.9868	.1644	6.084	1.013	6.166	40	1.4079
.1658	30	.1650	.9863	.1673	5.976	1.014	6.059	30	1.4050
.1687	40	.1679	.9858	.1703	5.871	1.014	5.955	20	1.4021
.1716	50	.1708	.9853	.1733	5.769	1.015	5.855	10	1.3992
.1745	10°00'	.1736	.9848	.1763	5.671	1.015	5.759	80°00'	1.3963
.1774	10	.1765	.9843	.1793	5.576	1.016	5.665	50	1.3934
.1804	20	.1794	.9838	.1823	5.485	1.016	5.575	40	1.3904
.1833	30	.1822	.9833	.1853	5.396	1.017	5.487	30	1.3875
.1862	40	.1851	.9827	.1883	5.309	1.018	5.403	20	1.3846
.1891	50	.1880	.9822	.1914	5.226	1.018	5.320	10	1.3817
.1920	11°00'	.1908	.9816	.1944	5.145	1.019	5.241	79°00'	1.3788
.1949	10	.1937	.9811	.1974	5.066	1.019	5.164	50	1.3759
.1978	20	.1965	.9805	.2004	4.989	1.020	5.089	40	1.3730
.2007	30	.1994	.9799	.2035	4.915	1.020	5.016	30	1.3701
.2036	40	.2022	.9793	.2065	4.843	1.021	4.945	20	1.3672
.2065	50	.2051	.9787	.2095	4.773	1.022	4.876	10	1.3643
.2094	12°00'	.2079	.9781	.2126	4.705	1.022	4.810	78°00'	1.3614
.2123	10	.2108	.9775	.2156	4.638	1.023	4.745	50	1.3584
.2153	20	.2136	.9769	.2186	4.574	1.024	4.682	40	1.3555
.2182	30	.2164	.9763	.2217	4.511	1.024	4.620	30	1.3526
.2211	40	.2193	.9757	.2247	4.449	1.025	4.560	20	1.3497
.2240	50	.2221	.9750	.2278	4.390	1.026	4.502	10	1.3468
.2269	13°00'	.2250	.9744	.2309	4.331	1.026	4.445	77°00'	1.3439
.2298	10	.2278	.9737	.2339	4.275	1.027	4.390	50	1.3410
.2327	20	.2306	.9730	.2370	4.219	1.028	4.336	40	1.3381
.2356	30	.2334	.9724	.2401	4.165	1.028	4.284	30	1.3352
.2385	40	.2363	.9717	.2432	4.113	1.029	4.232	20	1.3323
.2414	50	.2391	.9710	.2462	4.061	1.030	4.182	10	1.3294
.2443	14°00'	.2419	.9703	.2493	4.011	1.031	4.134	76°00'	1.3265
		cos t	sin t	cot t	tan t	csc t	sec t	degrees	t

TABLE 2. VALUES OF THE TRIGONOMETRIC FUNCTIONS   A3

t	degrees	csc t	sec t	cot t	tan t	cos t	sin t	degrees	t
1.2043	69°00'	2.790	1.071	2.605	.3839	.9336	.3584	21°00'	.3665
1.2014	50	2.769	1.072	2.583	.3872	.9325	.3611	10	.3694
1.1985	40	2.749	1.074	2.560	.3906	.9315	.3638	20	.3723
1.1956	30	2.729	1.075	2.539	.3939	.9304	.3665	30	.3752
1.1926	20	2.709	1.076	2.517	.3973	.9293	.3692	40	.3782
1.1897	10	2.689	1.077	2.496	.4006	.9283	.3719	50	.3811
1.1868	68°00'	2.669	1.079	2.475	.4040	.9272	.3746	22°00'	.3840
1.1839	50	2.650	1.080	2.455	.4074	.9261	.3773	10	.3869
1.1810	40	2.632	1.081	2.434	.4108	.9250	.3800	20	.3898
1.1781	30	2.613	1.082	2.414	.4142	.9239	.3827	30	.3927
1.1752	20	2.595	1.084	2.394	.4176	.9228	.3854	40	.3956
1.1723	10	2.577	1.085	2.375	.4210	.9216	.3881	50	.3985
1.1694	67°00'	2.559	1.086	2.356	.4245	.9205	.3907	23°00'	.4014
1.1665	50	2.542	1.088	2.337	.4279	.9194	.3934	10	.4043
1.1636	40	2.525	1.089	2.318	.4314	.9182	.3961	20	.4072
1.1606	30	2.508	1.090	2.300	.4348	.9171	.3987	30	.4102
1.1577	20	2.491	1.092	2.282	.4383	.9159	.4014	40	.4131
1.1548	10	2.475	1.093	2.264	.4417	.9147	.4041	50	.4160
1.1519	66°00'	2.459	1.095	2.246	.4452	.9135	.4067	24°00'	.4189
1.1490	50	2.443	1.096	2.229	.4487	.9124	.4094	10	.4218
1.1461	40	2.427	1.097	2.211	.4522	.9112	.4120	20	.4247
1.1432	30	2.411	1.099	2.194	.4557	.9100	.4147	30	.4276
1.1403	20	2.396	1.100	2.177	.4592	.9088	.4173	40	.4305
1.1374	10	2.381	1.102	2.161	.4628	.9075	.4200	50	.4334
1.1345	65°00'	2.366	1.103	2.145	.4663	.9063	.4226	25°00'	.4363
1.1316	50	2.352	1.105	2.128	.4699	.9051	.4253	10	.4392
1.1286	40	2.337	1.106	2.112	.4734	.9038	.4279	20	.4422
1.1257	30	2.323	1.108	2.097	.4770	.9026	.4305	30	.4451
1.1228	20	2.309	1.109	2.081	.4806	.9013	.4331	40	.4480
1.1199	10	2.295	1.111	2.066	.4841	.9001	.4358	50	.4509
1.1170	64°00'	2.281	1.113	2.050	.4877	.8988	.4384	26°00'	.4538
1.1141	50	2.268	1.114	2.035	.4913	.8975	.4410	10	.4567
1.1112	40	2.254	1.116	2.020	.4950	.8962	.4436	20	.4596
1.1083	30	2.241	1.117	2.006	.4986	.8949	.4462	30	.4625
1.1054	20	2.228	1.119	1.991	.5022	.8936	.4488	40	.4654
1.1025	10	2.215	1.121	1.977	.5059	.8923	.4514	50	.4683
1.0996	63°00'	2.203	1.122	1.963	.5095	.8910	.4540	27°00'	.4712
1.0966	50	2.190	1.124	1.949	.5132	.8897	.4566	10	.4741
1.0937	40	2.178	1.126	1.935	.5169	.8884	.4592	20	.4771
1.0908	30	2.166	1.127	1.921	.5206	.8870	.4617	30	.4800
1.0879	20	2.154	1.129	1.907	.5243	.8857	.4643	40	.4829
1.0850	10	2.142	1.131	1.894	.5280	.8843	.4669	50	.4858
1.0821	62°00'	2.130	1.133	1.881	.5317	.8829	.4695	28°00'	.4887
t	degrees	sec t	csc t	tan t	cot t	sin t	cos t		

Table 2. Values of the Trigonometric Functions (cont'd)

t	degrees	csc t	sec t	cot t	tan t	cos t	sin t	degrees	t
.2443	14°00'	4.134	1.031	4.011	.2493	.9703	.2419	76°00'	1.3265
.2473	10	4.086	1.031	3.962	.2524	.9696	.2447	50	1.3235
.2502	20	4.039	1.032	3.914	.2555	.9689	.2476	40	1.3206
.2531	30	3.994	1.033	3.867	.2586	.9681	.2504	30	1.3177
.2560	40	3.950	1.034	3.821	.2617	.9674	.2532	20	1.3148
.2589	50	3.906	1.034	3.776	.2648	.9667	.2560	10	1.3119
.2618	15°00'	3.864	1.035	3.732	.2679	.9659	.2588	75°00'	1.3090
.2647	10	3.822	1.036	3.689	.2711	.9652	.2616	50	1.3061
.2676	20	3.782	1.037	3.647	.2742	.9644	.2644	40	1.3032
.2705	30	3.742	1.038	3.606	.2773	.9636	.2672	30	1.3003
.2734	40	3.703	1.039	3.566	.2805	.9628	.2700	20	1.2974
.2763	50	3.665	1.039	3.526	.2836	.9621	.2728	10	1.2945
.2793	16°00'	3.628	1.040	3.487	.2867	.9613	.2756	74°00'	1.2915
.2822	10	3.592	1.041	3.450	.2899	.9605	.2784	50	1.2886
.2851	20	3.556	1.042	3.412	.2931	.9596	.2812	40	1.2857
.2880	30	3.521	1.043	3.376	.2962	.9588	.2840	30	1.2828
.2909	40	3.487	1.044	3.340	.2994	.9580	.2868	20	1.2799
.2938	50	3.453	1.045	3.305	.3026	.9572	.2896	10	1.2770
.2967	17°00'	3.420	1.046	3.271	.3057	.9563	.2924	73°00'	1.2741
.2996	10	3.388	1.047	3.237	.3089	.9555	.2952	50	1.2712
.3025	20	3.356	1.048	3.204	.3121	.9546	.2979	40	1.2683
.3054	30	3.326	1.049	3.172	.3153	.9537	.3007	30	1.2654
.3083	40	3.295	1.049	3.140	.3185	.9528	.3035	20	1.2625
.3113	50	3.265	1.050	3.108	.3217	.9520	.3062	10	1.2595
.3142	18°00'	3.236	1.051	3.078	.3249	.9511	.3090	72°00'	1.2566
.3171	10	3.207	1.052	3.047	.3281	.9502	.3118	50	1.2537
.3200	20	3.179	1.053	3.018	.3314	.9492	.3145	40	1.2508
.3229	30	3.152	1.054	2.989	.3346	.9483	.3173	30	1.2479
.3258	40	3.124	1.056	2.960	.3378	.9474	.3201	20	1.2450
.3287	50	3.098	1.057	2.932	.3411	.9465	.3228	10	1.2421
.3316	19°00'	3.072	1.058	2.904	.3443	.9455	.3256	71°00'	1.2392
.3345	10	3.046	1.059	2.877	.3476	.9446	.3283	50	1.2363
.3374	20	3.021	1.060	2.850	.3508	.9436	.3311	40	1.2334
.3403	30	2.996	1.061	2.824	.3541	.9426	.3338	30	1.2305
.3432	40	2.971	1.062	2.798	.3574	.9417	.3365	20	1.2275
.3462	50	2.947	1.063	2.773	.3607	.9407	.3393	10	1.2246
.3491	20°00'	2.924	1.064	2.747	.3640	.9397	.3420	70°00'	1.2217
.3520	10	2.901	1.065	2.723	.3673	.9387	.3448	50	1.2188
.3549	20	2.878	1.066	2.699	.3706	.9377	.3475	40	1.2159
.3578	30	2.855	1.068	2.675	.3739	.9367	.3502	30	1.2130
.3607	40	2.833	1.069	2.651	.3772	.9356	.3529	20	1.2101
.3636	50	2.812	1.070	2.628	.3805	.9346	.3557	10	1.2072
.3665	21°00'	2.790	1.071	2.605	.3839	.9336	.3584	69°00'	1.2043
t	degrees	sec t	csc t	tan t	cot t	sin t	cos t		

Table 2. Values of the Trigonometric Functions (cont'd)

t	degrees	csc t	sec t	cot t	tan t	cos t	sin t	degrees	t
.9599	55°00'	1.743	1.221	1.428	.7002	.8192	.5736	35°00'	.6109
.9570	50	1.736	1.223	1.419	.7046	.8175	.5760	10	.6138
.9541	40	1.729	1.226	1.411	.7089	.8158	.5783	20	.6167
.9512	30	1.722	1.228	1.402	.7133	.8141	.5807	30	.6196
.9483	20	1.715	1.231	1.393	.7177	.8124	.5831	40	.6225
.9454	10	1.708	1.233	1.385	.7221	.8107	.5854	50	.6254
.9425	54°00'	1.701	1.236	1.376	.7265	.8090	.5878	36°00'	.6283
.9396	50	1.695	1.239	1.368	.7310	.8073	.5901	10	.6312
.9367	40	1.688	1.241	1.360	.7355	.8056	.5925	20	.6341
.9338	30	1.681	1.244	1.351	.7400	.8039	.5948	30	.6370
.9308	20	1.675	1.247	1.343	.7445	.8021	.5972	40	.6400
.9279	10	1.668	1.249	1.335	.7490	.8004	.5995	50	.6429
.9250	53°00'	1.662	1.252	1.327	.7536	.7986	.6018	37°00'	.6458
.9221	50	1.655	1.255	1.319	.7581	.7969	.6041	10	.6487
.9192	40	1.649	1.258	1.311	.7627	.7951	.6065	20	.6516
.9163	30	1.643	1.260	1.303	.7673	.7934	.6088	30	.6545
.9134	20	1.636	1.263	1.295	.7720	.7916	.6111	40	.6574
.9105	10	1.630	1.266	1.288	.7766	.7898	.6134	50	.6603
.9076	52°00'	1.624	1.269	1.280	.7813	.7880	.6157	38°00'	.6632
.9047	50	1.618	1.272	1.272	.7860	.7862	.6180	10	.6661
.9018	40	1.612	1.275	1.265	.7907	.7844	.6202	20	.6690
.8988	30	1.606	1.278	1.257	.7954	.7826	.6225	30	.6720
.8959	20	1.601	1.281	1.250	.8002	.7808	.6248	40	.6749
.8930	10	1.595	1.284	1.242	.8050	.7790	.6271	50	.6778
.8901	51°00'	1.589	1.287	1.235	.8098	.7771	.6293	39°00'	.6807
.8872	50	1.583	1.290	1.228	.8146	.7753	.6316	10	.6836
.8843	40	1.578	1.293	1.220	.8195	.7735	.6338	20	.6865
.8814	30	1.572	1.296	1.213	.8243	.7716	.6361	30	.6894
.8785	20	1.567	1.299	1.206	.8292	.7698	.6383	40	.6923
.8756	10	1.561	1.302	1.199	.8342	.7679	.6406	50	.6952
.8727	50°00'	1.556	1.305	1.192	.8391	.7660	.6428	40°00'	.6981
.8698	50	1.550	1.309	1.185	.8441	.7642	.6450	10	.7010
.8668	40	1.545	1.312	1.178	.8491	.7623	.6472	20	.7039
.8639	30	1.540	1.315	1.171	.8541	.7604	.6494	30	.7069
.8610	20	1.535	1.318	1.164	.8591	.7585	.6517	40	.7098
.8581	10	1.529	1.322	1.157	.8642	.7566	.6539	50	.7127
.8552	49°00'	1.524	1.325	1.150	.8693	.7547	.6561	41°00'	.7156
.8523	50	1.519	1.328	1.144	.8744	.7528	.6583	10	.7185
.8494	40	1.514	1.332	1.137	.8796	.7509	.6604	20	.7214
.8465	30	1.509	1.335	1.130	.8847	.7490	.6626	30	.7243
.8436	20	1.504	1.339	1.124	.8899	.7470	.6648	40	.7272
.8407	10	1.499	1.342	1.117	.8952	.7451	.6670	50	.7301
.8378	48°00'	1.494	1.346	1.111	.9004	.7431	.6691	42°00'	.7330
t	degrees	sec t	csc t	tan t	cot t	sin t	cos t	degrees	t

t	degrees	sin t	cos t	tan t	cot t	sec t	csc t	degrees	t
.4887	28°00'	.4695	.8829	.5317	1.881	1.133	2.130	62°00'	1.0821
.4916	10	.4720	.8816	.5354	1.868	1.134	2.118	50	1.0792
.4945	20	.4746	.8802	.5392	1.855	1.136	2.107	40	1.0763
.4974	30	.4772	.8788	.5430	1.842	1.138	2.096	30	1.0734
.5003	40	.4797	.8774	.5467	1.829	1.140	2.085	20	1.0705
.5032	50	.4823	.8760	.5505	1.816	1.142	2.074	10	1.0676
.5061	29°00'	.4848	.8746	.5543	1.804	1.143	2.063	61°00'	1.0647
.5091	10	.4874	.8732	.5581	1.792	1.145	2.052	50	1.0617
.5120	20	.4899	.8718	.5619	1.780	1.147	2.041	40	1.0588
.5149	30	.4924	.8704	.5658	1.767	1.149	2.031	30	1.0559
.5178	40	.4950	.8689	.5696	1.756	1.151	2.020	20	1.0530
.5207	50	.4975	.8675	.5735	1.744	1.153	2.010	10	1.0501
.5236	30°00'	.5000	.8660	.5774	1.732	1.155	2.000	60°00'	1.0472
.5265	10	.5025	.8646	.5812	1.720	1.157	1.990	50	1.0443
.5294	20	.5050	.8631	.5851	1.709	1.159	1.980	40	1.0414
.5323	30	.5075	.8616	.5890	1.698	1.161	1.970	30	1.0385
.5352	40	.5100	.8601	.5930	1.686	1.163	1.961	20	1.0356
.5381	50	.5125	.8587	.5969	1.675	1.165	1.951	10	1.0327
.5411	31°00'	.5150	.8572	.6009	1.664	1.167	1.942	59°00'	1.0297
.5440	10	.5175	.8557	.6048	1.653	1.169	1.932	50	1.0268
.5469	20	.5200	.8542	.6088	1.643	1.171	1.923	40	1.0239
.5498	30	.5225	.8526	.6128	1.632	1.173	1.914	30	1.0210
.5527	40	.5250	.8511	.6168	1.621	1.175	1.905	20	1.0181
.5556	50	.5275	.8496	.6208	1.611	1.177	1.896	10	1.0152
.5585	32°00'	.5299	.8480	.6249	1.600	1.179	1.887	58°00'	1.0123
.5614	10	.5324	.8465	.6289	1.590	1.181	1.878	50	1.0094
.5643	20	.5348	.8450	.6330	1.580	1.184	1.870	40	1.0065
.5672	30	.5373	.8434	.6371	1.570	1.186	1.861	30	1.0036
.5701	40	.5398	.8418	.6412	1.560	1.188	1.853	20	1.0007
.5730	50	.5422	.8403	.6453	1.550	1.190	1.844	10	.9977
.5760	33°00'	.5446	.8387	.6494	1.540	1.192	1.836	57°00'	.9948
.5789	10	.5471	.8371	.6536	1.530	1.195	1.828	50	.9919
.5818	20	.5495	.8355	.6577	1.520	1.197	1.820	40	.9890
.5847	30	.5519	.8339	.6619	1.511	1.199	1.812	30	.9861
.5876	40	.5544	.8323	.6661	1.501	1.202	1.804	20	.9832
.5905	50	.5568	.8307	.6703	1.492	1.204	1.796	10	.9803
.5934	34°00'	.5592	.8290	.6745	1.483	1.206	1.788	56°00'	.9774
.5963	10	.5616	.8274	.6787	1.473	1.209	1.781	50	.9745
.5992	20	.5640	.8258	.6830	1.464	1.211	1.773	40	.9716
.6021	30	.5664	.8241	.6873	1.455	1.213	1.766	30	.9687
.6050	40	.5688	.8225	.6916	1.446	1.216	1.758	20	.9657
.6080	50	.5712	.8208	.6959	1.437	1.218	1.751	10	.9628
.6109	35°00'	.5736	.8192	.7002	1.428	1.221	1.743	55°00'	.9599
t	degrees	cos t	sin t	cot t	tan t	csc t	sec t	degrees	t

TABLE 3.   TRIGONOMETRIC FUNCTIONS OF RADIANS AND REAL NUMBERS   **A5**

Table 3. Trigonometric Functions of Radians and Real Numbers

t	sin t	cos t	tan t	cot t	sec t	csc t
.00	.0000	1.0000	.0000	—	1.000	—
.01	.0100	1.0000	.0100	99.997	1.000	100.00
.02	.0200	.9998	.0200	49.993	1.000	50.00
.03	.0300	.9996	.0300	33.323	1.000	33.34
.04	.0400	.9992	.0400	24.987	1.001	25.01
.05	.0500	.9988	.0500	19.983	1.001	20.01
.06	.0600	.9982	.0601	16.647	1.002	16.68
.07	.0699	.9976	.0701	14.262	1.002	14.30
.08	.0799	.9968	.0802	12.473	1.003	12.51
.09	.0899	.9960	.0902	11.081	1.004	11.13
.10	.0998	.9950	.1003	9.967	1.005	10.02
.11	.1098	.9940	.1104	9.054	1.006	9.109
.12	.1197	.9928	.1206	8.293	1.007	8.353
.13	.1296	.9916	.1307	7.649	1.009	7.714
.14	.1395	.9902	.1409	7.096	1.010	7.166
.15	.1494	.9888	.1511	6.617	1.011	6.692
.16	.1593	.9872	.1614	6.197	1.013	6.277
.17	.1692	.9856	.1717	5.826	1.015	5.911
.18	.1790	.9838	.1820	5.495	1.016	5.586
.19	.1889	.9820	.1923	5.200	1.018	5.295
.20	.1987	.9801	.2027	4.933	1.020	5.033
.21	.2085	.9780	.2131	4.692	1.022	4.797
.22	.2182	.9759	.2236	4.472	1.025	4.582
.23	.2280	.9737	.2341	4.271	1.027	4.386
.24	.2377	.9713	.2447	4.086	1.030	4.207
.25	.2474	.9689	.2553	3.916	1.032	4.042
.26	.2571	.9664	.2660	3.759	1.035	3.890
.27	.2667	.9638	.2768	3.613	1.038	3.749
.28	.2764	.9611	.2876	3.478	1.041	3.619
.29	.2860	.9582	.2984	3.351	1.044	3.497
.30	.2955	.9553	.3093	3.233	1.047	3.384
.31	.3051	.9523	.3203	3.122	1.050	3.278
.32	.3146	.9492	.3314	3.018	1.053	3.179
.33	.3240	.9460	.3425	2.920	1.057	3.086
.34	.3335	.9428	.3537	2.827	1.061	2.999
.35	.3429	.9394	.3650	2.740	1.065	2.916
.36	.3523	.9359	.3764	2.657	1.068	2.839
.37	.3616	.9323	.3879	2.578	4.073	2.765
.38	.3709	.9287	.3994	2.504	1.077	2.696
.39	.3802	.9249	.4111	2.433	1.081	2.630

Table 2. Values of the Trigonometric Functions (cont'd)

t	t degrees	sin t	cos t	tan t	cot t	sec t	csc t	t degrees	t
.7330	42°00'	.6691	.7431	.9004	1.111	1.346	1.494	48°00'	.8378
.7359	10	.6713	.7412	.9057	1.104	1.349	1.490	50	.8348
.7389	20	.6734	.7392	.9110	1.098	1.353	1.485	40	.8319
.7418	30	.6756	.7373	.9163	1.091	1.356	1.480	30	.8290
.7447	40	.6777	.7353	.9217	1.085	1.360	1.476	20	.8261
.7476	50	.6799	.7333	.9271	1.079	1.364	1.471	10	.8232
.7505	43°00'	.6820	.7314	.9325	1.072	1.367	1.466	47°00'	.8203
.7534	10	.6841	.7294	.9380	1.066	1.371	1.462	50	.8174
.7563	20	.6862	.7274	.9435	1.060	1.375	1.457	40	.8145
.7592	30	.6884	.7254	.9490	1.054	1.379	1.453	30	.8116
.7621	40	.6905	.7234	.9545	1.048	1.382	1.448	20	.8087
.7650	50	.6926	.7214	.9601	1.042	1.386	1.444	10	.8058
.7679	44°00'	.6947	.7193	.9657	1.036	1.390	1.440	46°00'	.8029
.7709	10	.6967	.7173	.9713	1.030	1.394	1.435	50	.7999
.7738	20	.6988	.7153	.9770	1.024	1.398	1.431	40	.7970
.7767	30	.7009	.7133	.9827	1.018	1.402	1.427	30	.7941
.7796	40	.7030	.7112	.9884	1.012	1.406	1.423	20	.7912
.7825	50	.7050	.7092	.9942	1.006	1.410	1.418	10	.7883
.7854	45°00'	.7071	.7071	1.0000	1.0000	1.414	1.414	45°00'	.7854
		cos t	sin t	cot t	tan t	csc t	sec t	t degrees	t

Table 3.   Trigonometric Functions of Radians and Real Numbers (cont'd)

t	sin t	cos t	tan t	cot t	sec t	csc t
.80	.7174	.6967	1.030	.9712	1.435	1.394
.81	.7243	.6895	1.050	.9520	1.450	1.381
.82	.7311	.6822	1.072	.9331	1.466	1.368
.83	.7379	.6749	1.093	.9146	1.482	1.355
.84	.7446	.6675	1.116	.8964	1.498	1.343
.85	.7513	.6600	1.138	.8785	1.515	1.331
.86	.7578	.6524	1.162	.8609	1.533	1.320
.87	.7643	.6448	1.185	.8437	1.551	1.308
.88	.7707	.6372	1.210	.8267	1.569	1.297
.89	.7771	.6294	1.235	.8100	1.589	1.287
.90	.7833	.6216	1.260	.7936	1.609	1.277
.91	.7895	.6137	1.286	.7774	1.629	1.267
.92	.7956	.6058	1.313	.7615	1.651	1.257
.93	.8016	.5978	1.341	.7458	1.673	1.247
.94	.8076	.5898	1.369	.7303	1.696	1.238
.95	.8134	.5817	1.398	.7151	1.719	1.229
.96	.8192	.5735	1.428	.7001	1.744	1.221
.97	.8249	.5653	1.459	.6853	1.769	1.212
.98	.8305	.5570	1.491	.6707	1.795	1.204
.99	.8360	.5487	1.524	.6563	1.823	1.196
1.00	.8415	.5403	1.557	.6421	1.851	1.188
1.01	.8468	.5319	1.592	.6281	1.880	1.181
1.02	.8521	.5234	1.628	.6142	1.911	1.174
1.03	.8573	.5148	1.665	.6005	1.942	1.166
1.04	.8624	.5062	1.704	.5870	1.975	1.160
1.05	.8674	.4976	1.743	.5736	2.010	1.153
1.06	.8724	.4889	1.784	.5604	2.046	1.146
1.07	.8772	.4801	1.827	.5473	2.083	1.140
1.08	.8820	.4713	1.871	.5344	2.122	1.134
1.09	.8866	.4625	1.917	.5216	2.162	1.128
1.10	.8912	.4536	1.965	.5090	2.205	1.122
1.11	.8957	.4447	2.014	.4964	2.249	1.116
1.12	.9001	.4357	2.066	.4840	2.295	1.111
1.13	.9044	.4267	2.120	.4718	2.344	1.106
1.14	.9086	.4176	2.176	.4596	2.395	1.101
1.15	.9128	.4085	2.234	.4475	2.448	1.096
1.16	.9168	.3993	2.296	.4356	2.504	1.091
1.17	.9208	.3902	2.360	.4237	2.563	1.086
1.18	.9246	.3809	2.427	.4120	2.625	1.082
1.19	.9284	.3717	2.498	.4003	2.691	1.077

t	sin t	cos t	tan t	cot t	sec t	csc t
.40	.3894	.9211	.4228	2.365	1.086	2.568
.41	.3986	.9171	.4346	2.301	1.090	2.509
.42	.4078	.9131	.4466	2.239	1.095	2.452
.43	.4169	.9090	.4586	2.180	1.100	2.399
.44	.4259	.9048	.4708	2.124	1.105	2.348
.45	.4350	.9004	.4831	2.070	1.111	2.299
.46	.4439	.8961	.4954	2.018	1.116	2.253
.47	.4529	.8916	.5080	1.969	1.122	2.208
.48	.4618	.8870	.5206	1.921	1.127	2.166
.49	.4706	.8823	.5334	1.875	1.133	2.125
.50	.4794	.8776	.5463	1.830	1.139	2.086
.51	.4882	.8727	.5594	1.788	1.146	2.048
.52	.4969	.8678	.5726	1.747	1.152	2.013
.53	.5055	.8628	.5859	1.707	1.159	1.978
.54	.5141	.8577	.5994	1.668	1.166	1.945
.55	.5227	.8525	.6131	1.631	1.173	1.913
.56	.5312	.8473	.6269	1.595	1.180	1.883
.57	.5396	.8419	.6410	1.560	1.188	1.853
.58	.5480	.8365	.6552	1.526	1.196	1.825
.59	.5564	.8309	.6696	1.494	1.203	1.797
.60	.5646	.8253	.6841	1.462	1.212	1.771
.61	.5729	.8196	.6989	1.431	1.220	1.746
.62	.5810	.8139	.7139	1.401	1.229	1.721
.63	.5891	.8080	.7291	1.372	1.238	1.697
.64	.5972	.8021	.7445	1.343	1.247	1.674
.65	.6052	.7961	.7602	1.315	1.256	1.652
.66	.6131	.7900	.7761	1.288	1.266	1.631
.67	.6210	.7838	.7923	1.262	1.276	1.610
.68	.6288	.7776	.8087	1.237	1.286	1.590
.69	.6365	.7712	.8253	1.212	1.297	1.571
.70	.6442	.7648	.8423	1.187	1.307	1.552
.71	.6518	.7584	.8595	1.163	1.319	1.534
.72	.6594	.7518	.8771	1.140	1.330	1.517
.73	.6669	.7452	.8949	1.117	1.342	1.500
.74	.6743	.7385	.9131	1.095	1.354	1.483
.75	.6816	.7317	.9316	1.073	1.367	1.467
.76	.6889	.7248	.9505	1.052	1.380	1.452
.77	.6961	.7179	.9697	1.031	1.393	1.437
.78	.7033	.7109	.9893	1.011	1.407	1.422
.79	.7104	.7038	1.009	.9908	1.421	1.408

TABLE 4.   POWERS AND ROOTS   **A7**

Table 4.   *Powers and Roots*

n	n²	√n	n³	∛n
51	2,601	7.141	132,651	3.708
52	2,704	7.211	140,608	3.733
53	2,809	7.280	148,877	3.756
54	2,916	7.348	157,464	3.780
55	3,025	7.416	166,375	3.803
56	3,136	7.483	175,616	3.826
57	3,249	7.550	185,193	3.849
58	3,364	7.616	195,112	3.871
59	3,481	7.681	205,379	3.893
60	3,600	7.746	216,000	3.915
61	3,721	7.810	226,981	3.936
62	3,844	7.874	238,328	3.958
63	3,969	7.937	250,047	3.979
64	4,096	8.000	262,144	4.000
65	4,225	8.062	274,625	4.021
66	4,356	8.124	287,496	4.041
67	4,489	8.185	300,763	4.062
68	4,624	8.246	314,432	4.082
69	4,761	8.307	328,509	4.102
70	4,900	8.367	343,000	4.121
71	5,041	8.426	357,911	4.141
72	5,184	8.485	373,248	4.160
73	5,329	8.544	389,017	4.179
74	5,476	8.602	405,224	4.198
75	5,625	8.660	421,875	4.217
76	5,776	8.718	438,976	4.236
77	5,929	8.775	456,533	4.254
78	6,084	8.832	474,552	4.273
79	6,241	8.888	493,039	4.291
80	6,400	8.944	512,000	4.309
81	6,561	9.000	531,441	4.327
82	6,724	9.055	551,368	4.344
83	6,889	9.110	571,787	4.362
84	7,056	9.165	592,704	4.380
85	7,225	9.220	614,125	4.397
86	7,396	9.274	636,056	4.414
87	7,569	9.327	658,503	4.431
88	7,744	9.381	681,472	4.448
89	7,921	9.434	704,969	4.465
90	8,100	9.487	729,000	4.481
91	8,281	9.539	753,571	4.498
92	8,464	9.592	778,688	4.514
93	8,649	9.644	804,357	4.531
94	8,836	9.695	830,584	4.547
95	9,025	9.747	857,375	4.563
96	9,216	9.798	884,736	4.579
97	9,409	9.849	912,673	4.595
98	9,604	9.899	941,192	4.610
99	9,801	9.950	970,299	4.626
100	10,000	10.000	1,000,000	4.642

n	n²	√n	n³	∛n
1	1	1.000	1	1.000
2	4	1.414	8	1.260
3	9	1.732	27	1.442
4	16	2.000	64	1.587
5	25	2.236	125	1.710
6	36	2.449	216	1.817
7	49	2.646	343	1.913
8	64	2.828	512	2.000
9	81	3.000	729	2.080
10	100	3.162	1,000	2.154
11	121	3.317	1,331	2.224
12	144	3.464	1,728	2.289
13	169	3.606	2,197	2.351
14	196	3.742	2,744	2.410
15	225	3.873	3,375	2.466
16	256	4.000	4,096	2.520
17	289	4.123	4,913	2.571
18	324	4.243	5,832	2.621
19	361	4.359	6,859	2.668
20	400	4.472	8,000	2.714
21	441	4.583	9,261	2.759
22	484	4.690	10,648	2.802
23	529	4.796	12,167	2.844
24	576	4.899	13,824	2.884
25	625	5.000	15,625	2.924
26	676	5.099	17,576	2.962
27	729	5.196	19,683	3.000
28	784	5.292	21,952	3.037
29	841	5.385	24,389	3.072
30	900	5.477	27,000	3.107
31	961	5.568	29,791	3.141
32	1,024	5.657	32,768	3.175
33	1,089	5.745	35,937	3.208
34	1,156	5.831	39,304	3.240
35	1,225	5.916	42,875	3.271
36	1,296	6.000	46,656	3.302
37	1,369	6.083	50,653	3.332
38	1,444	6.164	54,872	3.362
39	1,521	6.245	59,319	3.391
40	1,600	6.325	64,000	3.420
41	1,681	6.403	68,921	3.448
42	1,764	6.481	74,088	3.476
43	1,849	6.557	79,507	3.503
44	1,936	6.633	85,184	3.530
45	2,025	6.708	91,125	3.557
46	2,116	6.782	97,336	3.583
47	2,209	6.856	103,823	3.609
48	2,304	6.928	110,592	3.634
49	2,401	7.000	117,649	3.659
50	2,500	7.071	125,000	3.684

Table 3.   *Trigonometric Functions of Radians and Real Numbers (cont'd)*

t	sin t	cos t	tan t	cot t	sec t	csc t
1.20	.9320	.3624	2.572	.3888	2.760	1.073
1.21	.9356	.3530	2.650	.3773	2.833	1.069
1.22	.9391	.3436	2.733	.3659	2.910	1.065
1.23	.9425	.3342	2.820	.3546	2.992	1.061
1.24	.9458	.3248	2.912	.3434	3.079	1.057
1.25	.9490	.3153	3.010	.3323	3.171	1.054
1.26	.9521	.3058	3.113	.3212	3.270	1.050
1.27	.9551	.2963	3.224	.3102	3.375	1.047
1.28	.9580	.2867	3.341	.2993	3.488	1.044
1.29	.9608	.2771	3.467	.2884	3.609	1.041
1.30	.9636	.2675	3.602	.2776	3.738	1.038
1.31	.9662	.2579	3.747	.2669	3.878	1.035
1.32	.9687	.2482	3.903	.2562	4.029	1.032
1.33	.9711	.2385	4.072	.2456	4.193	1.030
1.34	.9735	.2288	4.256	.2350	4.372	1.027
1.35	.9757	.2190	4.455	.2245	4.566	1.025
1.36	.9779	.2092	4.673	.2140	4.779	1.023
1.37	.9799	.1994	4.913	.2035	5.014	1.021
1.38	.9819	.1896	5.177	.1931	5.273	1.018
1.39	.9837	.1798	5.471	.1828	5.561	1.017
1.40	.9854	.1700	5.798	.1725	5.883	1.015
1.41	.9871	.1601	6.165	.1622	6.246	1.013
1.42	.9887	.1502	6.581	.1519	6.657	1.011
1.43	.9901	.1403	7.055	.1417	7.126	1.010
1.44	.9915	.1304	7.602	.1315	7.667	1.009
1.45	.9927	.1205	8.238	.1214	8.299	1.007
1.46	.9939	.1106	8.989	.1113	9.044	1.006
1.47	.9949	.1006	9.887	.1011	9.938	1.005
1.48	.9959	.0907	10.983	.0910	11.029	1.004
1.49	.9967	.0807	12.350	.0810	12.390	1.003
1.50	.9975	.0707	14.101	.0709	14.137	1.003
1.51	.9982	.0608	16.428	.0609	16.458	1.002
1.52	.9987	.0508	19.670	.0508	19.695	1.001
1.53	.9992	.0408	24.498	.0408	24.519	1.001
1.54	.9995	.0308	32.461	.0308	32.476	1.000
1.55	.9998	.0208	48.078	.0208	48.089	1.000
1.56	.9999	.0108	92.620	.0108	92.626	1.000
1.57	1.0000	.0008	1255.8	.0008	1255.8	1.000

ANSWERS TO ODD-NUMBERED EXERCISES

CHAPTER 1

Section 1.2, page 8

1 Commutative   **3** Associative   **5** Identity   **7** Identity   **9** Inverse
11 $ab + 3a + 2b + 6$ **13** $ab + 3a + 2b + 6$ **15** $2xy + 4x - 3y - 6$ **17** $2xy + 4x - 3y - 6$
19 $12rs + 15s + 24r + 30$ **21** $29/12$ **23** $5/9$ **25** $11/12$ **27** $13/10$ **29** $10/21$
31 Two examples are $2 - 3 \neq 3 - 2$ and $2 - (3 - 4) \neq (2 - 3) - 4$.
33 If $a = -a$, then $a + a = 0$, or $2a = 0$. Multiplying both sides by $1/2$ produces $a = 0$. Conversely, if $a = 0$, then also $(-1)a = (-1)0 = 0$ and hence $-a = 0$. Thus $a = 0 = -a$.
35 If $ab = 0$ and $a \neq 0$, then $(1/a)ab = (1/a)0 = 0$, and hence $b = 0$.
37 Since $(a/b)(b/a) = ab/ba = 1$, b/a is the multiplicative inverse of a/b; that is, $b/a = (a/b)^{-1}$
39 We shall use the fact that if $x + y = 0$, then $y = -x$.
   (i) Since $-a + a = 0$ we may take $x = -a$ and $y = a$ above, obtaining $a = -(-a)$.
   (ii) Since $ab + (-a)b = [a + (-a)]b = (0)b = 0$, we take $x = ab$ and $y = (-a)b$, obtaining $(-a)b = -(ab)$. Similarly, $a(-b) = -(ab)$.
   (iii) Using (ii) twice and then (i) we obtain $(-a)(-b) = -[a(-b)] = -[-(ab)] = ab$.
   (iv) Using (ii), $(-1)a = -(1 \cdot a) = -a$.

Section 1.3, page 13

1 (a) $<$ (b) $>$ (c) $=$ **3** (a) $>$ (b) $<$ (c) $>$ **5** $-8 < -5$ **7** $0 > -1$ **9** $x < 0$
11 $3 < a < 5$ **13** $b \geq 2$ **15** $c \leq 1$ **17** (a) 5   (b) -5 (c) 13   **19** (a) 0   (b) $4 - \pi$ (c) -1
21 (a) 4   (b) 6   (c) 6   (d) 10   **23** (a) 12   (b) 3   (c) 3   (d) 9
25 (a) No. For example, $2 - 5$ is not positive.   (b) Yes. If a and b are positive, so is a/b.
27 $x - 5$ **29** $4 + x^2$

Section 1.4, page 19

1 $-27/125$ **3** 15   **5** $1/10$ **7** 1   **9** $27/625$ **11** $12a^8$ **13** $1/2c^3$ **15** $48x^8$
17 16   **19** $4t^4u^2$ **21** $-10ab^4c$ **23** $2x^2$ **25** $4a^2$ **27** $64a^6/b^3$ **29** $-8y^5/5u^5$

31 $-a^{13}b^4/3$ **33** $25v^4/t^2$ **35** 1   **37** $y^6/9x^4$ **39** $4/v^2$ **41** $3a + 2b$ **43** y^6/x^{12}
45 $1/a$ **47** $-200s^8v^7/r^3p$ **49** 1

Section 1.5, page 25

1 9   **3** $-2\sqrt[3]{2}$ **5** $\sqrt[3]{4}/2$ **7** $8\sqrt{3}$ **9** $15\sqrt{5}$ **11** $3y^3/x^2$ **13** $2a^2/b$
15 $(3a^2/b)\sqrt[3]{2ab}$ **17** $\sqrt{3uv}/3u^2v$ **19** $3x^5/y^2$ **21** $(2x/y^2)\sqrt[5]{x^2y^4}$ **23** $-3tv^2$ **25** $2pyz$
27 $\sqrt[3]{9x}/3x$ **29** $(2x^2/y^2)\sqrt{3xy}$ **31** $|a - b|$ **33** 1   **35** $(rs^6/4)\sqrt[4]{4r^2}$ **37** $3a\sqrt{b}$
39 $(x^2y^2/z)\sqrt[7]{x^2y^4z^3}$ **41** $2a^2b\sqrt{5b}$ **43** $a^2\sqrt[6]{b}$ **45** $\sqrt{ab}/ab$ **47** $(5x^2 + 3x - 1)\sqrt{x}$

Calculator Exercises

5 >   **7** >   **9** >

Section 1.6, page 29

1 $x^{3/4}$ **3** $(a + b)^{2/3}$ **5** $(x^2 + y^2)^{1/2}$
7 (a) $4x\sqrt{x}$ (b) $8x\sqrt{x}$ (c) $4 + x\sqrt{x}$ (d) $(4 + x)\sqrt{4 + x}$
9 16   **11** 0.000027   **13** 1/64   **15** $15x^2$ **17** $18a^{5/6}$ **19** $1/64t^6$ **21** $24c^2\sqrt[5]{c}$
23 $8a^6p^3$ **25** p^2r^4/s^6 **27** $1/s^9$ **29** $u^{3/4}$ **31** 1   **33** $(18bc^5/a^3)\sqrt{c}$ **35** $625x^8/y^4$
37 $a + 1$ **39** $\sqrt{x}$ **41** $\sqrt{cd}$ **43** $\sqrt[4]{a^2b}$ **45** $2\sqrt[6]{2}$ **47** $x\sqrt[6]{xy^5}$ **49** $\sqrt[6]{u^5v^4}/uv$
51 $\sqrt[6]{b^5}$ **53** 1.7×10^{-24} **55** (a) 4.27×10^5 (b) 9.8×10^{-8} (c) 8.1×10^8
57 (a) 830,000 (b) 0.0000000000029 (c) 563,000,000   **59** 6.93   **61** 1.125×10^{24} **63** 3.6×10^9
65 5.87×10^{12} miles

Calculator Exercises

5 -3.050 **7** 0.139   **9** <   **11** >

Section 1.7, page 35

1 $5x^3 - x^2 - 6x + 6$ **3** $5x^4 - 2x^3 - 9x^2 + 17x - 4$ **5** $-12y^2 - 9$ **7** $3a^4 - 4a + 5$
9 $6x^3 - 5x^2 - 19x + 20$ **11** $3r^4 + 4r^3 + 9r^2 + 2r + 12$ **13** $12x^5 + 18x^4 + 16x^3 - 5x^2 + 23x + 35$
15 $r^3 - t^3$ **17** $6x^5 - x^4 + 29x^3 - 2x^2 + 20x + 8$ **19** $4y^2 - 5x$ **21** $3v^2 - 2u^2 + uv^2$
23 $2x^2 - 5x - 3$ **25** $8s^2 - 38ts + 35t^2$ **27** $15x^4 - 29x^2y - 14y^2$ **29** $36t^2 - 25v^2$
31 $9r^2 + 60rs + 100s^2$ **33** $16x^4 - 40x^2y^2 + 25y^4$ **35** $x^2 - 2 + (1/x^2)$ **37** $a - b$
39 $x^3 - 6x^2y + 12xy^2 - 8y^3$ **41** $27r^3 + 108r^2s + 144rs^2 + 64s^3$ **43** $x^6 + 3x^4y^2 + 3x^2y^4 + y^6$
45 $a - 3a^{2/3}b^{1/3} + 3a^{1/3}b^{2/3} - b$ **47** $x^2 + 2xy + y^2 - z^2$ **49** $9x^2 + 12xy + 6xz + 4y^2 + 4yz + z^2$

Section 1.8, page 40

1 $s(r + 4t)$ **3** $3a^2b(b - 2)$ **5** $3xy^2(3x + 5y^2)$ **7** $(4x - 3)(x + 2)$ **9** $(2x - 5)(2x + 3)$
11 $(8y + 5)(2y - 3)$ **13** $(3z - 2)^2$ **15** $(3c + 8)(2c + 3)$ **17** $(2r + 5t)(2r - 5t)$
19 $(10x - 3y)(5x + 6y)$ **21** $(4w^2 + 3s)(4w^2 - 3s)$ **23** $(6z + 5)^2$ **25** $(3x - y)(9x^2 + 3xy + y^2)$
27 $(3a + 4b)(9a^2 - 12ab + 16b^2)$ **29** $(2x^2 - 5)(4x^4 + 10x^2 + 25)$ **31** $(3a + b)(2x - y)$
33 $(5z + x^2)(8w - 7)$ **35** $(a^2 + b^2)(a - b)$ **37** $(a - b)(a^2 + ab + b^2)(a + b)(a^2 - ab + b^2)$
39 Irreducible   **41** $6(x + 2)(x + 5)$ **43** $4(x - 3)^2$ **45** $3(5x + 4)^2$ **47** $9(2 + x)(2 - x)$
49 $(2x + 1)(xy - z)$ **51** $(3x^2 + 2y^2)(4z - 5w)$ **53** $(y + 2)(y^2 - 2y + 4)(y - 1)(y^2 + y + 1)$
55 $(x + y - 3)(x^2 + 2xy + y^2 + 3x + 3y + 9)$ **57** Irreducible
59 $(x^8 + 1)(x^4 + 1)(x^2 + 1)(x + 1)(x - 1)$

Section 1.9, page 45

1 $(x + 2)/(x + 3)$ **3** $3y/(5y + 1)$ **5** $-(a + 2)/(2a + 1)$ **7** $(2x - 3)/x(5x - 2)$
9 $34r/(3r - 1)(2r + 5)$ **11** $6x/(2x - 1)(x + 1)$ **13** $3/2(4t^2 + 6t + 9)$ **15** $a/(a^2 + 4)(5a + 2)$
17 $(6x - 7)/(3x + 1)^2$ **19** $(3 - 2c)/c^3$ **21** $(5x^2 - 2x - 6)/(x - 1)^3$
23 $(18x^3 - 10x^2 - 66x + 63)/x^2(2x - 3)^2$ **25** $(p^2 + 2p + 4)/(p - 3)$ **27** $-2/(2x + 2h + 3)(2x + 3)$
29 $-(3x^2 + 3xh + h^2)/x^3(x + h)^3$ **31** $11/(7x - 3)(2x + 1)$ **33** $a - b$ **35** $(x^2 + xy + y^2)/(x + y)$
37 $(2x^2 + 7x + 15)/(x^2 + 10x + 7)$ **39** $a - b$ **41** $(\sqrt{a} - \sqrt{b})/(a - b)$ **43** $(y\sqrt{x} + x\sqrt{y})/xy$
45 $c(1 + \sqrt{c})/(1 - c)$ **47** $(a + b - \sqrt{c})/(a^2 + 2ab + b^2 - c)$ **49** $-1/(x + h)x$
51 $(3x + 1)^5(7x - 9)(2x - 5)^{1/2}/(2x - 5)$ **53** $(x^2 + 12)(x^2 + 4)^{2/3}/3(x^2 + 4)^2$

Section 1.10, Review Exercises, page 47

1 (a) $-5/12$ (b) $39/20$ (c) $-13/56$ (d) $5/8$
3 (a) $x < 0$ (b) $1/3 < a < 1/2$ (c) $|x| \leq 4$ **5** (a) 5 (b) 5 (c) 7 **7** $18a^5b^5$ **9** $xy^5/9$
11 $-p^8/2q$ **13** x^3z/y^{10} **15** b^6/a^2 **17** $s + r$ **19** x^8/y^2 **21** $\sqrt[3]{2}/2$ **23** $2x^2y\sqrt[3]{x}$
25 $(1 - \sqrt{t})/t$ **27** $2x/y^2$ **29** $x^{12}\sqrt[3]{x^5}\sqrt[6]{y^5}$ **31** $(\sqrt{a} + \sqrt{b})/(a - b)$
33 (a) 9.37×10^{10} (b) 4.02×10^{-6} **35** $x^4 + x^3 - x^2 + x - 2$ **37** $-x^2 + 18x + 7$
39 $3y^5 - 2y^4 - 8y^3 + 10y^2 - 3y - 12$ **41** $a^4 - b^4$ **43** $6a^2 + 11ab - 35b^2$ **45** $169a^4 - 16b^2$
47 $8a^3 + 12a^2b + 6ab^2 + b^3$ **49** $81x^4 - 72x^2y^2 + 16y^4$ **51** $10w(6x + 7)$ **53** $(14x + 9)(2x - 1)$
55 $(2w + 3x)(y - 4z)$ **57** $8(x + 2y)(x^2 - 2xy + 4y^2)$ **59** $(p^4 + q^4)(p^2 + q^2)(p + q)(p - q)$
61 $(w^2 + 1)(w^4 - w^2 + 1)$ **63** $(3x - 5)/(2x + 1)$ **65** $(3x + 2)/x(x - 2)$
67 $(5x^2 - 6x - 20)/x(x + 2)^2$ **69** $-(2x^2 + x + 3)/x(x + 1)(x + 3)$

CHAPTER 2

Section 2.1, page 53

1 $-7/4$ **3** $5\sqrt{2}/2$ **5** $5/2$ **7** $8/13$ **9** $-12/19$ **11** -6 **13** $34/25$ **15** $10/9$
17 1 **19** $31/79$ **21** $9/38$ **23** $s = -20/39$ **25** 3 **27** $89/48$ **29** No solution
31 No solution **33** $5/9$ **35** $-2/3$ **37** No solution **45** $5/7$
47 Choose any a and b such that $b = -5a/3$. For example, take $a = 3$, $b = -5$, etc.
49 (a) $x + 1 = x + 2$ (b) 2

Section 2.2, page 63

1 $P = I/rt$ **3** $h = 2A/b$ **5** $w = (P - 2l)/2$ **7** $z = (d - ax - by)/c$ **9** $I = E/R$
11 $h = 3V/\pi r^2$ **13** $r = (S - a)/S$ **15** $R_2 = RR_1R_3/(R_1R_3 - RR_3 - RR_1)$ **17** $P = S/(1 + rt)$
19 $r = (3V + \pi h^3)/3\pi h^2$ **21** $r = (a - Sl)/(l - S)$ **23** $m = Ft/(v_1 - v_2)$ **25** $b_1 = (2A - hb_2)/h$
27 $v_1 = v_2 - at$ **29** $n = -rI/(IR - E)$ **31** 43 quarters, 27 dimes **33** 136, 137, 138, 139
35 -40 **37** 88 **39** \$12000 at 13%; \$18000 at $15\frac{1}{2}$% **41** \$1820 at 6%; \$1180 at 5.5%
43 64 seconds; 96 meters and 128 meters, respectively **45** 3 cm **47** After an additional 50 games
49 $20/3$ liters **51** 3 liters **53** 225 miles **55** 36 minutes **57** $500/3$ gm **59** $d(b - a)/b$ gm
61 400 **63** (a) 4050 square feet (b) 2592 square feet (c) 3600 square feet **65** 36 minutes
67 (a) $5/9$ mph (b) $2\frac{2}{9}$ miles **69** (a) 125 (b) 21

Section 2.3, page 70

1 $2/3, -5/2$ **3** $7/5, 1/4$ **5** $-5/4, -6$ **7** $3/4$ **9** $5/6, -2/3$ **11** $3/2, -1$
13 $-1 \pm \sqrt{7}$ **15** $(2 \pm \sqrt{14})/2$ **17** $5/2$ **19** $(-9 \pm \sqrt{21})/10$ **21** $(3 \pm \sqrt{13})/2$ **23** $0, -4/9$

25 $r = \sqrt{3\pi h V / \pi h}$ **27** $d = \sqrt{g m_1 m_2 F} / F$ **29** $t = (-v_0 + \sqrt{v_0^2 + 2gs})/g$ **31** $y = \pm (b/a)\sqrt{x^2 - a^2}$
33 $h = 4R \pm \sqrt{16R^2 - C^2}$ **35** 15, 17   **37** 2 cm decrease   **39** 2 ft
41 (a) $d = \sqrt{(400t)^2 + (200t + 100)^2} = 100\sqrt{20t^2 + 4t + 1}$ (b) 3:30 P.M.   **43** 24 in., 76 in.
45 $\sqrt{150/\pi} \approx 6.9\,\text{cm}$ **47** Until 9:24 A.M.   **49** 7 mph
51 1.5 inches at sides and top, 3 inches at bottom
53 (a) After $25 - 5\sqrt{17}$ seconds and also after $25 + 5\sqrt{17}$ seconds   (b) After 50 seconds   (c) 10,000 feet
55 $-3/5$ **59** $x = (y \pm \sqrt{2y^2 - 1})/2$; $y = -2x \pm \sqrt{8x^2 + 1}$

Calculator Exercises

1 $-0.166, -3.728$ **3** $0.689, -3.524$

Section 2.4, page 76

1 $-3, -3/2, 3/2$ **3** $0, -5/3, \pm\sqrt{2}/2$ **5** 0, 16   **7** 1   **9** $-57/5$ **11** 9/5   **13** $\pm\sqrt{62}/2$
15 6   **17** -1 **19** $-5/4$ **21** 3   **23** 0, 4   **25** $1/2, -1/2, 3, -3$
27 $\pm(1/6)\sqrt{30 + 6\sqrt{13}}, \pm(1/6)\sqrt{30 - 6\sqrt{13}}$ **29** $1/27, -27$ **31** 16/9, 25   **33** 3, 1/2
35 $-5/6, -3/2$ **37** $h = \sqrt{S^2 - \pi^2 r^4}/\pi r$ **39** $x = \pm(a/b)\sqrt{b^2 - y^2}$ **41** $y = (a^{2/3} - x^{2/3})^{3/2}$
43 $2\sqrt[3]{432/\pi} \approx 10.3\,\text{cm}$

Calculator Exercises

1 0, 106.17   **3** $3.13, -0.02$

Section 2.5, page 84

1 (a) $-1 < 1$ (b) $-11 < -9$ (c) $-3 < -2$ (d) $3 > 2$
3 (2, 5)   **5** $(-1, 3]$ **7** $[1, 4]$ **9** $(-1, \infty)$ **11** $(-\infty, 2]$

13 $-1 < x < 7$ **15** $8 < x \le 9$ **17** $5 < x$ **19** $(17/5, \infty)$ **21** $[-2, \infty)$ **23** $(5/2, \infty)$
25 $(-\infty, 44/3]$ **27** $(-3, 1)$ **29** $[1, 5]$ **31** $[2, 8/3)$ **33** $(6/38, \infty)$ **35** $(-2/3, \infty)$
37 $(-7/3, \infty)$ **39** $(1/4, \infty)$ **41** $(-\infty, 1)$ and $(1, \infty)$ **43** $140/9 \le c \le 80/3$ **45** $20/9 \le x \le 4$
47 The result is false if $a < 0$ (consider, for example, $a = -2$ and $b = 2$).
49 If $a < b$ and $c < d$, then $b - a$ and $d - c$ are positive and hence $(b - a) + (d - c)$ is positive. Consequently, $(b + d) - (a + c)$ is positive; that is, $a + c < b + d$.

Section 2.6, page 91

1 $(-2, 2)$ **3** $(6, \infty) \cup (-\infty, -6)$ **5** $[-10, 10]$ **7** $(9.95, 10.05)$ **9** $[-10, 2]$ **11** $(-2/3, 4)$
13 $(-\infty, -6) \cup (2, \infty)$ **15** $(-\infty, 1] \cup [4, \infty)$ **17** $(-2, 3)$ **19** $(-\infty, -5/2) \cup (1, \infty)$
21 $[0, 10]$ **23** $(-\infty, -1/2) \cup (10/3, \infty)$ **25** $(-3, 3)$ **27** $(4, \infty) \cup (-4, 0]$
29 $[-5, -3/2) \cup (3/2, 5]$ **31** $(-\infty, 1/2) \cup (7/5, \infty)$ **33** $(-\infty, -1) \cup (0, 1)$
35 $[-2, 1] \cup [2, \infty)$ **37** $(-\infty, -1] \cup (1, 2] \cup (3, \infty)$ **39** $[1/2, 4]$

Section 2.7, Review Exercises, page 92

1 $-5/6$ **3** -32 **5** All $x > 0$ **7** $(-2 \pm \sqrt{19})/3$ **9** $(1 \pm \sqrt{21})/2$ **11** $125, -27$
13 1/4, 1/9   **15** 2   **17** 5   **19** The interval $(-11/4, 9/4)$ **21** $(-\infty, -3/10)$
23 $(-\infty, 11/3] \cup [7, \infty)$ **25** $(-\infty, -3/2) \cup (2/5, \infty)$ **27** $(-\infty, -3/2) \cup (2, 9)$ **29** $(1, \infty)$
31 $R = (S/\pi s) - r$ **33** $R = \sqrt[4]{8nVL/\pi P}$ **35** $R_2 = (n - 1)R_1 R / (R_1 - R(n - 1))$

37 $36 - 25\sqrt{2}$ cm ≈ 0.645 cm   **39** 64 mph   **41** 16 ml   **43** 9 in., 11 in.
45 $d = \sqrt{4 - 200t + 2900t^2}$; $t = (1 + \sqrt{30})/29 \approx 0.22$ hr after 10:00 A.M.

CHAPTER 3

Section 3.1, page 100

1   **3** The line bisecting quadrants I and III.

5 (a) The line parallel to the y-axis which intersects the x-axis at $(3, 0)$
    (b) The line parallel to the x-axis which intersects the y-axis at $(0, -1)$
    (c) All points to the right of, and on, the y-axis
    (d) All points in quadrants I and III
    (e) All points under the x-axis
7 (a) $\sqrt{29}$ (b) $(5, -1/2)$ **9** (a) $\sqrt{13}$ (b) $(-7/2, -1)$ **11** (a) 4 (b) $(5, -3)$ **13** Area $= 28$
17 $(13, -28)$ **19** $d(A, P) = d(B, P)$
21 $\sqrt{x^2 + y^2} = 5$. A circle of radius 5 with center at the origin
23 $(0, 3 + \sqrt{11}), (0, 3 - \sqrt{11})$ **25** $(-2, -1)$ **27** $a > 4$ and $a < 2/5$

Section 3.2, page 108

1   **3**   **5**   **7**   **9**

11   **13**   **15**   **17**   **19**

21   **23**   **25**   **27**

29 Circle of radius 4, center at the origin
31 Circle of radius 1/3, center at the origin
33 Circle of radius 2, center at $(2, -1)$
35 Circle of radius 3, center at $(0, 3)$
37 $(x - 3)^2 + (y + 2)^2 = 16$ **39** $(x - 1/2)^2 + (y + 3/2)^2 = 4$ **41** $x^2 + y^2 = 34$
43 $(x + 4)^2 + (y - 2)^2 = 4$ **45** $(x - 1)^2 + (y - 2)^2 = 34$ **47** $(-1, 5); 4$ **49** $(0, 3); 2$
51 $(5, -4); \sqrt{41}$ **53** $(-1, 1); 1/2,$ **55** $(1/2, -1/3); 1/6$

Calculator Exercises

1   **3**   **5**   **7**

Section 3.3, page 115

1 3, 9, 4, 6   **3** 2, $\sqrt{2} + 6$, 12, 23
5 (a) $5a - 2$ (b) $-5a - 2$ (c) $-5a + 2$ (d) $5a + 5h - 2$ (e) $5a + 5h - 4$ (f) 5
7 (a) $2a^2 - a + 3$ (b) $2a^2 + a + 3$ (c) $-2a^2 + a - 3$ (d) $2a^2 + 4ah + 2h^2 - a - h + 3$
  (e) $2a^2 - a + 2h^2 - h + 6$ (f) $4a + 2h - 1$
9 (a) $3/a^2$ (b) $1/3a^2$ (c) $3a^4$ (d) $9a^4$ (e) $3a$ (f) $\sqrt{3a^2}$
11 (a) $2a/(a^2 + 1)$ (b) $(a^2 + 1)/2a$ (c) $2a^2/(a^4 + 1)$ (d) $4a^2/(a^4 + 2a^2 + 1)$ (e) $2\sqrt{a}/(a + 1)$
  (f) $\sqrt{2a^3 + 2a}/(a^2 + 1)$
13 $[5/3, \infty)$ **15** $[-2, 2]$
17 All real numbers except 0, 3, and -3
19 All nonnegative real numbers except 4 and 3/2
21 $9/7$, $(a + 5)/7$, $\mathbb{R}$ **23** 19, $a^2 + 3$, all nonnegative real numbers   **25** $\sqrt[3]{4}$, $\sqrt[3]{a}$, $\mathbb{R}$
27 Yes   **29** No   **31** Yes   **33** No   **35** Odd   **37** Even   **39** Even   **41** Neither
43 Neither
45 The equation $y = f(x)$ is not changed if $-x$ is substituted for x.
47 $r = C/2\pi$; $6/\pi \approx 1.9$ inches   **49** $V = 4x^3 - 100x^2 + 600x$ **51** $P = 4\sqrt{A}$ **53** $d = 2\sqrt{t^2 + 2500}$
55 $A = 20x$ if $x < 50$ and $A = 20x - 0.02x^2$ if $50 \le x \le 600$ **57** Yes   **59** No   **61** Yes   **63** No

Section 3.4, page 127

1 Increasing on $\mathbb{R}$ **3** Decreasing on $\mathbb{R}$ **5** Increasing on $(-\infty, 0]$ **7** Decreasing on $(-\infty, 0]$
                                                                              Decreasing on $[0, \infty)$ Increasing on $[0, \infty)$

9 Increasing on $[-4, \infty)$ **11** Increasing on $[0, \infty)$ **13** Decreasing on $(-\infty, 0)$
                                                                                      and on $(0, \infty)$

15 Decreasing on $(-\infty, 2]$ **17** Decreasing on $(-\infty, 0]$ **19** Neither increasing   **21**
    Increasing on $[2, \infty)$ Increasing on $[0, \infty)$ nor decreasing

23   **25**   **27**   **29**

31

33   **35**   **37**

39 (a) (b) (c)   **41**   **43** $f(x) = -\sqrt{1 - x^2}$

45   If $a < b$, then $f(a) > f(b)$; and if $b < a$, then $f(b) > f(a)$. Hence, if $a \ne b$, then $f(a) \ne f(b)$.

Section 3.5, page 137

1 $m = 8$ **3** $m = 1/11$ **5** $m = -1/4$ **9** $x - 3y - 11 = 0$ **11** $x + 5y + 23 = 0$
13 $4x + y - 7 = 0$ **15** $7x - 3y + 21 = 0$ **17** (a) $y = -3$ (b) $x = 7$ **19** $y = -x$
21 $13x - 8y + 132 = 0$ **23** $8x - 15y - 13 = 0, x + 6y + 1 = 0, 10x - 3y - 11 = 0$
25 $m = 2/5, b = 2$ **27** $m = -3/5, b = 0$ **29** $m = 0, b = -2$ **31** $m = -5/6, b = 10/3$
33 $m = 0, b = 0$ **35** $k = -3$
37

41 $-2/5$ **43** $f(x) = -6x - 2$ **47** $\dfrac{x}{(3/2)} + \dfrac{y}{(-3)} = 1$ **49** $r < 1$ or $r > 2$

51 $5x + 2y - 29 = 0$ **53** $5x - 7y + 15 = 0$
55 $x + 6y - 9 = 0; 4x + y - 4 = 0; 3x - 5y + 5 = 0$

Section 3.6, page 144

1 $6x - 1, 6x + 3$ **3** $36x^2 - 5, 12x^2 - 15$ **5** $12x^2 - 8x + 1, 6x^2 + 4x - 1$
7 $x^3 - 1, x^3 - 3x^2 + 3x - 1$ **9** $x + 9 + 9\sqrt{x + 9}, \sqrt{x^2 + 9x + 9}$ **11** $x^2/(2 - 5x^2), 4x^2 - 20x + 25$
13 $5, -5$ **15** $1/x^4, 1/x^4$ **17** x, x **19** and **21**   Show $f(g(x)) = x = g(f(x))$
23 $f^{-1}(x) = (x + 3)/4$ **25** $f^{-1}(x) = (1 - 5x)/2x, x > 0$ **27** $f^{-1}(x) = \sqrt{9 - x}, x \le 9$

29 $f^{-1}(x) = \sqrt[3]{(x+2)/5}$ **31** $f^{-1}(x) = (x^2 + 5)/3, x \geq 0$ **33** $f^{-1}(x) = (x-8)^3$ **35** $f^{-1}(x) = x$
37 (a) $f^{-1}(x) = (x-b)/a$ (b) No (not one-to-one)
39 If g and h are both inverse functions of f, then $f(g(x)) = x = f(h(x))$ for all x. Since f is one-to-one this implies that $g(x) = h(x)$ for all x, that is, $g = h$.

Section 3.7, page 147

1 $a = kv, k = 2/5$ **3** $r = ks/t, k = -14$ **5** $y = kx^2/z^3, k = 27$ **7** $c = ka^2b^3, k = -2/49$
9 $F = km_1m_2/d^2, k = 225/8$ **11** 295 pounds per square foot   **13** 20 pounds
15 $V = 20T/3P; 640/9$ **17** 50/9 ohms   **19** $3\sqrt{3}/2$ seconds

Section 3.8, Review Exercises, page 149

1 Area $= 10$ **3** The points in quadrants II and IV   **7** $(x+5)^2 + (y+1)^2 = 81$
9 (a) $18x + 6y - 7 = 0$ (b) $2x - 6y - 3 = 0$
11   **13**   **15**   **17**

19   **21** Decreasing on $\mathbb{R}$ **23** Decreasing on $[-1, \infty)$ **25**

27 (a) $1/2$ (b) $-\sqrt{2}/2$ (c) 0 (d) $-x/\sqrt{3-x}$ (e) $-x/\sqrt{x+3}$ (f) $x^2/\sqrt{x^2+3}$ (g) $x^2/(x+3)$
29

31 $18x^2 + 9x - 1, 6x^2 - 15x + 5$ **33** $f^{-1}(x) = (10-x)/15$ **35** $V = C^3/8\pi^2$

CHAPTER 4

Section 4.1, page 156

1   **3** Vertex: $(0, -9)$ **5** Vertex: $(0, 9)$

7 $3, -1/4$ **9** $2/3$ **11** $(-1 \pm \sqrt{41})/4$ **13** $(-9 \pm \sqrt{85})/2$

15 $2(x-4)^2 - 9$; minimum: $f(4) = -9$ **17** $-5(x+1)^2 + 8$; maximum: $f(-1) = 8$

19 $4(x - \frac{1}{2})^2 + 1$; minimum: $f(1/2) = 1$

21 min: $f(-5/2) = -9/4$ **23** max: $f(4) = 4$ **25** min: $f(-1/2) = 11/4$ **27** min: $f(2) = 4$

29 max: $f(-1) = 1$ **31** $10, 2$ **33** $-1/2$ **35** $20, -20$

39 125 yards by 250 yards, with intermediate fences parallel to the short side

Section 4.2, page 164

1   **3** $f(x) > 0$ if $x > 2$
$f(x) < 0$ if $x < 2$

5 $f(x) > 0$ for all x **7** $f(x) > 0$ if $-3 < x < 0$ or $x > 3$ **9** $f(x) > 0$ if $x < -2$ or $0 < x < 1$
$f(x) < 0$ if $x < -3$ or $0 < x < 3$ $f(x) < 0$ if $-2 < x < 0$ or $x > 1$

11 $f(x) > 0$ if $-4 < x < 1$ or $x > 5$
$f(x) < 0$ if $x < -4$ or $1 < x < 5$

13 $f(x) > 0$ if $x < -2$ or $x > 2$
$f(x) < 0$ if $-2 < x < 2$

15 $f(x) > 0$ if $-1 < x < 1$
$f(x) < 0$ if $x < -1$ or $x > 1$

17 $f(x) > 0$ if $x < -3$, $-1 < x < 0$, or $x > 2$
$f(x) < 0$ if $-3 < x < -1$ or $0 < x < 2$

19 $k = -4/3$

Calculator Exercises

In each of the following there is a zero between the indicated numbers.
3 2.51 and 2.52   **5** -1.10 and -1.09 **7** 1.331 and 1.332   **9** -1.733 and -1.732

Section 4.3, page 176

1

3

5

7

9

11

13

15

17

19

Section 4.4, page 182

1 $3/(x-2) + 5/(x+3)$ **3** $5/(x-6) - 4/(x+2)$ **5** $2/(x-1) + 3/(x+2) - 1/(x-3)$
7 $(3/x) + 2/(x-5) - 1/(x+1)$ **9** $2/(x-1) + 5/(x-1)^2$ **11** $(-7/x) + (5/x^2) + 2/(3x-5)$
13 $(-23/25)/(2x-1) + (24/25)/(x+2) + (2/5)/(x+2)^2$ **15** $(5/x) - 2/(x+1) + 3/(x+1)^3$
17 $-2/(x-1) + (3x+4)/(x^2+1)$ **19** $(4/x) + (5x-3)/(x^2+2)$ **21** $(4x-1)/(x^2+1) + 3/(x^2+1)^2$
23 $2x + 3x/(x^2+1) + 1/(x-1)$

Section 4.5, page 192

1   **3**   **5**   **7**

9   **11**   **13**   **15**

17   **19**   **21**   **23**

25   **27**   **29**   **31**

33 Ellipse; center $(-5, 4)$, vertices $(-9, 4)$ and $(-1, 4)$, endpoints of minor axis $(-5, 7)$ and $(-5, 1)$
35 Hyperbola; center $(2, -7)$, vertices $(-4, -7)$ and $(8, -7)$, endpoints of conjugate axis $(2, 0)$ and $(2, -14)$
37 Ellipse; center $(-2, 2)$, vertices $(-2, 3)$ and $(-2, 1)$, endpoints of minor axis $(-5/2, 3)$ and $(-3/2, 3)$

33   **35**   **37**

39 Hyperbola; center $(3, 0)$, vertices $(3, \pm 10)$, endpoints of conjugate axis $(-1, 0)$ and $(7, 0)$
41 Ellipse; center $(-3, 5)$, endpoints of major axis $(-3, 9)$ and $(-3, 1)$, endpoints of minor axis $(-5, 5)$ and $(-1, 5)$
43 Ellipse; center $(6, 2)$, vertices $(6, 5)$ and $(6, -1)$, endpoints of minor axis $(5, 2)$ and $(7, 2)$

39   **41**   **43**

45 Hyperbola; center $(-5, -3)$, vertices $(-5, 1)$ and $(-5, -7)$, endpoints of conjugate axis $(-3, -3)$ and $(-7, -3)$
47 Hyperbola; center $(2, 6)$, vertices $(0, 6)$ and $(4, 6)$, endpoints of conjugate axis $(2, 0)$ and $(2, 12)$

45   **47**

49 $x^2 + 4y^2 = 100$ **51** $2\sqrt{21}$ feet   **53** $A = 4a^2b^2/(a^2 + b^2)$
55 The graphs have the same asymptotes.

Section 4.6, Review Exercises, page 194

1   **3**   **5**   **7**   **9**

11   **13**   **15**   **17**

19   **21** $f(x) = 5(x + 3)^2 + 4$; min: $f(-3) = 4$

23 Radius of semicircle is $1/8\pi$ miles; length of rectangle is $1/8$ miles.
25 $8/(x - 1) - 3/(x + 5) - 1/(x + 3)$

27 Ellipse; center $(-3, 2)$, vertices $(-6, 2)$ and $(0, 2)$, endpoints of minor axis $(-3, 0)$ and $(-3, 4)$

29 Hyperbola; center $(2, -3)$, vertices $(2, -1)$ and $(2, -5)$, endpoints of conjugate axis $(2 \pm \sqrt{2}, -3)$

27   **29**

CHAPTER 5

Section 5.1, page 201

1   **3**   **5**   **7**   **9**

11   **13**   **15**   **17**   **19**

21   **23**   **25**   **27** (a) 1039, 3118, 5400

(b)

29 (a) 50 gm, 25 gm, 18 gm

(b)

31 $-1/1600$ **33** \$1010.00, \$1020.10, \$1061.52, \$1126.83   **35** (a) \$7800 (b) \$4790 (c) \$2942
37 a^x is not always real if $a < 0$.   **39** Reflection through the x-axis.

Calculator Exercises

1 4.7288, 4.6555, 4.7070, 4.7277, 4.7287   **3** <   **7**

9 (a) $1,061.36 (b) $1,126.49 (c) $1,346.86 (d) $1,814.01
11 (a) 5 years (b) 8 years (c) 12.5 years

13

15   **17** 2.7048, 2.7169, 2.7181, 2.7183, 2.7183; $f(x)$ gets closer to e as x gets closer to 0.

19 7.44   **21**

$G(t)$ approaches 10

Section 5.2, page 209

1 $\log_4 64 = 3$ **3** $\log_2 128 = 7$ **5** $\log_{10}(0.001) = -3$ **7** $\log_t s = r$ **9** $10^3 = 1000$
11 $3^{-5} = 1/243$ **13** $7^0 = 1$ **15** $t^p = r$ **17** -2 **19** 2   **21** 5   **23** 1/3   **25** -1
27 -3 **29** 16   **31** 0.3   **33** 2.4   **35** -0.25 **37** 1.8   **39** 1.25   **41** 13   **43** 27
45 $1/5, -1/5$ **47** 2, 3   **49** $\sqrt[3]{7}$ **51** 7/2   **53** No solution   **55** 1   **57** $2 + \sqrt{8}$
59 $2\log_a x + \log_a y - 3\log_a z$ **61** $(1/2)\log_a x + 2\log_a z - 4\log_a y$
63 $(2/3)\log_a x - (1/3)\log_a y - (5/3)\log_a z$ **65** $(1/2)\log_a x + (1/4)\log_a y + (3/4)\log_a z$
67 $\log_a(x^2\sqrt[3]{x} - 2/(2x + 3)^5)$ **69** $\log_a x^4/y^{5/3}$

Calculator Exercises

7 1.8928   **9** 2.1252

Section 5.3, page 214

1   **3**

5

  (a)   (b)

7

  (a)   (b)

9

  (a)   (b)

11 They are reflections of one another through the line $y = x$. They are inverse functions of one another.
13 $t = -1600 \log_2 (q/q_0)$ **15** $t = (1/k)(\ln Q - \ln Q_0)$ **17** (a) 10 (b) 30 (c) 40
19 (a) 2 (b) 4 (c) 5

Calculator Exercises

1

  (a)   (b)   (c)

5 (a) 1.2619 (b) 0.3074 (c) 0.1781 (d) 2.5821 (e) -0.1203

Section 5.4, page 220

1 2.5403, 7.5403 − 10, 0.5403   **3** 9.7324 − 10, 2.7324, 5.7324   **5** 1.7796, 5.7796 − 10, 2.7796
7 3.3044, 0.8261, 6.6956 − 10   **9** 9.9235 − 10, 0.0765, 9.9915 − 10   **11** 4.0428   **13** 2.0878
15 8.1462   **17** 4,240   **19** 8.85   **21** 162,000   **23** 0.543   **25** 0.0677   **27** 189
29 0.0237   **31** (a) 2.2 (b) 5 (c) 8.3   **33** Acidic if pH < 7, basic if pH > 7   **35** $\alpha \approx 54.5$
37 $10^{4.5} I_0$ (approximately 31,623 times that of I_0)

Section 5.5, page 223

1 $\log 7/\log 10$ **3** $\log 3/\log 4$ **5** $4 - (\log 5/\log 3)$ **7** $(\log 2 - \log 81)/\log 24$ **9** $-\log 8/\log 2$
11 5   **13** $(2/3)\sqrt{1001/111}$ **15** 100, 1   **17** 10^{100} **19** 10,000   **21** $x = \log(y \pm \sqrt{y^2 - 1})$
23 $x = (1/2)\log(1 + y)/(1 - y)$ **25** $x = \ln(y + \sqrt{y^2 + 1})$ **27** $x = (1/2)\ln(1 + y)/(1 - y)$
29 $t = -(L/R)\ln(1 - Ri/E)$ **31** (a) $I = I_0 10^{x/10}$

Calculator Exercises

1 $x \approx 1.2091$ **3** $x \approx \pm 1.347$ **5** 139 months   **7** 110 days   **9** 0.807

Section 5.6, page 227

1 1.4062   **3** 3.7294   **5** 7.1000 − 10   **7** 5.0913   **9** 9.8913 − 10   **11** 2.5851
13 9.9760 − 10   **15** 4.8234   **17** 8.6356 − 10   **19** 0.4776   **21** 27.78   **23** 49,330
25 0.1467   **27** 0.006536   **29** 234.7   **31** 1.367   **33** 0.09445   **35** 109,900   **37** 0.001345
39 0.2442

Section 5.7, page 230

1 36,500   **3** 4.26   **5** 1,750,000   **7** 0.876   **9** 6.20   **11** 22.0   **13** 22.1   **15** 0.000724
17 1.37   **19** 1.99   **21** 90.27   **23** 0.9356   **25** 3.068   **27** 88.6 square units
29 1.84 seconds

Section 5.8, Review Exercises, page 231

1 -4 **3** 4   **5** 6
7   **9**   **11**   **13**   **15**

17 9   **19** 1   **21** $\log(16/3)/\log 2$ **23** $\log(3/8)/\log(32/9)$ **25** $4\log x + (2/3)\log y - (1/3)\log z$
27 $x = (1/2)\log(y+1)/(y-1)$ **29** 1.6796   **31** 5.4780   **33** 315.9   **35** 0.06266   **37** 25,800
39 31.4

CHAPTER 6

Section 6.1, page 238

1 $(-1, 0)$ **3** $(0, -1)$ **5** $(1, 0)$ **7** $(-1, 0)$ **9** $(0, -1)$ **11** $(-\sqrt{2}/2, -\sqrt{2}/2)$
13 $(\sqrt{2}/2, -\sqrt{2}/2)$ **15** $(-\sqrt{2}/2, \sqrt{2}/2)$ **17** $(1/2, -\sqrt{3}/2)$ **19** $(-\sqrt{3}/2, -1/2)$
21 $(-1/2, \sqrt{3}/2)$ **23** (a) $(-3/5, -4/5)$ (b) $(-3/5, -4/5)$ (c) $(3/5, -4/5)$ (d) $(-3/5, 4/5)$
25 (a) $(8/17, -15/17)$ (b) $(8/17, -15/17)$ (c) $(-8/17, -15/17)$ (d) $(8/17, 15/17)$
27 (a) $(1, 0)$ (b) $(1, 0)$ (c) $(-1, 0)$ (d) $(1, 0)$
29 (a) $(-a/\sqrt{a^2 + b^2}, -b/\sqrt{a^2 + b^2})$ (b) $(-a/\sqrt{a^2 + b^2}, -b/\sqrt{a^2 + b^2})$ (c) $(a/\sqrt{a^2 + b^2}, -b/\sqrt{a^2 + b^2})$
    (d) $(-a/\sqrt{a^2 + b^2}, b/\sqrt{a^2 + b^2})$
31 $\sqrt{3}/2$ **33** $-\sqrt{5}/3$ **35** $-1/2$ **37** $\sqrt{0.9999}$

Calculator Exercises

1 IV   **3** III   **5** III   **7** III   **9** -0.2419

Section 6.2, page 243

1 Divide both sides of $\sin^2 t + \cos^2 t = 1$ by $\sin^2 t$.
3 $\csc t = -5/3$, $\sec t = 5/4$, $\tan t = -3/4$, $\cot t = -4/3$
5 $\csc t = \sqrt{a^2 + b^2}/a$, $\sec t = \sqrt{a^2 + b^2}/b$, $\tan t = a/b$, $\cot t = b/a$

Values in Exercises 7–17 are given in the order $\sin t$, $\cos t$, $\tan t$, $\csc t$, $\sec t$, $\cot t$; a dash means the value is undefined.

7 $1, 0, -, 1, -, 0$ **9** $\sqrt{2}/2, -\sqrt{2}/2, -1, \sqrt{2}, -\sqrt{2}, -1$ **11** $0, -1, 0, -, -1, -$
13 $-1, 0, -, -1, -, 0$ **15** $0, 1, 0, -, 1, -$ **17** $\sqrt{2}/2, -\sqrt{2}/2, -1, \sqrt{2}, -\sqrt{2}, -1$

19 $\sqrt{3}/2$, $1/2$, $\sqrt{3}$, $2\sqrt{3}/3$, 2, $\sqrt{3}/3$ **21** $1/2$, $-\sqrt{3}/2$, $-\sqrt{3}/3$, 2, $-2\sqrt{3}/3$, $-\sqrt{3}$
23 $-\sqrt{3}/2$, $-1/2$, $\sqrt{3}$, $-2\sqrt{3}/3$, -2, $\sqrt{3}/3$ **25** IV   **27** III   **29** II   **31** III   **33** I

Values in Exercises 35–41 are given in the order $\sin t$, $\cos t$, $\tan t$, $\csc t$, $\sec t$, $\cot t$.

35 $3/5$, $-4/5$, $-3/4$, $5/3$, $-5/4$, $-4/3$ **37** $-5/13$, $12/13$, $-5/12$, $-13/5$, $13/12$, $-12/5$
39 $-2\sqrt{2}/3$, $-1/3$, $2\sqrt{2}$, $-3\sqrt{2}/4$, -3, $\sqrt{2}/4$ **41** $\sqrt{15}/4$, $-1/4$, $-\sqrt{15}$, $4\sqrt{15}/15$, -4, $-\sqrt{15}/15$
43 No; $|\sin t| \le 1$ **45** (a) $\sqrt{2}/2$ (b) $-1/4$ **47** $(\cos t)(\sec t) = (\cos t)(1/\cos t) = 1$

Calculator Exercises

1 $\tan t = -0.3313$, $\cot t = -3.018$, $\sec t = -1.054$, $\csc t = 3.180$
3 $\sec t = 1.122$, $\csc t = 2.203$, $\cot t = 1.963$, $\sin t = 0.4540$
5 $\sin t = 0.9925$, $\tan t = -8.142$, $\sec t = -8.203$, $\csc t = 1.008$, $\cot t = -0.1228$

Section 6.3, page 247

1

$$\begin{array}{c}\!\!\!\!+\!\!\bullet\!\!+\!\!\bullet\!\!+\!\!\bullet\!\!+\!\!\bullet\!\!+\!\!\bullet\!\!+\!\!\bullet\!\!+\!\!\bullet\!\!+\!\!\bullet\!\!+\!\!\bullet\!\!+\!\!\blacktriangleright \\ -4\pi\ -3\pi\ -2\pi\ -\pi\quad 0\quad \pi\ \ 2\pi\ 3\pi\ 4\pi\end{array}$$

3 We shall give an indirect proof. Suppose there is a positive number k less than 2π such that $\sin(t + k) = \sin t$ for all t. Letting $t = 0$, we obtain $\sin k = \sin 0 = 0$. Since $0 < k < 2\pi$ it follows that $P(k)$ has coordinates $(-1, 0)$. Consequently, $k = \pi$, and we may write $\sin(t + \pi) = \sin t$ for all t. In particular, if $t = \pi/2$, then $\sin(3\pi/2) = \sin \pi/2$, or $-1 = 1$, an absurdity. This completes the proof.
5 If $y/x = a$, where $x^2 + y^2 = 1$, then $\pm\sqrt{1 - x^2}/x = a$. Hence $(1 - x^2)/x^2 = a^2$, and $1 - x^2 = a^2 x^2$ or $1 = a^2 x^2 + x^2 = (a^2 + 1)x^2$. Consequently $x^2 = 1/(a^2 + 1)$ or $x = \pm 1/\sqrt{a^2 + 1}$. If $a > 0$, choose the point $P(x, y)$ on U where $x = 1/\sqrt{a^2 + 1}$. If $a < 0$ choose $P(x, y)$ such that $x = -1/\sqrt{a^2 + 1}$. Thus there is always a point $P(x, y)$ on U such that $y/x = a$.

Section 6.4, page 253

1 $480°$, $840°$, $-240°$, $-600°$ **3** $495°$, $855°$, $-225°$, $-585°$ **5** $330°$, $690°$, $-390°$, $-750°$
7 $260°$, $980°$, $-100°$, $-460°$ **9** $17\pi/6$, $29\pi/6$, $-7\pi/6$, $-19\pi/6$ **11** $7\pi/4$, $15\pi/4$, $-9\pi/4$, $-17\pi/4$
13 $5\pi/6$ **15** $-\pi/3$ **17** $5\pi/4$ **19** $5\pi/2$ **21** $2\pi/5$ **23** $5\pi/9$ **25** $120°$ **27** $330°$
29 $135°$ **31** $-630°$ **33** $1260°$ **35** $20°$ **37** $114°35'30''$ **39** $286°28'44''$
41 (a) 1.75 (b) $100.27°$ **43** 6.98 meters   **45** 8.59 km
47 Approximations (in miles): (a) 4189 (b) 3142 (c) 2094 (d) 698 (e) 70
49 $1/8$ radian $= (45/2\pi)° \approx 7°10'$

Calculator Exercises

1 1.2811 **3** 8.4173 **5** $149.7651°$
7 Approximations (in miles): (a) 4150.26 (b) 3112.69 (c) 2075.13 (d) 691.71 (e) 69.17
9 Approximately 0.126 radians, or $7.219°$

Section 6.5, page 261

Answers to Exercises 1–29 are in the order sin, cos, tan, csc, sec, cot.

1 $-3/5$, $4/5$, $-3/4$, $-5/3$, $5/4$, $-4/3$ **3** $-5\sqrt{29}/29$, $-2\sqrt{29}/29$, $5/2$, $-\sqrt{29}/5$, $-\sqrt{29}/2$, $2/5$
5 $4\sqrt{17}/17$, $-\sqrt{17}/17$, -4, $\sqrt{17}/4$, $-\sqrt{17}$, $-1/4$ **7** $-7\sqrt{53}/53$, $-2\sqrt{53}/53$, $7/2$, $-\sqrt{53}/7$, $-\sqrt{53}/2$, $2/7$
9 $4/5$, $3/5$, $4/3$, $5/4$, $5/3$, $3/4$ **11** 1, 0, —, 1, —, 0 **13** $\sqrt{2}/2$, $-\sqrt{2}/2$, -1, $\sqrt{2}$, $-\sqrt{2}$, -1
15 $-\sqrt{2}/2$, $\sqrt{2}/2$, -1, $-\sqrt{2}$, $\sqrt{2}$, -1 **17** 0, 1, 0, —, 1, —   **21** $4/5$, $3/5$, $4/3$, $5/4$, $5/3$, $3/4$

23 $2\sqrt{13}/13, 3\sqrt{13}/13, 2/3, \sqrt{13}/2, \sqrt{13}/3, 3/2$ **25** $2/5, \sqrt{21}/5, 2\sqrt{21}/21, 5/2, 5\sqrt{21}/21, \sqrt{21}/2$
27 $a/\sqrt{a^2+b^2}, b/\sqrt{a^2+b^2}, a/b, \sqrt{a^2+b^2}/a, \sqrt{a^2+b^2}/b, b/a$
29 $b/c, \sqrt{c^2-b^2}/c, b/\sqrt{c^2-b^2}, c/b, c/\sqrt{c^2-b^2}, \sqrt{c^2-b^2}/b$

Calculator Exercises

1 Table A is exact. Calculator answers are approximations.   **3** 0.743, 0.669, 1.110, 1.346, 1.494, 0.901

Section 6.6, page 271

1 (a) $\pi/4$ (b) $\pi/3$ (c) $\pi/6$ **3** (a) $\pi/4$ (b) $\pi/6$ (c) $\pi/3$ **5** (a) 1.5 (b) $2\pi-5$
7 (a) 60° (b) 20° (c) 70°   **9** (a) 49°20′ (b) 45° (c) 80°35′   **11** (a) $\sqrt{3}/2$ (b) $-\sqrt{3}/2$
13 (a) -1 (b) -1 **15** (a) $-2\sqrt{3}/3$ (b) -2 **17** 0.2644   **19** 1.540   **21** 1.431
23 0.7826   **25** 6.197   **27** -1.019 **29** -0.3035 **31** 0.4440   **33** -0.1426 **35** 1.032
37 0.7911   **39** 0.5217   **41** 3.436   **43** 0.5315   **45** 1.3521   **47** 0.7703
49 21°33′, 158°27′   **51** 26°45′, 206°45′   **53** 69°44′, 290°16′   **55** 43°48′, 316°12′

Section 6.7, page 278

1 (c) $[-2\pi, -3\pi/2), (-3\pi/2, -\pi], [0, \pi/2), (\pi/2, \pi]$ (d) $[-\pi, -\pi/2), (-\pi/2, 0], [\pi, 3\pi/2), (3\pi/2, 2\pi]$
3 (a) $\csc(-t) = 1/\sin(-t) = 1/(-\sin t) = -(1/\sin t) = -\csc t$
7   **9**   **11**   **13**

15

Section 6.8, page 284

1 The amplitudes and periods are: (a) $4, 2\pi$ (b) $1, \pi/2$ (c) $1/4, 2\pi$ (d) $1, 8\pi$ (e) $2, 8\pi$ (f) $1/2, \pi/2$
(g) $4, 2\pi$ (h) $1, \pi/2$
3 The amplitudes and periods are (a) $3, 2\pi$ (b) $1, 2\pi/3$ (c) $1/3, 2\pi$ (d) $1, 6\pi$ (e) $2, 6\pi$ (f) $1/3, \pi$
(g) $3, 2\pi$ (h) $1, 2\pi/3$
5   **7**   **9**   **11**

13   **15**   **17**   **19**

21   **23**   **25**   **27**

29   **31**   **33**   **35**

37   **39**   **41**   **43**

Section 6.9, page 288

1   **3**   **5**   **7**

9   **11**   **13**   **15**

17   **19**   **21**   **23**

25   **27**   **29**

Section 6.10, page 293

1 $\beta = 60°$, $a = 20\sqrt{3}/3 \approx 12$, $c = 40\sqrt{3}/3 \approx 23$ **3** $\alpha = 38°$, $b \approx 19$, $c \approx 24$
5 $\beta = 72°20'$, $b \approx 14.1$, $c \approx 14.8$ **7** $\alpha = 18°9'$, $a \approx 78.67$, $c \approx 252.6$ **9** $\alpha \approx 29°$, $\beta \approx 61°$, $c \approx 51$
11 $\alpha \approx 69°$, $\beta \approx 21°$, $a \approx 5.4$ **13** $\beta = 52°14'$, $a \approx 396.7$, $c \approx 647.7$ **15** $\alpha \approx 49°40'$, $\beta \approx 40°20'$, $b \approx 522$
17 $51°$ **19** 70.6 meters   **21** 20.2 meters   **23** 29.7 km   **25** $d \approx 160$ meters

27 $192(\sin 22°30') \approx 73.5$ cm   **29** $9{,}659$ ft   **31** 126 mph   **33** $28{,}800$ ft   **35** 55 miles
37 325 miles   **39** $h = d(\tan \beta - \tan \alpha)$

Calculator Exercises

1 $\beta = 48.73°$, $b \approx 358.5$, $c \approx 476.9$ **3** $\alpha \approx 52.94°$, $b \approx 0.3484$, $c \approx 0.5781$
5 $\alpha \approx 87.29°$, $a \approx 151{,}000$, $c \approx 151{,}200$ **7** $c \approx 48.67$, $\alpha \approx 74.36°$, $\beta \approx 15.64°$
9 $\alpha \approx 60.97°$, $\beta \approx 29.03°$, $a \approx 4437$

Section 6.11, page 299

1 (a) $\omega = 200\pi$ radians per min.   (b) In cm: $x = 40 \cos 200\pi t$, $y = 40 \sin 200\pi t$
3 Amplitude 10 cm, period 1/3 sec, frequency 3 osc/sec. The point is at the origin at $t = 0$. It then moves upward with decreasing speed, reaching the point with coordinate 10 at $t = 1/12$. It then reverses direction and moves downward, gaining speed until it reaches the origin at $t = 1/6$. It continues downward, with decreasing speed, reaching the point with coordinate -10 at $t = 1/4$. It then reverses direction and moves upward with increasing speed, returning to the origin at $t = 1/3$.
5 Amplitude 4 cm, period 4/3 sec, frequency 3/4 osc/sec. The motion is similar to that in Exercise 3; however, the point starts 4 units above the origin and moves downward, reaching O at $t = 1/3$, and the point with coordinate -4 at $t = 2/3$. It then reverses direction and moves upward, reaching O at $t = 1$ and its initial point at $t = 4/3$.
7 $d = 5 \cos (2\pi/3)t$ **9** Current lags emf by $1/1440$ sec.

Section 6.12, Review Exercises, page 301

1 $(-1, 0), (0, -1), (0, 1), (-\sqrt{2}/2, -\sqrt{2}/2), (1, 0), (\sqrt{3}/2, 1/2)$ **3** (a) II   (b) III   (c) IV
5 The order will be sin, cos, tan, csc, sec, cot:
   (a) $1, 0, \text{—}, 1, \text{—}, 0$ (b) $\sqrt{2}/2, -\sqrt{2}/2, -1, \sqrt{2}, -\sqrt{2}, -1$ (c) $0, 1, 0, \text{—}, 1, \text{—}$
   (d) $-1/2, \sqrt{3}/2, -\sqrt{3}/3, -2, 2\sqrt{3}/3, -\sqrt{3}$ **7** $810°, -120°, 315°, 900°, 36°$ **9** $65°, 42°50', 8°$
11 The order will be sin, cos, tan, csc, sec, cot:
   (a) $-4/5, 3/5, -4/3, -5/4, 5/3, -3/4$ (b) $2\sqrt{13}/13, -3\sqrt{13}/13, -2/3, \sqrt{13}/2, -\sqrt{13}/3, -3/2$
   (c) $-1, 0, \text{—}, -1, \text{—}, 0$ **13** (a) 0.8500 (b) 2.972 (c) 1.112

15   **17**   **19**   **21**

23   **25**   **27**   **29**

CHAPTER 7

Section 7.1, page 305

The following are arranged in the order $\sin t, \cos t, \tan t, \csc t, \sec t, \cot t$:

1 $-\sqrt{3}/2, 1/2, -\sqrt{3}, -2\sqrt{3}/3, 2, -\sqrt{3}/3$ **3** $\sqrt{2}/2, -\sqrt{2}/2, -1, \sqrt{2}, -\sqrt{2}, -1$

5 $-1/2, \sqrt{3}/2, -\sqrt{3}/3, -2, 2\sqrt{3}/3, -\sqrt{3}$ **7** $0, -1, 0, \text{---}, -1, \text{---}$

9 $-3/5, -4/5, 3/4, -5/3, -5/4, 4/3$ **11** $12/13, -5/13, -12/5, 13/12, -13/5, -5/12$

13 $-2/3, \sqrt{5}/3, -2\sqrt{5}/5, -3/2, 3\sqrt{5}/5, -\sqrt{5}/2$ **15** $1/8, -\sqrt{63}/8, -\sqrt{63}/63, 8, -8\sqrt{63}/63, -\sqrt{63}$

33 $\cos t = \pm\sqrt{1 - \sin^2 t}$, $\tan t = \pm\sin t/\sqrt{1 - \sin^2 t}$, $\csc t = 1/\sin t$, $\sec t = \pm 1/\sqrt{1 - \sin^2 t}$, $\cot t = \pm\sqrt{1 - \sin^2 t}/\sin t$ **35** $\sin t = \pm\tan t/\sqrt{1 + \tan^2 t}$ **37** $\sec t = \pm\csc t/\sqrt{\csc^2 t - 1}$

Section 7.3, page 315

In the following, n denotes any integer.

1 $(2\pi/3) + 2\pi n, (4\pi/3) + 2\pi n$ **3** $(\pi/4) + (\pi/2)n$ **5** $2\pi n, (3\pi/2) + 2\pi n$ **7** $(\pi/3) + \pi n, (2\pi/3) + \pi n$

9 $(4\pi/3) + 2\pi n, (5\pi/3) + 2\pi n$ **11** $(\pi/6) + \pi n, (5\pi/6) + \pi n$ **13** $(7\pi/6) + 2\pi n, (11\pi/6) + 2\pi n$

15 $(\pi/12) + n\pi, (5\pi/12) + n\pi$ **17** $\pi/3, 2\pi/3, 4\pi/3, 5\pi/3; 60°, 120°, 240°, 300°$

19 $\pi/6, 5\pi/6, 3\pi/2; 30°, 150° 270°$ **21** $0, \pi, \pi/4, 3\pi/4, 5\pi/4, 7\pi/4; 0°, 180°, 45°, 135°, 225°, 315°$

23 $\pi/2, 3\pi/2, 2\pi/3, 4\pi/3; 90°, 270°, 120°, 240°$ **25** No solutions   **27** $11\pi/6, \pi/2; 330°, 90°$

29 $0, \pi/2; 0°, 90°$ **31** $0; 0°$ **33** $7\pi/6, 11\pi/6, \pi/2, 3\pi/2; 210°, 330°, 90°, 270°$

35 $3\pi/4, 7\pi/4; 135°, 315°$ **37** $15°30', 164°30'$ **39** $41°50', 138°10', 194°30', 345°30'$

Section 7.4, page 322

1 (a) $\cos 43°23'$ (b) $\sin 16°48'$ (c) $\cot \pi/3$ **3** (a) $\sin 3\pi/20$ (b) $\cos(2\pi - 1)/4$ (c) $\cot(\pi - 2)/2$

5 (a) $(\sqrt{2} + 1)/2$ (b) $(\sqrt{2} - \sqrt{6})/4$ **7** (a) $\sqrt{3} + 1$ (b) $-2 - \sqrt{3}$ **9** (a) $(\sqrt{2} - 1)/2$

(b) $(\sqrt{6} + \sqrt{2})/4$ **11** $\cos 25°$ **13** $\sin(-5°)$ **15** $-\sin 5$ **17** $36/85, 77/85; \text{I}$

19 $3/5, 4/5, 3/4, -117/125, 44/125, -117/44$

41 $\sin u \cos v \cos w + \cos u \sin v \cos w + \cos u \cos v \sin w - \sin u \sin v \sin w$

49 $0, \pi/3, 2\pi/3, \pi, 4\pi/3, 5\pi/3; 0°, 60°, 120°, 180°, 240°, 300°$ **51** $f(x) = 2\cos(2x - \pi/6); 2, \pi, \pi/12$

53 $f(x) = 2\sqrt{2}\cos(3x + \pi/4); 2\sqrt{2}, 2\pi/3, -\pi/12$

Section 7.5, page 329

1 $24/25, -7/25, -24/7$ **3** $-4\sqrt{2}/9, -7/9, 4\sqrt{2}/7$ **5** $\sqrt{10}/10, 3\sqrt{10}/10, 1/3$

7 $-\sqrt{2 + \sqrt{2}}/2, \sqrt{2 - \sqrt{2}}/2, -\sqrt{2} - 1$ **9** (a) $\sqrt{2 - \sqrt{2}}/2$ (b) $\sqrt{2 - \sqrt{3}}/2$ (c) $\sqrt{2} + 1$

29 $\frac{3}{8} + \frac{1}{2}\cos\theta + \frac{1}{8}\cos 2\theta$ **31** $0, 2\pi/3, \pi, 4\pi/3; 0°, 120°, 180°, 240°$ **33** $\pi/3, \pi, 5\pi/3; 60°, 180°, 300°$

35 $0, \pi; 0°, 180°$ **37** $0, \pi/3, 5\pi/3; 0°, 60°, 300°$

Calculator Exercises

1 -0.8217

Section 7.6, page 334

1 $\sin 12\theta + \sin 6\theta$ **3** $(1/2)\cos 4t - (1/2)\cos 10t$ **5** $(1/2)\cos 10u + (1/2)\cos 2u$

7 $(3/2)\sin 3x + (3/2)\sin x$ **9** $2\cos 2\theta \sin 4\theta$ **11** $-2\sin 4x \sin x$ **13** $-2\cos 5t \sin 2t$

15 $2\cos(3x/2)\cos(x/2)$ **25** $(1/2)\sin(a + b)x + (1/2)\sin(a - b)x$ **27** $n\pi/4$ where n is any integer

29 $n\pi/2$ where n is any integer.

Section 7.7, page 342

1 (a) $\pi/6$ (b) $\pi/3$ **3** (a) 0 (b) $\pi/2$ **5** (a) $\pi/2$ (b) 0   **7** (a) $\pi/3$ (b) $-\pi/3$
9 (a) $-\pi/4$ (b) $-\pi/6$ **11** -0.7069 **13** 1.1403 **15** $\sqrt{3}/2$ **17** $3/5$ **19** $-\pi/4$ **21** 0
23 $-77/36$ **25** $-24/25$ **27** $u\sqrt{1+u^2}/(1+u^2)$ **29** $\sqrt{2+2u}/2$
39 $\cot^{-1} u = v$ if and only if $\cot v = u$.
43   **45**   **47**   **49**   **51**

55 $\arctan(-9 \pm \sqrt{57})/4$ **57** $\arccos(\pm\sqrt{15}/5), \arccos(\pm\sqrt{3}/3)$ **59** $\arcsin(\pm\sqrt{30}/6)$

Calculator Exercises

1 $-0.3478, -1.3336$ **3** $0.6847, 2.4569, 0.9553, 2.1863$ **5** ± 1.1503

Section 7.8, page 349

1 $\beta \approx 62°, b \approx 14.1, c \approx 15.6$ **3** $\gamma \approx 100°10', b \approx 55.1, c \approx 68.7$ **5** $\alpha \approx 58°40', a \approx 487, b \approx 442$
7 $\beta \approx 53°40', \gamma \approx 61°10', c \approx 20.6$ **9** $\alpha \approx 77°30', \beta \approx 49°10', b \approx 108; \alpha \approx 102°30', \beta \approx 24°10', b \approx 58.7$
11 $\alpha \approx 20°30', \gamma \approx 46°20', a \approx 94.5$ **13** 219   **15** 50 feet   **17** 2.7 miles
19 3.7 miles from A and 5.4 miles from B **21** 627 meters

Calculator Exercises

1 $\alpha = 119.7°, a \approx 371.5, c \approx 243.5$ **3** $\beta = 49.36°, a \approx 49.78, c \approx 23.39$
5 $\alpha \approx 25.993°, \gamma \approx 32.383°, a \approx 0.146$

Section 7.9, page 352

1 $a \approx 26, \beta \approx 41°, \gamma \approx 79°$ **3** $b \approx 177, \alpha \approx 25°10', \gamma \approx 4°50'$ **5** $c \approx 2.75, \alpha \approx 21°10', \beta \approx 43°40'$
7 $\alpha \approx 29°, \beta \approx 47°, \gamma \approx 104°$ **9** $\alpha \approx 12°30', \beta \approx 136°30', \gamma \approx 31°$ **11** 63, 87 inches   **13** 92 feet
15 24 miles   **17** 39 miles   **19** Approximately 2.3 miles   **21** Approximately N55°31'E

Calculator Exercises

1 $a \approx 40.8, \beta \approx 26.9°, \gamma \approx 104.8°$ **3** $c \approx 0.487, \beta \approx 5.12°, \alpha \approx 168.03°$
5 $\alpha \approx 157°16', \beta \approx 7°49', \gamma \approx 14°55'$

Section 7.10, page 362

1 $\langle 3,1 \rangle, \langle 1,-7 \rangle, \langle 13,8 \rangle, \langle 3,-32 \rangle$ **3** $\langle -15,6 \rangle, \langle 1,-2 \rangle, \langle -68,28 \rangle, \langle 12,-12 \rangle$
5 $4\mathbf{i} - 3\mathbf{j}, -2\mathbf{i} + 7\mathbf{j}, 19\mathbf{i} - 17\mathbf{j}, -11\mathbf{i} + 33\mathbf{j}$ **7** $-2\mathbf{i} - 5\mathbf{j}, -6\mathbf{i} + 7\mathbf{j}, -6\mathbf{i} - 26\mathbf{j}, -26\mathbf{i} + 34\mathbf{j}$
9 $-3\mathbf{i} + 2\mathbf{j}, 3\mathbf{i} + 2\mathbf{j}, -15\mathbf{i} + 8\mathbf{j}, 15\mathbf{i} + 8\mathbf{j}$ **31** $3\sqrt{2}, 7\pi/4$ **33** $5, \pi$ **35** $\sqrt{41}, \arccos(-4\sqrt{41}/41)$
37 $18, 3\pi/2$ **39** 89 kg, S66°W   **41** 5.8lb, 129°   **43** 56°, 232 mph   **45** 420 mph, 244°
47 N20°30'W

Section 7.11, Review Exercises, page 364

17 $\pi/2, 3\pi/2, \pi/4, 3\pi/4, 5\pi/4, 7\pi/4; 90°, 270°, 45°, 135°, 225°, 315°$ **19** $0, \pi; 0°, 180°$
21 $0, \pi; 0°, 180°$ **23** $\pi/2, 7\pi/6, 11\pi/6; 90°, 210°, 330°$ **25** $\pi/6, 5\pi/6, \pi/3, 5\pi/3; 30°, 150°, 60°, 300°$

27 $\pi/3, 5\pi/3; 60°, 300°$ **29** $\sqrt{2}-\sqrt{3}/2$ or $(\sqrt{2}-\sqrt{6})/4$ **31** $\sqrt{2}-\sqrt{3}/2$ or $(\sqrt{2}-\sqrt{6})/4$
33 $84/85$ **35** $-36/77$ **37** $240/289$ **39** $24/7$ **41** $1/3$
43 (a) $(1/2)\cos 3t - (1/2)\cos 11t$ (b) $(1/2)\cos(u/12)+(1/2)\cos(5u/12)$ (c) $3\sin 8x - 3\sin 2x$
45 $5\pi/6$ **47** π **49** $1/2$ **51** $-7/25$ **53** $\pi/2$ **55**

57 $a = 50\sqrt{6}, c = 50(1+\sqrt{3}), \gamma = 75°$ **59** $a = \sqrt{43}, \beta = \arccos(11/16), \gamma = \arccos(-1/4)$

CHAPTER 8

Section 8.1, page 371

1 $(3,5), (-1,-3)$ **3** $(1,0), (-3,2)$ **5** $(0,0), (1/8,1/128)$ **7** $(3,-2)$ **9** No solutions
11 $(-4,3), (5,0)$ **13** $(-2,2)$ **15** $((-6-\sqrt{86})/10, (2-3\sqrt{86})/10), ((-6+\sqrt{86})/10, (2+3\sqrt{86})/10)$
17 $(-4,0), (12/5, 16/5)$ **19** $(0,1), (4,-3)$ **21** $(\pm 2,5), (\pm\sqrt{5},4)$ **23** $(\sqrt{2}, \pm 2\sqrt{3}), (-\sqrt{2}, \pm 2\sqrt{3})$
25 $(2\sqrt{2}, \pm 2), (-2\sqrt{2}, \pm 2)$ **27** $x=3, y=-1, z=2$ **29** $(\log 3/\log 5, 7)$ **31** $r=3, s=-4$
33 $x=35/22, y=43/44$ **35** $13, 9$ **37** 8 inches, 12 inches **39** $5-\sqrt{19}, 5+\sqrt{19}, 10$

Section 8.2, page 378

1 $(4,-2)$ **3** $(8,0)$ **5** $(-1,3/2)$ **7** $(76/53, 28/53)$ **9** $(51/13, 96/13)$
11 $(8/7, -3\sqrt{6}/7)$ **13** No solution **15** All ordered pairs (m, n) such that $3m - 4n = 2$
17 $(0,0)$ **19** $(-22/7, -11/5)$ **21** $a=2, b=4$ **23** 137 adults, 313 children **25** 68
27 25 dimes, 18 quarters **29** \$6,650 at 6.5% and \$3,325 at 8%
31 40 gm of 35% alloy, 60 gm of 60% alloy **33** 540 mph, 60 mph **35** $v_0 = 10, a = 3$

Section 8.3, page 385

1 $(2,3,-1)$ **3** $(-2,4,5)$ **5** No solution **7** $(2/3, 31/21, 1/21)$

There are other forms for the answers in Exercises 9–15.

9 $(2c, -c, c)$ where c is any real number **11** $(0, -c, c)$ where c is any real number
13 $((12-9c)/7, (8c-13)/14, c)$, where c is any real number
15 $((7c+5)/10, (19c-15)/10, c)$ where c is any real number **17** $(1,3,-1,2)$ **19** $(1/11, 31/11, 3/11)$
21 $(-2,-3)$ **23** No solution **25** $(2,1,2)$ **27** 24 **29** 17 1 of 10%, 11 1 of 30%, 22 1 of 50%
31 4 hr for A, 2 hr for B, 5 hr for C **33** 380 lb of G_1, 60 lb of G_2, 160 lb of G_3
35 $f(x) = (-\frac{1}{2})x^2 + x + (\frac{5}{2})$ **37** $x^2 + y^2 - x + 3y - 6 = 0$

Section 8.4, page 392

29 $x=2, y=-1, z=3, s=4, t=1$

Section 8.5, page 400

1 $\begin{bmatrix} 9 & -1 \\ -2 & 5 \end{bmatrix}, \begin{bmatrix} 1 & -3 \\ 4 & 1 \end{bmatrix}, \begin{bmatrix} 10 & -4 \\ 2 & 6 \end{bmatrix}, \begin{bmatrix} -12 & -3 \\ 9 & -6 \end{bmatrix}$

3 $\begin{bmatrix} 9 & 0 \\ 1 & 5 \\ 3 & 4 \end{bmatrix}, \begin{bmatrix} 3 & -2 \\ 3 & -5 \\ -9 & 4 \end{bmatrix}, \begin{bmatrix} 12 & -2 \\ 4 & 0 \\ -6 & 8 \end{bmatrix}, \begin{bmatrix} -9 & -3 \\ 3 & -15 \\ -18 & 0 \end{bmatrix}$

5 $[11 \quad -3 \quad -3], [-3 \quad -3 \quad 7], [8 \quad -6 \quad 4], [-21 \quad 0 \quad 15]$

7 $\begin{bmatrix} -3 & 4 & 1 & 6 \\ 3 & 2 & 7 & -7 \end{bmatrix}, \begin{bmatrix} 3 & 4 & -1 & 0 \\ -1 & 2 & -7 & -3 \end{bmatrix}, \begin{bmatrix} 0 & 8 & 0 & 6 \\ 2 & 4 & 0 & -10 \end{bmatrix}, \begin{bmatrix} 9 & 0 & -3 & -9 \\ -6 & 0 & -21 & 6 \end{bmatrix}$

9 $\begin{bmatrix} 16 & 38 \\ 11 & -34 \end{bmatrix}, \begin{bmatrix} 4 & 38 \\ 23 & -22 \end{bmatrix}$ **11** $\begin{bmatrix} 3 & -14 & -3 \\ 16 & 2 & -2 \\ -7 & -29 & 9 \end{bmatrix}, \begin{bmatrix} 3 & -20 & -11 \\ 2 & 10 & -4 \\ 15 & -13 & 1 \end{bmatrix}$

13 $\begin{bmatrix} 4 & 8 \\ -18 & 11 \end{bmatrix}, \begin{bmatrix} 3 & -4 & 4 \\ -5 & 2 & 2 \\ -51 & 26 & 10 \end{bmatrix}$ **15** $\begin{bmatrix} 1 & 2 & 3 \\ 4 & 5 & 6 \\ 7 & 8 & 9 \end{bmatrix}, \begin{bmatrix} 1 & 2 & 3 \\ 4 & 5 & 6 \\ 7 & 8 & 9 \end{bmatrix}$ **17** $[15], \begin{bmatrix} -3 & 7 & 2 \\ -12 & 28 & 8 \\ 15 & -35 & -10 \end{bmatrix}$

19 $\begin{bmatrix} 4 \\ 12 \\ -1 \end{bmatrix}$ **21** $\begin{bmatrix} 18 & 0 & -2 \\ -40 & 10 & -10 \end{bmatrix}$ **31** $\frac{1}{10}\begin{bmatrix} 3 & 4 \\ -1 & 2 \end{bmatrix}$ **33** Does not exist. **35** $\frac{1}{8}\begin{bmatrix} 2 & 1 & 0 \\ -2 & 3 & 0 \\ 0 & 0 & 2 \end{bmatrix}$

37 $\frac{1}{3}\begin{bmatrix} -4 & -5 & 3 \\ -4 & -8 & 3 \\ 1 & 2 & 0 \end{bmatrix}$ **39** $\begin{bmatrix} \frac{1}{2} & 0 & 0 \\ 0 & \frac{1}{4} & 0 \\ 0 & 0 & \frac{1}{6} \end{bmatrix}$ **41** $\frac{1}{6}\begin{bmatrix} -8 & -7 & 4 & -9 \\ -8 & -4 & 4 & -6 \\ -4 & -5 & 2 & -3 \\ 6 & 3 & 0 & 3 \end{bmatrix}$ **43** $ab \neq 0; \begin{bmatrix} 1/a & 0 \\ 0 & 1/b \end{bmatrix}$

47 $(13/10, -1/10)$ **49** $(-25/3, -34/3, 7/3)$

Section 8.6, page 407

1 $M_{11} = -14, M_{21} = 7, M_{31} = 11, M_{12} = 10, M_{22} = -5, M_{32} = 4, M_{13} = 15, M_{23} = 34, M_{33} = 6;$
$A_{11} = -14, A_{21} = -7, A_{31} = 11, A_{12} = -10, A_{22} = -5, A_{32} = -4, A_{13} = 15, A_{23} = -34, A_{33} = 6$
3 $M_{11} = 0, M_{12} = 5, M_{21} = -1, M_{22} = 7; A_{11} = 0, A_{12} = -5, A_{21} = 1, A_{22} = 7$ **5** -83 **7** 5
9 2 **11** 0 **13** -125 **15** 48 **17** -216 **19** $abcd$

Section 8.7, page 412

Answers to Exercises 1, 3, 5, 9, and 13 refer to the Theorem on Row and Column Transformations of a Determinant.

1 (i) **3** (iii) **5** (ii) **7** Two rows are identical **9** (ii)
11 Every number in column 2 is 0; also two rows are identical. **13** (iii) **15** -10 **17** -142
19 -183 **21** 44 **23** 359

Section 8.9, page 422

1 The region above the graph of $2y = 3x - 6$
3 The region above on the graph of $y = 1 - 2x$
5 The region under the graph of $y = x^2 - 2$
7 The region above and on the graph of $y = x^2 + 1$
9 The region above and on the graph of $y = 1/x^2$
11 The region under the graph of $3x + y = 3$ and also above the graph of $y = 4 - 2x$
13 The region under the graph of $y = x$ and also under the graph of $2x + 5y = 10$
15 The quadrilateral bounded by the graphs of $3x + y = 6$, $y - 2x = 1$, $x = -2$, and $y = 4$

17 The region above and on the line $x + y = 1$ which is also inside or on the circle $x^2 + y^2 = 4$
19 The region under or on the graph of $y = 1 - x^2$ and also under or on the line $x = 1 + y$
21 The region in the first quadrant between the graphs of $y = 2^x$ and $y = 3^x$
23 The region bounded by the graphs of $y = \log x$, $y + x = 1$, and $x = 10$
25 Let y denote the number of bats and x the number of balls. Then $x \geq 2$, $y \geq 3$, and $5x + 7y \leq 80$.
27 If x and y denote the number of brand A and brand B, respectively, then $x \geq 20$, $y \geq 10$, $x \geq 2y$, and $x + y \leq 100$.
29 If x and y denote the amounts in the first and second accounts, respectively, then $x \geq 2000$, $y \geq 2000$, $y \geq 3x$, and $x + y \leq 15,000$.

Section 8.10, page 428

1 50 Double Fault and 30 Set Point   **3** 3.5 pounds of S and 1 pound of T
5 Send 25 from W_1 to X and 0 from W_1 to Y. Send 10 from W_2 to X and 60 from W_2 to Y.
7 45 acres of crop A and 50 acres of crop B
9 Minimum cost: 16 ounces X, 4 ounces Y, 0 ounces Z; maximum cost: 0 ounces X, 8 ounces Y, 12 ounces Z

Section 8.11, Review Exercises, page 430

1 $(19/23, -18/23)$ **3** $(-3, 5), (1, -3)$ **5** $(2\sqrt{3}, \pm\sqrt{2}), (-2\sqrt{3}, \pm\sqrt{2})$ **7** $(14/17, 14/27)$
9 $(6/11, -7/11, 1)$ **11** $(-2c, -3c, c)$ where c is any real number
13 $(5c - 1, (-19c + 5)/2, c)$ where c is any real number   **15** $(-1, 1/2, 1/3)$
17 The region inside the circle $x^2 + y^2 = 16$ which is also above the graph of $y = x^2$
19 The triangular region bounded by the graphs of $x - 2y = 2$, $y - 3x = 4$, and $2x + y = 4$ **21** -6
23 48   **25** -84 **27** 120   **29** 0   **31** $a_{11}a_{22}a_{33} \cdots a_{nn}$

33 $\left(-\dfrac{1}{2}\right)\begin{bmatrix} 2 & 4 \\ 3 & 5 \end{bmatrix}$ **35** $\dfrac{1}{7}\begin{bmatrix} 2 & 1 & 0 & 0 \\ -1 & 3 & 0 & 0 \\ 0 & 0 & 3 & 2 \\ 0 & 0 & -5 & -1 \end{bmatrix}$ **37** $\begin{bmatrix} 4 & -5 & 6 \\ 4 & -11 & 5 \end{bmatrix}$ **39** $\begin{bmatrix} 0 & 4 & -6 \\ 16 & 22 & 1 \\ 12 & 11 & 9 \end{bmatrix}$

41 $\begin{bmatrix} -12 & 4 & -11 \\ 6 & -11 & 5 \end{bmatrix}$ **43** $\begin{bmatrix} a & 3a \\ 2b & 4b \end{bmatrix}$ **45** $\begin{bmatrix} 5 & 9 \\ 13 & 19 \end{bmatrix}$

CHAPTER 9

Section 9.1, page 437

1 $-2 + 6i$ **3** $-2 + 4i$ **5** $7 - 5i$ **7** $-4 - i$ **9** $4 + 7i$ **11** $-6 + 3i$
13 $-10 + 5i$ **15** $20 - 10i$ **17** 25   **19** $-72 - 36i$ **21** $32 - 44i$ **23** 15   **25** 10
27 $5 + 12i$ **29** $7 + 17i$ **31** 1   **33** -1 **35** i **37** $x = -8/5, y = 3$
39 $x = -1/2, y = -1/7$

Section 9.2, page 441

1 $(3/13) - (2/13)i$ **3** $(35/61) + (42/61)i$ **5** $(-1/5) - (11/10)i$ **7** $(-7/13) + (17/13)i$
9 $-7 - 21i$ **11** $(12/17) - (54/17)i$ **13** $(42/5) - (9/5)i$ **15** $(-1/4) - (1/4)i$ **17** $-2 - i$
19 $(1/125)i$ **21** 5   **23** $\sqrt{85}$ **25** 8   **27** 1
The geometric representations in Exercises 29–37 are the following points:
29 $P(4, 2)$ **31** $P(3, -5)$ **33** $P(-3, 6)$ **35** $P(-6, 4)$ **37** $P(0, 2)$ **39** $(26/29) - (7/29)i$
41 $(-1/2)i$

Section 9.3, page 444

1 -5 **3** $-3 + 7i$ **5** 54 **7** $-2 + 2i$ **9** $-10\sqrt{10}i$ **11** $(11/10) - (13/10)i$

13 $(3 \pm \sqrt{31}i)/2$ **15** $-1 \pm 2i$ **17** $(-1 \pm \sqrt{47}i)/8$ **19** $5, (-5 \pm 5\sqrt{3}i)/2$

21 $2, -2, -1 \pm \sqrt{3}i, 1 \pm \sqrt{3}i$ **23** $\pm 2i, \pm(3/2)i$ **25** $0, (-3 \pm \sqrt{7}i)/2$ **27** $x^2 - 8x + 17 = 0$

29 $x^2 + 6x + 13 = 0$ **31** $x^2 + 25 = 0$ **33** $x^2 - (10 + i)x + 10i = 0$ **35** $x^2 + 1 = 0$

37 $x^2 - 5ix - 6 = 0$

Section 9.4, page 448

1 $\sqrt{2}(\cos 7\pi/4 + i \sin 7\pi/4)$ **3** $8(\cos 5\pi/6 + i \sin 5\pi/6)$ **5** $20(\cos 3\pi/2 + i \sin 3\pi/2)$

7 $10(\cos 4\pi/3 + i \sin 4\pi/3)$ **9** $7(\cos \pi + i \sin \pi)$ **11** $\sqrt{5}(\cos \theta + i \sin \theta)$, where $\theta = \arctan(1/2)$

13 $4\sqrt{2}(\cos 5\pi/4 + i \sin 5\pi/4)$ **15** $6(\cos \pi/2 + i \sin \pi/2)$ **17** $8(\cos 5\pi/6 + i \sin 5\pi/6)$

19 $12(\cos 0 + i \sin 0)$ **21** $-2, i$ **23** $10\sqrt{3} - 10i, (-2\sqrt{3}/5) + (2/5)i$ **25** $40, 5/2$

27 $8 - 4i, (8/5) + (4/5)i$ **29** $z_1 z_2 z_3 = r_1 r_2 r_3 [\cos(\theta_1 + \theta_2 + \theta_3) + i \sin(\theta_1 + \theta_2 + \theta_3)],$
$$z_1 z_2 \cdots z_n = (r_1 r_2 \cdots r_n)[\cos(\theta_1 + \theta_2 + \cdots + \theta_n) + i \sin(\theta_1 + \theta_2 + \cdots + \theta_n)]$$

Section 9.5, page 452

1 $-972 - 972i$ **3** $-32i$ **5** -8 **7** $-(\sqrt{2}/2) - (\sqrt{2}/2)i$ **9** $(-1/2) - (\sqrt{3}/2)i$

11 $-64\sqrt{3} - 64i$ **13** $(\sqrt{6}/2) + (\sqrt{2}/2)i, -(\sqrt{6}/2) - (\sqrt{2}/2)i$

15 $(\sqrt[4]{2}/2) + (\sqrt[4]{18}/2)i, -(\sqrt[4]{18}/2) + (\sqrt[4]{4}/2)i, (\sqrt[4]{18}/2) - (\sqrt[4]{2}/2)i, -(\sqrt[4]{2}/2) - (\sqrt[4]{18}/2)i$

17 $3i, -(3\sqrt{3}/2) - (3/2)i, (3\sqrt{3}/2) - (3/2)i$ **19** $\pm 1, (1/2) \pm (\sqrt{3}/2)i, (-1/2) \pm (\sqrt{3}/2)i$

21 $\sqrt[5]{2}[\cos \theta + i \sin \theta]$, where $\theta = 9°, 81°, 153°, 225°, 279°$

23 $\pm 2, \pm 2i$ **25** $\pm 2i, \pm \sqrt{3} + i, \pm \sqrt{3} - i$ **27** $2i, -\sqrt{3} - i, \sqrt{3} - i$

29 $3\cos \theta + 3\sin \theta i$, where $\theta = 0, 2\pi/5, 4\pi/5, 6\pi/5, 8\pi/5$

Section 9.6, Review Exercises, page 453

1 $-1 + 8i$ **3** $-28 + 6i$ **5** $40 + 56i$ **7** 97 **9** $16 + 15i$ **11** $(7/13) + (17/13)i$

13 $(5/82) + (2/41)i$ **15** $2\sqrt{34}$ **17** $-19 - 7i$ **19** $-i$ **21** $(1 \pm \sqrt{14}i)/5$

23 $\pm(\sqrt{14}/2)i, \pm(2\sqrt{3}/3)i$ **25** $x^2 - 14x + 50 = 0$ **27** $x^2 + 2x + 17 = 0$

29 $10\sqrt{2}[\cos 3\pi/4 + i \sin 3\pi/4]$ **31** $17[\cos \pi + i \sin \pi]$ **33** $10[\cos 7\pi/6 + i \sin 7\pi/6]$ **35** $-512i$

37 $-972 + 972i$ **39** $-3, (3/2) \pm (3\sqrt{3}/2)i$

CHAPTER 10

Section 10.1, page 458

1 $x^2 + x + 2, -2x + 13$ **3** $(5/2)x, (-9/2)x$ **5** $0, 7x^3 - 5x + 2$ **7** 95 **9** -23 **11** 0

13 $k = 17/12$ **15** $f(2) = 0$ **17** $f(c) > 0$

19 If $f(x) = x^n - y^n$, then $f(y) = 0$. If n is even, then $f(-y) = 0$.

Section 10.2, page 462

1 $2x^2 + x + 6, 7$ **3** $x^2 - 3x + 1, -8$ **5** $3x^4 - 6x^3 + 12x^2 - 18x + 36, -65$

7 $4x^3 + 2x^2 - 4x - 2, 0$ **9** $x^{n-1} + x^{n-2} + \cdots + x + 1, 0$ **11** $-23, 53$ **13** 3277 **15** $8 + 7\sqrt{3}$

Section 10.3, page 468

1 $(-1/15)x^3 + (19/5)x + 2$ **3** $-2x^3 + 12x^2 - 22x + 12$ **5** $(3/10)x^3 - (33/10)x + 6$
7 $x^4 - 2x^3 - 11x^2 + 12x + 36$ **9** $(2/9)x^8 - (4/3)x^7 + (8/3)x^6 - (16/9)x^5$
11 -4 (multiplicity 3), $4/3$ (multiplicity 1)   **13** 0 (multiplicity 3), 5, -1 (each of multiplicity 1)
15 $\pm 5/3$ (each of multiplicity 4), $\pm 4i$ (each of multiplicity 1)
17 -2 (multiplicity 3); 1 (multiplicity 2), 2 (multiplicity 1)   **19** $(x + 3)^2(x + 2)(x - 1)$
21 $(x - 1)^5(x + 1)$ **23** Positive: 3 or 1; Negative: 0   **25** Positive: 0; Negative: 1
27 Positive: 2 or 0; Negative: 2 or 0   **29** Positive: 2 or 0; Negative: 3 or 1   **31** Upper 5, lower -2
33 Upper 2, lower -2 **35** Upper 3, lower -3

Section 10.4, page 473

1 $x^2 - 8x + 17$ **3** $x^3 - 10x^2 + 37x - 52$ **5** $x^4 - 6x^3 + 30x^2 - 78x + 85$
7 $x^5 - 2x^4 + 27x^3 - 50x^2 + 50x$
9 No. If i is a root, then $-i$ is also a root. Hence, the polynomial would have factors $x - 1$, $x + 1$, $x - i$, $x + i$ and, therefore, would be of degree greater than 3.
11 $-1, -2, 4$ **13** $2, -3, 5/2$ **15** $4, -7, \pm\sqrt{2}$ **17** $4, -2, 3/2$ **19** $1/2, -2/3, -3$
21 $-3/4, (-3 \pm 3\sqrt{7}i)/4$ **23** $4, -3, \pm\sqrt{3}i$ **25** $1, -2, -4/3, 2/3$ **27** $3, -4, -2, 1/2, -1/2$
37 Complex zeros occur in conjugate pairs.

Section 10.5, Review Exercises, page 475

1 $3x^2 + 2$, $-21x^2 + 5x - 9$ **3** $9/2, 53/2$ **5** -132 **7** $6x^4 - 12x^3 + 24x^2 - 52x + 104$, -200
9 $(2/41)x^3 + (14/41)x^2 + (80/41)x + (68/41)$ **11** $x^7 + 6x^6 + 9x^5$
13 1 (multiplicity 5), -3 (multiplicity 1)
15 (a) Positive: 3 or 1; negative: 1   (b) Upper bound 3; lower bound -1 **19** $-1/2, 1/4, 3/2$

CHAPTER 11

Section 11.2, page 489

1 $a^6 + 6a^5b + 15a^4b^2 + 20a^3b^3 + 15a^2b^4 + 6ab^5 + b^6$
3 $a^8 - 8a^7b + 28a^6b^2 - 56a^5b^3 + 70a^4b^4 - 56a^3b^5 + 28a^2b^6 - 8ab^7 + b^8$
5 $81x^4 - 540x^3y + 1350x^2y^2 - 1500xy^3 + 625y^4$
7 $u^{10} + 20u^8v + 160u^6v^2 + 640u^4v^3 + 1280u^2v^4 + 1024v^5$
9 $r^{-12} - 12r^{-9} + 60r^{-6} - 160r^{-3} + 240 - 192r^3 + 64r^6$
11 $1 + 10x + 45x^2 + 120x^3 + 210x^4 + 252x^5 + 210x^6 + 120x^7 + 45x^8 + 10x^9 + x^{10}$
13 $(3^{25})c^{10} + 25(3^{24})c^{52/5} + 300(3^{23})c^{54/5} + 2300(3^{22})c^{56/5}$ **15** $60(3^{14})b^{13} - (3^{15})b^{15}$ **17** $30618a^{10}b^2$
19 $840u^4v^6$ **21** $924x^2y^2$ **23** -61236 **25** $24x^2y^6$ **27** $210c^3d^2$
29 $1 + 0.2 + 0.018 + 0.00096 = 1.21896$. If $n = (1.02)^{10}$, then $\log n = 10 \log(1.02) \approx 0.0860$ and $n \approx 1.2190$.

Section 11.3, page 495

1 $9, 6, 3, 0, -3; -12$ **3** $1/2, 4/5, 7/10, 10/17, 13/26; 22/65$ **5** $9, 9, 9, 9, 9; 9$
7 $1.9, 2.01, 1.999, 2.0001, 1.99999; 2.00000001$ **9** $4, -9/4, 5/3, -11/8, 6/5; -15/16$
11 $2, 0, 2, 0, 2; 0$ **13** $2/3, 2/3, 8/11, 8/9, 32/27; 128/33$ **15** $4, 9, 25, 49, 121; 289$
17 $7, 19, 29, 37, 47; 79$ **19** $1, 2, 3, 4, 5; 8$ **21** $2, 1, -2, -11, -38$ **23** $-3, 3^2, 3^4, 3^8, 3^{16}$
25 $5, 5, 10, 30, 120$ **27** $2, 2, 4, 4^3, 4^{12}$

29 $a_n = 2n + (1/24)(n-1)(n-2)(n-3)(n-4)(a-10)$. There are many other answers. **31** -5
33 10 **35** 25 **37** $-17/15$ **39** 61 **41** $10,000$ **43** $(2n^3 + 12n^2 + 40n)/6$

45 $(4n^3 - 12n^2 + 11n)/3$ **47** $\sum_{k=1}^{5}(4k-3)$ **49** $\sum_{k=1}^{4}k/(3k-1)$ **51** $1 + \sum_{k=1}^{n}(-1)^k \dfrac{x^{2k}}{2k}$

53 $\sum_{k=1}^{7}(-1)^{k-1}/k$ **55** $\sum_{n=1}^{99}1/n(n+1)$

Section 11.4, page 500

1 $18, 38, 4n-2$ **3** $1.8, 0.3, 3.3 - 0.3n$ **5** $5.4, 20.9, (3.1)n - (10.1)$ **7** $\log 3^5, \log 3^{10}, \log 3^n$
9 -8.5 **11** -9.8 **13** -105 **15** 30 **17** 530 **19** $423/2$ **21** $60; 12,780$ **23** 24
25 $10/3, 14/3, 6, 22/3, 26/3$ **27** $2, 2, 2^2, 2^6, 2^{24}$ **29** $23.25, 22.5, 21.75, 21, 20.25, 19.5, 18.75$

Section 11.5, page 506

1 $1/2, 1/6; 8(1/2)^{n-1} = 2^{4-n}$ **3** $0.03, -0.00003; 300(-0.1)^{n-1}$ **5** $3125, 5^8; 5^n$
7 $81/4, -3^7/2^5; 4(-3/2)^{n-1}$ **9** $x^8, -x^{14}; (-1)^{n-1}x^{2n-2}$ **11** $2^{4x+1}, 2^{7x+1}; 2^{nx-x+1}$ **13** $243/8$
15 $a_1 = 1/81, S_5 = 211/1296$ **17** $(-3/2)(1 - 3^{10}) = 88,572$ **19** $(-1/3)(1 - 2^{-10})$ **21** $25/256\%$
23 $10,000(6/5)^r, 10,000(6/5)^{10}$ **25** $2/3$ **27** $50/33$ **29** Sum does not exist ($|r| = \sqrt{2} > 1$).
31 $23/99$ **33** $2393/990$ **35** $5141/999$ **37** $16123/9999$ **39** 30 meters

Section 11.6, page 511

1 (a) 60 (b) 125 **3** 64 **5** 336 **7** 24 **9** (a) $(2.34)10^6$ (b) $(2.16)10^6$
11 $151,200; 5,760$ **13** 1024 **15** 210 **17** 60480 **19** 120 **21** 6 **23** $40,320$
25 120 **27** (a) 60 (b) 125 **29** $9,000,000$

Section 11.7, page 516

1 35 **3** 9 **5** n **7** 1 **9** $166,320$ **11** $151,200$ **13** 252 **15** $28,56$
17 $8!5!4!3!$ **19** $4,082,400$

Section 11.8, page 521

1 (a) $1/13$ (b) $2/13$ (c) $3/13$ **3** (a) $1/6$ (b) $1/6$ (c) $1/3$ **5** (a) $1/18$ (b) $5/36$ (c) $7/36$
7 $1/36$ **9** $3/8$ **11** $624/_{52}C_5 \approx 0.00024$ **13** $(13)_{13}C_4/_{52}C_5 \approx 0.0036$
15 $(4)_{13}C_5/_{52}C_5 \approx 0.002$ **17** (a) $1/256$ (b) $_8C_7/256 = 1/32$ (c) $_8C_6/256 = 7/64$ (d) $37/256$
19 (a) Consider $P(E \cup E')$ (b) $1 - (_{48}C_5/_{52}C_5) \approx 0.34$

Section 11.9, Review Exercises, page 522

7 $x^{12} - 18x^{10}y + 135x^8y^2 - 540x^6y^3 + 1215x^4y^4 - 1458x^2y^5 + 729y^6$ **9** $-(63/16)b^{12}c^{10}$
11 $5, -2, -1, -20/29, -7/19$ **13** $1, 1/2, 5/4, 7/8, 65/64$ **15** $10, 11/10, 21/11, 32/21, 53/32$
17 $9, 3, \sqrt{3}, \sqrt[4]{3}, \sqrt[8]{3}$ **19** 75 **21** 1000 **23** $\sum_{k=1}^{5}3k$ **25** $\sum_{k=0}^{4}(-1)^k5(20-k)$
27 $13 + 10\sqrt{3}, 85 + 55\sqrt{3}$ **29** $-31, 50$ **31** 64 **33** $4\sqrt{2}$ **35** 530 **37** -2041
39 $5/7$ **41** $_{52}P_{13}; (_{13}P_5)(_{13}P_3)(_{13}P_3)(_{13}P_2)$ **43** $495, 126$ **45** (a) $1/1000$ (b) $1/100$ (c) $1/20$
47 $(_6C_6 + _6C_5 + _6C_4)/2^6 = 11/32$

INDEX

The Trigonometric Functions

I. *Of Real Numbers*

$$\sin t = y \qquad\qquad \csc t = \frac{1}{y}$$

$$\cos t = x \qquad\qquad \sec t = \frac{1}{x}$$

$$\tan t = \frac{y}{x} \qquad\qquad \cot t = \frac{x}{y}$$

II. *Of Arbitrary Angles*

$$\sin \theta = \frac{b}{r} \qquad\qquad \csc \theta = \frac{r}{b}$$

$$\cos \theta = \frac{a}{r} \qquad\qquad \sec \theta = \frac{r}{a}$$

$$\tan \theta = \frac{b}{a} \qquad\qquad \cot \theta = \frac{a}{b}$$

III. *Of Acute Angles*

$$\sin \theta = \frac{\text{opp}}{\text{hyp}} \qquad\qquad \csc \theta = \frac{\text{hyp}}{\text{opp}}$$

$$\cos \theta = \frac{\text{adj}}{\text{hyp}} \qquad\qquad \sec \theta = \frac{\text{hyp}}{\text{adj}}$$

$$\tan \theta = \frac{\text{opp}}{\text{adj}} \qquad\qquad \cot \theta = \frac{\text{adj}}{\text{opp}}$$